NON-PERTURBATIVE FIELD THEORY

From Two-Dimensional Conformal Field Theory
to QCD in Four Dimensions

Providing a new perspective on quantum field theory, this book gives a pedagogical and up-to-date exposition of non-perturbative methods in relativistic quantum field theory and introduces the reader to modern research work in theoretical physics.

It describes in detail non-perturbative methods in quantum field theory, and explores two-dimensional and four-dimensional gauge dynamics using those methods. The book concludes with a summary emphasizing the interplay between two- and four-dimensional gauge theories.

Aimed at graduate students and researchers, this book covers topics from two-dimensional conformal symmetry, affine Lie algebras, solitons, integrable models, bosonization and 't Hooft model, to four-dimensional conformal invariance, integrability, large N expansion, Skyrme model, monopoles and instantons. Applications, first to simple field theories and gauge dynamics in two dimensions, and then to gauge theories in four dimensions and quantum chromodynamics (QCD) in particular, are thoroughly described. This title, first published in 2010, has been reissued as an Open Access publication on Cambridge Core.

YITZHAK FRISHMAN is a Professor Emeritus at the Weizmann Institute, Israel, where he has served as Head of the Einstein Centre for Theoretical Physics and Head of the Department of Particle Physics.

JACOB SONNENSCHEIN is a Professor of Physics at Tel Aviv University, Israel, where he was Head of the Particle Physics Department from 2003 to 2007.

CAMBRIDGE MONOGRAPHS ON MATHEMATICAL PHYSICS

General Editors: P. V. Landshoff, D. R. Nelson, S. Weinberg

I. Montvay and G. Münster *Quantum Fields on a Lattice*[†]
L. O'Raifeartaigh *Group Structure of Gauge Theories*[†]
T. Ortín *Gravity and Strings*[†]
A. M. Ozorio de Almeida *Hamiltonian Systems: Chaos and Quantization*[†]
L. Parker and D. J. Toms *Quantum Field Theory in Curved Spacetime: Quantized Fields and Gravity*
R. Penrose and W. Rindler *Spinors and Space-Time Volume 1: Two-Spinor Calculus and Relativistic Fields*[†]
R. Penrose and W. Rindler *Spinors and Space-Time Volume 2: Spinor and Twistor Methods in Space-Time Geometry*[†]
S. Pokorski *Gauge Field Theories, 2nd edition*[†]
J. Polchinski *String Theory Volume 1: An Introduction to the Bosonic String*
J. Polchinski *String Theory Volume 2: Superstring Theory and Beyond*
V. N. Popov *Functional Integrals and Collective Excitations*[†]
R. J. Rivers *Path Integral Methods in Quantum Field Theory*[†]
R. G. Roberts *The Structure of the Proton: Deep Inelastic Scattering*[†]
C. Rovelli *Quantum Gravity*[†]
W. C. Saslaw *Gravitational Physics of Stellar and Galactic Systems*[†]
M. Shifman and A. Yung *Supersymmetric Solitons*
H. Stephani, D. Kramer, M. MacCallum, C. Hoenselaers and E. Herlt *Exact Solutions of Einstein's Field Equations, 2nd edition*
J. Stewart *Advanced General Relativity*[†]
T. Thiemann *Modern Canonical Quantum General Relativity*
D. J. Toms *The Schwinger Action Principle and Effective Action*
A. Vilenkin and E. P. S. Shellard *Cosmic Strings and Other Topological Defects*[†]
R. S. Ward and R. O. Wells, Jr *Twistor Geometry and Field Theory*[†]
J. R. Wilson and G. J. Mathews *Relativistic Numerical Hydrodynamics*

[†] Issued as a paperback.

Non-Perturbative Field Theory

From Two-Dimensional Conformal Field Theory to QCD in Four Dimensions

YITZHAK FRISHMAN
The Weizmann Institute of Science

JACOB SONNENSCHEIN
Tel Aviv University

CAMBRIDGE
UNIVERSITY PRESS

Shaftesbury Road, Cambridge CB2 8EA, United Kingdom

One Liberty Plaza, 20th Floor, New York, NY 10006, USA

477 Williamstown Road, Port Melbourne, VIC 3207, Australia

314–321, 3rd Floor, Plot 3, Splendor Forum, Jasola District Centre, New Delhi – 110025, India

103 Penang Road, #05-06/07, Visioncrest Commercial, Singapore 238467

Cambridge University Press is part of Cambridge University Press & Assessment,
a department of the University of Cambridge.

We share the University's mission to contribute to society through the pursuit of
education, learning and research at the highest international levels of excellence.

www.cambridge.org
Information on this title: www.cambridge.org/9781009401647

DOI: 10.1017/9781009401654

First published 2010
Reissued as OA 2023

A catalogue record for this publication is available from the British Library.

ISBN 978-1-009-40164-7 Hardback
ISBN 978-1-009-40161-6 Paperback

To my wife Yehudith,
mother Faiga
and daughter Einat

Yitzhak Frishman

To my mother Hilda,
wife Nava
and children Nir, Ori and Tal

Jacob Sonnenschein

Contents

Contents

Preface

Field theory is the framework with which one describes the theory of the standard model of elementary particles and their interactions. The electromagnetic sector (QED) of the standard model is understood extremely well using perturbation theory, but the color interaction (QCD) which is responsible for hadron physics can only be accounted for perturbatively for a limited set of observational data. Due to the fact that at long distances the color interaction is strongly coupled, one cannot reliably apply perturbative methods to extract, for instance, the spectrum of the hadrons. The arsenal of tools to handle strongly coupled systems is obviously much more limited than the one used for weakly coupled ones. Nevertheless, several methods to handle non-perturbative field theories have been developed. The main goal of this book is to expose the reader to those techniques and to describe their applications in two-dimensional and four-dimensional field theories and finally in QCD in four dimensions.

The topic of non-perturbative field theory is by itself very rich and it is clear that one cannot cover it in a non superficial manner in one book. Thus we had to make certain decisions about the flow of the book and about the topics that should be addressed. As for the former issue we have decided to present the book in three parts. In the first part we describe, in detail, the most important non-perturbative techniques of two-dimensional field theory. The reason for this is obvious since physical systems with one space dimension and one time dimension are the simplest and hence it is easier to grasp the non-perturbative tools when applied to these systems. In the second part of the book we study two-dimensional gauge theories with the emphasis on employing the techniques developed in the first part. The third part is devoted to the non-perturbative aspects of gauge dynamics in four dimensions. In this part we elevate the techniques of the first part to four dimensions and we examine to what extent gauge theories in four dimensions behave like their two-dimensional simplified analogs.

There are several books on the shelves discussing non-perturbative methods in general such as [66] and [182], there are books describing one particular method, like conformal field theory in two dimensions for instance [77], there are books that describe two-dimensional QCD, [2] and books that study various aspects of four-dimensional QCD, for example [151] and of course there are books on the basics of field theory, for example [37], [130], [173] and [215]. The aim of this book is three-fold, to review a package of non-perturbative methods, to present a picture which is close to the state-of-the-art in the topics described and to demonstrate application of the methods in addressing several questions of gauge dynamics.

The particular methods we explore in Part 1 of the book associate with conformal field theory, with affine Lie algebras, with topological properties of fields, solitons and integrable models, with bosonization and with the large N approximation.

In Part 2 we first present the basics of gauge field theories in two dimensions and in particular the bosonized version of them, we then describe the seminal large N solution of 't Hooft of the mesonic spectrum of two-dimensional QCD; we address the mesonic spectrum using current algebra methods, we describe the discrete light-cone quantization of QCD with quarks in the fundamental representation and also adjoint quarks, we compute the spectrum of baryons and their properties in the strong coupling limit, we discuss the issue of confinement versus screening behavior, we analyze QCD_2 using coset model and BRST techniques, and finally we digress and devote a chapter to generalized Yang–Mills theory on Riemann surfaces and their stringy nature.

In Part 3 we demonstrate the applications in four-dimensional gauge dynamics of conformal invariance, techniques of integrable models, of large N expansion and of topology. In particular we devote chapters to Skyrmions, magnetic monopoles and gauge theory instantons.

As we have mentioned above we had to take decisions about what topics related to non-perturbative field theory we should not include. We decided not to address string theories, supersymmetric field theories and the holographic string (gravity)/gauge duality. The main reason for this decision was that to cover each of these topics requires a book in itself, or even more than one book. In fact certain subjects that we do cover in the book, like conformal field theory, magnetic monopoles or instantons would require a full book to cover properly. What we have tried to achieve is to describe the basic ideas of each topic and to demonstrate its application. We have also not treated subjects like anomalies, lattice formulations, sigma models, chiral Lagrangians and other non-perturbative topics.

Some topics described in the book are "fully established topics", in the sense that presumably the most important developments in those have been already achieved, for instance conformal field theory in two dimensions and bosonization in two dimensions. On the other hand some other topics of the book are under current intensive investigation and are certainly still not fully established. An example of the latter is integrability in four-dimensional gauge dynamics. The reason we have decided to include topics of the latter kind is that we wanted the book to be fairly up to date and useful to researchers investigating "modern" topics.

In the more basic issues we have made an effort to present the material in a pedagogical manner and to be self contained. For instance our discussion started from a free massless scalar field theory in two dimensions and gradually evolved into general conformal field theories. In dealing with more advanced topics, like for instance instantons in four dimensions, the reader will need to consult with specialized references to obtain a more complete and wider picture of the topic.

Some of the content of the book, mainly in Part 2, is based on the research work of the authors, but most of the material is a review of the work of many researchers in the field.

The book is aimed for advanced Ph.D. students, post-docs and other newcomers to the arena of non-perturbative methods in field theory. The reader should definitely be equipped with a basic knowledge of field theory, group theory and algebra, differential equations, geometry and topology.

Throughout the book we refer to only a limited list of references. The number of scientific contributions to the topics discussed in this book is enormous and since we could not cover all of them we have referred to papers that initiated the various topics, and to review papers and books where a much more exhaustive list of references can be found.

We have made an attempt to keep the same notations throughout the book. However in certain instances we have changed notations during the course of the book, mainly to be in accordance with relevant literature. In these cases we specified explicitly the change in notation made.

Acknowledgements

We would like to thank O. Aharony, A. Armoni, M. Bernstein, S. J. Brodsky, E. Cohen, R. Dashen, G. Date, G. F. Dell-Antonio, J. R. Ellis, O. Ganor, D. Gepner, E. Gimon, A. Hanany, G. P. Lepage, M. Karliner, L. A. Pando Zayas, C. T. Sachrajda, N. Sochen, M. J. Strassler, S. Yankielowicz, W. J. Zakrzewski and D. Zwanziger who collaborated with us in the research works that are covered in this book.

We would like to thank Ori Sonnenschein for drawing the figures of the book.

We would like to thank O. Aharony, M. Karliner and S. Theisen for their remarks on the manuscript.

The work of Jacob Sonnenschein was supported in part by the Albert Einstein Minerva Center of The Weizmann Institute of Science.

PART I

Non-perturbative methods in two-dimensional
field theory

1

From massless free scalar field to conformal field theories

In this chapter we analyze the simplest field theory, which is the theory of a free massless scalar field in two space-time dimensions, one space and one time.[1] The rich symmetry and algebraic structure of this theory encapsulates the basic concepts of two-dimensional conformal field theory, which will be the topic of the next chapter.

1.1 Complex geometry

It is convenient for the discussion of two-dimensional free scalar theory and later conformal field theories to introduce complex coordinates as follows:[2]

$$\xi = x^0 + ix^1 \quad \bar{\xi} = x^0 - ix^1.$$ (1.1)

We now take x^0 and x^1 to be in Euclidean space. Correspondingly we define the derivatives

$$\partial_\xi = \frac{1}{2}(\partial_0 - i\partial_1) \quad \partial_{\bar{\xi}} = \frac{1}{2}(\partial_0 + i\partial_1),$$ (1.2)

which is a special case of the decomposition to components of vectors, namely

$$A_\xi = \tfrac{1}{2}(A^0 - iA^1) \quad A_{\bar{\xi}} = \frac{1}{2}(A^0 + iA^0)$$
$$A^\xi = (A^0 + iA^1) \quad A^{\bar{\xi}} = (A^0 - iA^1).$$ (1.3)

The metric of the flat Euclidean space-time $ds^2 = dx^{0^2} + dx^{1^2}$ translates into $ds^2 = d\xi d\bar{\xi}$, namely

$$g_{\xi\bar{\xi}} = g_{\bar{\xi}\xi} = \frac{1}{2}, \quad g^{\xi\bar{\xi}} = g^{\bar{\xi}\xi} = 2, \quad g_{\xi\xi} = g_{\bar{\xi}\bar{\xi}} = g^{\xi\xi} = g^{\bar{\xi}\bar{\xi}} = 0.$$ (1.4)

With this metric at hand the scalar product of two vectors takes the form

$$A^\mu B_\mu = A^\xi B_\xi + A^{\bar{\xi}} B_{\bar{\xi}} = \frac{1}{2}(A^\xi B^{\bar{\xi}} + A^{\bar{\xi}} B^\xi).$$ (1.5)

Complex components of higher-order tensors relate in a similar manner to the real components, in particular for a symmetric two-tensor (like the

---

[1] The content of this chapter comprises the basics of massless scalar fields in two dimensions. This is covered in many textbooks.
[2] The use of complex coordinates in the context of the bosonic string theory is described by Polyakov in [177].

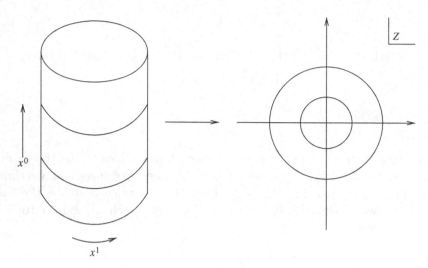

Fig. 1.1. The map between ξ and z.

energy-momentum tensor),

$$T \equiv T_{\xi\xi} = \frac{1}{4}(T_{00} - 2iT_{10} - T_{11})$$

$$\bar{T} \equiv T_{\bar{\xi}\bar{\xi}} = \frac{1}{4}(T_{00} + 2iT_{10} - T_{11})$$

$$T_{\bar{\xi}\xi} = T_{\xi\bar{\xi}} = \frac{1}{4}(T_{00} + T_{11}). \tag{1.6}$$

Often, especially in the context of string theory, the space direction is no longer $\mathcal{R}$, but rather is compactified on S^1 so that $x^1 \equiv x^1 + 2\pi$. For such a geometry it is convenient to introduce the following conformal map:

$$\xi \to z = e^\xi = e^{x^0 + ix^1},$$

which maps the cylinder to the complex plane (see Fig. 1.1).

In particular the past $x^0 = -\infty$ is mapped into the origin and the future $x^0 = \infty$ into a circle with an infinite radius. It is clear that the relations between $(\xi, \bar{\xi})$ and (x^0, x^1) derived above hold also between $(z, \bar{z})$ and $(Real(z), Im(z))$. The holomorphic and anti-holomorphic derivatives with respect to z will be denoted by $\partial \equiv \partial_z$ and $\bar{\partial} \equiv \partial_{\bar{z}}$.

1.2 Free massless scalar field

The action S of the free massless scalar field $\hat{X}(z, \bar{z})$ is

$$S = \int d^2x \mathcal{L} = \frac{1}{8\pi} \int d^2x \partial_\nu \hat{X} \partial^\nu \hat{X}$$

$$= \frac{1}{4\pi} \int d^2\xi \partial_\xi \hat{X} \partial_{\bar{\xi}} \hat{X} = \frac{1}{4\pi} \int d^2z \partial \hat{X} \bar{\partial} \hat{X}, \tag{1.7}$$

where $\mathcal{L}$ is the Lagrangian density. The factor $\frac{1}{4\pi}$ is used to match the normalization of the bosonic string theory (with $\alpha' = 2$). In the complex coordinate notation $(\xi, \bar{\xi})$ and $(z, \bar{z})$ the measure of the integral is $d^2\xi = (i/2)d\xi \wedge d\bar{\xi}$ and $d^2z = (i/2)dz \wedge d\bar{z}$, respectively. Note that $\mathcal{L}$ is a local expression and thus is the same for the Euclidean plane or for any compact two-surface.

Varying the scalar field $\hat{X}(z, \bar{z}) \to \hat{X}(z, \bar{z}) + \delta\hat{X}(z, \bar{z})$ induces a variation in the action of the form

$$\delta S = -\frac{1}{2\pi} \int d^2z (\partial \bar{\partial} \hat{X}) \delta \hat{X}. \tag{1.8}$$

The action is thus extremized by configurations that solve the corresponding equation of motion

$$\partial \bar{\partial} \hat{X} = 0. \tag{1.9}$$

It is thus clear that $\partial \hat{X}$ is a holomorphic function and $\bar{\partial} \hat{X}$ is an anti-holomorphic function, and the most general solution takes the form

$$\hat{X}(z, \bar{z}) = [X(z) + \bar{X}(\bar{z})]. \tag{1.10}$$

1.3 Symmetries of the classical action

By construction the action is invariant under translations and $SO(2)$ rotations. Translations in x^0 and x^1 translate in complex coordinates to

$$z \to z + a; \quad \bar{z} \to \bar{z} + \bar{a}, \tag{1.11}$$

where a is a constant complex number, and the $SO(2)$ rotations, in infinitesimal form, to

$$\delta z = -i\epsilon z; \quad \delta \bar{z} = i\epsilon \bar{z}, \tag{1.12}$$

where ϵ is an infinitesimal real parameter.

When we go back to Minkowski space, the $SO(2)$ rotations turn into $SO(1, 1)$ transformations. In addition it is easy to realize that a shift of the field by a constant A,

$$\hat{X}(z, \bar{z}) \to \hat{X}(z, \bar{z}) + A, \tag{1.13}$$

leaves the Lagrangian invariant. It is a special feature of two dimensions that the symmetry group of the action is in fact much richer since one can replace the constant A with $A(z)$ and the constant $\bar{A}$ with $\bar{A}(\bar{z})$, which are arbitrary holomorphic and anti-holomorphic functions, respectively,

$$\hat{X}(z, \bar{z}) \to \hat{X}(z, \bar{z}) + A(z); \quad \hat{X}(z, \bar{z}) \to \hat{X}(z, \bar{z}) + \bar{A}(\bar{z}). \tag{1.14}$$

These are the *affine current algebra transformations*.[3]

---

[3] Affine Lie algebras describing a physical system were first discussed in [27]. More references will be given in the next two chapters.

In a similar manner the space-time translations (1.11) can also be elevated to holomorphic and anti-holomorphic transformations,

$$z \to f(z); \quad \bar{z} \to \bar{f}(\bar{z}), \tag{1.15}$$

referred to as two-dimensional *conformal transformations*. Affine current algebra transformations and conformal transformations will be further discussed in Sections 1.10 and 1.11

1.4 Mode expansion

The mode expansion of the classical solution depends on the boundary conditions. For the case where the underlying two-dimensional manifold is the infinite plane, a standard Fourier transform is used:

$$\hat{X}(x^0, x^1) = \int \frac{dk^1}{\sqrt{2\pi}\sqrt{k^0}} [a(k^1)e^{-ik \cdot x} + a^\dagger(k^1)e^{ik \cdot x}]. \tag{1.16}$$

If the range of the space coordinate is bounded, one may impose two types of boundary conditions, associated with closed and open strings. In the case of *closed strings* the boundary conditions

$$\hat{X}(x^0, x^1) = \hat{X}(x^0, x^1 + 2\pi) \tag{1.17}$$

are automatically obeyed by $\hat{X}(z, \bar{z})$. For this case the mode expansion is expressed in terms of a Laurent series,

$$\partial X = -i \sum_{n=-\infty}^{\infty} \frac{\alpha_n}{z^{n+1}} \quad \bar{\partial}\bar{X} = -i \sum_{n=-\infty}^{\infty} \frac{\bar{\alpha}_n}{\bar{z}^{n+1}}. \tag{1.18}$$

Integrating this expansion we get

$$\hat{X}(z, \bar{z}) = \mathcal{X} - i\mathcal{P}\ln(z\bar{z}) + i \sum_{m=-\infty,\ m \neq 0}^{\infty} \left(\frac{\alpha_m}{m} z^{-m} + \frac{\bar{\alpha}_m}{m} \bar{z}^{-m} \right), \tag{1.19}$$

with $\mathcal{X}$ a constant and

$$\mathcal{P} = \alpha_0 = \bar{\alpha}_0. \tag{1.20}$$

For *open strings* the boundary conditions are of Neumann type, namely

$$\partial_1 \hat{X}(x^0, x^1 = 0) = \partial_1 \hat{X}(x^0, x^1 = \pi) = 0 \implies \partial\hat{X}(z, \bar{z} = z) = \bar{\partial}\hat{X}(z, \bar{z} = z). \tag{1.21}$$

The corresponding mode expansion takes the form

$$\hat{X}(z, \bar{z}) = \mathcal{X} - i\mathcal{P}\ln(z\bar{z}) + i \sum_{m=-\infty,\ m \neq 0}^{\infty} \frac{\alpha_m}{m} \left(z^{-m} + \bar{z}^{-m} \right). \tag{1.22}$$

1.5 Noether currents and charges

Associated with the symmetries (1.14) and (1.15) are conserved Noether currents and charges. In the Noether procedure one is instructed to elevate the global parameters of transformations into local ones and extract the associated currents from the variation of the action, namely $\delta S \sim \int d^2 x J_\mu \partial^\mu \epsilon$. Let us apply this procedure first to the affine current algebra transformations so that we vary the action with respect to $\delta \hat{X}(z, \bar{z}) = \epsilon(z, \bar{z})$ yielding

$$\delta S = \frac{1}{4\pi} \int d^2 z [\partial \epsilon(z, \bar{z}) \bar{\partial} \hat{X}(z, \bar{z}) + \bar{\partial} \epsilon(z, \bar{z}) \partial \hat{X}(z, \bar{z})]. \tag{1.23}$$

Unlike the situation in more than two dimensions, and due to the fact that the symmetries (1.14) are in fact not only global ones but rather "half local", the currents

$$J \equiv \partial X; \quad \bar{J} \equiv \bar{\partial} \bar{X} \tag{1.24}$$

are holomorphic and anti-holomorphic conserved,

$$\bar{\partial} J \equiv \bar{\partial} \partial X = 0; \quad \partial \bar{J} \equiv \partial \bar{\partial} \bar{X} = 0. \tag{1.25}$$

The classical currents are determined up to an overall constant.

A similar situation occurs with respect to the conformal transformation. Replacing in the infinitesimal version of (1.15) $\delta z \to \epsilon(z, \bar{z})$ and $\delta \bar{z} \to \bar{\epsilon}(z, \bar{z})$ one finds,

$$\delta S = \frac{1}{2\pi} \int d^2 z [\partial \bar{\epsilon}(z, \bar{z}) \bar{\partial} \hat{X}(z, \bar{z}) \bar{\partial} \hat{X}(z, \bar{z}) + \bar{\partial} \epsilon(z, \bar{z}) \partial \hat{X}(z, \bar{z}) \partial \hat{X}(z, \bar{z})]. \tag{1.26}$$

The associated holomorphic and anti-holomorphic conserved energy-momentum tensor components are

$$T = -\frac{1}{2} \partial X \partial X; \quad \bar{T} = -\frac{1}{2} \bar{\partial} \bar{X} \bar{\partial} \bar{X}, \tag{1.27}$$

where the coefficients were chosen in a way that will turn out to be convenient when discussing the corresponding quantum generators.

1.6 Canonical quantization

Prior to imposing the canonical quantization condition one has to identify the time direction. There are several options. Using x^0 as the time direction, the corresponding conjugate momentum of $\hat{X}(z, \bar{z})$ is

$$\Pi = \frac{\delta \mathcal{L}}{\delta x_0 \hat{X}} = \frac{1}{4\pi} \partial_0 \hat{X},$$

and the standard quantization conditions are

$$[\hat{X}(x^0, x^1), \Pi(y^0, y^1)]_{x^0 = y^0} = i\delta(x^1 - y^1)$$

$$[\hat{X}(x^0, x^1), \hat{X}(y^0, y^1)]_{x^0 = y^0} = 0$$

$$[\Pi(x^0, x^1), \Pi(y^0, y^1)]_{x^0 = y^0} = 0. \tag{1.28}$$

These conditions yield the standard algebra of the creation and annihilation operators for (1.16),

$$[a(k^1), a^\dagger(p^1)] = \delta(k^1 - p^1); \quad [a(k^1), a(p^1)] = [a^\dagger(k^1), a^\dagger(p^1)] = 0. \tag{1.29}$$

Substituting the mode expansion (1.16) into the expressions of the Noether charges associated with the symmetries of the action (1.7) one finds that the energy-momentum operators are proportional to $a^\dagger(k)a(k) + a(k)a^\dagger(k)$ and hence their vacuum expectation values are proportional to $\delta(0) \sim L$, where L is the size of the space direction. It is thus clear that for the infinite Euclidean plane (or a Minkowski space-time with space $\mathcal{R}$) these expectation values diverge. One then defines the **normal ordered** operators:

$$: \mathcal{O} : \equiv \mathcal{O} - <0|\mathcal{O}|0> . \tag{1.30}$$

For free fields this is equivalent to ordering annihilation operators to the right of creation operators, and sufficient to make $: \mathcal{O} :$ finite.

Using the algebra of the creation and annihilation operators and the normal ordered Hamiltonian, the construction of the Fock space is standard. One defines the vacuum state $|0>$ such that

$$a(k^1)|0> = 0. \tag{1.31}$$

The states in the Fock space are

$$\prod_i a^\dagger(k_i)^{n_i}|0>, \tag{1.32}$$

and their energies, by applying the Hamiltonian,

$$H| \prod a^\dagger(k_i)^{n_i}|0> = \sum_i (k_j^0) n_i(k_i) \prod a^\dagger(k_i)^{n_i}|0> . \tag{1.33}$$

The canonical quantization for the scalar field on a compact space direction, with the boundary conditions of open or closed string, (1.21) and (1.17), respectively, follows very similar steps. Imposing the quantization conditions (1.28) above implies the following algebra for the α_n operators of the open string and for the α_n and $\bar{\alpha}_n$ operators for the closed string:

$$[\alpha_m, \alpha_n] = m\delta_{m+n}$$
$$[\bar{\alpha}_m, \bar{\alpha}_n] = m\delta_{m+n}$$
$$[\alpha_m, \bar{\alpha}_n] = 0. \tag{1.34}$$

It is thus clear that α_n operators are related to the $a(k)$ operators as

$$\alpha_m = \sqrt{m}a(m), m > 0; \quad \alpha_{-m} = \sqrt{m}a^\dagger(m), m > 0. \tag{1.35}$$

1.7 Radial quantization

For the case of a cylinder-like two-dimensional manifold, namely, where the space direction is compactified so that $x^1 \equiv x^1 + 2\pi$, it is natural to use the $z = e^{x^0 + ix^1}$ coordinates. Space translations $x^1 \to x^1 + a$ take the form of multiplying by a phase factor $z \to e^{ia}z$, and time translations $x^0 \to x^0 + a$ turn into dilatations $z \to e^a z$. Rotations $(x^0 + ix^1) \to (c + is)(x^0 + ix^1)$, go into $z \to z^{(c+is)}$, with $(c + is) = e^{i\theta}$, θ the rotation angle. Correspondingly the generators of these transformations change their geometrical operation. For instance the Hamiltonian obviously goes into the dilatation generator. Moreover, generators which are Noether charges transform into contour integrals. Recall that the Noether charge is $Q = \int dx^1 J_0(x^1)$ which in the new coordinates reads $Q = \int d\theta J_r(\theta)$ so that we can write,

$$Q = \frac{1}{2\pi i} \oint [dz J(z) + d\bar{z}\bar{J}(\bar{z})], \tag{1.36}$$

where the contour integral is performed at some radius and the sign convention we adopt is that both the dz and $d\bar{z}$ integral are taken to be positive for the counter-clockwise sense.

The infinitesimal transformation of an operator generated by the Noether charge Q is given by:

$$\delta_{\epsilon,\bar{\epsilon}}\mathcal{O} = \frac{1}{2\pi i} \oint [dz J(z)\epsilon(z), \mathcal{O}(w, \bar{w})] + d\bar{z}[\bar{J}(\bar{z})\bar{\epsilon}(\bar{z}), \mathcal{O}(w, \bar{w})]. \tag{1.37}$$

Define a product R of two operators $A(z)B(w)$ as taken radially, namely[4]

$$R(A(z)B(w)) = A(z)B(w), |z| > |w|; \quad B(w)A(z), |w| > |z|. \tag{1.38}$$

In Fig. 1.2 we show the two contour integrals that lead to a contour integral around w, the location of the operator $\mathcal{O}$, so that the infinitesimal transformation is given by,

$$\delta_{\epsilon,\bar{\epsilon}}\mathcal{O} = \frac{1}{2\pi i} \oint [dz\epsilon(z)R(J(z)\mathcal{O}(w, \bar{w})) + d\bar{z}\bar{\epsilon}(\bar{z})R(J(\bar{z})\mathcal{O}(w, \bar{w}))]. \tag{1.39}$$

---

[4] The notion of radial quantization was introduced in [104]. This construction was used in the context of complex geometry in [93].

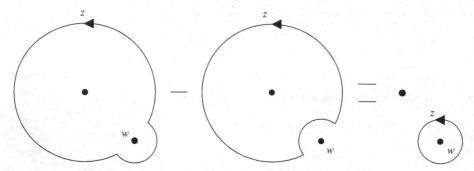

Fig. 1.2. A contour around w from the commutator.

We now apply this formulation to the symmetry generators (discussed in Section 2.1):

(i) The infinitesimal affine current algebra transformation $\hat{X}(z,\bar{z}) \to \hat{X}(z,\bar{z}) - \epsilon(z)$ is generated by the holomorphic current $J(z) = \partial X$ via

$$\delta_\epsilon \hat{X}(w,\bar{w}) = \frac{1}{2\pi i} \oint dz\epsilon(z)R(\partial X(z)\hat{X}(w,\bar{w}))$$
$$= \frac{1}{2\pi i} \oint dz \frac{-1}{z-w}\epsilon(z) = -\epsilon(w), \tag{1.40}$$

where we have used for the product of operators,

$$R(X(z)X(w)) = -\log(z-w) + \text{finite terms.} \tag{1.41}$$

This is an example of the concept of operator product expansion, which is addressed in the next section.

(ii) In a similar manner we can compute the transformation of ∂X generated by the energy momentum tensor T

$$\delta_\epsilon \partial X(w) = \frac{1}{2\pi i} \oint dz\epsilon(z)R\left(-\frac{1}{2} : \partial X(z)\partial X(z) : \partial X(w)\right)$$
$$= \frac{1}{2\pi i} \oint dz\frac{1}{(z-w)^2}\partial X(z)\epsilon(z) = \partial\epsilon(w)\partial X(w) + \epsilon(w)\partial^2 X(w), \tag{1.42}$$

which is indeed the infinitesimal transformation of the holomorphic current $J = \partial X(z)$. The generator T is normal ordered using the following expression:

$$T(w) = -\frac{1}{2} : \partial X(z)\partial X(w) := -\frac{1}{2}\lim_{z\to w}\left[\partial X(z)\partial X(w) + \frac{1}{(z-w)^2}\right]. \tag{1.43}$$

1.8 Operator product expansion

In computing the contour integrals associated with infinitesimal transformations we have made use of the operator product expansions of pairs of operators.[5] The singularities that occur when the points are taken to approach one another are captured in the notion of *operator product expansion (OPE)*,

$$\mathcal{O}_i(x)\mathcal{O}_j(y) = \sum_k c_{ij}^k(x - y)\mathcal{O}_k(y), \tag{1.44}$$

where $c_{ij}^k(x - y)$ are the *coefficient functions* which are singular in the limit of $x \to y$. Such expansions were proven to hold in renormalizable field theories. The OPEs are an essential tool in exploring quantum field theories. Recall that all of the information on the QFT is encoded into the values of all possible correlation functions of the complete set of local operators $\mathcal{O}_i(x)$, namely, $< \mathcal{O}_1(x_1)...\mathcal{O}_n(x_n) >$. In particular, one is interested in the behavior of these correlation functions when two or more points approach each other, which is encapsulated in the OPEs. For all applications discussed here the OPEs are treated as asymptotic expansions and only their singular terms will be specified. For the present case of two-dimensional free massless scalar field theory the OPE converges and in fact, as will be discussed in Section 3.7.2, a similar situation occurs in all 2d CFTs.

The OPEs of the free massless scalar can be deduced from its propagator, which can be evaluated from the solution. It takes the form:

$$< \hat{X}(z\bar{z})\hat{X}(w\bar{w}) >= -log|z - w|^2. \tag{1.45}$$

In terms of the separation of the solution into holomorphic and anti-holomorphic parts the two propagators read:

$$< X(z)X(w) >= -\log(z - w); \quad < \bar{X}(\bar{z})\bar{X}(\bar{w}) >= -\log(\bar{z} - \bar{w}). \tag{1.46}$$

By differentiating the last relation with respect to z and to w one finds the short distance expansion of other operators like $J(z)$, $T(z)$ etc. In particular the OPE of the currents is

$$J(z)J(w) = \partial X(z)\partial X(w) = -\frac{1}{(z - w)^2} + \text{finite terms}, \tag{1.47}$$

with a similar result for the anti-holomorphic currents.

A different, though equivalent, approach is to write the OPE as a Taylor expansion in $(z - w)$ and $(\bar{z} - \bar{w})$ in the following form:

$$\hat{X}(z, \bar{z})\hat{X}(w\bar{w}) = -log|z - w|^2 + \sum_{k=1}^{\infty} \frac{1}{k!}[(z - w)^k : (\partial^k \hat{X}(w, \bar{w}))\hat{X}(w, \bar{w}) :$$
$$+(\bar{z} - \bar{w})^k : (\bar{\partial}^k \hat{X}(w, \bar{w}))\hat{X}(w, \bar{w}) :]. \tag{1.48}$$

---

[5] Wilson introduced for the first time the concept of an operator product expansion [219]. It was used for two-dimensional conformal field theories in [33].

This form of expansion is based on the property that the normal ordered product of the scalar fields,

$$: \hat{X}(z, \bar{z})\hat{X}(w\bar{w}) := \hat{X}(z, \bar{z})\hat{X}(w, \bar{w}) + log|z - w|^2, \tag{1.49}$$

obeys the equation of motion, namely,

$$\partial\bar{\partial} : \hat{X}(z, \bar{z})\hat{X}(w, \bar{w}) := 0, \tag{1.50}$$

and hence can be decomposed to holomorphic and anti-holomorphic functions and thus is non singular.

In the previous subsection we used two OPEs to determine the symmetry transformation of X and ∂X. We will work out now two additional examples of OPEs, involving the operator which will later be found to be very useful $: e^{i\alpha X(w)} :$.

(i) The conformal properties of $: e^{i\alpha X(w)}$ are being determined by its OPE with $T(z)$ which takes the form

$$T(z) : e^{i\alpha X(w)} := -\frac{1}{2}(: \partial X(z)\partial X(z) :)(: e^{i\alpha X(w)} :)$$

$$= \frac{(\frac{\alpha^2}{2})}{(z - w)^2}e^{i\alpha X(w)} + \frac{1}{(z - w)}\partial e^{i\alpha X(w)}. \tag{1.51}$$

In language that will be developed in Section 2.2 this result will mean that $: e^{i\alpha X(w)} :$ has a conformal dimension of $\frac{\alpha^2}{2}$.

(ii) The OPE of two operators of the form $: e^{i\alpha X(w)}$ is

$$(: e^{i\alpha X(z)} :)(: e^{-i\beta X(w)} :) = \frac{: e^{i\alpha X(w)}e^{-i\beta X(w)} :}{(z - w)^{\alpha\beta}}. \tag{1.52}$$

1.9 Path integral quantization

So far we have been using canonical quantization. Before proceeding to the general structure of affine current algebra and Virasoro algebra we introduce the quantization of a free massless scalar field using the Euclidean path integral approach. As usual the functional integration $D\hat{X}(z, \bar{z})$ can be approximated by discretizing the two-dimensional space and representing the functional integral by products of ordinary integrals. Expectation values of operators $\mathcal{O}(X)$ constructed from X are given by,

$$< \mathcal{O}(\hat{X}) >= \int D\hat{X}(z, \bar{z})\mathcal{O}(\hat{X})e^{-S} = \int D\hat{X}(z, \bar{z})\mathcal{O}(\hat{X})e^{-\frac{1}{2\pi}\int d^2 z\partial\hat{X}\bar{\partial}\hat{X}}. \tag{1.53}$$

Correlation functions have to obey the equation

$$\partial\bar{\partial} < \hat{X}(z, \bar{z})\hat{X}(w, \bar{w}) >= -2\pi\delta^2(z - w, \bar{z} - \bar{w}), \tag{1.54}$$

as can be deduced by using the fact that the path integral of a total derivative vanishes:

$$
\begin{aligned}
0 &= \int D\hat{X}(z,\bar{z}) \frac{\delta}{\delta \hat{X}(z,\bar{z})} [e^{-S}\hat{X}(w,\bar{w})] \\
&= \int D\hat{X}(z,\bar{z}) e^{-S} \left[-\frac{\delta S}{\delta \hat{X}(z,\bar{z})} \hat{X}(w,\bar{w}) + \delta^2(z-w,\bar{z}-\bar{w}) \right] \\
&= \frac{1}{2\pi} < \partial\bar{\partial}\hat{X}(z,\bar{z})\hat{X}(w,\bar{w}) > + < \delta^2(z-w,\bar{z}-\bar{w}) > .
\end{aligned}
\tag{1.55}
$$

Alternatively one can use (1.45) and (1.46) directly. Note that in that case care must be exercised, as naively we would get zero rather than the delta function, since the expression is a sum of two terms, one depending on z only and the other on $\bar{z}$ only. The point is that the expressions (1.46) cannot be taken over at the origin. A working rule is:

$$
\bar{\partial}\left(\frac{1}{z}\right) = \partial\left(\frac{1}{\bar{z}}\right) = (2\pi)\delta^2(z,\bar{z})
$$

This can be derived by going over from $\frac{1}{z}$ to $\frac{\bar{z}}{z\bar{z}+\epsilon^2}$, to regulate the singularity at the origin.

1.10 Affine current algebra

As was shown in Section 1.3 the classical action is invariant under both affine current algebra transformations and conformal transformations. We would like to study the algebraic structure of the generators of these symmetries. We start with the invariance under affine current algebra transformations. Recall that the corresponding generators are the holomorphic and anti-holomorphic currents J and $\bar{J}$, given in (1.24). Expanding the currents in Laurant series,

$$
J = \sum_{n=-\infty}^{\infty} J_n z^{-(n+1)}; \quad \bar{J} = \sum_{n=-\infty}^{\infty} \bar{J}_n \bar{z}^{-(n+1)},
\tag{1.56}
$$

it is obvious from the mode expansion (1.18) that the algebra of the currents is related to that of the α_n operators, namely

$$
[J_m, J_n] = -m\delta_{m+n}; \quad [\bar{J}_m, \bar{J}_n] = -m\delta_{m+n}.
\tag{1.57}
$$

This form of algebra will be shown in the next chapter to be associated with level $k=1$ abelian affine current algebra (or Kac–Moody algebra as it is sometimes referred to).

This algebra translates into the following algebra of the currents:

$$
[J(z), J(w)] = \delta'(z-w).
\tag{1.58}
$$

Using the technique developed in Section 1.7 we can derive this result also from the operator product expansion of two currents,

$$J(z)J(w) = \frac{1}{(z-w)^2} + \text{finite terms.} \tag{1.59}$$

1.11 Virasoro algebra

Next we address the algebraic structure of the generators of conformal transformations (1.27). Upon inserting (1.18) into the Laurent expansion of the energy momentum tensor,

$$T = \sum_{n=-\infty}^{\infty} L_n z^{-(n+2)}; \quad \bar{T} = \sum_{n=-\infty}^{\infty} \bar{L}_n \bar{z}^{-(n+2)}, \tag{1.60}$$

one finds that for L_n with $n \neq 0$,

$$L_n = 1/2 \sum_{m=-\infty}^{\infty} : \alpha_{n-m}\alpha_m : . \tag{1.61}$$

For $n \neq 0$ the operators α_{n-m} and α_m commute, and so the product equals the normal ordered one. The situation is different for L_0. Here one encounters an infinity in the product of chiral fields, which normal ordering removes, resulting in,

$$L_0 = 1/2\mathcal{P}^2 + \sum_{1}^{\infty} \alpha_{-m}\alpha_m. \tag{1.62}$$

We shall later see that it is sometimes necessary to shift L_0 by a constant. Using the commutation relation of α_n one finds the following "naive" expression for the commutator of L_n operators:

$$\begin{aligned}
[L_m, L_n] &= \frac{1}{4} \sum_{k,l} [\alpha_{m-k}\alpha_k, \alpha_{n-l}\alpha_l] \\
&= \frac{1}{2} \sum_k k\alpha_{m-k}\alpha_{k+n} + \frac{1}{2} \sum_k (m-k)\alpha_{m-k+n}\alpha_k \\
&= (m-n)L_{m+n},
\end{aligned} \tag{1.63}$$

where to get to the third line we have changed a variable in the first sum from $k \to k - n$. This is the *classical Virasoro algebra*,[6] and in fact in the quantum theory it is further corrected. The correction appears only for the case $m + n = 0$, so for $m \neq -n$ the classical form (1.63) is exact. For generators with $m + n = 0$ the two sums in the second line of (1.63) have to be brought to normal order. As re-ordering means using the commutator, one gets divergent series for which, in the case at hand, one cannot shift the variable of summation without changing

---

[6] The Virasoro algebra was presented in [212]. More references will be given in the next two chapters.

the result. Taking this into account, one gets a c-number shift in the commutation rule,

$$[L_m, L_n] = (m - n)L_{m+n} + \mathcal{A}(m)\delta_{m+n}.$$ (1.64)

To compute the anomaly term $\mathcal{A}(m)$ we introduce a cutoff function $f_\Lambda(k)$, which tends to 1 in the limit of infinite regulator Λ for any k, but for every finite Λ goes to zero sufficiently rapidly at infinite k. Thus we view the operators L_n as regularized sums,

$$L_n = 1/2 \sum_{m=-\infty}^{\infty} : \alpha_{n-m}\alpha_m : f_\Lambda(m),$$ (1.65)

to replace (1.61). With this regularized expression, a direct computation gives for the anomaly,

$$\mathcal{A}(m) = 1/4 \sum_{k=1}^{\infty} \{k(m-k)f_\Lambda(m-k)[f_\Lambda(k-m) + f_\Lambda(-k)]$$
$$+ k(m+k)f_\Lambda(-k)[f_\Lambda(k) + f_\Lambda(-m-k)]\}.$$ (1.66)

If we now take $f_\Lambda(k)$ to 1, without being careful, we get the divergent sum,

$$\mathcal{A}(m) \to m \sum_{k=1}^{\infty} k.$$ (1.67)

Using ζ-function regularization, namely replacing k by k^{-s}, we get a convergent sum for any $s > 1$, and then we continue analytically to $s = -1$, to get $-m/12$ for the right-hand side of the last equation.

To compute $\mathcal{A}(m)$ with the regulators f_Λ, we now look at,

$$\mathcal{A}(m) + \frac{m}{12} = 1/4 \sum_{k=1}^{\infty} \{k(m-k)f_\Lambda(m-k)[f_\Lambda(k-m) + f_\Lambda(-k)]$$
$$+ k(m+k)f_\Lambda(-k)[f_\Lambda(k) + f_\Lambda(-m-k)] - 4mk\}.$$ (1.68)

Only large k is relevant now, as for any finite k we can take Λ to infinity first, obtaining zero on the right-hand side. We now take,

$$f_\Lambda(q) \approx |q|^{-p} \quad |q| \to \infty,$$ (1.69)

with $p \to 0$ as $\Lambda \to \infty$. Expanding in powers of $\frac{m}{k}$, and recalling that $\zeta(s)$ has a pole only at $s = 1$, we get by summing first and then letting $p \to 0$, the result,

$$\mathcal{A}(m) + \frac{m}{12} = \frac{m^3}{12}.$$ (1.70)

The anomaly term $\mathcal{A}(m)$ can also be determined using the Jacobi identity $[L_k, [L_m, L_n]] + [L_m, [L_n, L_k]] + [L_n, [L_k, L_m]] = 0$. One finds that for $k + m + n = 0$ the anomaly term obeys $(m-n)\mathcal{A}(k) + (n-k)\mathcal{A}(m) + (k-n)\mathcal{A}(m) = 0$. Recall also that $\mathcal{A}(0) = 0$ and $\mathcal{A}(m) = -\mathcal{A}(-m)$ so it is enough to determine $\mathcal{A}(m)$ for positive m. The relation derived from the Jacobi identity can be used to get a recursion relation which is determined by values of $\mathcal{A}(1)$ and $\mathcal{A}(2)$. In fact the general solution is of the form $\mathcal{A}(n) = b_3 n^3 + b_1 n$. The coefficient b_1 is correlated with the normal ordering ambiguity constant of L_0. One can determine the coefficients b_1 and b_3 by computing the vacuum expectation values of $<0|[L_1, L_{-1}]|0> = 0$ and $<0|[L_2, L_{-2}]|0> = \frac{1}{2}$, so that altogether one finds $\mathcal{A}(n) = \frac{1}{12} n(n^2 - 1)$ and the full Virasoro algebra associated with the massless free scalar field is,

$$[L_m, L_n] = (m - n)L_{m+n} + \frac{1}{12} m(m^2 - 1)\delta_{m+n}. \tag{1.71}$$

In the next chapter the Virasoro algebra will be discussed in a broader perspective. In that context it will become clear that the algebra of (1.71) associated with the massless free scalar is characterized by a $c = 1$ Virasoro anomaly.

2

Conformal field theory

Conformal invariance of two-dimensional massless scalar field theory was shown in the previous chapter to associate with the infinite algebra of conserved charges, the Virasoro algebra. In this chapter we describe the basic building blocks of any two-dimensional conformal field theory (CFT). The notions of primary and descendant operators will be introduced and the structure of the Hilbert space of states will be described. We will discuss and classify certain classes of unitary CFTs. Crossing symmetry, duality and bootstrap equations will be defined and applied to computing correlators of CFTs. We then discuss the Verlinde formula which relates the fusion rules and the S transformation. We will end up with two examples of CFTs that demonstrate all of the concepts that have been introduced before. The first one is the theory of a Majorana fermion and the second is the $m = 3$ unitary minimal model, which is shown to be the continuum limit of the two-dimensional Ising model.

Conformal field theory in two dimensions is covered by many review articles and books. The former include [109] which we use intensively in this chapter, also [25], [13], [59], [233] and many others.

Among the books that discuss 2d CFT is [140] and books on string theories [113], [154], [174], [138], [237], [142], [30].

The most complete book on the topic is [77].

The basics of conformal field theory were stated in the seminal paper by Belavin, Polyakov and Zamolodchikov [33]. This includes the introduction of primary fields, the behavior of the energy-momentum tensor and the central charge. Conformal Ward identity and the use of OPEs appears in [93], [95] and [94].

2.1 Conformal symmetry in two dimensions

The theory of the free massless scalar field in two dimensions was shown to be invariant under the holomorphic and anti-holomorphic coordinate transformations

$$z \rightarrow z' = f(z); \quad \bar{z} \rightarrow \bar{z}' = \bar{f}(\bar{z}). \tag{2.1}$$

Under such a transformation the metric transforms as

$$ds^2 = dz d\bar{z} \rightarrow dz' d\bar{z}' = \frac{\partial z'}{\partial z} \frac{\partial \bar{z}'}{\partial \bar{z}} dz d\bar{z}. \tag{2.2}$$

At this point we can understand why we referred to these transformations as conformal transformations. In general in d space-time dimensions the conformal group is the subgroup of coordinate transformations that leaves the metric invariant up to a scale, namely,

$$g_{\mu\nu}(x) \rightarrow g'_{\mu\nu}(x') = \Omega(x)g_{\mu\nu}(x). \tag{2.3}$$

It is obvious from (2.2) that the 2d conformal transformations (2.1) indeed produce such a variation of the metric. An important property of conformal transformations in any dimension is that they preserve the angle $\frac{\vec{A}\cdot\vec{B}}{\sqrt{A^2 B^2}}$ between two vectors $\vec{A}$ and $\vec{B}$.

Starting from flat space, the general infinitesimal coordinate transformations $x^\mu \rightarrow x^\mu + \epsilon^\mu(x)$ induces a change of the metric $ds^2 \rightarrow ds^2 + (\partial_\mu \epsilon_\nu + \partial_\nu \epsilon_\mu)dx^\mu dx^\nu$, so that the condition for conformal transformations reads,

$$\partial_\mu \epsilon_\nu + \partial_\nu \epsilon_\mu = \frac{2}{d}(\partial \cdot \epsilon)g_{\mu\nu}, \tag{2.4}$$

where $g_{\mu\nu}$ is $\eta_{\mu\nu}$ or $\delta_{\mu\nu}$ for a Minkowskian signature, or Euclidean signature, respectively.

It is thus obvious that for two-dimensional Euclidean space-time $\epsilon = \epsilon(z)$ and $\bar{\epsilon} = \bar{\epsilon}(\bar{z})$ are the unique solutions of (2.4), which reduces to the Cauchy–Riemann equation $\partial_1 \epsilon_1 = \partial_2 \epsilon_2$ and $\partial_1 \epsilon_2 = -\partial_2 \epsilon_1$.

We would like now to put aside scalar field theory and explore the general properties of conformal field theories in two dimensions. Any theory with a vanishing trace of the energy-momentum tensor $T^\mu_\mu = 0$, or in complex coordinates $T_{z\bar{z}} = 0$, has necessarily an independent holomorphically (and anti-holomorphically) conserved energy-momentum tensor components, namely,

$$\bar{\partial}T \equiv \bar{\partial}T_{zz} = 0 \quad \partial\bar{T} \equiv \partial T_{\bar{z}\bar{z}} = 0. \tag{2.5}$$

This follows trivially from the usual conservation law $\bar{\partial}T_{zz} + \partial T_{z\bar{z}} = 0$, and its complex conjugation. It is also clear that in fact there are infinitely many conserved currents, since $g(z)T(z)$ for any analytic function $g(z)$ is also a holomorphically conserved current (we sometimes call any conserved tensor "current"). We show in the following section that indeed the energy-momentum tensor $T(z)$ and $\bar{T}(\bar{z})$ generate the conformal transformations given in (2.1).

2.2 Primary fields

Conformal invariance constrains the OPEs of the theory. In particular, since T is holomorphic, the OPE of $T(z)$ with a general operator can be expanded in terms of a Laurent expansion in integer powers of z. The singular part of the OPE takes the form,

$$T(z)\tilde{\mathcal{O}}(w,\bar{w}) = \sum_{n=0}^{\infty} \frac{1}{(z-w)^{n+1}} \tilde{\mathcal{O}}^{(n)}(w,\bar{w}), \tag{2.6}$$

where the sum is usually finite, and the operators $\tilde{O}^{(n)}(w, \bar{w})$ have to be determined. Using radial quantization as in Section 1.7 and the OPE above, we get for the transformation generated by $T(z)$,

$$\delta_\epsilon \tilde{O}(w, \bar{w}) = \sum_n \frac{1}{n!} \left[(\partial^n \epsilon) \tilde{O}^{(n)}(w, \bar{w}) \right]. \tag{2.7}$$

We now consider operators that transform under conformal transformation in a way that generalizes the transformation of the metric, (2.2),

$$\mathcal{O}(z, \bar{z}) \to \mathcal{O}'(z'\bar{z}') = \left(\frac{\partial z'}{\partial z} \right)^h \left(\frac{\partial \bar{z}'}{\partial \bar{z}} \right)^{\bar{h}} \mathcal{O}(z'\bar{z}'). \tag{2.8}$$

An operator with such conformal transformations is a *primary field* or a *tensor operator* with *conformal weights* $(h, \bar{h})$, which are sometimes referred to as the holomorphic and anti-holomorphic conformal dimensions.[1] The sum of the weights $h + \bar{h}$ is the *total dimension* that determines the behavior under scaling, whereas $h - \bar{h}$ is the *spin* that controls the behavior under rotations. The infinitesimal transformations that correspond to (2.8) are,

$$\delta_{\epsilon, \bar{\epsilon}} \mathcal{O}(z, \bar{z}) = \left[(h \partial \epsilon + \epsilon \partial) + (\bar{h} \bar{\partial} \bar{\epsilon} + \bar{\epsilon} \bar{\partial}) \right] \mathcal{O}(z\bar{z}). \tag{2.9}$$

This form of transformation implies that the singular part of the OPE of T and $\mathcal{O}(w, \bar{w})$ reduces to,

$$T(z)\mathcal{O}(w, \bar{w}) = \frac{h}{(z-w)^2} \mathcal{O}(w, \bar{w}) + \frac{1}{(z-w)} \partial \mathcal{O}(w, \bar{w}). \tag{2.10}$$

Applying these notions to the free scalar field we find that $\partial X(z)$ has $(1, 0)$ weights, $\bar{\partial} \bar{X}(\bar{z})$ has $(0, 1)$ and the weights of $: e^{i\alpha X(z, \bar{z})}:$ are $(\frac{\alpha^2}{2}, \frac{\alpha^2}{2})$.

In Chapter 1 the notion of OPE was discussed in the context of scalar field theory. The generalization to any CFT is straightforward. Normalize the operators with fixed conformal weights as,

$$\langle \mathcal{O}_i(z, \bar{z}) \mathcal{O}_j(w, \bar{w}) \rangle = \delta_{ij} \frac{1}{(z-w)^{2h_i}} \frac{1}{(\bar{z} - \bar{w})^{2\bar{h}_i}}, \tag{2.11}$$

then, for a complete set, the OPE of any pair of such operators is, to leading singularity,

$$\mathcal{O}_i(z, \bar{z}) \mathcal{O}_j(w, \bar{w}) \sim \sum_k C_{ijk} (z-w)^{h_k - h_i - h_j} (\bar{z} - \bar{w})^{\bar{h}_k - \bar{h}_i - \bar{h}_j} \mathcal{O}_k(w, \bar{w}), \tag{2.12}$$

where C_{ijk} are the *product coefficients* of the theory.

---

[1] The notion of conformal primary field and its descendants was introduced in [33] and further discussed in [236].

2.3 Conformal properties of the energy-momentum tensor

For the free massless scalar field we found that the OPE of $T(z)T(w)$ is not of the form shown as (2.6), due to the anomaly term as in (1.71). The form of $T(z)T(w)$ OPE for any CFT is rather,

$$T(z)T(w) = \frac{c/2}{(z-w)^4} + \frac{2T(w)}{(z-w)^2} + \frac{\partial_w T(w)}{(z-w)}, \tag{2.13}$$

where c is the *central charge* (or the Virasoro anomaly), a constant that characterizes the theory. The second term represents the dimensions and the third the property of translations under T. For theories with positive semi-definite Hilbert space $c \geq 0$, as follows from,

$$< T(z)T(w) >= \frac{c/2}{(z-w)^4}.$$

This type of OPE implies the following infinitesimal transformation of T:

$$\delta_{\epsilon(z)} T(z) = \frac{c}{12} \partial^3 \epsilon(z) + 2(\partial \epsilon(z))T(z) + \epsilon(z)\partial T(z). \tag{2.14}$$

The corresponding finite transformation $T(z) \to T'(z')$ takes the form,

$$T'(z') = (\partial z')^2 T(z) + \frac{c}{12}\{z', z\}, \tag{2.15}$$

where $\{z', z\}$ is the Schwarzian derivative,

$$\{f, z\} = \frac{2\partial^3 f \partial f - 3\partial^2 f \partial^2 f}{2\partial f \partial f}. \tag{2.16}$$

To derive (2.16), we first note that by applying a second transformation $f \to w$ we get,

$$\{w, z\} = (\partial_z f)^2 \{w, f\} + \{f, z\}. \tag{2.17}$$

Then, we take $w = f + \delta f$, thus obtaining a functional equation,

$$\delta f \frac{\delta}{\delta f}\{f, z\} = (\partial_z f)^2 \frac{\partial^3 \delta f}{\partial^3 f}. \tag{2.18}$$

Expressing the right-hand side as derivatives with respect to z,

$$\frac{1}{f'}(\delta f)''' - \frac{3f''}{(f')^2}(\delta f)'' + \left[\frac{3(f'')^2}{(f')^3} - \frac{f'''}{(f')^2}\right](\delta f)',$$

we can integrate the equation to get (2.16). The first term suggests integrating to f'''/f', the variation of which gives $1/f'(\delta f)''' - f'''/(f')^2(\delta f)'$, while the second term suggests $-3(f'')^2/2(f')^2$, the variation of which gives $-3f''/(f')^2(\delta f)'' + 3(f'')^2/(f')^3(\delta f)'$.

For the massless scalar case T can be written as $T(z) = -\frac{1}{2} : J(z)J(z) :$, as we saw in (1.5). In fact, as will be discussed in Chapter 3, there is a large class of theories that share this so-called *Sugawara form*. For this type of theory the proof that the finite transformation is of the form of (2.15) is as follows. Recall that as a primary field of weights $(1,0)$, $J(z) \to \frac{\partial z'}{\partial z} J(z')$. If we write $T(z) = -\frac{1}{2}\lim_{z \to w} (J(z)J(w) + \frac{1}{(z-w)^2})$ and substitute the transformation of the currents we end up after some lengthy but straightforward calculation with (2.15).

2.4 Virasoro algebra for CFT

Let us use the Laurent expansion of T for CFT, following (1.60),

$$T = \sum_{n=-\infty}^{\infty} L_n z^{-(n+2)} \qquad \bar{T} = \sum_{n=-\infty}^{\infty} \frac{\bar{L}_n}{\bar{z}^{-(n+2)}}, \tag{2.19}$$

so that,

$$L_n = \frac{1}{2\pi i} \oint dz\, z^{n+1} T(z). \tag{2.20}$$

The expansion is chosen such that L_n has scale dimension n under $z \to z/a$, namely, $L_n \to a^n L_n$.

The Virasoro algebra[2] can now be derived using the OPE of $T(z)T(w)$ given in (2.13),

$$[L_n, L_m] = \left(\frac{1}{2\pi i}\right)^2 \oint dz \oint dw [z^{n+1}w^{m+1} - z^{m+1}w^{n+1}]T(z)T(w). \tag{2.21}$$

The double integral is performed by fixing w and transforming the difference of the two $\oint dz$ integrals into one integral around w,

$$[L_n, L_m] = \left(\frac{1}{2\pi i}\right)^2 \oint dz \oint dw \left[z^{n+1}w^{m+1} - z^{m+1}w^{n+1}\right]$$

$$\left[\frac{c/2}{(z-w)^4} + \frac{2T(w)}{(z-w)^2} + \frac{\partial_w T(w)}{(z-w)}\right]$$

$$= \left(\frac{1}{2\pi i}\right) \oint dw \left[c/12(n^3 - n)w^{n+m-1}\right.$$

$$\left. + [2(n+1) - (n+m+2)]\, w^{n+m+1}T(w)\right]$$

$$= \frac{c}{12}(n^3 - n)\delta(n+m) + (n-m)L_{n+m}. \tag{2.22}$$

Performing identical steps for $\bar{L}_n$ we get that $\bar{L}_n$ obeys the same infinite algebra, with some central charge $\bar{c}$, and that $[L_n, \bar{L}_m] = 0$.

Any CFT is a representation of the Virasoro algebra characterized by c and $\bar{c}$. It is straightforward to identify the following properties of the algebra:

- The generators $(L_{\pm 1}, L_0)$ span an $SL(2, \mathcal{R})$ algebra,

$$[L_{+1}, L_{-1}] = 2L_0 \quad [L_0, L_{\pm 1}] = \mp L_\pm \tag{2.23}$$

---

[2] The first use of the Virasoro algebra was by M. Virasoro in the context of the dual resonance model [212]. Its application to two-dimensional CFT was presented in [33].

Table 2.1. *The conformal family*

Level	Weight	Fields
0	h	ϕ
1	$h+1$	$L_{-1}\phi$
2	$h+2$	$L_{-2}\phi, L_{-1}^2\phi$
3	$h+3$	$L_{-3}\phi, L_{-2}L_{-1}\phi, L_{-1}^3\phi$
$\vdots$	$\vdots$	$\vdots$
N	$h+N$	$P(N)$ fields

- For $n > 0$, L_{-n} is a raising operator and L_n is a lowering one, since $[L_0, L_n] = -nL_n$ so that if $|\psi>$ is an eigenstate of L_0, $L_0|\psi> = h|\psi>$ then $L_0|L_n\psi> = (h - n)|L_n\psi>$.

2.5 Descendant operators

From every primary operator $\phi(z, \bar{z})$ one can construct an infinite tower of Virasoro descendant operators,

$$(L_{-n}\phi(w, \bar{w})) = \frac{1}{2\pi i} \oint dz \frac{1}{z^{n-1}} T(z)\phi(w, \bar{w}). \tag{2.24}$$

A distinguished descendant operator is the energy momentum tensor $T(z)$ since,

$$L_{-2}\mathbf{1} = \frac{1}{2\pi i} \oint \frac{dz}{z} T(z)\mathbf{1} = T(0). \tag{2.25}$$

The set containing the primary field $\phi(z, \bar{z})$ and all its descendant operators is called a *conformal family* and it is denoted by $[\phi]$. A conformal family is a tower of operators where each layer is characterized by its level as shown in Table 2.1, where $P(N)$ is the number of partitions of N into positive integer parts, which can be written in terms of the generating function $\prod_{n=1} \frac{1}{(1-q^n)} = \sum_{N=0}^{\infty} P(N)q^N$.

We can now use the conformal family to rewrite the expression of the OPE (2.12) of two primary fields,

$$\phi_i(z, \bar{z})\phi_j(w, \bar{w})$$
$$= \sum_{k\{l\bar{l}\}} C_{ijk}^{\{l\bar{l}\}} (z - w)^{h_k - h_i - h_j + \sum_n l_n} (\bar{z} - \bar{w})^{\bar{h}_k - \bar{h}_i - \bar{h}_j + \sum_n \bar{l}_n} \phi_k^{l\bar{l}}(w, \bar{w}),$$

$$\tag{2.26}$$

where we denote by $\phi_k^{l\bar{l}}(w, \bar{w})$ the descendants $L_{-l_1} \ldots L_{-l_n} \bar{L}_{-\bar{l}_1} \ldots \bar{L}_{-\bar{l}_n} \phi_k(w, \bar{w})$ with the normalization given in (2.11). The product coefficients $C_{ijk}^{\{l\bar{l}\}}$ are given in terms of those of (2.12) C_{ijk} as,

$$C_{ijk}^{\{l\bar{l}\}} = C_{ijk} \beta_{ij}^{k\{l\}} \bar{\beta}_{ij}^{k\{\bar{l}\}}, \tag{2.27}$$

where $\beta_{ij}^{k\{l\}}$ are determined by conformal invariance and are functions of c and h_i, h_j, h_k, and similarly for $\bar{\beta}_{ij}^{k\{\bar{l}\}}$. This follows from a detailed analysis that we do not show here.

The OPEs of any pair of descendant fields can also be deduced from (2.12) which implies in fact that all the information about the OPE is encoded in the product coefficients C_{ijk}. Moreover since the structure of (2.26) holds for all the primaries and their descendants, one can write the so-called *fusion algebra* for conformal, families, which takes the form,

$$[\phi_i][\phi_j] = \sum_k N_{ij}^k [\phi_k]. \tag{2.28}$$

2.6 Hilbert space of states

Our next task is to construct the Hilbert space of states. First we define the ground state $|0>$ by,

$$L_n|0> = 0 \quad n \geq 0. \tag{2.29}$$

The next step in this program is to build the highest weight states (hws). Consider the state generated from the vacuum by a primary field $\phi(z)$ of dimension h,

$$|h> = \phi(0)|0> . \tag{2.30}$$

It is easy to check that for $n > 0$, $[L_n, \phi(0)] = 0$ since,

$$[L_n, \phi(w)] = \frac{1}{2\pi i} \oint dz z^{n+1} T(z)\phi(w) = h(n+1)w^n \phi(w) + w^{n+1}\partial\phi(w). \tag{2.31}$$

Hence the *highest weight state* $|h>$ obeys

$$L_0|h> = h|h> \quad L_n|h> = 0 \quad n > 0. \tag{2.32}$$

Expanding the primary field $\phi(z)$ in a Laurent series $\sum_n \phi_n z^{(n-h)}$, one can write the highest weight state symbolically as $\phi_h|0>$.

Descendant states are generated by applying the descendant operators $L_{-n}\phi$ on the vacuum or alternatively by applying L_{-n} on highest weight states, namely,

$$L_{-n}|h> = L_{-n}\phi(0)|0> = (L_{-n}\phi)|0> . \tag{2.33}$$

It is thus clear that the highest weight states, or equivalently the primary operators, play a major role in constructing representations of the Virasoro algebra. In fact one can show that every representation is characterized by a primary operator. Consider an eigenstate of L_0, $L_0|\psi> = h_\psi|\psi>$. Now act on it with the lowering operator L_n with $n > 0$. The L_0 eigenvalue of the new state $L_n|\psi>$ is $h_\psi - n$. Since we require that the Hamiltonian is bounded from below, L_0 has to be also bounded. This implies that after repeating the lowering process one finally hits a state that is annihilated by L_n for every $n > 0$ and hence an hws.

It is thus clear that any state in a positive Hilbert space is a linear combination of hws, and their descendants. The representation given in Table 2.1 is referred to as the *Verma module*. Denoting it by $\mathcal{V}(c, h)$ and its analogous representation for the anti-holomorphic Virasoro algebra by $\bar{\mathcal{V}}(\bar{c}, \bar{h})$, the Hilbert space of the theory is a direct sum of the products $\mathcal{V}(c, h) \otimes \bar{\mathcal{V}}(\bar{c}, \bar{h})$, namely,

$$\mathcal{H} = \sum_{h,\bar{h}} \mathcal{V}(c, h) \otimes \bar{\mathcal{V}}(\bar{c}, \bar{h}). \tag{2.34}$$

The Verma module may be *reducible* in the sense that there is a submodule that is by itself a Verma module. Such a submodule whose states transform amongst themselves under any conformal transformation, is built from a $|h_{\text{null}}\rangle$. The latter is both an hws., namely $L_n|h_{\text{null}}\rangle = 0$ for $n > 0$, as well as a descendant. Such a state is called *null state* or *null vector*, motivated by what follows. It generates its own Verma module which is included in the parent module. It is orthogonal to the whole Verma module as well as to itself $\langle h_{\text{null}}|h_{\text{null}}\rangle = 0$, since $\langle h_{\text{null}}|L_{-k_1} \ldots L_{-k_n}|h\rangle = \langle h|L_{k_n} \ldots L_{k_1}|h_{\text{null}}\rangle^* = 0$, and in particular it has a zero norm $\langle h_{\text{null}}|h_{\text{null}}\rangle = 0$ and similarly also its descendants. The null state corresponds to a null operator which is simultaneously a primary and a secondary field.

Let us now demonstrate the construction of a null vector. Consider a general linear combination of the states of level 2,

$$L_{-2}|h\rangle + aL_{-1}^2|h\rangle, \tag{2.35}$$

we would like to check whether for certain values of the mixing coefficient a, this state is a null state. If indeed it is $|\text{null}\rangle$, then so is the state $[L_n|\text{null}\rangle]$. In fact it is easy to verify that at level 2, one has to check these consistency conditions only for L_1 and L_2. Now using the Virasoro algebra we find that,

$$[L_1, L_{-2}]|h\rangle + a[L_1, L_{-1}^2]|h\rangle = (3 + 2a(2h + 1))L_{-1}|h\rangle,$$
$$[L_2, L_{-2}]|h\rangle + a[L_2, L_{-1}^2]|h\rangle = \left(4h + \frac{c}{2} + 6ah\right)|h\rangle. \tag{2.36}$$

It is thus clear that for the following values of a and c,

$$a = -\frac{3}{2(2h + 1)} \qquad c = \frac{2h}{2h + 1}(5 - 8h), \tag{2.37}$$

the linear combination state (2.35) is a null state. In the unitary case we have h and c positive (see next section). Hence in this example $h < \frac{5}{8}$.

An irreducible representation of the Virasoro algebra can be constructed from a Verma module that contains a null vector by a quotient procedure, taking out of the Verma module the null module. In the next section we discuss this construction.

2.7 Unitary CFT and Kac determinant

Unitarity is obviously lost if there are negative norm states in the Verma module. Hence, our task is to derive the conditions for having a negative norm state. In the basis of the Verma module,

$$L_{-k_1} \dots L_{-k_i} |h> \equiv |s> \quad (1 \le k_1 \le \dots \le k_i), \tag{2.38}$$

the matrix of inner products $\mathbf{I}_{ss'} = <s|s'>$ is block diagonal with blocks $\mathbf{I}^{(N)}$ for states at level $N(\sum_i k_i = N)$. For a given Verma module the elements of $\mathbf{I}$ are functions of (h, c). It is easy to realize that unitarity dictates $c > 0$ and $h > 0$. This follows from $<h|L_n L_{-n}|h> = [2nh + 1/12cn(n^2 - 1)] <h|h>$, which is positive for $n = 1$ only if $h > 0$ and for large enough n only for $c > 0$. To determine the full set of constraints for unitarity let us analyze further the properties of $\mathbf{I}$. A general state $|\hat{s}> = \sum_k c_k |s>$ has a norm $<\hat{s}|\hat{s}> = \hat{c}^\dagger \mathbf{I} \hat{c}$, with $\hat{c}$ the vector of the c_k. Now since $\mathbf{I}$ is hermitian it can be diagonalized by a unitary matrix U so that the norm can be written as $<\hat{s}|\hat{s}> = \sum_k l_k |t_k|^2$ where $t = U\hat{c}$ and l_k are the eigenvalues of $\mathbf{I}$, which are real. It is thus clear that there are negative norm states if and only if $\mathbf{I}$ has negative eigenvalues. A vanishing eigenvalue indicates a null vector which means a reducible Verma module.

For the low lying levels these matrices take the following form:

$$\begin{aligned}
\mathbf{I}^{(0)} &= 1 \\
\mathbf{I}^{(1)} &= 2h \\
\mathbf{I}^{(2)} &= \begin{pmatrix} 4h(2h+1) & 6h \\ 6h & 4h + c/2 \end{pmatrix}.
\end{aligned} \tag{2.39}$$

The derivation of the various elements is straightforwad, for instance,

$$\begin{aligned}
\mathbf{I}^{(2)}_{11} &= <h|L_1 L_1 L_{-1} L_{-1}|h> = <h|L_1 L_{-1} L_1 L_{-1}|h> + 2 <h|L_1 L_0 L_{-1}|h> \\
&= 4 <h|L_1 L_{-1} L_0|h> + 2 <h|L_1 L_{-1}|h> = 8h^2 + 4h.
\end{aligned} \tag{2.40}$$

The determinant of $\mathbf{I}^{(2)}$ is given by

$$\det[\mathbf{I}^{(2)}] = 32(h - h_{1,1})(h - h_{1,2})(h - h_{2,1}), \tag{2.41}$$

where $h_{1,1} = 0$ and $h_{1,2}, h_{2,1}$ are $(1/16)[(5 - c) \pm \sqrt{(1 - c)(25 - c)}]$. The trace of $\mathbf{I}^{(2)}$ is $\text{Tr}[\mathbf{I}^{(2)}] = 8h(h + 1) + c/2$. Since the trace and the determinant are the sum and product of the two eigenvalues, unitarity is lost if either the trace or the determinant is negative.

The determinant for $\mathbf{I}^{(N)}$ at general level N, which is referred to as the *Kac determinant*,[3] has the form

$$\det[\mathbf{I}^{(N)}] = \alpha_N \prod_{pq \le N} [h - h_{p,q}(c)]^{P(N-pq)}, \tag{2.42}$$

---

[3] The proof of the Kac determinant is detailed in [89], [206] and [95].

where α_N are constants independent of (c, h) and $h_{p,q}(c)$ can be expressed in terms of $m = -\frac{1}{2} \pm \frac{1}{2}\sqrt{\frac{25-c}{1-c}}$ as,

$$h_{p,q}(c) = \frac{[(m+1)p - mq]^2 - 1}{4m(m+1)}. \qquad (2.43)$$

Note that we can choose either the plus or the minus sign in the expression for m, as their interchange is like interchanging p with q, which does not change the determinant. Note also that $h_{p,q}$ is invariant under $p \to m - p, q \to m + 1 - q$. Let us also mention that for $N = 2$ the result is identical to (2.41).

In the (h, c) plane the determinant vanishes along the curves $h = h_{p,q}(c)$ which are therefore called the *vanishing curves*. If the determinant (2.42) is negative it means that there is an odd number of negative eigenvalues and hence the corresponding Virasoro representation is not unitary. If the determinant is vanishing or positive one needs to further analyze the determinant as follows:

- For $c > 1$ and $h > 0$ it is straightforward to show that the determinant does not vanish.

  In the domain $1 < c < 25$ the value for m has an imaginary part. Thus $h_{p,q}$ are complex for $p \neq q$, and as they come in complex conjugate pairs the product of the appropriate two factors in the determinant is positive. For $p = q$ the value of $h_{p,q}$ is negative. Thus the determinant is positive in that domain.

  For $c > 25$ the $h_{p,q}$ are negative.

  For large h the matrix is dominated by its diagonal elements.

  Since these elements are positive, the eigenvalues for large h are all positive. Now since the determinant never vanishes in the region considered ($h > 0$, $c > 1$) all the eigenvalues have to be positive on the entire region.

  Note that in $\mathbf{I}^{(2)}$ the off-diagonal element is larger at large h than the 22 element, but still the determinant is dominated at large h by the diagonal elements, and thus also the eigenvalues, as a 2×2 matrix.

- For $c = 1$ we have $h_{p,q} = (p - q)^2/4$, and so the determinant is never negative. However, it vanishes when $h = n^2/4$ for some integer n.

- For $0 < c < 1$, $h > 0$ a closer look at the determinant is required. We draw $h_{p,q}(c)$ in Fig. 2.1.

  By expanding the curves around $c = 1$ one can show that any point in the region can be connected to the right of $c = 1$ by crossing a single vanishing curve. The vanishing of the determinant is due to one eigenvalue that reverses its sign which implies that there are negative norm states at any point in the region that are not on the vanishing curve. In fact it turns out that there are additional negative norm states at points along the vanishing curve except at

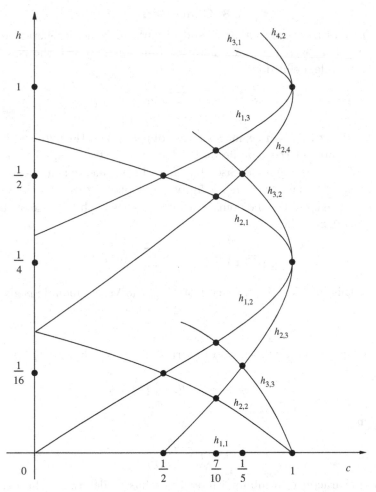

Fig. 2.1. $h_{p,q}(c)$ as a function of c for various values of (p,q).

certain points where they intersect. On these points the central charge c is a solution of $m = -\frac{1}{2} + \frac{1}{2}\sqrt{\frac{25-c}{1-c}}$ for the cases of m an integer from 3 up,

$$c = 1 - \frac{6}{m(m+1)}, \qquad m = 3, 4, \ldots \qquad (2.44)$$

For each such *unitary minimal model*[4] there are $m(m-1)/2$ primary fields with h given by (2.43) where p, q are integers satisfying $1 \leq p \leq m-1$, $1 \leq q \leq p$. The simplest of those models is the Ising model, given in the section $m = 3$, $c = 1/2$ with $h_{1,1} = 0$, $h_{2,1} = 1/2$, $h_{2,2} = 1/16$. It will be described in Section 2.13.

---

[4] The minimal models were presented in [33] and discussed in [95].

2.8 Characters

The structure of the Verma module, and in particular the degeneracy of states at each level, is captured in the generating function $\chi_{(c,h)}(\tau)$, the *character* of the Verma module, defined by,

$$\chi_{(c,h)}(\tau) = \text{Tr}[q^{L_0 - \frac{c}{24}}] = \sum_{n=0}^{\infty} \dim(h+n) q^{h+n-\frac{c}{24}}, \quad (2.45)$$

where $q \equiv e^{2\pi i \tau}$, τ is a complex number, and $\dim(n+h)$ is the number of linearly independent states of the module at level n. The latter is equal to $P(n)$ the partitions of n in the generic case, but may be smaller when there are null states. For $|q| < 1$, namely, τ in the upper half plane, the series is uniformly convergent, since $|q| < 1$ is the domain of convergence of the inverse of the Euler function $\varphi(q)$ defined by,

$$\frac{1}{\varphi(q)} = \prod_{n=1}^{\infty} \frac{1}{1 - q^n} = \sum_{n=0}^{\infty} P(n) q^n. \quad (2.46)$$

In terms of this function the character of a generic Verma module is given by,

$$\chi_{(c,h)}(\tau) = \frac{q^{h - \frac{c}{24}}}{\varphi(q)}. \quad (2.47)$$

The character can be expressed also in terms of the Dedekind $\eta(\tau)$ function,

$$\eta(\tau) \equiv q^{\frac{1}{24}} \varphi(q) = q^{\frac{1}{24}} \prod_{n=1}^{\infty} (1 - q^n), \quad (2.48)$$

in the form

$$\chi_{(c,h)}(\tau) = \frac{q^{h + \frac{1-c}{24}}}{\eta(\tau)}. \quad (2.49)$$

To get the character of a minimal model one has to determine the irreducible Verma module using the quotient procedure discussed in the previous section. We do not give the derivation here, just the final result, which is,

$$\chi(c(p,p'), h_{rs}(p,p')) = \frac{q^{h - \frac{c}{24}}}{\varphi(q)} = \sum_{n \in \mathbb{Z}} \left[q^{\frac{(2pp'n + pr - p's)^2}{4pp'}} - q^{\frac{(2pp'n + pr + p's)^2}{4pp'}} \right], \quad (2.50)$$

where

$$c(p,p') = 1 - 6\frac{(p - p')^2}{pp'}, \quad (2.51)$$

and

$$h_{rs}(p,p') = \frac{(pr - p's)^2 - (p - p')^2}{4pp'}. \quad (2.52)$$

Note that these are the non-unitary minimal models, except for the cases $p - p' = \pm 1$, which coincide with the cases of the previous section with the identification of $p = m$ or $p' = m$.

2.9 Correlators and the conformal Ward identity

Now that the Hilbert space of states has been analyzed we would like to determine the correlation functions of all possible operators of a given CFT. Naturally, we first investigate correlators of primary fields and then those also involving descendents.

A very useful tool for determining correlators are the symmetries of the system. In the present case we obviously implement conformal invariance. In particular we first determine the consequences of the $SL(2,C)$ *Ward identities*. Recall that the vacuum is annihilated by $L_{0,\pm 1}$ and $\bar{L}_{0,\pm 1}$, and hence is invariant under $SL(2,C)$, namely, $U|0> = |0>$ for $U \in SL(2,C)$. It thus follows that,

$$<0|U^{-1}\phi_1(z_1,\bar{z}_1)U\ldots U^{-1}\phi_n(z_n,\bar{z}_n)U|0> = <0|\phi_1(z_1,\bar{z}_1)\ldots\phi-n(z_n,\bar{z}_n)|0> .$$
(2.53)

Recall that by definition a primary field of dimension h transforms under an $SL(2,C)$ transformation $z \to f(z) = \frac{az+b}{cz+d}$ (with $ad - bc = 1$), as,

$$U^{-1}\phi(z,\bar{z})U = (\partial f(z))^h \phi(f(z),\bar{z}).$$
(2.54)

Let us mention that $SL(2,C)$ invariance holds for CFT in any dimension.

The invariance of the vacuum implies, in infinitesimal form,

$$<0|[L_k,\phi_1(z_1,\bar{z}_1)]\ldots\phi_n(z_n,\bar{z}_n)|0> +\ldots <0|\phi_1(z_1,\bar{z}_1)\ldots[L_k,\phi_n(z_n,\bar{z}_n)]|0> = 0,$$
(2.55)

for $k = 0,\pm 1$. Using $[L_k,\phi(z,\bar{z})] = h(k+1)z^k\phi(z,\bar{z}) + z^{k+1}\partial\phi(z,\bar{z})$ we get Ward identities in terms of differential equations:

$$k = -1: \quad \sum_i \partial_i <0|\phi_1(z_1,\bar{z}_1)\ldots\phi_n(z_n,\bar{z}_n)|0> = 0$$

$$k = \quad 0: \quad \sum_i (z_i\partial_i + h_i) <0|\phi_1(z_1,\bar{z}_1)\ldots\phi_n(z_n,\bar{z}_n)|0> = 0$$

$$k = +1: \quad \sum_i (z_i^2\partial_i + 2z_i h_i) <0|\phi_1(z_1,\bar{z}_1)\ldots\phi_n(z_n,\bar{z}_n)|0> = 0. \quad (2.56)$$

These Ward identities are associated with the invariance under translations, dilations and special conformal transformations. Applying these equations to the two point function one finds that,

$$G_2(z_1,\bar{z}_1,z_2,\bar{z}_2) \equiv <0|\phi_1(z_1,\bar{z}_1)\phi_1(z_2,\bar{z}_2)|0> = \frac{c_2}{(z_1 - z_2)^{2h_1}(\bar{z}_1 - \bar{z}_2)^{2\bar{h}_1}},$$
(2.57)

where c_2 is a constant, to be put to 1 in the normalization (2.11). Note also that when taking two different fields ϕ_1 and ϕ_2, $SL(2,C)$ implies that $h_1 = h_2$ is necessary for a non-zero two-point function.

In a similar manner the three-point function is given by,

$$G_3(z_i,\bar{z}_i) = c_{123}\left(\frac{1}{z_{12}^{h_1+h_2-h_3}z_{13}^{h_1+h_3-h_2}z_{23}^{h_2+h_3-h_1}}\right)(z \to \bar{z}, h \to \bar{h}), \quad (2.58)$$

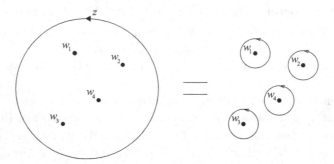

Fig. 2.2. Integration along C that bounds all the operators.

where $z_{ij} = z_i - z_j$ and c_{123} is the correpsonding product coefficient defined in (2.12). Using the $SL(2C)$ invariance one can set the points z_1, z_2, z_3 at $\infty, 1, 0$, respectively so that the constant c_{123} is determined from the corresponding correlator via $\lim_{z_1, \bar{z}_1 \to \infty} \left[z_1^{2h} \bar{z}_1^{2\bar{h}} G_3 \right] = c_{123}$. For G_n with $n > 3$ the global conformal transformations do not fully determine the correlator. For instance the four-point function G_4 can be written using these transformations as,

$$G_4(z_i, \bar{z}_i) = f(\mathcal{Z}, \bar{\mathcal{Z}}) \left[\left(\prod_{i<j} z_{ij}^{-(h_i + h_j) + h/3} \right) (z \to \bar{z}, h \to \bar{h}) \right], \qquad (2.59)$$

where $h = \sum_{i=1}^{4} h_i$ and the cross ratio $\mathcal{Z}$ is defined as $\mathcal{Z} = \frac{z_{12} z_{34}}{z_{13} z_{24}}$, which is an $SL(2, C)$ invariant.

For a general n-point function, denoting the power of z_{ij} by $-h_{ij}$, we get,

$h_{ij} = \left[\frac{2}{n-2}(h_i + h_j) - \frac{2}{(n-1)(n-2)} h \right]$ for $n \geq 3$.

So far we have implemented the global Ward identities. To get the *local Ward identity* one performs a conformal transformation of an n-point function of primary fields G_n. This is achieved by integrating $\epsilon(z) T(z)$ along a contour C which bounds a region that includes all the operators (see Fig. 2.2)

Now using analyticity one can deform the contour into a sum of countours each of which encircles one operator. The result of the integral is therefore,

$$\left\langle \oint \frac{dz}{2\pi i} \epsilon(z) T(z) \phi(w_1, \bar{w}_1) \dots \phi(w_n, \bar{w}_n) \right\rangle$$

$$= \sum_{i=1}^{n} \left\langle \phi(w_1, \bar{w}_1) \dots \oint \frac{dz}{2\pi i} \epsilon(z) T(z) \phi(w_i, \bar{w}_i) \dots \phi(w_n, \bar{w}_n) \right\rangle$$

$$= \sum_{i=1}^{n} \left\langle \phi(w_1, \bar{w}_1) \dots \delta_\epsilon \phi(w_i, \bar{w}_i) \dots \phi(w_n, \bar{w}_n) \right\rangle. \qquad (2.60)$$

Using (2.9) we substitute now for $\delta_\epsilon \phi(w_i, \bar{w}_i) = \epsilon(w_i)\partial + h\partial\epsilon(w_i)\phi(w_i, \bar{w}_i)$. Since this holds for arbitrary ϵ we can get a local form of the Ward identity,

$$\langle T(z)\phi(w_1, \bar{w}_1)\ldots\phi(w_n, \bar{w}_n)\rangle = \sum_{i=1}^{n}\left(\frac{h_i}{(z-w_i)^2} + \frac{1}{(z-w_i)}\frac{\partial}{\partial w_i}\right)$$

$$\langle\phi(w_1, \bar{w}_1)\ldots\phi(w_n, \bar{w}_n)\rangle, \qquad (2.61)$$

similar to the transition from (2.9) to (2.10). It is thus clear that the correlation function above is a meromorphic function of z with singularities at the positions of the operators.

A useful tool for computing correlators is the use of null vectors. Rather than discussing this for a general null vector we demonstrate this procedure on a level two null vector. Recall that in models with a primary of weight h such that $c = \frac{2h}{2h+1}(5-8h)$ there is a null vector at level two of the form $(L_{-2} + aL_{-1}^2)\Phi^{(h)} = 0$ where $a = -\frac{3}{2(2h+1)}$. As $L_{-1}\phi^{(h)}(z) = \partial\phi^{(h)}(z)$ one can trade $L_{-2}\phi^{(h)}(z)$ with $-a\partial^2\phi^{(h)}(z)$. Now $L_{-2}\phi^{(h)}(w)$ is given by,

$$L_{-2}\phi^{(h)}(w) = \lim_{z\to w}\left[T(z)\phi^{(h)}(w) - \frac{h\phi^{(h)}(w)}{(z-w)^2} - \frac{\partial_w\phi^{(h)}(w)}{(z-w)}\right]. \qquad (2.62)$$

Substituting this into (2.61) one finds the following differential equation,

$$-a\partial_{w_1}^2\langle\phi(w_1, \bar{w}_1)\ldots\phi(w_n, \bar{w}_n)\rangle$$

$$= \sum_{i\neq 1}^{n}\left(\frac{h_i}{(w_1-w_i)^2} + \frac{1}{(w_1-w_i)}\frac{\partial}{\partial w_i}\right)\langle\phi(w_1, \bar{w}_1)\ldots\phi(w_n, \bar{w}_n)\rangle. \qquad (2.63)$$

This exact differential equation will enable us to compute the four-point function for the Ising model as we discuss in Section (2.13).

Next we would like to deduce the implications of the associativity on correlation functions of primaries and descendant operators.

2.10 Crossing symmetry, duality and bootstrap

The complete package of information that specifies a CFT is its Virasoro anomaly c, the set of primary fields $\phi_i(z, \bar{z})$, with their weights $(h_i, \bar{h}_i)$ and the operator product coefficients C_{ijk}. Hence, to determine all consistent CFTs one has to find all the allowed sets of such data. The latter have to comply with the constraints that follow from conformal symmetry as well as with the associativity of the operator algebra. To study the implications of associativity it is useful to consider the four-point function,

$$\langle\phi_i(w_1, \bar{w}_1)\phi_j(w_2, \bar{w}_2)\phi_k(w_3, \bar{w}_3)\phi_l(w_4, \bar{w}_4)\rangle. \qquad (2.64)$$

The idea is to compare the computation of this correlator using the OPE of ϕ_i and ϕ_j and of ϕ_k and ϕ_l, with those of ϕ_i and ϕ_k and of ϕ_j and ϕ_l, namely calculation where $(z_1 \to z_2), (z_3 \to z_4)$ versus one in which $(z_1 \to z_3), (z_2 \to z_4)$.

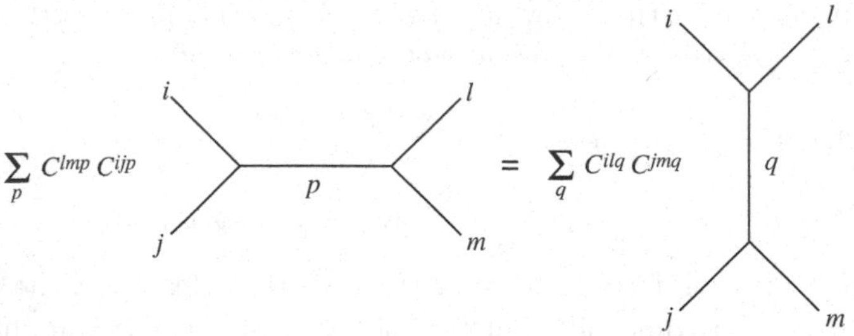

Conformal field theory

Fig. 2.3. Crossing symmetry.

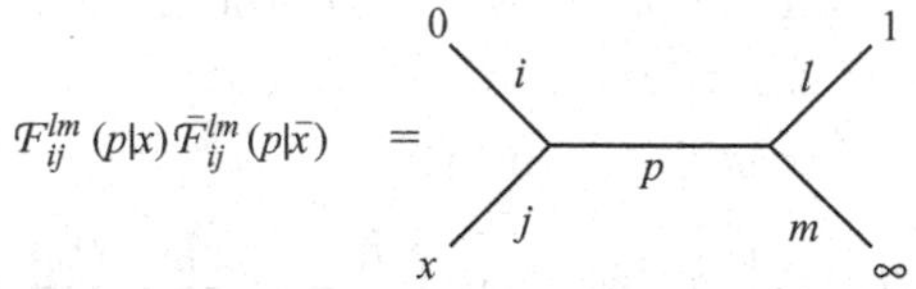

Fig. 2.4. Single channel amplitude.

The requirement that the two ways of computing coincide, referred to as *crossing symmetry*, is expressed in Fig. 2.3.[5]

Using conformal transformations, we can relate the diagram on the left-hand side of Fig. 2.3 to the diagram drawn in Fig. 2.4, which corresponds to the sum of the contributions of intermediate states belonging to the conformal family $[\phi_p]$ with the four-point function of operators located at $(w_1, w_2, w_3, w_4) = (0, z, 1, \infty)$. Note that in such a situation, z is actually also the cross ratio $\mathcal{Z}$. We denote this amplitude by the *conformal block* $\mathcal{F}_{ij}^{kl}(m|z)\bar{\mathcal{F}}_{ij}^{kl}(m|\bar{z})$ which depends on the Virasoro anomaly of the theory and the dimensions of all the operators involved. In terms of conformal blocks the crossing symmetry condition takes the form,

$$\sum_m C_{ijm} C_{klm} \mathcal{F}_{ij}^{kl}(m|z)\bar{\mathcal{F}}_{ij}^{kl}(m|\bar{z})$$

$$= \sum_n C_{ijn} C_{kln} \mathcal{F}_{ik}^{jl}(n|1-z)\bar{\mathcal{F}}_{ik}^{jl}(n|1-\bar{z}). \tag{2.65}$$

For a given set of conformal blocks (2.65) is a set of equations that determine the C_{ijk} and the weights. The general set of solutions of these equations is not known, but for a particular class of theories like the minimal models these equations can be solved.

---

[5] Crossing symmetry, duality and bootstrap was discussed in [33].

2.11 Verlinde's formula

The fusion rules (2.28), namely,

$$[\phi_i][\phi_j] = \sum_k N_{ij}^k [\phi_k],$$

constitute a commutative associative algebra. The commutativity implies that $N_{ij}^k = N_{ji}^k$ and the associativity means that,

$$\sum_k N_{ij}^k N_{kl}^m = \sum_k N_{ik}^m N_{jl}^k. \tag{2.66}$$

Using matrix notation in which $N_{ij}^k = (N_i)_j^k$ the associativity translates into the commutativity of the matrices, namely $N_i N_l = N_l N_i$. Thus, the matrices N_i are also members of an associative commutative algebra. Hence they can be diagonalized simultaneously to form a one-dimensional representation. This implies that there is a common matrix $\tilde{S}$,

$$N_{ij}^k = \sum_{lm} \tilde{S}_j^l \lambda_i^{(l)} \delta_l^m (\tilde{S}^{-1})_m^k = \sum_l \tilde{S}_j^l \lambda_i^{(l)} (\tilde{S}^{-1})_l^k, \tag{2.67}$$

where we denote the eigenvalues of N_i by $\lambda_i^{(l)}$. If j is the vacuum state $j = 0$ then $N_{i0}^k = \delta_i^k$, if all the representations labeled by i are irreducible. We now multiply from the right by $\tilde{S}_k^n$ to get,

$$N_{i0}^k \tilde{S}_k^n = \sum_l \tilde{S}_0^l \lambda_i^{(l)} (\tilde{S}^{-1})_l^k \tilde{S}_k^n$$

$$\tilde{S}_i^n = \sum_l \tilde{S}_0^l \lambda_i^{(l)} \delta_l^n = \tilde{S}_0^n \lambda_i^{(n)}, \tag{2.68}$$

which means that $\lambda_i^{(n)} = \frac{\tilde{S}_i^n}{\tilde{S}_0^n}$ and therefore,

$$N_{ij}^k = \sum_l \frac{\tilde{S}_j^l \tilde{S}_i^l (\tilde{S}^{-1})_l^k}{\tilde{S}_0^l}. \tag{2.69}$$

Now, for the reader who knows about the τ parameter and the characters (discussed in Section 2.8), we recall that under the S-transformation $\tau \to -\frac{1}{\tau}$ the characters of a given CFT transform as,

$$\chi_j\left(-\frac{1}{\tau}\right) = \sum_k S_j^k \chi_k(\tau). \tag{2.70}$$

Verlinde's formula[6] states that the matrix $\tilde{S}$ above is identical to the S-transformation matrix,

$$S = \tilde{S}. \tag{2.71}$$

This is a remarkable relation.

---

[6] The Verlinde formula was introduced in the seminal paper [210].

2.12 Free Majorana fermions – an example of a CFT

The theory of free massless fermions in two dimensions is an example of a 2d conformal theory of the utmost importance. In this section we describe this theory in detail following the steps taken in the general analysis of conformal field theories. The well-known Dirac action of a massless free fermion in two Euclidean dimensions is,

$$S = \frac{1}{4\pi} \int d^2x\, \bar{\Psi}\, \not{\partial}\Psi. \tag{2.72}$$

Expressing the Dirac fermion in terms of *chiral (or Weyl)* fermions, a left ψ and a right $\tilde{\psi}$, with $\Psi \equiv (\psi, \tilde{\psi})$, and using the fact that in two dimensions one can take $\gamma^0 = \sigma^2$ and $\gamma^1 = \sigma^1$, we rewrite the action as,

$$S = \frac{1}{4\pi} \int d^2z (\psi^\dagger \bar{\partial}\psi + \tilde{\psi}^\dagger \partial\tilde{\psi}). \tag{2.73}$$

We remind the reader that $\partial \equiv \partial_z$ and $\bar{\partial} \equiv \partial_{\bar{z}}$. The equations of motion are,

$$\bar{\partial}\psi = 0 \quad \partial\tilde{\psi} = 0 \;\rightarrow\; \psi = \psi(z) \quad \tilde{\psi} = \tilde{\psi}(\bar{z}). \tag{2.74}$$

In analogy to the symmetries of the scalar field it is straightforward to realize that the action is invariant under left holomorphic chiral and right anti-holomorphic transformations,

$$\psi \to \psi' = e^{i\alpha(z)}\psi \quad \tilde{\psi} \to \tilde{\psi}' = e^{i\tilde{\alpha}(\bar{z})}\tilde{\psi}. \tag{2.75}$$

The corresponding "affine current algebra" currents, given by,

$$J = i\psi^\dagger\psi \quad \bar{J} = i\tilde{\psi}^\dagger\tilde{\psi}, \tag{2.76}$$

are holomorphically and anti-holomorphically conserved.

In addition the theory is obviously invariant under conformal transformations $z \to f(z)$, $\bar{z} \to \bar{f}(\bar{z})$.

Dirac (or Weyl) fermions can be further decomposed into Majorana (or Weyl–Majorana) fermions as $\Psi = \frac{1}{\sqrt{2}}(\Psi_1 + i\Psi_2)$ (or $\psi = \frac{1}{\sqrt{2}}(\psi_1 + i\psi_2)$). Substituting these, the action reads,

$$S = \frac{1}{8\pi} \int d^2z \left(\sum_{i=1}^{2} \psi_i\bar{\partial}\psi_i + \tilde{\psi}_i\partial\tilde{\psi}_i \right). \tag{2.77}$$

From this point on we discuss the theory of single Majorana fermions, namely χ and $\tilde{\chi}$ are a left and a right Weyl–Majorana fermion with the action,

$$= \frac{1}{8\pi} \int d^2z (\chi\bar{\partial}\chi + \tilde{\chi}\partial\tilde{\chi}). \tag{2.78}$$

The equations of motion are still as in (2.74) so that χ is a holomorphic function and $\tilde{\chi}$ is an anti-holomorphic one (their extensions, as they are real on the real line).

Before spelling out the conformal structure of the theory we pause for a moment with the complex coordinate formulation and discuss canonical quantization in two dimensions with a Minkowski signature. The conjugate momentum to χ is $\pi_\chi = \frac{\partial \mathcal{L}}{\partial_0 \chi} = \frac{1}{2}\chi$, and as we are dealing with a real field $\{\pi_\chi, \chi\}$ has a factor $\frac{1}{2}$ multiplying a delta function, which gives,

$$\{\chi(x_1, x_0), \chi(y_0, y_1)\}|_{x_0=y_0} = \delta(x_1 - y_1). \tag{2.79}$$

Combining a pair of two Majorana fermions (each consisting of two Weyl–Majorana) into a Dirac fermion, one finds for the latter the usual anti-commutation relations, namely,

$$\{\Psi^\dagger(x_1, x_0), \Psi(y_1, y_0)\}|_{x_0=y_0} = \delta(x_1 - y_1) \quad \{\Psi(x_1, x_0), \Psi(y_1, y_0)\}|_{x_0=y_0} = 0. \tag{2.80}$$

The Noether currents associated with conformal transformations, namely the components of the energy-momentum tensor, are given by,

$$T(z) = -\frac{1}{2} : \chi\partial\chi : \quad \bar{T}(\bar{z}) = -\frac{1}{2} : \tilde{\chi}\bar{\partial}\tilde{\chi} :, \tag{2.81}$$

where $: \chi\chi :$ the normal ordered product, stands for the product with the subtraction of its OPE. The latter is given by,

$$\chi(z)\chi(w) = \frac{1}{z-w} \quad \tilde{\chi}(z)\tilde{\chi}(w) = \frac{1}{\bar{z}-\bar{w}}. \tag{2.82}$$

Using this basic OPE in $T(z)\chi(w)$ one finds,

$$T(z)\chi(w) = \frac{1}{2}\frac{\chi(w)}{(z-w)^2} + \frac{\partial\chi(w)}{z-w}, \tag{2.83}$$

which implies that χ is a *primary* field of conformal dimensions of $(\frac{1}{2}, 0)$, and similarly $\tilde{\chi}$ with $(0, \frac{1}{2})$. The Virasoro anomaly, which comes as usual from $T(z)T(w)$, is $c = \frac{1}{2}$, and $\bar{c} = \frac{1}{2}$ from $\bar{T}(z)\bar{T}(w)$.

Recall that the energy-momentum tensor of the scalar field (1.27) takes the form of a bilinear of the "current algebra" currents (1.24). We want to examine now if such a construction can be applied also for the fermionic fields. Since for Weyl–Majorana fermions there are no such currents it is left only to check for the $T(z)$ of Weyl fermions. Let us note first the OPE of the currents and the Weyl fermions that read,

$$J(z)\psi(w) = -i\frac{\psi(z)}{(z-w)} \quad J(z)\psi^\dagger(w) = i\frac{\psi^\dagger(z)}{(z-w)}, \tag{2.84}$$

where $J = i : \psi^\dagger\psi :$ with our conventions.

Using this OPE one finds,

$$T(z) = -\frac{1}{2} : J(z)J(z) := -\frac{1}{2}\lim_{z \to w}\left[J(z)J(w) + \frac{1}{(z-w)^2}\right]$$

$$= -\frac{1}{2}\lim_{z \to w}\lim_{\epsilon \to 0}\left[iJ(z)\left(\psi^\dagger(w-\epsilon)\psi(w+\epsilon) + \frac{1}{2\epsilon}\right) + \frac{1}{(z-w)^2}\right]$$

$$= -\frac{1}{2}\lim_{z \to w}\lim_{\epsilon \to 0}\left[-\frac{\psi^\dagger(z)\psi(w+\epsilon)}{[z-(w-\epsilon)]} + \frac{\psi^\dagger(w-\epsilon)\psi(z)}{[z-(w+\epsilon)]} + \frac{1}{(z-w)^2}\right]$$

$$= -\frac{1}{2}\left[\psi^\dagger\partial\psi - \partial\psi^\dagger\psi\right]. \tag{2.85}$$

This construction of the energy-momentum tensor in terms of a normal ordered product of two currents, which is known as the *Sugawara construction*, will play a key role in the discussion in Chapter 4.

The mode expansion of the Weyl–Majorana fermion takes the form,

$$\psi = \sum_{r \in \mathbb{Z}+\nu} \frac{\psi_r}{z^{r+\frac{1}{2}}} \qquad \psi_r = \frac{1}{2\pi i}\oint dz z^{r-\frac{1}{2}}\psi(z), \tag{2.86}$$

with $z = e^{-iw}$, ν is related to the boundary conditions as

$$\psi(w+2\pi) = e^{2\pi i\nu}\psi(w), \tag{2.87}$$

so that there are two types of fermions:

Ramond fermions $\nu = \frac{1}{2}$ $\leftrightarrow$ periodic boundary condition
Neveu–Schwarz fermions $\nu = 0$ $\leftrightarrow$ anti-periodic boundary condition.

The anti-commutation relations of the fermionic modes follow straightforwardly upon using (2.86) and the OPEs (2.82), namely,

$$\{\psi_r, \psi_s\} = \delta_{r+s}. \tag{2.88}$$

The form (2.82) holds for the periodic case. For the anti-periodic case there is an extra factor of $\frac{1}{2}(\sqrt{\frac{z}{w}} + \sqrt{\frac{w}{z}})$, which tends to 1 as $z \to w$.

The canonical quantization conditions in terms of real space-time coordinates take the following form,

$$\{\psi(x^1), \psi(y^1)\}|_{x^0=y^0} = \frac{1}{2}\delta(x^1 - y^1), \tag{2.89}$$

since ψ is the conjugate momentum of itself. Combining two Majorana fermions into a Dirac one yields the following anti-commutation relations for the Dirac fermions,

$$\{\Psi^\dagger(x^1), \Psi(y^1)\}|_{x^0=y^0} = \delta(x^1 - y^1) \quad \{\Psi(x^1), \Psi(y^1)\}|_{x^0=y^0} = 0, \tag{2.90}$$

so that now $\Psi^\dagger$ is the conjugate momentum of Ψ.

2.13 The Ising model – the $m = 3$ unitary minimal model

The first unitary minimal model with $m = 3$ has $c = 1/2$, just like the Majorana fermion discussed above. We now analyze the primaries of this model, their fusion rules and their correlators. Comparing the latter with correlation functions of the Ising model,[7] we show that in fact the $m = 3$ unitary minimal model is the continuum limit of the Ising model. Recall that the set of primaries of the $m = 3$ model are characterized by the following conformal weights:

$$h_{1,1} = 0 \quad h_{2,1} = \frac{1}{2} \quad h_{2,2} = \frac{1}{16}, \tag{2.91}$$

which determine the two-point functions,

$$\left\langle \Phi^{(1/2)}(z, \bar{z}) \Phi^{(1/2)}(w, \bar{w}) \right\rangle = \frac{1}{|z - w|^2}$$

$$\left\langle \Phi^{(1/16)}(z, \bar{z}) \Phi^{(1/16)}(w, \bar{w}) \right\rangle = \frac{1}{|z - w|^{1/4}}, \tag{2.92}$$

where $\Phi^{(h)}(z, \bar{z}) = \phi^{(h)}(z) \bar{\phi}^{(h)}(\bar{z})$. It turns out that at the critical point of the Ising model, the two-point function of the spin operator σ at a lattice point n and at the origin behaves like $<\sigma_n \sigma_0> \sim \frac{1}{|n|^{1/4}}$. Thus in the continuum, it has the same "critical exponent" as that of $\Phi^{(1/16)}$, and similarly the Greens function of the energy density falls like $<\epsilon_n \epsilon_0> \sim \frac{1}{|n|^2}$, namely like the two-point function of $\Phi^{(1/2)}$.

There are additional properties of the $m = 3$ unitary model that can be shown to match those of the Ising model. Here we demonstrate this with the determination of the four-point function of $\Phi^{(1/16)}$, namely,

$$\left\langle \Phi^{(1/16)}(z_1, \bar{z}_1) \dots \Phi^{(1/16)}(z_4, \bar{z}_4) \right\rangle. \tag{2.93}$$

From equation (2.63) we have that

$$\left[\frac{4}{3} \partial_{w_1}^2 - \sum_{i \neq 1}^{4} \left(\frac{1/16}{(w_1 - w_i)^2} + \frac{1}{(w_1 - w_i)} \frac{\partial}{\partial w_i} \right) \right] \langle \phi(w_1, \bar{w}_1) \dots \phi(w_4, \bar{w}_4) \rangle = 0, \tag{2.94}$$

where ϕ denotes $\Phi^{(1/16)}$.

Using the global Ward identities we express G_4 as in (2.59),

$$G_4(z_1, \bar{z}_1 \dots z_4, \bar{z}_4) = \tilde{f}(\mathcal{Z}, \bar{\mathcal{Z}}) \left[(z_{12} z_{13} z_{14} z_{23} z_{24} z_{34})^{-1/24} (\text{C.C.}) \right], \tag{2.95}$$

where $\mathcal{Z} = \frac{z_{12} z_{34}}{z_{13} z_{24}}$. Using also $\frac{z_{14} z_{23}}{z_{13} z_{24}} = 1 - \mathcal{Z}$, we can rewrite as,

$$G_4(z_1, \bar{z}_1 \dots z_4, \bar{z}_4) = \tilde{f}(\mathcal{Z}, \bar{\mathcal{Z}}) \left[(z_{13} z_{24})^{-1/8} [\mathcal{Z}(1 - \mathcal{Z})]^{-1/24} (\text{C.C.}) \right]. \tag{2.96}$$

---

[7] The two-dimensional Ising model has a long history. It was discussed in [137]. The relation to Majorana fermions was discussed in [187].

Anticipating the result of the Ising model, we actually write,

$$G_4(z_1, \bar{z}_1 \ldots z_4, \bar{z}_4) = f(\mathcal{Z}, \bar{\mathcal{Z}}) \cdot \left[[z_{13} z_{24} \mathcal{Z} (1 - \mathcal{Z})]^{-1/8} (C.C.) \right]. \qquad (2.97)$$

If we now substitute this ansatz into (2.94) we find the following differential equation for f,

$$\left[\mathcal{Z}(1 - \mathcal{Z}) \partial^2 + (1/2 - \mathcal{Z}) \partial + 1/16 \right] f(\mathcal{Z}, \bar{\mathcal{Z}}) = 0. \qquad (2.98)$$

A similar equation holds for $\bar{\mathcal{Z}}$. The solutions of this differential equation are $f_{1,2}(\mathcal{Z}) = \left(1 \pm \sqrt{1 - \mathcal{Z}} \right)^{1/2}$ and so finally the four-point function takes the form,

$$G_4(z_1, \bar{z}_1 \ldots z_4, \bar{z}_4) = \left| \left(\frac{z_{13} z_{24}}{z_{12} z_{23} z_{34} z_{34}} \right) \right|^{1/4} \left(|f_1(\mathcal{Z})|^2 + |f_2(\mathcal{Z})|^2 \right), \qquad (2.99)$$

where the unique combination is dictated by the requirement for a single value. This is identical to the result found in the Ising model for G_4.

Note also that $f(1 - \mathcal{Z}, 1 - \bar{\mathcal{Z}})$ is a solution, a result of the symmetry under the interchange of z_1 with z_3.

Note also that although $\Phi^{(1/2)}$ is a free fermion, $\Phi^{(1/16)}$ cannot be constructed in a local way from the fermion.

3
Theories invariant under affine current algebras

In addition to being conformal invariant, it was shown in Chapter 1 that the theory of a free massless scalar field admits also affine algebra currents $J(z)$ and $\bar{J}(\bar{z})$ which are holomorphically and anti-holomorphically conserved, namely $\bar{\partial}J(z,\bar{z}) = \partial\bar{J}(z,\bar{z}) = 0$. The existence of these currents, as was the case with the energy-momentum tensor, implies that the theory is invariant under an infinite-dimensional group of transformations. Inspired by this invariance of the free scalar theory we would like to identify and investigate field theories equipped with an affine current algebra (ACA), which is often refered to as Kac–Moody algebra or affine Lie algebra (ALA). Conformal field theories (CFT) are characterized by the Virasoro anomaly, the set of primary fields and the corresponding structure constants. It will be shown in this chapter that theories with ACA admit a similar algebraic structure and moreover that they are necessarily also CFTs. Thus every ACA theory will be characterized by the Virasoro anomaly as well as its ACA analog, the ACA level. Primary fields that have been defined so far via their operator product expansion (OPE) with the $T(z)$ and $\bar{T}(\bar{z})$ will have to obey certain OPE also with currents $J(z)$ and $\bar{J}(\bar{z})$. The zero modes of the free scalar affine currents J_0 and $\bar{J}_0$ (1.57) commute, namely, they generate an abelian group. For theories with "ordinary" non-affine currents the generalization of the abelian group to non-abelian ones led (in four dimensions) to the standard model of the basic interactions and in fact to an enormously rich spectrum of interesting theories. It is thus very natural to explore the generalization of the abelian ACA to non-abelian affine current algebras. The investigation of two-dimensional theories which are invariant under transformations generated by such affine currents is the subject of this chapter. We start with a brief reminder of the properties of finite dimensional Lie algebras.

The topics included in this chapter are covered in several books and review papers. In particular we have made use of the famous review by Goddard and Olive [111], and the book by Di Francesco, Mathieu and Senechal [77].

3.1 Simple finite-dimensional Lie algebras

Consider the Lie algebra $\mathcal{G}$,

$$[T^a, T^b] = if^{ab}_c T^c, \qquad (3.1)$$

associated with a group G, namely, the set T^a are the generators of the group G.[1] We will consider simple groups, namely those that do not contain invariant subgroups. Denote the maximal set of commuting Hermitian generators by $H^i, i = 1, \ldots, r$ so that

$$[H^i, H^j] = 0 \quad i, j = 1, \ldots, r. \tag{3.2}$$

This abelian subalgebra of $\mathcal{G}$ is referred to as the *Cartan subalgebra*. It can be shown that any two such abelian subalgebras with generators H^i and $\tilde{H}^i$ are conjugate under the action of the group, namely, $\tilde{H}^i = gH^ig^{-1}$ for some $g \in G$. The dimension of the Cartan subalgebra, which is the maximal number of commuting generators is defined as the *rank* of the algebra $\mathcal{G}$, rank $(\mathcal{G}) = r$.

A basis of the full algebra $\mathcal{G}$ constitutes H^i and the *step operators* or *ladder operators* E^α defined by,

$$[H^i, E^\alpha] = \alpha^i E^\alpha, \quad i = 1, \ldots, r. \tag{3.3}$$

The r-dimensional vector α is called a *root* associated with the step operator E^α. The roots are real and up to multiplication with a scalar there is a single E^α associated with α via (3.3). No multiple of a given root α is a root apart from $-\alpha$ which is the root paired with $E^{-\alpha} = E^{\alpha\dagger}$. The number of roots is obviously $(\dim\mathcal{G} - \mathrm{rank}\mathcal{G})$.

The rest of the algebra are the commutation relations $[E^\alpha, E^\beta]$ which follow from the Jacobi identity,

$$[H^i, [E^\alpha, E^\beta]] = (\alpha^i + \beta^i)[E^\alpha, E^\beta], \tag{3.4}$$

so that if $(\alpha^i + \beta^i) \neq 0$ and is not a root, then $[E^\alpha, E^\beta] = 0$. If on the other hand $(\alpha^i + \beta^i)$ is a root then $[E^\alpha, E^\beta]$ must be a multiple of $E^{\alpha+\beta}$. If $(\alpha^i + \beta^i) = 0$ it follows that $[E^\alpha, E^{-\alpha}] \sim \alpha \cdot H$.

To summarize the full algebra reads,

$$\begin{aligned}
[H^i, H^j] &= 0 \quad i, j = 1, \ldots, \mathrm{rank}(\mathcal{G}) \\
[H^i, E^\alpha] &= \alpha^i E^\alpha \quad \alpha = 1, \ldots, (\dim G - \mathrm{rank}(\mathcal{G})) \\
[E^\alpha, E^\beta] &= \epsilon(\alpha\beta)E^{\alpha+\beta} \quad \text{if } \alpha + \beta \text{ is a root} \\
&= \frac{2\alpha \cdot H}{\alpha^2} \quad \text{if } \alpha + \beta = 0 \\
&= 0 \quad \text{otherwise.}
\end{aligned} \tag{3.5}$$

This basis of the algebra is a modified version of the *Cartan–Weyl basis*. The constants $\epsilon(\alpha\beta)$ can be chosen to be ± 1 if all the root vectors have the same length.

---

[1] Finite-dimensional Lie algebra is covered in many books, for instance Cahn [54].

It is straightforward to realize that the triplet of generators E^α, $E^{-\alpha}$, and $\frac{\alpha \cdot H}{\alpha^2}$ is isomorphic to J_+, J_-, J_3 of an $SU(2)$ algebra, namely,

$$\left[\frac{\alpha \cdot H}{\alpha^2}, E^{\pm\alpha}\right] = \pm E^{\pm\alpha}, \quad [E^{+\alpha}, E^{-\alpha}] = 2\frac{\alpha \cdot H}{\alpha^2}. \tag{3.6}$$

Consequently the eigenvalues of $2\frac{\alpha \cdot H}{\alpha^2}$, just like those of $2J_3$, are integers in any unitary representation. The eigenvalues associated with each root β is given by $\frac{2\alpha \cdot \beta}{\alpha^2} \in \mathcal{Z}$. It is natural to define the notion of coroot $\alpha^\wedge = \frac{2\alpha}{\alpha^2}$.

3.1.1 The Weyl group

Consider a root β such that $2\alpha \cdot \beta/\alpha^2 \neq 0$ is its eigenvalue under the operation of $2\frac{\alpha \cdot H}{\alpha^2}$. There must be another step operator $E^{\beta+m\alpha}$ which is a member of the $SU(2)$ multiplet with the opposite eigenvalue, namely,

$$2\alpha \cdot \beta/\alpha^2 + 2m = -2\alpha \cdot \beta/\alpha^2. \tag{3.7}$$

Then $m = -2\alpha \cdot \beta/\alpha^2$, and

$$\beta + m\alpha = \beta - 2\frac{(\alpha \cdot \beta)\alpha}{\alpha^2} = \beta - (\alpha^\wedge, \beta)\alpha \equiv \sigma_\alpha(\beta) \tag{3.8}$$

is a root for each pair of roots α and β; $\sigma_\alpha(\beta)$ is a reflection in the hyperplane perpendicular to α. The set of these reflections that permute the roots, generate a finite group $W(\mathcal{G})$, the *Weyl group* of $\mathcal{G}$.

3.1.2 Cartan matrix and Dynkin diagrams

It is convenient to define the notion of *simple roots* as follows. Select a $\text{rank}(\mathcal{G})$ dimensional basis of the roots that consists of $\alpha_{(i)}$, $i = 1, \ldots, \text{rank}(\mathcal{G})$ in such a way that any root α can be written as,

$$\alpha = \sum_i^{\text{rank } \mathcal{G}} n_i \alpha_{(i)}, \tag{3.9}$$

where the n_i are integers which are either all $n_i \geq 0$ or all $n_i \leq 0$. In the former case α is *positive*, while in the latter it is *negative*. This base is called the basis of *simple roots*. Associated with the simple roots one defines the *simple Weyl reflections* $\sigma_{\alpha_{(i)}}$ that generate the entire Weyl group.

The scalar products of simple roots define the *Cartan matrix* as follows:

$$A_{ij} = \frac{2\alpha_{(i)} \cdot \alpha_{(j)}}{\alpha_{(j)}^2}. \tag{3.10}$$

The Cartan matrix is a $\text{rank}(\mathcal{G}) \times \text{rank}(\mathcal{G})$ matrix with integer components and with diagonal elements which take the value of 2. The off diagonal elements are either negative or vanishing.

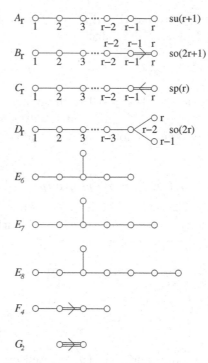

Fig. 3.1. Dynkin diagrams for all the simple Lie algebras.

It can be shown that the roots of a simple Lie algebra can have at most two different lengths, a long root and a short one. The ratio between their lengths are either 2 or 3. When all the roots have the same length the algebra is a *simply laced algebra*. From the Cartan matrix A_{ij} one can reconstruct a basis of simple roots up to a scale and overall orientation. In fact constructing all the roots from the simple roots, one finds that full information on $\mathcal{G}$ is encoded in A_{ij}.

The information contained in the Cartan matrix A_{ij} may be encoded in a planar diagram, the *Dynkin diagram*. The construction of such a diagram is as follows:

- To each simple root $\alpha_{(i)}$ assign a node in the diagram.
- If a node represents a short root mark it by a black dot, and if a long one by a white dot.
- Join the points $\alpha_{(i)}$ and $\alpha_{(j)}$ by $A_{ij}A_{ji}$ lines. For $i \neq j$, $A_{ij}A_{ji}$ can take the values of $0, 1, 2, 3$. In fact since $A_{ij}A_{ji} = 4\cos^2\theta_{ij}$, where θ_{ij} is the angle between the two roots, then orthogonal simple roots are not connected, and those with an angle of $120, 135, 150$ degrees are connected with one, two, or three lines.
- In some conventions one does not separate between black and white dots, but rather one draws an arrow pointing from $\alpha_{(i)}$ to $\alpha_{(j)}$ when $\alpha_{(i)}^2 > \alpha_{(j)}^2$.

The Dynkin diagrams for all simple Lie algebras are given in Fig. [3.1].

3.1.3 Highest weight states

We now consider finite-dimensional representations of $\mathcal{G}$ other than the adjoint representation that has been analyzed so far, the latter being the same as the generators. We can always choose a basis $\{|\mu\rangle\}$ for which,

$$H^i|\mu\rangle = \mu^i|\mu\rangle. \tag{3.11}$$

The rank($\mathcal{G}$) dimensional vector μ of eigenvalues of the Cartan subalgebra generators is called the *weight vector*. A root is a weight of the adjoint representation. In a similar manner to their action as roots, the triplet $E^\alpha, E^{-\alpha}$ and $2\alpha \cdot H/\alpha^2$ form an $SU(2)$ algebra and hence $\{|\mu\rangle\}$ must be an $SU(2)$ multiplet, and in particular,

$$2\alpha \cdot \mu/\alpha^2 \in \mathcal{Z}, \tag{3.12}$$

for any root α.

This property of any given weight defines a lattice $\Lambda_W(\mathcal{G})$, the *weight lattice* of the algebra $\mathcal{G}$. The weights associated with a representation must be mapped into one another under the operation of σ_α and in fact the whole Weyl group. One can choose a basis for the weight lattice $\Lambda_W(\mathcal{G})$ consisting of *fundamental weights* $\lambda_{(i)}$ such that,

$$2\lambda_{(i)} \cdot \alpha_{(j)}/\alpha_{(j)}^2 = \delta_{ij}. \tag{3.13}$$

Any weight λ can then be expanded as $\lambda = \sum n_i \lambda_{(i)}$ with integer coefficients n_i. If all $n_i \geq 0$, the weight λ is called a *dominant weight*. Every weight can be mapped into a unique dominant weight by action of the Weyl group. The dominant weight $\rho = \sum_i \lambda_{(i)}$, where $i = 1, \ldots, \text{rank}\mathcal{G}$, is characterized by $\rho \cdot \alpha > 0$ for any positive α and $\rho \cdot \alpha < 0$ for any negative α. In fact $\rho = 1/2 \sum \alpha$ where the sum is over all the positive roots.

For any given finite-dimensional representation of $\mathcal{G}$ one defines the *highest weight state* $|\mu_0\rangle$ for which $\rho \cdot \mu_0$ is the largest. Such a state is annihilated by all the raising operators,

$$E^\alpha|\mu_0\rangle = 0, \tag{3.14}$$

for every $\alpha > 0$. All the states of a given irreducible representation can be built by acting on the highest weight state with lowering operators, namely, each state takes the form,

$$E^{-\beta_1}\ldots.E^{-\beta_n}|\mu_0\rangle. \tag{3.15}$$

In fact every irreducible representation has a unique highest weight state $|\mu_0\rangle$ and the other weights μ have the property that $\mu_0 - \mu$ is a sum of positive roots. The highest weight state is a dominant weight. In the opposite direction for each dominant weight there is a unique irreducible representation for which it is the highest weight state. As was mentioned for the adjoint representation the weights are

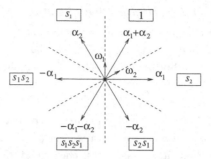

Fig. 3.2. Root system and Weyl chambers of $SU(3)$.

the roots so that the corresponding highest weight state is the *highest root*. The difference between the highest root and any other root is a sum of positive roots.

We end this subsection with an example. Consider the $SU(3)$ algebra. The Cartan matrix for this algebra takes the form,

$$\begin{pmatrix} 2 & -1 \\ -1 & 2 \end{pmatrix}. \tag{3.16}$$

The simple roots are related to the fundamental weights as $\alpha_{(1)} = 2\lambda_{(1)} - \lambda_{(2)}$ and $\alpha_{(2)} = -\lambda_{(1)} + 2\lambda_{(2)}$. The scalar products between the fundamental weights are $(\lambda_{(1)}, \lambda_{(1)}) = (\lambda_{(2)}, \lambda_{(2)}) = 2/3$ and $(\lambda_{(1)}, \lambda_{(2)}) = 1/3$, using the standard normalization of $(\alpha_{(i)}, \alpha_{(i)}) = 2$. The full Weyl group is given by $W = \{1, \sigma_1, \sigma_2, \sigma_1\sigma_2, \sigma_2\sigma_1, \sigma_1\sigma_2\sigma_1\}$. The action of the different elements of the Weyl group on the two simple roots gives all roots.

The root system and the Weyl chambers are given in Fig. 3.2. The Weyl chambers are separated by the dashed lines, and are specified here by the elements of the Weyl group, with the latter denoted in the figure by s_i. The Weyl chambers are defined by,

$$C_\omega = \{\lambda | (\omega\lambda, \alpha_i) \geq 0, i = 1 \dots r\}.$$

3.2 Affine current algebra

In the previous subsection we acquired a certain familiarity with the notions of roots, highest weights, Cartan matrices, Dynkin diagrams etc., in the context of a simple Lie algebra. As was explained in the introduction to this chapter, two-dimensional CFTs are characterized by an extended algebraic structure, that of affine Lie algebra.[2] We now describe the basic properties of the affine Lie algebra using the notions of the previous subsection.

---

[2] Affine Lie algebras were introduced into the physics literature by Bardacki and Halpern in [27]. Independently V. Kac [136] and R.V. Moody [164] introduced them in the mathematical literature.

The basic ALA is given by,

$$[J_m^a, J_n^b] = i f_c^{ab} J_{m+n}^c + \hat{k} m \delta_{ab} \delta_{m+n,0}, \tag{3.17}$$

where the central element $\hat{k}$ commutes with all the generators J_m^a, namely, $[J_m^a, \hat{k}] = 0$.

We now use a generalization of the Cartan–Weyl basis to the affine algebra as follows:

$$[H_m^i, H_n^j] = \hat{k} m \delta^{ij} \delta_{m+n,0}$$
$$[H_m^i, E_n^\alpha] = \alpha^i E_{m+n}^\alpha,$$
$$[E_m^\alpha, E_n^\beta] = \epsilon(\alpha\beta) E_{m+n}^{\alpha+\beta} \quad \text{if } \alpha + \beta \text{ is a root}$$
$$= \frac{2}{\alpha^2} (\alpha \cdot H_{m+n} + \hat{k} m \delta_{m+n}) \quad \text{if } \alpha + \beta = 0$$
$$= 0 \quad \text{otherwise.}$$
$$[H_m^i, \hat{k}] = [E_n^\alpha, \hat{k}] = 0. \tag{3.18}$$

We have used the normalization $(H^i, H^j) = \delta^{ij}$, $(E^\alpha, E^\beta) = \frac{2}{\alpha^2} \delta_{\alpha+\beta}$, where (X, Y) denotes the Killing form, defined as the trace of the product in the adjoint representation, $Tr(adX, adY)$. The hermiticity properties are,

$$H_m^{i\,\dagger} = H_{-m}^i, \quad E_m^{\alpha\,\dagger} = E_{-m}^{-\alpha}, \quad \hat{k}^\dagger = \hat{k} \tag{3.19}$$

Unlike the simple Lie algebra, here we have an $r + 1$ dimensional abelian subalgebra consisting of $[H_0^1, \ldots, H_0^r, \hat{k}]$. With respect to these generators, E_m^α are step operators,

$$[H_0^i, E_m^\alpha] = \alpha^i E_m^\alpha \quad [\hat{k}, E_n^\alpha] = 0. \tag{3.20}$$

Each of the eigenvectors $(\alpha^1, \ldots, \alpha^r, 0)$ is infinitely degenerate since it is independent of m. Moreover this abelian subalgebra is not maximal since $[H_0^i, H_n^j] = 0$. Thus one has to extend the algebra by adding a grading operator which can be taken to be L_0 such that,

$$[L_0, J_n^a] = -n J_n^a \quad [L_0, \hat{k}] = 0. \tag{3.21}$$

Using the generators $(H_0^i, \hat{k}, -L_0)$ as in the Cartan subalgebra we have as the step operators E_n^α corresponding to a root $(\alpha, 0, n)$ and H_n^i corresponding to $(0, 0, n)$.

The root system of ALA is thus infinite but spans an $(r + 1)$ dimensional space. We can divide the roots into positive and negative according to the following rule:

$$(\alpha, 0, n) > 0 \quad \text{if } n > 0, \text{ or if } n = 0 \text{ and } \alpha > 0 \tag{3.22}$$

The basis of the simple roots can therefore be taken as,

$$\alpha_{(i)} = (\alpha_i, 0, 0), \quad 1 \leq i \leq r$$
$$\alpha_{(0)} = (-\theta, 0, 1), \tag{3.23}$$

where α_i is the basis of simple roots of the Lie algebra, and θ is the highest root. Thus an arbitrary root of the ALA can be expressed as,

$$\alpha = \sum_{i=0}^{r} n_i \alpha_{(i)}. \tag{3.24}$$

It is positive if $n_i \geq 0$ and negative when $n_i \leq 0$, and these are the only two posibilities.

3.2.1 Cartan matrix and Dynkin diagrams

The first step is to define the scalar product $\langle X, Y \rangle$ which has to be symmetric and obey the relation,

$$\langle X, [Y, Z] \rangle + \langle Y, [X, Z] \rangle = 0$$

for $X, Y, Z \in \hat{g}$, the ALA. Upon using a convenient normalization one can bring the basic scalar products to the following form,

$$\langle T_m^a, T_n^b \rangle = \delta^{ab} \delta_{m+n}$$
$$\langle T_m^a, -L_0 \rangle = 0$$
$$\langle T_m^a, \hat{k} \rangle = 1$$
$$\langle \hat{k}, \hat{k} \rangle = 1$$
$$\langle \hat{k}, -L_0 \rangle = 1$$
$$\langle -L_0, -L_0 \rangle = 0. \tag{3.25}$$

The last relation is actually a choice, following on from the invariance of the algebra under a shift of L_0 by a multiple of $\hat{k}$. Here we use T instead of δ used previously.

In the following Hermitian basis,

$$T_0^i, \quad \frac{(T_m^i + T_{-m}^i)}{\sqrt{2}}, \quad \frac{(\hat{k} - L_0)}{\sqrt{2}}, \quad \frac{(\hat{k} + L_0)}{\sqrt{2}}, \tag{3.26}$$

the scalar product is Lorentzian since the norm of all the first three basis vectors is $+1$ while that of $\frac{(\hat{k}+L_0)}{\sqrt{2}}$ is -1.

The Lorentzian signature holds also for the Cartan sub-algebra (CSA) generators. One can define the scalar product of two vectors of simultaneous eigenvalues of the CSA,

$$m^i = (\mu^i, \mu_k^i, \mu_{-L_0}^i), \quad m^j = (\mu^j, \mu_k^j, \mu_{-L_0}^j)$$

to be,

$$m^i \cdot m^j = \mu^i \cdot \mu^j + \mu_k^i \cdot \mu_{-L_0}^j + \mu_{-L_0}^i \cdot \mu_k^j. \tag{3.27}$$

In particular the scalar product of two roots,

$$a^i = (\alpha^i, 0, n^i), \quad a^j = (\alpha^j, 0, n^j)$$

is

$$a^i \cdot a^j = \alpha^i \cdot \alpha^j. \tag{3.28}$$

The root that corresponds to E_n^α, $a = (\alpha, 0, n)$ has a norm $a^2 = \alpha^2 > 0$ and hence is referred to as a space-like root, whereas the root that is associated with H_n^i, $n\delta = (0, 0, n)$ has zero norm (light-like) and is orthogonal to all other roots. We have used the "unit" of $\delta = (0, 0, 1)$.

The *Cartan matrix* of $\hat{g}$, which is an $(r + 1) \times (r + 1)$ matrix, is defined in a similar way to one of the Lie algebra, namely,

$$A_{ij} = \frac{2a_{(i)} \cdot a_{(j)}}{a_{(j)}^2} \quad 0 \leq i, j \leq r. \tag{3.29}$$

We add to the Cartan matrix of the Lie algebra, the extra row and column A_{i0} and A_{0i} which can be found using (3.10) with $\alpha_0 = -\theta$. Now from the definition of the fundamental weight it follows that $\theta = -\sum_{i=0}^{r} A_{0i}\lambda_i$. Since θ is a long root of g, namely, $\theta^2 \geq \alpha_{(i)}^2$ one gets that,

$$-A_{i0} = 1 \quad \text{if } A_{0i} \neq 0$$
$$-A_{i0} = 0 \quad \text{if } A_{0i} = 0, \tag{3.30}$$

provided that θ is not itself a simple root, as happens for $SU(2)$. The Dynkin diagram of $\hat{g}$ is obtained using that of g appended with an extra point that corresponds to α_0 connected by $-A_{0i}$ lines to the points $a_{(i)}$. If $-A_{0i} > 1$ an arrow is drawn which points toward $a_{(0)}$. We demonstrate the construction in the following example:

$\hat{SU}(2)$ - There are only two simple roots $a_{(0)} = (-\alpha, 0, 1)$ and $a_{(1)} = (\alpha, 0, 0)$ so that $A_{0i} = A_{i0} = -2$ and the Cartan matrix is

$$\begin{pmatrix} 2 & -2 \\ -2 & 2 \end{pmatrix}. \tag{3.31}$$

Thus, there are two roots of equal length connected by four lines with arrows pointing in both directions to indicate that $a_{(0)}^2$ is equal to $a_{(1)}^2$.

The Dynkin diagrams of the affine simple algebra are shown in Fig. 3.3. The point that corresponds to $\alpha_{(0)}$ is marked by a zero. The black dots relate to the notion of twisted affine Lie algebra which we do not discuss here (see for example [111]).

3.2.2 The Weyl group

The Weyl group of $\hat{g}$ is defined in a similar way to that of g, namely it is generated by reflections in the hyperplanes normal to a,

$$\sigma_a(b) = b - 2\left(\frac{b \cdot a}{a^2}\right) a = (\sigma_\alpha(\beta), 0, n_a - 2\left(\frac{\alpha \cdot \beta}{\alpha^2}\right) n_\beta), \tag{3.32}$$

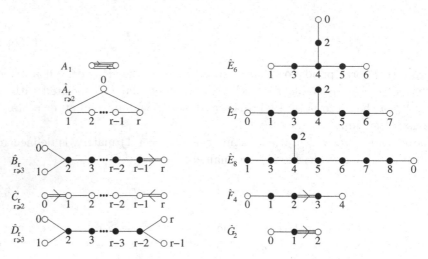

Fig. 3.3. Dynkin diagrams for all the affine simple Lie algebras.

for $a = (\alpha, 0, n_\alpha)$ and $b = (\beta, 0, n_\beta)$ a space-like root. Light-like roots are invariant under such reflections $\sigma_a(n\delta) = n\delta$. In fact the Weyl group of $\hat{g}$ is a semidirect product of the Weyl group of g and the coroot lattice of g which is the lattice generated by the coroots $\alpha^\nu = \frac{\alpha}{\alpha^2}$. The coroots form the root system of the Lie algebra dual to g obtained by interchanging the root lengths. The simply laced algebras A_n, D_n, E_n are obviously self-dual, as are F_4 and G_2, whereas $B_n^\nu = C_n$ and $C_n^\nu = B_n$.

3.2.3 Highest weight representations

A highest weight state $|\hat{\mu}_0\rangle$ is a state that is annihilated by all the raising operators for positive roots, namely,

$$E_0^\alpha|\hat{\mu}_0\rangle = E_n^{\pm\alpha}|\hat{\mu}_0\rangle = H_n^i|\hat{\mu}_0\rangle = 0, \tag{3.33}$$

for $n > 0, \alpha > 0$. The eigenvalue of this state is the highest weight vector $\hat{\mu}_0 = (\mu_0^i, k, h)$ given by,

$$H_0^i|\hat{\mu}_0\rangle = \mu_0^i|\hat{\mu}_0\rangle, \quad \hat{k}|\hat{\mu}_0\rangle = k|\hat{\mu}_0\rangle, \quad L_0|\hat{\mu}_0\rangle = h|\hat{\mu}_0\rangle. \tag{3.34}$$

We can set h to zero as a matter of convention. A highest weight representation is characterized by a unique highest weight state. To have a unitary highest weight representation the following necessary and sufficient conditions have to be obeyed:

$$\frac{2k}{\theta^2} \in \mathcal{Z} \quad k \geq \theta \cdot \mu_0 \geq 0. \tag{3.35}$$

The non-negative integer $\frac{2k}{\theta^2}$ is the *level* of the representation. Any state in the representation is characterized by a weight vector $\hat{\mu} = (\mu^i, k, h)$ such that $\mu_0 - \mu$ is a sum of positive roots. Introduce now a set of fundamental weights $l_{(i)}$ for $\hat{g}$, $i = 0, \ldots r$, such that $2l_{(i)} \cdot a_{(j)}/a_{(j)}^2 = \delta_{ij}$. The general solution of the condition $2a \cdot \hat{\mu}/\theta^2 \in \mathbb{Z}$, which is equivalent to the condition Eq. (3.35) for $a = a_{(0)} = (-\theta, 0, 1)$, now takes the form $\hat{\mu}_0 = \sum_{i=0}^{r} n_i l_{(i)}$, where n_i are non-negative integers, apart from the indeterminate component in the L_0 direction. For $l_{(i)}$ one finds,

$$l_{(i)} = \left(\lambda_{(i)}, \frac{1}{2}m_i\theta^2, 0\right) \quad l_{(0)} = \left(0, \frac{1}{2}m_0\theta^2, 0\right),\tag{3.36}$$

where $m_0 = 1$ and where the integers m_i are defined via $\theta/\theta^2 = \sum_{i=0}^{r} m_i \alpha_{(i)}/\alpha_{(i)}^2$. The corresponding level is given by,

$$\text{level} = \sum_{i=0}^{r} n_i m_i.\tag{3.37}$$

Level 1 representations are thus associated with highest weights $l_{(i)}$ with all $m_i = 1$. Those are indicated by open points in Fig. 3.3.

From the definition of m_i it follows that,

$$\sum_{i=0}^{r} A_{ij} m_j = 0.\tag{3.38}$$

Since the Cartan matrix has the basic symmetry of the extended Dynkin diagram, also the positions of the open dots have to preserve this symmetry. For the classical groups A_r, B_r, C_r, D_r the values of m_i for the closed dots is 2. For the exceptional groups the vector $(m_0, \ldots, m_r)$ is as follows

$$
\begin{array}{ll}
\hat{E}^6 & (1, 1, 2, 2, 3, 2, 1) \\
\hat{E}^7 & (1, 2, 2, 3, 4, 3, 2, 1) \\
\hat{E}^8 & (1, 2, 4, 6, 5, 4, 3, 2, 1) \\
\hat{F}^4 & (1, 2, 3, 2, 1) \\
\hat{G}^2 & (1, 2, 1)
\end{array}\tag{3.39}
$$

3.3 Current OPEs and the Sugawara construction

In Section 1.8 for the free scalar theory it was shown that the OPE of two currents $J(z)J(w)$ takes the form of $J(z)J(w) = \frac{1}{(z-w)^2} +$ finite terms. This type of OPE, which corresponds to the abelian ALA, is generalized following the discussion in Section 3.2 to,

$$J^a(z)J^b(w) = \frac{k\delta^{ab}}{(z-w)^2} + i\frac{f_c^{ab}J^c(w)}{(z-w)} + \text{finite terms}.\tag{3.40}$$

We can now use the OPE to evaluate the infinitesimal transformation of the current under ALA transformations,

$$\delta_\epsilon J^a(w) = \frac{1}{2\pi i} \oint_w dz \epsilon^b(z) J^b(z) J^a(w)$$

$$= \frac{1}{2\pi i} \oint_w dz \epsilon^b(z) \left[\frac{k\delta^{ab}}{(z-w)^2} + i \frac{f_c^{ab} J^c(w)}{(z-w)} \right] = i f_{bc}^a \epsilon^b J^c - k\partial\epsilon^a. \quad (3.41)$$

The same structure also holds for $\bar{J}^a(\bar{z})$.

The OPE form of the ALA can be transformed into a commutator form of the algebra. We introduce a Laurent expansion of the currents,

$$J^a(z) = \sum_n J_n^a z^{-(n+1)} \quad J_n^a = \frac{1}{2\pi i} \oint dz z^n J^a(z). \quad (3.42)$$

Substituting the OPE into the expression of the commuation relation we indeed find the ALA of (3.17), namely

$$[J_m^a, J_n^b] = i f_c^{ab} J_{m+n}^c + \hat{k} m \delta_{ab} \delta_{m+n,0}. \quad (3.43)$$

In free scalar theory there are two "currents" which are holomorphically conserved, J and T, and moreover the energy-momentum tensor is bilinear in J, as was shown in Section 1.5. We now elevate this special case into a general construction of T for theories which admit ALA structure. The construction is known as the *Sugawara construction*. One writes T as a normal ordered product of the currents,

$$T(z) = \frac{1}{2\kappa} : J^a(z) J_a(z) : \quad (3.44)$$

with a coefficient κ that has to be determined quantum mechanically. In fact, one way to determine κ is by requiring that J^a is a primary field of weight 1, namely,

$$T(z) J^a(w) = \frac{J^a}{(z-w)^2} + \frac{\partial J^a(w)}{(z-w)}. \quad (3.45)$$

Using the OPE (3.40) and the relation $-f_c^{ab} f_{bd}^c = 2C\delta_d^a$, where C is the dual Coxeter number, we find that $\frac{(k+C)}{\kappa} = 1$ so that the form of the Sugawara constructed T is,

$$T(z) = \frac{1}{2(k+C)} : J^a(z) J_a(z) : \quad (3.46)$$

Note that the Casimir of the adjoint is $2C$. Note also that in Section 1.5, for the free scalar case, we had a relative minus sign, due to a difference of factor i in defining the currents there.

In the WZW models discussed in the next section, classically one has T with a coefficient of $\frac{1}{2k}$. It is thus clear that for those models quantum mechanically, due to the double contraction, we get a finite renormalization of the level $k \to k + C$.

Here we have used currents which are in an orthonormal basis. If instead we express the current in the Cartan–Weyl basis used in the previous sections, the form of T in terms of the Cartan sub-algebra generators H^i and the step operators E^α is,

$$T(z) = \frac{1}{2(k+C)} \left[: H^i(z)H^i(z) : + \frac{|\alpha|^2}{2}(E^\alpha E^{-\alpha} + E^{-\alpha} E^\alpha) \right]. \qquad (3.47)$$

The OPEs (3.45) and (3.40) also enable us to determine c, the Virasoro anomaly of the model, via the computation,

$$T(z)T(w) = T(z)\frac{1}{2(k+C)} : J^a J^a : (w)$$

$$= \frac{1}{(k+C)} \left\{ \frac{J^a(z)J^a(w)}{(z-w)^2} + \frac{\partial J^a(z)J^a(w)}{(z-w)} \right\}$$

$$= \frac{1}{(k+C)} \left\{ \frac{k(dim\ G)}{(z-w)^4} + \frac{: J^a J^a : (w)}{(z-w)^2} + \frac{1}{2}\frac{\partial : J^a J^a : (w)}{(z-w)} \right\}$$

$$= \frac{1}{(k+C)}\frac{k(dim\ G)}{(z-w)^4} + 2\frac{T(w)}{(z-w)^2} + \frac{\partial T(w)}{(z-w)}. \qquad (3.48)$$

We thus read off the Virasoro anomaly

$$c = \frac{k\,dim\ G}{k+C}. \qquad (3.49)$$

The construction of T in terms of a normal ordered product of two currents calls for combining together the ALA and the Virasoro algebra. Substituting into (3.46) the mode expansions, of $T(z)$ in terms of L_n and of $J(z)$ in terms of J_n, one finds that,

$$L_n = \frac{1}{2(k+C)} \sum_m : J^a_{n-m} J^a_m : \qquad (3.50)$$

where here the normal ordering implies putting the currents with positive m to the right. In fact normal ordering is required only for L_0.

Using this relation, we write down the full Virasoro algebra and ALA,

$$[L_n, L_m] = (n-m)L_{n+m} + \frac{c}{12}(n-1)n(n+1)\delta(n+m)$$

$$[L_n, J^a_m] = -mJ^a_{m+n}$$

$$[J^a_m, J^b_n] = if^{ab}_c J^c_{m+n} + \hat{k}m\delta_{ab}\delta_{m+n,0}. \qquad (3.51)$$

In mathematical terminology the Virasoro algebra belongs to the enveloping algebra of the ALA.

3.4 Primary fields

Recall that the operators of any CFT were shown to be either Virasoro primaries or descendants. The former were defined by their OPE with T. In a similar manner ALA primaries $\Phi_{l,\bar{l}}(z,\bar{z})$ are defined via their OPE with J,

$$J^a(z)\Phi_{l,\bar{l}}(w,\bar{w}) = \frac{T_l^a \Phi_{l,\bar{l}}(w,\bar{w})}{(z-w)}$$

$$\bar{J}^a(z)\Phi_{l,\bar{l}}(w,\bar{w}) = \frac{T_{\bar{l}}^a \Phi_{l,\bar{l}}(w,\bar{w})}{(\bar{z}-\bar{w})}, \tag{3.52}$$

where $T_l^a, T_{\bar{l}}^a$ are the matrix T^a in the $l,\bar{l}$ representations, for the holomorphic and antiholomorphic sectors, respectively. From here on we discuss only holomorphic properties. In terms of the Laurent components J_n^a the condition for a primary field reads,

$$J_n^a \Phi_{l,\bar{l}}(0,\bar{z}) = 0 \quad \text{for } n > 0; \quad J_0^a \Phi_{l,\bar{l}}(z,\bar{z}) = T_l^a \Phi_{l,\bar{l}}(z,\bar{z}). \tag{3.53}$$

In theories where the energy-momentum tensor can be constructed in a Sugawara construction it is easy to see that the ALA primaries are also Virasoro primaries. Indeed, using (3.50) we see that L_n for $n > 0$ annihilates the ALA primary. For L_0 acting on the primaries we get,

$$L_0 \Phi_l = \frac{1}{2(k+C)} J_0^a J_0^a \Phi_l = \frac{C_2(l)}{2(k+C)}\Phi_l. \tag{3.54}$$

Thus the primaries in theories equipped with the Sugawara construction, for instance the WZW models that will be discussed in the next section, obey (3.53) and also,

$$L_n \Phi_{l,\bar{l}}(0,\bar{z}) = 0 \quad \text{for } n > 0; \quad L_0^a \Phi_{l,\bar{l}}(z,\bar{z}) = \frac{C_2(l)}{2(k+C)}\Phi_{l,\bar{l}}(z,\bar{z}). \tag{3.55}$$

Recall that T is not a Virasoro primary but rather is a descendant of the identity $T(0) = L_{-2}I$. The same applies to $J(z)$. From the mode expansion (3.42) it is clear that,

$$J^a(0) = J_{-1}^a I. \tag{3.56}$$

Note however that $J^a(z)$ is a Virasoro primary. Apart from the distinguished descendant J there are descendant operators of all the primaries. In fact all the local operators can be written as,

$$J_{-n_1}^{a_1} \cdots J_{-n_N}^{a_N} \bar{J}_{-\bar{n}_1}^{\bar{a}_1} \cdots \bar{J}_{-\bar{n}_N}^{\bar{a}_N} \Phi_{l,\bar{l}}(z,\bar{z}), \tag{3.57}$$

and in the case of a Sugawara construction all the operators are of the form,

$$L_{-m_1} \cdots L_{-m_M} \bar{L}_{-\bar{m}_1} \cdots \bar{L}_{-\bar{m}_M} J_{-n_1}^{a_1} \cdots J_{-n_N}^{a_N} \bar{J}_{-\bar{n}_1}^{\bar{a}_1} \cdots \bar{J}_{-\bar{n}_N}^{\bar{a}_N} \Phi_{l,\bar{l}}(z,\bar{z}). \tag{3.58}$$

3.5 ALA characters

In Section 2.8 we introduced the notion of the Virasoro character (2.45) which characterizes the structure of the Virasoro Verma module. In a complete analogy let us now define a character of the CFT and ALA module $\hat{\lambda}$ as follows:

$$\chi_{\hat{\lambda}}(z^j, \tau) = e^{-im_{\hat{\lambda}}\delta} Tr_{\hat{\lambda}}[e^{2\pi i\tau L_0} e^{-2\pi i \sum_j z_j h^j}], \tag{3.59}$$

where $m_{\hat{\lambda}}$, δ and h^j are the generators of the Cartan subalgebra associated with the group, and z_j are complex numbers. The character can also be expressed in terms of the generalized theta function $\Theta_{\hat{\lambda}}$ in the following form:

$$\chi_{\hat{\lambda}}(z^j, \tau) = \frac{\sum_{w \in W} \epsilon(w)\Theta_{w(\hat{\lambda}+\hat{\rho})}}{\sum_{w \in W} \epsilon(w)\Theta_{w\hat{\rho}}}, \tag{3.60}$$

where the sums are over the elements of the finite Weyl group, $\epsilon(w) = (-1)^{l(w)}$ with $l(w)$ the length of w.

Rather than defining the generalized theta function for any ALA at any level, we define here only the function for $\hat{SU}(2)$ level k. For this case we have,

$$\Theta_{\lambda_1}^{(k)}(z; \tau; t) = e^{-2\pi kt} \sum_{n \in \mathbb{Z}} e^{-2\pi i[knz + \frac{1}{2}\lambda_1 z - kn^2\tau - \frac{\lambda_1^2 \tau}{4k}]}. \tag{3.61}$$

with $\Theta_{\lambda_1}^{(k)}(z; \tau; o) \equiv \Theta(k - \lambda_1, \lambda_1)$ (see [77] for details).

In terms of this function the character of $\hat{SU}(2)_k$ takes the form,

$$\chi_{\hat{\lambda}} = \frac{\Theta_{\lambda_1+1}^{(k+2)} - \Theta_{-\lambda_1-1}^{(k+2)}}{\Theta_1^{(2)} - \Theta_{-1}^{(2)}}, \tag{3.62}$$

where $\hat{\lambda} = [k - \lambda_1, \lambda_1]$. For the special point $(z = 0, t = 0)$ the character expressed in terms of $q = e^{2\pi i\tau}$ reads,

$$\chi_{\hat{\lambda}}(q) = q^{\frac{(\lambda_1+1)^2}{4(k+2)} - \frac{1}{8}} \frac{\sum_{n \in \mathbb{Z}}[\lambda_1 + 1 + 2n(k + 2)]q^{n[\lambda_1+1+(k+2)n]}}{\sum_{n \in \mathbb{Z}}[1 + 4n]q^{n[1+2n]}}. \tag{3.63}$$

For level one and for $k = \lambda_1 = 1$ we get,

$$\chi_{\hat{\lambda}}(q) = q^{\frac{5}{24}} \frac{(2 - 4q + 8q^5 - 10q^8 + \ldots)}{(1 - 3q + 5q^3 - 7q^6 + 9q^{10} + \ldots)}$$

$$= q^{\frac{5}{24}}(2 + 2q + 6q^2 + 8q^3 + \ldots). \tag{3.64}$$

The content of the four first grades of the module $[k - \lambda_1 = 0, \lambda_1 = 1]$ is $(1), (1), (3) \oplus (1), (3) \oplus 2(1)$, so that the number of the states in these different grades is indeed 2,2,6 and 8 as in the expression of the character.

3.6 Correlators, null vectors and the Knizhnik–Zamolodchikov equation

Correlators of Virasoro primaries were subjected to local and global Ward identities, (2.61) and (2.56), respectively. We now derive their ALA duals. Performing a group transformation of a given correlator,

$$
\left\langle \oint \frac{dz}{2\pi i} \epsilon^a(z) J^a(z) \phi_{l_1}(w_1, \bar{w}_1) \dots \phi_{l_n}(w_n, \bar{w}_n) \right\rangle
$$

$$
= \sum_{i=1}^{n} \left\langle \phi_{l_1}(w_1, \bar{w}_1) \dots \oint \frac{dz}{2\pi i} \epsilon^a(z) J^a(z) \phi_{l_i}(w_i, \bar{w}_i) \dots \phi_{l_n}(w_n, \bar{w}_n) \right\rangle
$$

$$
= \sum_{i=1}^{n} \left\langle \phi_{l_1}(w_1, \bar{w}_1) \dots \delta_\epsilon \phi_{l_i}(w_i, \bar{w}_i) \dots \phi_{l_n}(w_n, \bar{w}_n) \right\rangle. \tag{3.65}
$$

Now from the OPE (3.52) we know that $\delta_\epsilon \phi_{l_i}(w_i, \bar{w}_i) = \epsilon^a(w_i) T_{l_i}^a \phi_{l_i}(w_i, \bar{w}_i)$. Since this holds for arbitrary ϵ we can get a local form of the Ward identity in the form,

$$
\langle J^a(z) \phi_{l_1}(w_1, \bar{w}_1) \dots \phi_{l_n}(w_n, \bar{w}_n) \rangle = \sum_{i=1}^{n} \frac{T_{l_i}^a}{(z - w_i)} \langle \phi_{l_1}(w_1, \bar{w}_1) \dots \phi_{l_n}(w_n, \bar{w}_n) \rangle.
$$

$$
\tag{3.66}
$$

As for the global Ward identity, we use the fact that the correlator has to be invariant under global g transformations (constant ϵ^a), namely,

$$
\delta_\epsilon^a \langle \phi_{l_1}(w_1, \bar{w}_1) \dots \phi_{l_n}(w_n, \bar{w}_n) \rangle = 0,
$$

leading to

$$
\sum_{i=1}^{n} T_{l_i}^a \langle \phi_{l_1}(w_1, \bar{w}_1) \dots \phi_{l_n}(w_n, \bar{w}_n) \rangle = 0. \tag{3.67}
$$

Null vectors of CFTs were found to be useful in Section 2.9, since they lead to differential equations for certain correlators. In a similar manner one can write down null vectors of ALA. In the context of the Sugawara construction, due to the link between the Virasoro algebra generator T and the ALA generators J^a, there are null vectors that combine generators from both infinite algebras. We discuss now an important example of this class that leads to the *Knizhnik–Zamolodchikov* equations. Consider, at Virasoro level one, the following null vector,

$$
|\text{null}\rangle = \left\{ L_{-1} - \frac{1}{k + C} J_{-1}^a T_{l_i}^a \right\} |\Phi_{l_i}\rangle. \tag{3.68}
$$

It is easy to see that this is indeed a null state, following (3.52). If we insert the corresponding null operator into a correlation function of primary fields, like $\langle \Phi_1(z_1) \dots \text{null}(z_i) \dots \Phi_n(z_n) \rangle$, the latter must vanish and hence we get,

$$
\left\{ \partial_i - \frac{1}{k + C} \sum_{j \neq i} \frac{T_i^a T_j^a}{(z_i - z_j)} \right\} \langle \Phi_1(z_1) \dots \Phi_n(z_n) \rangle = 0. \tag{3.69}
$$

In the derivation, we use,

$$<\Phi_1(z_1)\ldots J^a_{-1}\Phi_i(z_i)\ldots\Phi_n(z_n)>$$

$$= \frac{1}{2\pi i}\oint_{z_i}\frac{dz}{z-z_i}<\Phi_1(z_1)\ldots J^a(z)\Phi_i(z_i)\ldots\Phi_n(z_n)>$$

$$= \frac{1}{2\pi i}\oint_{z_i,j\neq i}\frac{dz}{z-z_i}\sum_{j\neq i}\frac{T^a_j}{(z-z_j)}<\Phi_1(z_1)\ldots\Phi_j(z_j)\ldots\Phi_n(z_n)>$$

$$= \sum_{j\neq i}\frac{T^a_j}{(z_i-z_j)}<\Phi_1(z_1)\ldots\Phi_n(z_n)> . \tag{3.70}$$

For the case of four-point functions, as the correlator depends only on the cross-ratio coordinate $\mathcal{Z} = \frac{z_{12}z_{34}}{z_{13}z_{24}}$, the partial differential equations reduce to an ordinary differential equation. In Section 4.4 we will demonstrate a solution of the Knizhnik–Zamolochikov equation for a four-point function.

3.7 Free fermion realization

In the previous chapter the theories of massless free Dirac and Majorana fermions were analyzed as examples of CFTs. In particular it was shown that the Dirac fermion admits an abelian ALA structure. It is thus natural to expect that the theory of N fermions should be invariant under the transformations associated with non-abelian ALAs.[3] Indeed, it will be shown in this section that an $\hat{SO}(N)$ ALA, and a $\hat{U}(N)$ are the underlying algebraic structures of N free massless Majorana fermions and N Dirac fermions, respectively. We start with the former case.

3.7.1 Free Majorana fermions and $\widehat{SO}(N)$

Consider a generalization of the action given in Section 2.11 for N Majorana fermions,

$$S = \frac{1}{8\pi}\int d^2z\sum_{i=1}^N\{\psi_i\bar{\partial}\psi_i + \tilde{\psi}_i\partial\tilde{\psi}_i\}, \tag{3.71}$$

where ψ and $\tilde{\psi}$ are left and right Weyl–Majorana fermions, respectively. Note that this is possible in 2d, and in any other dimension that is 2 modulo 8. In 4d, for example, we do not have a Weyl–Majorana fermion, as in the case of a single Majorana fermion, due to the equations of motion,

$$\psi \equiv \psi_i(z) \quad \tilde{\psi} \equiv \tilde{\psi}_i(\bar{z}). \tag{3.72}$$

---

[3] The free fermion realization of ALA was presented for the first time in [27].

However, unlike the case of a single fermion, here the action is invariant under transformations associated with $\widehat{SO}(N)$ affine algebra generated by the following holomorphically (anti-holomorphically) conserved currents,

$$J^a(z) = \frac{1}{2}\psi^i T^a_{ij}\psi^j \qquad \bar{J}^a(\bar{z}) = \frac{1}{2}\tilde{\psi}^i T^a_{ij}\tilde{\psi}^j, \tag{3.73}$$

where T^a are $SO(N)$ matrices which can be expressed as,

$$T^a_{ij} \equiv t^{(kl)}_{ij} = i(\delta^k_i \delta^l_j - \delta^k_j \delta^l_i). \tag{3.74}$$

The coefficients (halfs) are not determined by the Noether procedure, but are chosen in a manner that will be explained below.

The T^a_{ij} matrices obey the relations,

$$Tr[T^a T^b] = 2\delta^{ab}$$

$$\sum_a T^a_{ij} T^a_{kl} = -\delta_{ik}\delta_{jl} + \delta_{il}\delta_{jk}$$

$$\sum_{ab} f_{abc} f_{abd} = 2(N-2)\delta_{cd}. \tag{3.75}$$

The anticommutation relation and the OPE generalize in an obvious way those of the single Majorana fermion, namely,

$$\{\psi^i(x_0, x_1)\psi^j(y_0, y_1)\}|_{x_0=y_0} = \frac{1}{2}\delta^{ij}\delta(x_1 - y_1) \tag{3.76}$$

and

$$\psi^i(z)\psi^j(w) = \frac{\delta^{ij}}{z-w} \qquad \tilde{\psi}^i(z)\tilde{\psi}^j(w) = \frac{\delta^{ij}}{\bar{z}-\bar{w}}. \tag{3.77}$$

Now using this OPE one can derive the OPE of two currents and verify that they take the form of (3.40),

$$J^a(z)J^b(w) = \frac{1}{4} : \psi^i(z)T^a_{ij}\psi^j(z) :: \psi^k(w)T^b_{kl}\psi^l(w) :$$

$$= \frac{1}{4}T^a_{ij}T^b_{kl}\left[-\left(: \psi^i(z)\psi^k(w) : +\frac{\delta^{ik}}{z-w}\right)\psi^j(z)\psi^l(w)\right.$$

$$\left.+\left(: \psi^i(z)\psi^l(w) : \frac{\delta^{il}}{z-w}\right)\psi^j(z)\psi^k(w)\right] = \frac{1}{4}T^a_{ij}T^b_{kl}\frac{1}{z-w}$$

$$[-\delta^{ik} : \psi^j(z)\psi^l(w) : +\delta^{il} : \psi^j(z)\psi^k(w) : +\delta^{jk} : \psi^i(z)\psi^l(w) :$$

$$-\delta^{jl} : \psi^i(z)\psi^k(w) :] + \frac{1}{4}T^a_{ij}T^b_{kl}\frac{1}{(z-w)^2}\left[-\delta^{ik}\delta^{jl} + \delta^{il}\delta^{jk}\right]. \tag{3.78}$$

By expanding the fields that are functions of z around w and using the relations above one finds that indeed the OPE of the two currents take the form of (3.40), namely,

$$J^a(z)J^b(w) = \frac{1\delta^{ab}}{(z-w)^2} + \frac{f^{ab}_c J^c(w)}{(z-w)} + \text{finite terms} \tag{3.79}$$

It is thus clear that this is a realization of an $\widehat{SO}(N)$ ALA of level $k = 1$.

The Noether currents associated with the conformal transformations, the energy-momentum tensor T ($\bar{T}$) is just the sum of T ($\bar{T}$) associated with each one of the N Majorana fermions, hence,

$$T(z) = -\frac{1}{2}\sum_i : \psi^i \partial \psi^i : \quad \bar{T}(\bar{z}) = -\frac{1}{2}\sum_i : \tilde{\psi}^i \bar{\partial} \tilde{\psi}^i : . \tag{3.80}$$

Since the Virasoro anomaly of a single Majorana fermion is $c = \frac{1}{2}$ it is clear that the theory of N fermions has $c = \frac{N}{2}$.

In Section 2.12 it was shown that T of a Dirac fermion could be transformed into a Sugawara form, $T(z) = -\frac{1}{2} : J(z)J(z) :$, where $J(z)$ was the $U(1)$ current. Since we will show below that the Sugawara form is the underlying structure of the important class of WZW models, it is a natural question to ask whether also in the present case for the N fermions T can be put into a Sugawara construction.

Now, using the expression for the Virasoro anomaly for a theory with $\widehat{SO}(N)_1$, we find, as we saw before, that,

$$c = \frac{\dim G}{k + C} = \frac{\frac{1}{2}N(N-1)}{1 + (N-2)} = \frac{N}{2}. \tag{3.81}$$

3.7.2 Primary fields

Similarly to the case of a single Majorana field, the OPE of $T(z)\psi^i(w)$ is,

$$T(z)\psi^i(w) = \frac{1}{2}\frac{\psi^i}{(z-w)^2} + \frac{\partial\psi^i}{z-w}, \tag{3.82}$$

which implies that ψ^i are N *primary* fields of conformal dimensions $(\frac{1}{2}, 0)$, and similarly $\tilde{\psi}^i$ has dimension $(0, \frac{1}{2})$.

Is the primary field $\Phi^{(1/2,1/2)}(z\bar{z}) = \psi(z)\tilde{\psi}(\bar{z})$ the only primary operator (in addition to the identity operator that corresponds to the vacuum state)? For the primaries of the ALA $\widehat{SO}(N)_1$ we find (see (2.13)) that there is also one primary operator with dimension $\frac{N}{16}$ for odd N, and two primary operators for even N. These additional primaries transform in the spinor representation of $\widehat{SO}(N)_1$.

Can one construct these primaries in terms of the fermionic fields ψ and $\tilde{\psi}$? The situation here is similar to the one in the Ising model. In fact, using the spin operator $\sigma(z, \bar{z})$ or its dual, one indeed gets from the N independent Majorana fermion theories, the dimension $\frac{N}{16}$ and the number of degrees of freedom 2^N, which are identical to the dimension of the spinor representation.

So far we have shown the free fermion construction of $\widehat{SO}(N)_1$, namely, of the ALA at level 1. We would now like to investigate the possibility of having free fermion realization also to the affine Lie algebra at higher levels. Going through our previous derivation it is clear that the ALA structure of the OPE of two currents (3.40), applies to fermions at any representation. For a given representaion ρ the corresponding level k is determined from the first term on the right-hand side, namely $Tr(T^a T^b) = 2k\delta^{ab}$. Now since for a representation ρ,

$Tr(T_\rho^a T_\rho^b) = 2D_2(\rho)\delta^{ab}$ where $D_2(\rho)$ is the Dynkin index of the representation, it is clear that free fermions constitute a realization of $\widehat{SO}(N)$ at level $D_2(\rho)$.

3.8 Free Dirac fermions and the $\widehat{U}(N)$

Consider the theory of N Dirac fermions described by the following action,

$$S = \frac{1}{4\pi} \int d^2 z \{ \psi^{i\dagger} \bar{\partial} \psi_i + \tilde{\psi}^{i\dagger} \partial \tilde{\psi}_i \}. \tag{3.83}$$

In terms of symmetries, the difference between this theory and the one of a single Dirac fermion, is that now there is an invariance under $U(N)$ left holomorphic and right anti-holomorphic transformations, namely,

$$\psi \to \psi' = g(z)\psi \quad \tilde{\psi} \to \tilde{\psi}' = \bar{g}(\bar{z})\tilde{\psi}, \tag{3.84}$$

where $g(z), \bar{g}(\bar{z}) \in U(N)$. The associated holomorphic currents are given by,

$$J^a = \psi^{i\dagger} T^{aj}{}_i \psi_j \quad J = \psi^{i\dagger} \psi_i, \tag{3.85}$$

where J is the $U(1)$ current, $J^a(z)$ are the $SU(N)$ currents and $T^{aj}{}_i$ are matrices in the adjoint of $SU(N)$, that obey

$$Tr[T^a T^b] = \delta^{ab}$$

$$\sum_a T_{ij}^a T_{kl}^a = \delta_{il}\delta_{jk} - \frac{1}{N}\delta_{ij}\delta_{kl}$$

$$\sum_{ab} f_{abc} f_{abd} = N\delta_{cd}. \tag{3.86}$$

Using the OPEs of the fermions, it is straightforward to realize that the currents indeed constitute the OPEs that correspond to a $\widehat{U}(N)$ of level $k = 1$,

$$J^a(z)J^b(w) = \frac{1\delta^{ab}}{(z-w)^2} + \frac{f_c^{ab} J^c(w)}{(z-w)} + \text{finite terms}$$

$$J(z)J(w) = \frac{1}{(z-w)^2} + \text{finite terms}$$

$$J^a(z)J(w) = \text{finite terms}. \tag{3.87}$$

Similar to the case of Majorana fermions, the Noether current T is given by,

$$T(z) = T(z)_{U(1)} + T(z)_{SU(N)} = -\frac{1}{2}[\psi^{i\dagger} \partial \psi_i - \partial \psi^{i\dagger} \psi_i], \tag{3.88}$$

and can be reexpressed in terms of a Sugawara form,

$$T(z)_{U(1)} = \frac{1}{4N} : \psi^{\dagger i} \psi_i \psi^{\dagger j} \psi_j :$$

$$T(z)_{SU(N)} = \frac{1}{2(N+1)} \sum_a : \psi^{\dagger i} T^a \psi_i \psi^{\dagger j} T^a \psi_j : . \tag{3.89}$$

Since a Dirac fermion has a $c = 1$ Virasoro anomaly, it is clear that the theory of N Dirac fermions has $c = N$. This is also the outcome of the Virasoro anomaly associated with the Sugawara form as follows,

$$c_{U(1)} + c_{SU(N)} = 1 + \frac{N^2 - 1}{N + 1} = N. \tag{3.90}$$

4

Wess–Zumino–Witten model and coset models

The two-dimensional Wess–Zumino–Witten (WZW) model was introduced in the seminal paper of Witten [224]. The model makes use of the WZ term that was introduced by Wess and Zumino in [217]. Sometimes the model is referred to as the WZWN model, where the N stands for Novikov, who independently invoked a similar model [170]. Here we follow only [224].

4.1 From free massless scalar theory to the WZW model

Consider the free massless scalar theory that was described in Section 1.2, but now with $\hat{X}(z, \bar{z})$ being an angle variable defined in the interval $[0, 2\pi]$. The action of the scalar field can now be re-written in the following form,

$$
\begin{aligned}
S = \int \mathrm{d}^2 x \mathcal{L} &= \frac{1}{8\pi} \int \mathrm{d}^2 z \partial_\nu \hat{X} \partial^\nu \hat{X} \\
&= \frac{1}{8\pi} \int \mathrm{d}^2 z \partial_\nu (e^{i\hat{X}}) \partial^\nu (e^{-i\hat{X}}) = \frac{1}{4\pi} \int \mathrm{d}^2 z \partial u \bar{\partial} u^{-1},
\end{aligned}
\tag{4.1}
$$

where $u = e^{i\hat{X}(z, \bar{z})}$ is an abelian group element. Recall that the theory is characterized by a Virasoro algebra and an abelian ALA structure. In terms of this variable the currents J and $\bar{J}$ can be written as,

$$
J = -iu^{-1}\partial u = iu\partial u^{-1}, \quad \bar{J} = -iu^{-1}\bar{\partial} u = iu\bar{\partial} u^{-1},
$$

with

$$
\bar{\partial} J = \partial \bar{J} = 0
$$

and

$$
T =: (u^{-1}\partial u)^2 :=: (u\partial u^{-1})^2 : .
$$

It is now tempting to replace the abelian u with a non-abelian group element,

$$
u \in G, \quad G = SO(N) \text{ or } SU(N),
\tag{4.2}
$$

and consider the action,

$$
S_{sm} = \frac{1}{4\pi} \int \mathrm{d}^2 z Tr[\partial u \bar{\partial} u^{-1}],
\tag{4.3}
$$

where the trace is taken in fundamental representation so that,

$$Tr[T^a T^b] = \frac{1}{2}\delta^{ab}.$$

The question here is whether this action admits a similar non-abelian affine Lie algebra and Virasoro algebra. Let us analyze the equations of motion, symmetries and the corresponding currents of this action. The variation of the action under $u \to u + \delta u$ is,

$$
\begin{aligned}
\delta S_{sm} &= \frac{1}{4\pi} \int d^2 z Tr[\partial(\delta u)\bar{\partial}u^{-1} - \partial u \bar{\partial}(u^{-1}\delta u u^{-1})] \\
&= \frac{1}{4\pi} \int d^2 z Tr\left[u^{-1}\delta u [u^{-1}\bar{\partial}\partial(u) - \partial\bar{\partial}(u^{-1})u]\right] \\
&= \frac{1}{4\pi} \int d^2 z Tr\left[u^{-1}\delta u \partial^\mu(u^{-1}\partial_\mu u)\right] \\
&= \frac{1}{4\pi} \int d^2 z Tr\left[\delta u u^{-1} \partial^\mu(u\partial_\mu u^{-1})\right],
\end{aligned}
\tag{4.4}
$$

were we use $\delta u^{-1} = -u^{-1}\delta u u^{-1}$ and $\partial u^{-1} u = -u^{-1}\partial u$, following from $\delta(u^{-1}u) = 0$ and $\partial(u^{-1}u) = 0$. It is easy to realize that for a constant group element g the action is invariant under,

$$u \to gu \ (u^{-1} \to u^{-1}g^{-1}), \quad u \to uh \ (u^{-1} \to h^{-1}u^{-1}), \tag{4.5}$$

and the currents corresponding to the left and right multiplications take the form,

$$J^\mu = \frac{1}{4\pi}u^{-1}\partial^\mu u \quad \tilde{J}^\mu = -\frac{1}{4\pi}\partial^\mu u u^{-1}. \tag{4.6}$$

Both currents are conserved. Note that the conservation of one implies the conservation of the other. However, unlike the massless free scalar theory, now we do not have an ALA structure associated with a separate holomorphic and anti-holomorphic conservation. The latter would have taken the form of $J_L(z)$ corresponding to left transformation of the form $u \to g(z)u$ and $J_R(\bar{z})$, corresponding to right transformation of the form $u \to ug(\bar{z})$. In a similar manner one finds that the energy-momentum tensor,

$$T_{\mu\nu} \sim Tr[J_\mu J_\nu] - 1/2 \, g_{\mu\nu} Tr[J^\alpha J_\alpha],$$

and there is only the overall conservation law $\partial_\mu T^{\mu\nu} = 0$, not $\bar{\partial}T = \partial\bar{T} = 0$, namely not an external product of two Virasoro algebras.

Can we modify the action (4.3) so that it does have the desired ALA and Virasoro algebraic structure? For that let us consider first the variation of the action we are looking for. If instead of (4.4) one assumes a variation of the form,

$$\delta S = \frac{1}{4\pi} \int d^2 z Tr\left[u^{-1}\delta u \partial(u^{-1}\bar{\partial}u)\right] = \frac{1}{4\pi} \int d^2 z Tr\left[\delta u u^{-1}\bar{\partial}(u\partial u^{-1})\right], \tag{4.7}$$

then the global transformations of (4.5) are elevated into

$$u \to g(z)u \quad u \to uh(\bar{z}),$$ (4.8)

with the corresponding currents,

$$J_L \equiv J = \frac{k}{4\pi}\partial u u^{-1} \quad J_R \equiv \bar{J} = -\frac{k}{4\pi}u^{-1}\bar{\partial}u,$$ (4.9)

which have the desired ALA property,

$$\partial\bar{J} = \bar{\partial}J = 0.$$ (4.10)

Moreover, it can be shown that for an action whose variation takes the form of (4.7) the energy-momentum takes the form,

$$T \sim Tr[JJ], \quad \bar{T} \sim Tr[\bar{J}\bar{J}],$$ (4.11)

and hence it also has the appropriate Virasoro behavior.

The next question is obviously what action has a variation of the form (4.7), and in particular can it be built from S_{sm} plus an additional term that has the standard form of $\tilde{S} = \frac{1}{4\pi}\int d^2z\mathcal{L}$. To address this question we rewrite the variation (4.7) in the form,

$$\delta S = \frac{1}{4\pi}\int d^2z Tr\left[u^{-1}\delta u\partial^\mu(g_{\mu\nu} + \epsilon_{\mu\nu})(u^{-1}\partial^\nu u)\right]$$

$$= \frac{1}{4\pi}\int d^2z Tr\left[\delta u u^{-1}\partial^\mu(g_{\mu\nu} - \epsilon_{\mu\nu})(u\partial^\nu u^{-1})\right].$$ (4.12)

Clearly the term, proportional to $g_{\mu\nu}$ in both forms, is exactly the variation δS_{sm}, so that we need to find what action $\tilde{S}$ has a variation that takes the form of the $\epsilon^{\mu\nu}$ term. It may seem that the action,

$$\tilde{S} = \frac{1}{4\pi}\int d^2z\epsilon^{\mu\nu}Tr[\partial^\mu u\partial_\nu u^{-1}]$$

does the job, but in fact it vanishes.

It was the proposal of Witten to take for $\tilde{S}$ the so-called WZ action, which for the present case takes the form of a three-dimensional integral over a ball whose boundary is an S^2, which is the two-dimensional space-time,

$$S_{WZ} = \frac{1}{12\pi}\int d^3\sigma\epsilon^{ijk}Tr[(u^{-1}\partial_i u)(u^{-1}\partial_j u)(u^{-1}\partial_k u)],$$ (4.13)

where σ_i with $i = 1, 2, 3$ are the coordinates of the ball. Using the fact that $\int d^3\sigma\epsilon^{ijk}\partial^k(\ldots) = \int d^2\sigma\epsilon^{ij}(\ldots)$, it is straightforward to show that indeed the variation of (4.13) yields the extra term to change (4.4) to (4.7).

The map u, from the Euclidean space-time that we now take to be S^2 to the group manifold (Fig. 4.1) can be extended into a map from the ball to the group manifold. This is based on the fact that the homotopy group associated with

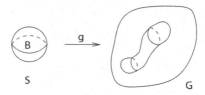

Fig. 4.1. The map between the space-time S^2 and the group manifold.

maps from S^2 to the group space G vanishes, namely, $\pi_2(G) = 0$[1] for any non-abelian group G.

On the other hand since $\pi_3(G) = \mathcal{Z}$, there are topologically inequivalent ways to extend the map u to a map from the ball to the group manifold. This implies that there is an ambiguity in S_{WZ} and it is well defined only modulo $S_{WZ} \rightarrow S_{WZ} + 2\pi$. Thus the coefficient of this term must be an integer k, and to have a variation of the form (4.12) it is clear that the sigma term has to have the same coefficient.

Let us now summarize. The classical action of the WZW model is,

$$S_{WZW} = \frac{k}{4\pi} \int d^2z Tr[\partial u \bar{\partial} u^{-1}]$$

$$+ \frac{k}{12\pi} \int d^3\sigma \epsilon^{ijk} Tr[(u^{-1}\partial_i u)(u^{-1}\partial_j u)(u^{-1}\partial_k u)]. \qquad (4.14)$$

The variation of this action is given by,

$$\delta S = \frac{k}{4\pi} \int d^2z Tr\left[u^{-1}\delta u \partial(u^{-1}\bar{\partial}u)\right] = \frac{k}{4\pi} \int d^2z Tr\left[\delta u u^{-1}\bar{\partial}(u\partial u^{-1})\right], \quad (4.15)$$

so that the equation of motion takes the form,

$$\partial(u^{-1}\bar{\partial}u) = \bar{\partial}(u\partial u^{-1}) = 0. \qquad (4.16)$$

The solutions of these equations of motion take the form,

$$u(z,\bar{z}) = u(z)\bar{u}(\bar{z}), \qquad (4.17)$$

where clearly $u \in G, \bar{u} \in G$.

We should state, that the form (4.14), with a term extended to one dimension higher, follows from general properties. Equations of motion that we want, in even space-time dimensions, imply a term in the action with one dimension higher, otherwise the action will involve singular terms, like the introduction of Dirac strings in the case of elementary monopoles.

The symmetries of the action are the ALA transformations,

$$u \rightarrow g(z)u \qquad u \rightarrow uh(\bar{z}), \qquad (4.18)$$

---

[1] $\pi_n(G)$ denotes the group of homotopy classes of maps f: $S^n \rightarrow G$

and the conformal transformations,

$$z \to f(z) \quad \bar{z} \to \bar{f}(\bar{z}). \tag{4.19}$$

The ALA currents are,

$$J = -\frac{k}{4\pi}\partial u u^{-1} \quad \bar{J} = \frac{k}{4\pi}u^{-1}\bar{\partial}u, \tag{4.20}$$

and the classical energy-momentum tensor takes the form,

$$T = \frac{1}{k}Tr[JJ] \quad \bar{T} = \frac{1}{k}Tr[\bar{J}\bar{J}]. \tag{4.21}$$

4.2 Perturbative conformal invariance

In the following section it will be shown in an exact way, based on algebraic properties, that the WZW model is a CFT. Prior to that we present now a perturbative computation, demonstrating that to a given order indeed the theory has a vanishing β function. Here we restrict ourselves to the one loop order. Of course this only serves as a motivation, as the CFT is at a finite coupling, and so exact demonstration is needed.

The idea is to use the background field method, expanding u around a solution of the equations of motion which we denote by u_0, so $u = u_0 e^{iT^a \pi^a}$. Substituting this ansatz into the action 4.14 one finds,

$$S_{WZW} = \frac{k}{4\pi}\int d^2z\{Tr[\partial u_0 \bar{\partial}u_0^{-1}] + \frac{1}{2}\partial_\mu \pi^a \partial^\mu \pi^a$$
$$+ \frac{1}{2}(\eta^{\mu\nu} - \epsilon^{\mu\nu})Tr\{(u_0^{-1}\partial_\mu u_0)[T^a \pi^a, T^b \partial_\nu \pi^b]\} + O(\pi^3)\} \tag{4.22}$$

The one loop renormalization diagram is shown in Fig. 4.2.

The non vanishing contributions are only when both vertices are proportional to $\eta^{\mu\nu}$ or to $\epsilon^{\mu\nu}$. The two contributions are the same apart from a sign since $\eta^{\rho\mu}\eta^\nu_\rho = \eta^{\mu\nu} = -\epsilon^{\rho\mu}\epsilon^\nu_\rho$, so that the one loop beta function vanishes. Obviously, this result relates to the choice of the coefficient of the sigma model term versus the WZ term, with the latter being fixed by topological arguments. The vanishing of the β function at this stage is an indication that we have chosen the coefficients in a way that is compatible with conformal invariance.

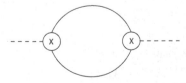

Fig. 4.2. Calculation of the one loop beta function.

4.3 ALA, Sugawara construction and the Virasoro algebra

An alternative approach to the quantization methods discussed in the previous section is the ALA and CFT approach. Notice first that under infinitesimal left transformation $\delta u = \epsilon(z)u$ the left current transforms as,

$$\delta_\epsilon J = \frac{k}{4\pi}[\partial(\epsilon u)u^{-1} - \partial u u^{-1}\epsilon] = \frac{k}{4\pi}(\partial\epsilon + [\epsilon, J]), \tag{4.23}$$

which translates into $\delta_\epsilon J^a = \frac{k}{4\pi}[\partial\epsilon^a + if^a_{bc}\epsilon^b J^c]$. Since the transformation of J is generated by,

$$\delta_\epsilon J^a(w) = \frac{1}{2\pi i}\oint_w \epsilon(z)J(z)J^a(w), \tag{4.24}$$

it is easy to realize that the OPE that is compatible with such a transformation is,

$$J^a(z)J^b(w) = \frac{k\delta^{ab}}{(z-w)^2} + \frac{if^{ab}_c J^c}{(z-w)}, \tag{4.25}$$

which is the OPE associated with the ALA discussed in Section 3.2.

Next we want to determine the conformal properties of u, in particular its confomal dimension. The classical form of the currents (4.20) is elevated to the quantum expression via the equations,

$$\kappa\partial u(z, \bar{z}) =: J^a T^a u(z, \bar{z}) :, \quad \kappa\bar{\partial}u(z, \bar{z}) =: \bar{J}^a u(z, \bar{z})T^a :, \tag{4.26}$$

where κ, which is a renormalized level, will be determined shortly, and the normal ordering refers as usual to subtracting the singular parts of the product. Assuming that u is an ALA primary field, the OPE takes the form,

$$J^a(z)T^a u(w, \bar{w}) = \frac{c_2}{z-w}u(w, \bar{w}) + \kappa\partial u(w, \bar{w}) + \sum_{n=2}^{\infty}(z-w)^{n-1}T^a J^a_{-n}u(w, \bar{w}), \tag{4.27}$$

where (4.26) was inserted as the $(z-w)^0$ term, and c_2 is the quadratic Casimir operator in the representation of u, $\sum_a T^a T^a = c_2$. If we assume that u is a Virasoro primary as well, then $L_{-1}u = \partial u$, so that combined with (4.27) one can write down the following null vector,

$$v_{\text{null}} \equiv (J^a_{-1}T^a - \kappa L_{-1})u = 0. \tag{4.28}$$

This is a special case of the degeneracy in the combined ALA–Virasoro algebra discussed in Section 3.3.

The null vectors obey,

$$L_0 V_{\text{null}} = (h+1)V_{\text{null}}$$
$$J^a_0 V_{\text{null}} = T^a V_{\text{null}}$$
$$L_n V_{\text{null}} = J^a_n V_{\text{null}} = 0 \quad \text{for } n > 0, \tag{4.29}$$

where $L_0 u = hu$. For $n = 1$ the conditions $L_1 V_{\text{null}} = J_1^a V_{\text{null}} = 0$ imply that the renormalized level and the conformal dimension of u take the form,

$$\kappa = \frac{1}{2}(C_2 + k) \quad h = \frac{c_2}{C_2 + k}, \tag{4.30}$$

where C_2 is the quadratic Casimir in the adjoint representation, defined as $f^{acd} f^{bcd} = C_2 \delta^{ab}$.

The use of null vectors and the differential equations that determine correlators of primary fields were introduced in the landmark paper of Knizhnik and Zamolodchikov [143]. An elaboration of the application of these equations appears in [77]. This direction was further developed by Gepner and Witten [108].

The WZW has an in-built Sugawara construction. In fact it is very often taken as the prototype model for this structure. According to the discussion in Section 3.3 the quantum version of the classical energy-momentum tensor (4.21) takes the form of (3.46),

$$T(z) = \frac{1}{2(k + C_2)} : J^a(z)J^a(z) :, \tag{4.31}$$

and the Virasoro anomaly of the model is,

$$c = \frac{k \dim G}{k + C_2}. \tag{4.32}$$

The Sugawara construction is described in [203]. This paper, however, does not have the correct expression that includes the finite renormalization. This was done later in the paper of Dashen and Frishman [73].

4.4 Correlation functions of primary fields

Primary fields of theories invariant under ALA were defined and discussed in Section 3.4. The group element of the WZW theory is an example of a primary field in the fundamental representation of $G \times G$. Indeed the transformation properties of $u(z, \bar{z})$ imply that it has the following OPE with the currents,

$$J^a(z)u(w, \bar{w}) = -\frac{t^a u(w, \bar{w})}{z - w} \quad \bar{J}^a(\bar{z})u(w, \bar{w}) = -\frac{u(w, \bar{w})t^a}{\bar{z} - \bar{w}}. \tag{4.33}$$

Next we would like to compute the n-point correlation function of the group element primary field of the WZW model. In Section 3.6 we presented the Knizhnik–Zamolodchikov equation which determines the correlators of theories invariant under ALA. We now demonstrate its use in determining the four-point correlation function of the primary field $u(z, \bar{z})$ of $SU(N)$ WZW model. We denote this correlator as,

$$G_4 = \langle u(z_1, \bar{z}_1)u^{-1}(z_2, \bar{z}_2)u^{-1}(z_3, \bar{z}_3)u(z_4, \bar{z}_4)\rangle. \tag{4.34}$$

Recall from (2.59) that in general due to the conformal Ward identity the four-point function can be written as,

$$G_4 = [z_{14}z_{23}\bar{z}_{14}\bar{z}_{23}]^{-2h}\mathcal{G}(x,\bar{x}), \tag{4.35}$$

where

$$x = \frac{z_{12}z_{34}}{z_{14}z_{32}} \qquad \bar{x} = \frac{\bar{z}_{12}\bar{z}_{34}}{\bar{z}_{14}\bar{z}_{32}}, \tag{4.36}$$

with $z_{ij} = z_i - z_j$ and h, the dimension of u, is given by $h = \frac{N^2-1}{2N(N+k)}$.

Now $\mathcal{G}(x,\bar{x})$ can be decomposed into a sum of terms, each one representing a conformal block, the latter having the form of a product of a holomorphic and anti-holomorphic function,

$$\mathcal{G}(x,\bar{x}) = \text{Sum of terms of the form } G(x)\bar{G}(\bar{x}). \tag{4.37}$$

Since $u(z,\bar{z})$ is in the fundamental representation of $SU(N)$, the four-point function is a product of two fundamentals and two anti-fundamentals, so each term in the last equation can be decomposed into,

$$G(x) = I_1 G_1 + I_2 G_2, \tag{4.38}$$

where I_1 and I_2 are the $SU(N)$ invariant factors,

$$I_1 \equiv \delta_{m_1,m_2}\delta_{m_3,m_4} \qquad I_2 \equiv \delta_{m_1,m_3}\delta_{m_2,m_4}. \tag{4.39}$$

If we now substitute this decomposed form of the four-point function into the Knizhnik–Zamolodchikov equation (3.69) we find,

$$\left(\partial_{z_i} + \frac{1}{k+N}\sum_{j\neq i}\frac{t_i^a \otimes t_j^a}{z_i - z_j}\right)[z_{14}z_{23}]^{-4h}(I_1 G_1 + I_2 G_2) = 0. \tag{4.40}$$

As was discussed in Section 2.9 conformal invariance allows us to fix three out of the four points. Using the standard convenient choice,

$$z_1 = x, \quad z_2 = 0, \quad z_3 = 1, \quad z_4 = \infty, \tag{4.41}$$

and the equation now reads,

$$\left(\partial_x + \frac{1}{k+N}\frac{t_1^a \otimes t_2^a}{x} + \frac{1}{k+N}\frac{t_1^a \otimes t_3^a}{x-1}\right)(I_1 G_1 + I_2 G_2) = 0, \tag{4.42}$$

After introducing the explicit expressions for the various group theoretical products $t_i^a \otimes t_j^a I_k$ and projecting to the I_1 and I_2 factors we get,

$$\partial_x G_1 = \frac{-1}{k+N}\left(\frac{(N^2-1)}{N}\frac{G_1}{x} + \frac{G_2}{x} - \frac{1}{N}\frac{G_1}{x-1}\right)$$

$$\partial_x G_2 = \frac{-1}{k+N}\left(\frac{(N^2-1)}{N}\frac{G_2}{x-1} + \frac{G_1}{x-1} - \frac{1}{N}\frac{G_2}{x}\right). \tag{4.43}$$

Extracting G_2 from the first equation and plugging it back into the second equation, the latter translates into a hypergeometric differential equation,

$$\frac{x(1-x)}{N^2}[N^2\kappa^2\partial_x^2 + A(x)\partial_x + B(x)]g_1(x) = 0, \qquad (4.44)$$

where $\kappa = k + N$ and with the following two possible values for $A(x)$, $B(x)$ and the relation between g_1 and G_1 as

$$A(x) = \left(\frac{N(N+\kappa)}{x} - \frac{N^2}{1-x}\right)N\kappa \quad B(x) = -\frac{N^4 - N^2 + 2}{x(1-x)},$$

$$G_1 = [x(1-x)]^{\frac{1}{\kappa N}}g_1^+$$

or

$$A(x) = \left(\frac{-N(N-\kappa)}{x} - \frac{N^2}{1-x}\right)N\kappa \quad B(x) = -\frac{2(N^2-1)}{x(1-x)},$$

$$G_1 = x^{-\frac{N^2-1}{\kappa N}}(1-x)^{\frac{1}{\kappa N}}g_1^- \qquad (4.45)$$

The solutions of the differential equations are the following hypergeometric functions,

$$g_1^- = F\left(\frac{1}{\kappa}, -\frac{1}{\kappa}, 1 - \frac{N}{\kappa}; x\right) \quad g_1^+ = F\left(\frac{N-1}{\kappa}, \frac{N+1}{\kappa}, 1 + \frac{N}{\kappa}; x\right). \qquad (4.46)$$

In a similar way the solutions for G_2 are found, defining an appropriate g_2.

To fully determine the correlator we still have to fix the linear combination of the solutions. This is done using crossing symmetry, as discussed in Section 2.10.[2] The latter implies that,

$$\mathcal{G}(x, \bar{x}) = \mathcal{G}(1 - x, 1 - \bar{x}). \qquad (4.47)$$

With parametrization,

$$\mathcal{G}(x, \bar{x}) = \sum_{i,j=1,2} I_i \bar{I}_j \mathcal{G}_{i,j}(x, \bar{x}) \quad \mathcal{G}_{i,j} = \sum_{n,m=+,-} \xi_{mn} G_i^{(m)} G_j^{(n)}. \qquad (4.48)$$

Crossing symmetry implies that,

$$\mathcal{G}_{i,j}(x, \bar{x}) = \mathcal{G}_{3-i,3-j}(1 - x, 1 - \bar{x}), \qquad (4.49)$$

which follows from the fact that under crossing symmetry $I_1 \leftrightarrow I_2$. Single valuedness implies that $\xi_{+-} = \xi_{-+} = 0$. To obey the crossing symmetry requirement

---

[2] Crossing symmetry to determine correlators of fermions in the fundamental representation of $SU(N)$ was used by Dashen and Frishman in [73].

we make use of the following property of hypergeometric functions:

$$F(a, b, c; x) = A_1 F(a, b, a + b - c; 1 - x)$$
$$+ A_2 (1 - x)^{c-a-b} F(c - a, c - b, c - a - b + 1; 1 - x), \quad (4.50)$$

where

$$A_1 = \frac{\Gamma(c)\Gamma(c - a - b)}{\Gamma(c - a)\Gamma(c - b)} \quad A_2 = \frac{\Gamma(c)\Gamma(a + b - c)}{\Gamma(a)\Gamma(b)}. \quad (4.51)$$

Finally we find that

$$\mathcal{G}_{ij} = G_i^{(-)}(x) G_j^{(-)}(\bar{x}) + \frac{c_{--}^2 - 1}{c_{+-}^2} G_i^{(+)}(x) G_j^{(+)}(\bar{x}), \quad (4.52)$$

where

$$c_{--} = N \frac{\Gamma(\frac{N}{\kappa})\Gamma(-\frac{N}{\kappa})}{\Gamma(\frac{1}{\kappa})\Gamma(-\frac{1}{\kappa})} \quad c_{+-} = -N \frac{[\Gamma(\frac{N}{\kappa})]^2}{\Gamma(\frac{N+1}{\kappa})\Gamma(\frac{N-1}{\kappa})}. \quad (4.53)$$

For $k = 1$ we have $c_{--} = -1$, and hence the second term in (4.52) vanishes. Using $F\left(\frac{1}{N+1}, -\frac{1}{N+1}, \frac{1}{N+1}; x\right) = (1 - x)^{\frac{1}{N+1}}$ the four-point function takes the form,

$$\mathcal{G}(x, \bar{x}) = [x\bar{x}(1 - x)(1 - \bar{x})]^{\frac{1}{N}} \left[I_1 \frac{1}{x} + I_2 \frac{1}{1 - x}\right] \left[\bar{I}_1 \frac{1}{x} + \bar{I}_2 \frac{1}{1 - x}\right]. \quad (4.54)$$

In Section 6.3 it will be shown that the WZW theory of $U(N)$ at level $k = 1$ is a bosonized equivalent to that of N Dirac fermions controlled by the action,

$$S_f = \int d^2 z \, [\psi_\alpha^\dagger \bar{\partial} \psi_\alpha + \tilde{\psi}_\beta^\dagger \partial \tilde{\psi}_\beta]. \quad (4.55)$$

In particular the fermion bilinear is equivalent to the group element as,

$$M \tilde{u}_\alpha^\beta(z, \bar{z}) =: \psi_\alpha(\tilde{\psi}^\dagger)^\beta : \quad M(\tilde{u}^{-1})_\beta^\alpha(z, \bar{z}) =: \tilde{\psi}_\beta \psi^{\dagger\alpha} :, \quad (4.56)$$

where M is a mass scale and where $\tilde{u}$ denotes the $U(N)$ group element. In the theory of free Dirac fermions the four-point function of the fermion bilinears takes the form,

$$\mathcal{G}(x, \bar{x}) = \left[I_1 \frac{1}{x} + I_2 \frac{1}{1 - x}\right] \left[\bar{I}_1 \frac{1}{x} + \bar{I}_2 \frac{1}{1 - x}\right]. \quad (4.57)$$

It is easy to see that by converting (4.54) to a similar correlator of $U(N)$ we find exactly the same answer. This is done as follows: define the $U(N)$ group element to be,

$$\tilde{u}(z, \bar{z}) = e^{i\sqrt{4\pi/N}\gamma\varphi(z,\bar{z})} u(z, \bar{z}). \quad (4.58)$$

Then the four-point function of group elements of $U(N)$ is,

$$\tilde{\mathcal{G}}(x, \bar{x}) = M^{-2\frac{\gamma^2}{N}} [x\bar{x}(1 - x)(1 - \bar{x})]^{-\frac{\gamma^2}{N}} \mathcal{G}(x, \bar{x}). \quad (4.59)$$

For $\gamma = 1$ we observe that indeed the correlator is identical to that of the fermion bilinears. For arbitrary γ the correlator corresponds to that of fermion bilinears of the Thirring model defined by,

$$S = S_f + \frac{\gamma^2 - 1}{2\gamma^2} \int d^2 z J(z) \bar{J}(\bar{z}). \tag{4.60}$$

This generalized bosonization will also be addressed in Section 6.2.

4.5 WZW models with boundaries – D branes

The WZW model described in Section 4.1 was shown to be based on a map from Σ, a compact two-dimensional manifold, in paticular an S^2, into a group manifold G. Let us now study the case where Σ has boundaries. For concreteness we take it to be the upper half-plane. In the bulk the theory is invariant under the holomorphic and anti-holomorphic ALA (4.18) and there is corresponding holomorphic and anti-holomorphic conservation of the associated currents (4.9). On the boundary the two types of modes mix, the symmetry is reduced to,

$$u \to g(\tau) u g(\tau)^{-1}, \tag{4.61}$$

where τ denotes the coordinate on the boundary, and accordingly there is a relation between J_L and J_R,

$$J(z) = \Omega_{\text{aut}} \bar{J}(\bar{z}) \quad \text{at } z = \bar{z}, \tag{4.62}$$

where Ω_{aut} is an automorphism of the ALA.

The notion of boundary conformal field theory was introduced in [58]. The gluing conditions used for D branes in the WZW model were introduced in [135]. From the many papers that have been written on the subject we have chosen to describe it following [11] and [85].

Let us first address the simplest case of a level k $S\hat{U}(2)$ WZW, for which $\Omega = -1$, and then later discuss the general case. In terms of ∂_t, ∂_x and the adjoint action of G on its Lie algebra,

$$Ad(g)u = gug^{-1}, \tag{4.63}$$

the gluing condition reads,

$$(1 - Ad(u))u^{-1}\partial_t u = (1 + Ad(u))u^{-1}\partial_x u. \tag{4.64}$$

The tangent space to the group G at the point u can be split into $T_u G = T_u^{\perp} G \oplus T_u^{\parallel} G$, where $T_u^{\parallel} G$ consists of vectors tangential to the orbit of Ad through u. On $T_u^{\perp} G$, $(1 - Ad(u)) = 0$ and $(1 + Ad(u)) = 2$, so that $(u^{-1}\partial_x u)^{\perp} = 0$ and the corresponding D branes, namely the submanifolds where the condition (4.62) is obeyed, coincide with the conjugacy classes. In the case that $(1 - Ad(u))$ is invertible, (4.64) can be written as,

$$u^{-1}\partial_t u = \frac{1 + Ad(u)}{1 - Ad(u)} u^{-1} \partial_x u. \tag{4.65}$$

We can now define a two form on the conjugacy class as,

$$\omega = \frac{k}{8\pi}\left(u^{-1}\mathrm{d}u\frac{1 + Ad(u)}{1 - Ad(u)}u^{-1}\mathrm{d}u\right). \tag{4.66}$$

Applying an exterior derivative to this form we find,

$$\mathrm{d}\omega = \frac{k}{12\pi}Tr(\mathrm{d}uu^{-1})^3, \tag{4.67}$$

namely, it is not closed. The submanifold $D \subset G$ on which the WZ term is exact, $(WZ) = \mathrm{d}\omega$ defines a D brane in G. There is a further restriction which follows from reasoning similar to that discussed in Section 4.1.

Consider the wave functional $\Psi(u(x))$ on the space of closed loops $u(x)$ in some conjugacy class C. The latter, for the group manifold $SU(2)$, are typically two-spheres so that C can be constructed in two different ways, and hence there is an ambiguity in the phase of the wave functional. It can be shown that the phase can take the values $2\pi j$ with $j = 1, \ldots, k - 1$. Thus there are $k - 1$ two-dimensional conjugacy classes or D_2 branes for the k level $S\hat{U}(2)$ WZW model. In addition there are two D_0 branes associated with the two points $\pm e$, where e is the identity on the group space.

To address the issue of the conjugacy classes in other groups it is convenient to rewrite the two form (4.66) as,

$$\omega = \frac{k}{8\pi}Tr[\tilde{k}^{-1}\mathrm{d}\tilde{k}h\tilde{k}^{-1}\mathrm{d}\tilde{k}h^{-1}], \tag{4.68}$$

for u that belongs to the conjugacy class,

$$C_h^G = \{u \in G|u = \tilde{k}h\tilde{k}^{-1}\}. \tag{4.69}$$

The WZW action that corresponds to the map from the two manifold with a boundary to the group space can be constructed as follows. Instead of considering the map from Σ that has a boundary, we take it from $\Sigma \cup D$ where D is an auxilliary disc that closes the hole in Σ, having a common boundary with it. The disc is mapped into the conjugacy class allowing for its boundary (4.69). The WZW action is now written as,

$$S = S_{WZW} + \int_D w, \tag{4.70}$$

where S_{WZW} is the ordinary WZW action with a three-dimensional WZ term defined now on a ball whose boundary is $\Sigma \cup D$. It can be explicitly checked that the new WZW action is invariant under (4.18). Similar to the topologically distinct D_2 branes of the k level $S\hat{U}(2)$ WZW model, there are different embeddings of the disc in a conjugacy class in a general group manifold G. This is related to the second homotopy group of the conjugacy class, which in general is non-trivial. The group element k of (4.69) is defined modulo a right multiplication with any element that commutes with h, and the group of such elements is isomorphic to the Cartan torus T_G generated by the generators in the Cartan subalgebra. Thus the conjugacy classes can be described as $\frac{G}{T_G}$ and the second homotopy group reads,

$$\Pi_2(C_h^G) = \Pi_1(T_G) \tag{4.71}$$

For a rank r algebra of G, the D_2 will be characterized by an r-dimensional vector in the coroot lattice of G. Namely, if two embeddings given by khk^{-1} and $k'hk'^{-1}$ then on the boundary of the world sheet they have to be related as,

$$k(\tau)k'(\tau)^{-1} = t(\tau), \tag{4.72}$$

where $t(\tau)$ is an element of the subgroup isomorphic to T_G which commutes with h. This relation determines a mapping from the boundary to T_G. Since the latter is R^r modulo $2\pi \times$ (coroot lattice), every such mapping belongs to the topological sector parameterized by a vector in the coroot lattice describing the winding of the boundary circle on the torus T_G. This lattice vector determines via (4.71) the element of $\Pi^2(C_h^G)$ corresponding to the union of the two embeddings. For the group element $h \in T_G$ of the form $h = e^{i\theta \cdot H}$, where H are the Cartan subalgebra generators, the change in the action resulting from the topological change of the embedding is $\Delta S = k(\theta \cdot s)$, where s is a coroot lattice vector. Consistency of the model then implies the condition,

$$\theta \cdot \alpha \in \frac{2\pi \mathcal{Z}}{k}. \tag{4.73}$$

This generalizes the condition that led to the set of $k - 1$ D_2 branes for level k $S\hat{U}(2)$. It implies that θ should be $\frac{2\pi}{k} \times$ (weight lattice vector). Since a point in T_G is defined modulo $2\pi \times$ (coroot lattice), the allowed conjugacy classes correspond to points in the weight lattice divided by k modulo the coroot lattice. This is also the characterization of the primary fields of the corresponding WZW model.

4.6 G/H coset models

The concept of coset models dates back to [110] or in fact even to as early as [27]. A Lagrangian formulation in terms of a gauged WZW model was introduced for instance in [12]. Here we follow the review of [111].

The WZW models constitute a large class of conformal field theories which are invariant under ALA. An even larger class of CFTs can be constructed by taking the quotient of two WZW models. Consider an ALA $\hat{g}$ at level k and a subalgebra of it $\hat{h}$ at level k_h. We denote the currents associated with the former as J^a and with the latter $J_h^{a_h}$ where $a = 1, \ldots, \dim G$ and $a_h = 1, \ldots, \dim H$. The currents associated with the subalgebra $\hat{h}$ can be expressed as a linear combination of $\hat{g}$ as,

$$J_h^{a_h} = \sum_a m_{a_h a} J^a. \tag{4.74}$$

Using the commutator of J^a and the corresponding generator of the Virasoro algebra constructed via the Sugawara construction,

$$[L_m, J_n^a] = -m J_{m+n}^a, \tag{4.75}$$

it follows that,

$$[L_m, J_{h\ n}^{a_h}] = -m J_{h\ m+n}^{a_h}. \tag{4.76}$$

It is also obvious that a similar relation holds with L_m^h which is the Virasoro generator built by a Sugawara construction from the currents of $\hat{h}$, namely,

$$[L_m^h, J_{h\ n}^{a_h}] = -m J_{h\ m+n}^{a_h}. \tag{4.77}$$

Thus it follows that,

$$[L_m - L_m^h, J_h^{a_h}] = 0. \tag{4.78}$$

Since L_m^h is a bilinear of $J_h^{a_h}$ it follows that,

$$[L_m - L_m^h, L_n^h] = 0 \quad \rightarrow \quad [L_m, L_n^h] = [L_m^h, L_n^h]. \tag{4.79}$$

We can now define,

$$L_m^{(g/h)} \equiv L_m - L_m^h. \tag{4.80}$$

The algebra of these coset generators is a Virasoro algebra, as follows from,

$$[L_m^{(g/h)}, L_n^{(g/h)}] = [L_m, L_n] - [L_m^h, L_n^h]$$
$$= (m-n)L_{m+n}^{(g/h)} + [c(\hat{g}_k) - c(\hat{h}_{k_h})]\frac{(m^3 - m)}{12}\delta_{m+n,0}. \tag{4.81}$$

Thus we have just found that the Virasoro generators of the coset $L_m^{(g/h)}$ obey a Virasoro algebra with a central charge of

$$c = \frac{k \, dimg}{k + C_2(g)} - \frac{k_h \, dimh}{k_h + C_2(h)}. \tag{4.82}$$

A special class of coset models are the *diagonal coset models* $\frac{\hat{g} \oplus \hat{g}}{\hat{g}}$. The generators of the coset in this case are given by $J_h^a = J_{(1)}^a + J_{(2)}^a$, namely the sum of the generators of each copy. It thus follows that the level of the coset must be the sum of the two levels since $[J_{(1)}^a, J_{(2)}^a] = 0$. The coset therefore takes the form,

$$\frac{\hat{g}_{k_1} \oplus \hat{g}_{k_2}}{\hat{g}_{k_1 + k_2}},$$

and its corresponding central charge is given by,

$$c = \dim g \left(\frac{k_1}{k_1 + C_2(g)} + \frac{k_2}{k_2 + C_2(g)} - \frac{k_1 + k_2}{k_1 + k_2 + C_2(g)} \right). \qquad (4.83)$$

Consider the case where $g = SU(2)$ and the coset is,

$$\frac{S\hat{U}(2)_k \oplus S\hat{U}(2)_1}{S\hat{U}(2)_{k+1}}, \qquad (4.84)$$

with the central charge,

$$c = \frac{3k}{k+2} + 1 - \frac{3(k+1)}{k+3} = 1 - \frac{6}{(k+2)(k+3)} = 1 - \frac{6}{p(p+1)}, \qquad (4.85)$$

where in the last step we introduced $p = k + 2 \geq 3$. This is exactly the central term of the unitary minimal models discussed in Section 2.7. In fact one can show that this is indeed an equivalence in the sense that the characters of the minimal models are the same as those of the coset model.

4.7 G/G coset models

The concept of the G/G model was introduced in [200] and [227]. Our description of the G/G model and in particular its BRST analysis follows [9] and [8].

A special class of the G/H models is the case where $H = G$, namely where we gauge the maximal anomaly-free diagonal group. Using the gauging procedure that will be discussed in Section 9.3.1 the classical action takes the form,

$$S_k(g, A^\mu) = S_k(g) - \frac{k}{2\pi} \int d^2z Tr(g^{-1}\partial g \bar{A}_{\bar{z}} + g\bar{\partial}g^{-1}A_z - \bar{A}_{\bar{z}}g^{-1}A_z g + A_z \bar{A}_{\bar{z}}). \qquad (4.86)$$

Next we introduce the following parameterization of the gauge fields, $A_z = ih^{-1}\partial_z h$, $\bar{A}_{\bar{z}} = ih^*\partial_{\bar{z}}h^{*-1}$ where $h(z) \in G^c$. In Section 15 we will elaborate more about this formulation. The action then reads,

$$S_k(g, A) = S_k(g) - S_k(hh^*). \qquad (4.87)$$

The Jacobian of the change of variables introduces a dimension $(1,0)$ system of anticommuting ghosts χ and ρ in the adjoint representation of the group. The quantum action thus takes the form of,

$$S_k(g, h, \rho, \chi) = S_k(g) - S_{k+2C_G}(hh^*) - i \int d^2z Tr[\rho\bar{\partial}\chi + c.c], \qquad (4.88)$$

where C_G is the second Casimir of the adjoint representation.[3] The second term can be viewed as $S_{-(k+2C_G)}(h)$. Since the Hilbert space of the model decomposes into holomorphic and anti-holomorphic sectors we restrict our discussion only to the former.

---

[3] This was $C_2(g)$ in the previous section; notation has changed according to the literature.

There are three sets of holomorphic G transformations which leave (4.88) invariant,

$$\delta_J g = i[\epsilon(z), g] \quad \delta_I h = i[\epsilon(z), h]$$
$$\delta_{J(gh)} \chi^a = i f^a_{bc} \epsilon^b \chi^c \quad \delta_{J(gh)} \rho^a = -i f^a_{bc} \epsilon^b \rho^c, \tag{4.89}$$

with ϵ in the algebra of G. The corresponding currents J^a, I^a and $J^{(gh)a} = i f^a_{bc} \chi_b \rho_c$ satisfy the G ALA with the levels $k, -(k + 2c_G)$ and $2c_G$, respectively. We now define $J^{(\text{tot})a}$

$$J^{(\text{tot})a} = J^a + I^a + J^{(gh)a} = J^a + I^a + i f^a_{bc} \chi_b \rho_c, \tag{4.90}$$

which obeys an affine Lie algebra of level,

$$k^{(\text{tot})} = k - (k + 2c_G) + 2c_G = 0. \tag{4.91}$$

The energy-momentum tensor $T(z)$ is a sum of Sugawara terms of the J^a and I^a currents and the usual contribution of a $(1, 0)$ ghost system, namely,

$$T(z) = \frac{1}{k + c_G} : J^a J^a : - \frac{1}{k + c_G} : I^a I^a : + \rho^a \partial \chi^a. \tag{4.92}$$

The corresponding Virasoro central charge vanishes,[4]

$$c^{(\text{tot})} = \frac{k d_G}{k + c_G} - \frac{(k + 2c_G) d_G}{-(k + 2c_G) + c_G} - 2d_G = 0 \tag{4.93}$$

The symmetry structure of the model is in fact richer. It is easy to realize that there are also two odd conserved currents, a dimension one current which is the BRST current $J^{(\text{BRST})}$ and a dimension two operator G. These holomorphic symmetry generators are given by

$$J^{(\text{BRST})} = \chi_a \left[J^a + I^a + \frac{1}{2} J^{(gh)a} \right],$$

$$G = \frac{1}{k + c_G} \rho_a [J^a - I^a]. \tag{4.94}$$

The reason we denote the dimension one current as a BRST current is that one can express both $T(z)$, $J^{(\text{tot})}{}_a$ and $J^{(\text{BRST})}$ itself in terms of its corresponding charge $Q = \int J^{(\text{BRST})}(z)$ as follows,

$$T(z) = \{Q, G(z)\},$$
$$J^{(\text{tot})a}(z) = \{Q, \rho^a\},$$
$$J^{(\text{BRST})} = \{Q, j^{\#}(z)\}, \tag{4.95}$$

where $j^{\#}$ is the ghost number current.

The fact that $T(z)$ is BRST exact namely $T(z) = \{Q, G(z)\}$ and that the total Virasoro anomaly vanishes, are indications that the $\frac{G}{G}$ model is a topological quantum field theory. These theories which were found to be very useful in dealing with various issues in physics and mathematics are beyond the scope

---

[4] d_G is what we called dim g in the previous section; changed according to the literature.

of this book. We thus do not discuss here the topological quantum field theory aspects of the $\frac{G}{G}$ models.

By construction of the BRST procedure the space of physical states of a $\frac{G}{G}$ model is given by the cohomology of Q. That is to say, a physical state |phys> has to be closed under Q, namely Q|phys> = 0 and not exact, namely |phys> ≠ Q|state> where |state> is any other state. It can be shown that taking the trace over those states one finds the torus partition function of the $\frac{G}{G}$ model which is based on the decomposition into WZW characters, discussed in Section 3.5. The torus partition function can be expressed as

$$Z_{\frac{G}{G}} = c\tau_2^{-r} \int du Z^g(\tau, u) Z^{hh^*}(\tau, u) Z^{gh}(\tau, u), \qquad (4.96)$$

where du is the measure over the flat gauge connections on the torus and r is the rank of G; $Z^g(\tau, u)$ is the torus partition function of the G_k WZW model,

$$Z^g(\tau, u) = (q\bar{q})^{\frac{-c}{24}} \sum_{\lambda_L, \lambda_R} \chi_{k,\lambda_L}(\tau, u) \bar{\chi}_{k,\lambda_R}(\tau, u) N_{\lambda_R, \lambda_L}, \qquad (4.97)$$

where $q = e^{2i\pi\tau}$, λ_R, λ_L denote the G_k highest weights, and for the diagonal invariant $N_{\lambda_R, \lambda_L} = \delta_{\lambda_R, \lambda_L}$. The character can be written as,

$$\chi_{k,\lambda}(\tau, u) = \frac{M_{k,\lambda}(\tau, u)}{M_{0,0}(\tau, u)}, \qquad (4.98)$$

with $M_{k,\lambda}$ defined explicitly for the $SU(2)$ case below. $Z^{hh^*}(\tau, u)$ in (4.96) is the contribution of $h \in \frac{G^c}{G}$ at level $k + 2C_G$ or equivalently $h \in G$ at level $-(k + 2c_G)$. This takes the form $Z^{hh^*}(\tau, u) \sim |M_{0,0}(\tau, u)|^{-2}$ indicating that $\frac{G^c}{G}$ contains just one conformal block. It is straightforward to calculate $Z^{gh}(\tau, u)$, the ghost contribution to the partiton function $Z^{gh}(\tau, u) \sim |M_{0,0}(\tau, u)|^4$. Thus there is a cancellation of the $|M_{0,0}(\tau, u)|$ factors and the resulting character is given by the numerator of the character of the "matter" sector. In the $\frac{G}{G}$ model it is $M_{k,\lambda}$. This cancellation property leads to an index interpretation for $M_{k,\lambda}$. For $G = SU(2)$ it was found that one can express,

$$M_{k,j}(\tau, \theta) = \sum_{l=-\infty}^{\infty} q^{(k+2)(l+\frac{j+\frac{1}{2}}{(k+2)})^2} \sin\left\{\pi\theta\left[(k+2)l + j + \frac{1}{2}\right]\right\}, \qquad (4.99)$$

as

$$M_{k,j}(\tau, \theta) = \frac{1}{2i} q^{\frac{(j+\frac{1}{2})^2}{(k+2)}} e^{i\pi\theta(j+\frac{1}{2})} Tr[(-)^{\hat{G}} q^{\hat{L}_0} e^{i\pi\theta \hat{J}_{(tot)}^0}], \qquad (4.100)$$

where $\theta = Reu$, $\hat{G}$ is the ghost number, $\hat{L}_0$ is the excitation level and $\hat{J}_{(tot)}^0$ is the $J_{(tot)}^0$ eigenvalue of the excitation. Note that $M_{k,j}(\tau, \theta)$ is obtained from the torus $M_{k,j}(\tau, u)$ by restricting to just one angle. This amounts to considering the propagation along a cylinder rather than around the torus. As long as we are interested in the spectrum it is sufficient to consider $M_{k,j}(\tau, \theta)$. This index interpretation enables us to read important information about the physical spectrum from (4.99). For a positive integer k, $2j = 0, \ldots, k$. Hence there are $k + 1$ zero ghost number primary states which correspond to the first term in the q expansion of the different $M_{k,j}$s, i.e. the term corresponding to $l = 0$ with $\hat{L}_0 = \frac{j(j+1)}{k+2}$. On each of these states there is a whole tower of descendant states correponding to the higher terms in the q expansion. For further discussion of the G/G model the reader is referred to [9].

5

Solitons and two-dimensional integrable models

5.1 Introduction

In the previous chapters we have addressed 2D field theories with no scale. As we discussed in Chapter 2, one cannot define an S-matrix for such theories. Generically physical systems are characterized by certain energy scales and the notion of S-matrix plays an important role. It is thus time to move forward and examine non-conformal field theories. Again we start our journey with the theory of a free scalar field, but now a massive one. We then move on and discuss interacting theories equipped with infinite numbers of conserved charges, the so-called integrable models, that resemble the free massive theory in a way that will be explained below.

5.2 From the theory of a massive free scalar field to integrable models

The classical action of a massive free scalar field is obviously the action of a massless scalar field with an additional mass term,

$$S = \int d^2x \left[\frac{1}{2} \partial^\mu \phi \partial_\mu \phi - \frac{1}{2} m^2 \phi^2 \right], \tag{5.1}$$

where m is the mass scale which momentarily will be shown to be the mass of the particle associated with the field ϕ. Unlike the analysis of CFTs there is no advantage here to the use of complex coordinates, so we will use real ones.

The corresponding equation of motion,

$$\partial^\mu \partial_\mu \phi + m^2 \phi = 0, \tag{5.2}$$

is solved for the case of uncompactified space-time by the following Fourier transform,

$$\phi(x^0, x^1) = \int \frac{dk^1}{\sqrt{2\pi}\sqrt{k^0}} \left[a(k^1) e^{-ik \cdot x} + a^\dagger(k^1) e^{ik \cdot x} \right], \tag{5.3}$$

where $(k^0)^2 - (k^1)^2 = m^2$.

A dramatic difference between the massless field discussed in Chapter 2, and the massive one we discuss here, shows up when analyzing the symmetries of the system.

The only transformations that leave the action invariant are the $ISO(1,1)$ Poincare transformations, namely, the space and time translations and a single Lorentz transformation. These are,

$$x^0 \to x^0 + a^0 \quad x^1 \to x^1 + a^1$$
$$x^0 \to x^0 + a^0_1 x^1 \quad x^1 \to x^1 + a^1_0 x^0, \tag{5.4}$$

where the transformation parameters are constants and $a^0_1 = a^1_0$. The fact that the parameters are constants, and not holomorphic and anti-holomorphic functions of the complex coordinates, has a tremendous impact, since it implies the absence of the powerful infinite-dimensional Virasoro algebra.

The corresponding Noether currents associated with the Poincare transformations are,

$$T_{\mu\nu} = \partial_\mu \phi \partial_\nu \phi - g_{\mu\nu} \mathcal{L},$$
$$J^\mu_{\text{Lor}} \equiv J^{\mu 01}_{\text{Lor}} = \epsilon_{\rho\nu} T^{\mu\rho} x^\nu. \tag{5.5}$$

However, since the space-time is two dimensional, there is an additional conserved current, the so-called *topological current*,

$$J^\mu_{\text{top}} = \epsilon^{\mu\nu} \partial_\nu \phi, \tag{5.6}$$

which is conserved regardless of the equations of motion, since obviously $\partial_\mu J^\mu_{\text{top}} = \epsilon^{\mu\nu} \partial_\mu \partial_\nu \phi = 0$. In fact this current is conserved for any interacting scalar field in 2d, and as we will see later on it plays an important role in the analysis of soliton solutions of integrable models.

The theory of a free massive scalar field, as well as other scalar theories that will be addressed in this chapter, are obviously invariant under the discrete symmetry of,

$$\phi \to -\phi. \tag{5.7}$$

The canonical quantization was described in Section 1.6. The normal ordered Hamiltonian and momentum expressed in terms of the creation and annihilation operators take the form,

$$H = \int dk^1 \sqrt{(k^1)^2 + m^2} a^\dagger(k^1) a(k^1) \quad \mathbf{P} = \int dk^1 k^1 a^\dagger(k^1) a(k^1). \tag{5.8}$$

The state $a^\dagger(k^1)|0\rangle$ is characterized by,

$$\mathbf{P} a^\dagger(p^1)|0\rangle = p^1 a^\dagger(p^1)|0\rangle \quad H a^\dagger(p)|0\rangle = \sqrt{(p^1)^2 + m^2} a^\dagger(p^1)|0\rangle, \tag{5.9}$$

and hence it is interpreted as a single free massive relativistic particle.

For the case of free single particles, the momentum and Hamiltonian can be generalized to an infinite set of conserved charges, like $Q_n \equiv P^n$ as,

$$Q_n \, a^\dagger(p)|0\rangle = (p^1)^n a^\dagger(p)|0\rangle, \tag{5.10}$$

and similarly for powers of p^0.

The conserved charge Q_n can also be represented as an integral in space. For odd $n = 2k + 1$,

$$Q_{2k+1} = \int dx [\phi^{(2k+1)}(t, x)\dot{\phi}(t, x) + \text{hermitian conjugate}], \qquad (5.11)$$

where $\phi^{(n)}$ is n derivatives of space on ϕ, t is x^0 and x is x^1. For even n it is a bit more complicated, but can be evaluated similarly. Note that the expression is local.

We elevate the field theory associated with a free massive scalar particle into a non-trivial interacting integrable model by replacing the mass term with a potential for the scalar field. It will be shown that identifying in such interacting field theories, an infinite set of conserved charges similar to the one of the free theory, will be a key ingredient in constructing integrable models. This will be discussed in Section 5.10.

A more general construction of integrable models is based on perturbing conformal field theories, which were discussed in Chapter 3, with relevant primary fields, namely, those that have conformal dimension $\Delta + \bar{\Delta} < 2$.[1] This class of models, which will include in particular the integrable minimal models, will be discussed in Section 5.9.

A very basic notion in scalar theories with interacting potential is the solitonic classical configurations, which will be the topic of the next section.

5.3 Classical solitons

We now let the massive particles interact with each other. The interaction is introduced in the form of a potential added to the Lagrangian of the free scalar field theory. Our first task in analyzing this type of field theories is to determine the solutions of the classical equations of motion. We start first with *solitons*, which are static solutions of finite energy, and then move on to time-dependent solutions.

The "classical" material about solitons is described in great detail in several books, in particular in [182], [66] and [183]. For "nontopological solitons" see [149].

Consider a two-dimensional scalar field described by the following action,

$$
\begin{aligned}
S &= \int d^2 x \left[\frac{1}{2} \partial_\mu \phi \partial^\mu \phi - V(\phi) \right] \\
&= \int d^2 x \left[\frac{1}{2} (\dot{\phi})^2 - \frac{1}{2} (\phi')^2 - V(\phi) \right],
\end{aligned} \qquad (5.12)
$$

---

[1] So far in the context of conformal field theory we have denoted the conformal dimension by h. Here in the chapter on integrable models it will be denoted by Δ.

where $\dot{}$ and $'$ refer to time and space derivatives and $V(\phi)$ is a positive semi-definite function of ϕ. The corresponding equation of motion is given by,

$$\partial_\mu \partial^\mu \phi + \partial_\phi V(\phi) = \ddot{\phi} - \phi'' + \partial_\phi V(\phi) = 0. \tag{5.13}$$

The energy associated with a given configuration of ϕ is,

$$E = \int \mathrm{d}x \left[\frac{1}{2}(\dot{\phi})^2 + \frac{1}{2}(\phi')^2 + V(\phi) \right]. \tag{5.14}$$

Let us assume that the potential has a set of N absolute minima at which it vanishes, namely $V(\phi^i) = 0$ for $i = 1, \ldots, N$. If ϕ^i are constants independent of space-time, then the corresponding energy vanishes, and in fact $E(\phi) = 0$ if and only if $\phi(x,t) = \phi^i$.

Static solutions of the equation of motion are determined by,

$$\phi'' - \partial_\phi V(\phi) = 0. \tag{5.15}$$

Solitons which have finite energy, must have ϕ' and $V(\phi)$ vanish rapidly enough at $\pm\infty$, and thus must approach asymptotically one of the configurations ϕ^i that minimizes the potential, namely

$$\lim_{x \to \infty} \phi(x) \to \phi^i, \quad \lim_{x \to -\infty} \phi(x) \to \phi^j. \tag{5.16}$$

Solving (5.15) is equivalent to solving a mechanical system where x becomes the time, ϕ the coordinate of a point particle of a unit mass subjected to a potential $-V(\phi)$, and $E_{\mathrm{mech}} = \frac{1}{2}\phi'^2 - V(\phi)$ is the conserved energy of the system. The boundary conditions where at $x \to \pm\infty$ $V(\phi) \to 0$ and $\phi' \to 0$ implies that $E_{\mathrm{mech}} = 0$. The energy of the field theory (5.14) translates into the action of the mechanical system. The particle trajectory is therefore characterized by having finite action and vanishing mechanical energy. The virial theorem for the particle system has the form,

$$\frac{1}{2}(\phi')^2 = V(\phi), \tag{5.17}$$

which is also easily derivable in field theory language by multiplying (5.15) by ϕ', integrating over x and using the boundary conditions.

From the mechanical analog it is clear that:

(i) there is no non-trivial solution for a potential with a single minimum.
(ii) For a potential with n minima there are $2(n-1)$ solutions associated with trajectories starting at $x \to -\infty$ at ϕ^i and ending at $x \to \infty$ at ϕ^{i+1} and vice versa. Trajectories where instead the particle ends at $\phi^{j>i+1}$ or back to ϕ^i are impossible, since all the derivatives $\frac{d^n \phi}{dx^n}$ vanish at ϕ^{i+1} so the particle that gets to this point will not be able to leave it.

The equation of motion (5.15) has solutions of the form,

$$x - x_0 = \pm \int_{\phi(x_0)}^{\phi(x)} \frac{d\tilde{\phi}}{\sqrt{2V(\tilde{\phi})}},$$

(5.18)

where x_0 the integration constant is any arbitrary point where the field has the value of $\phi(x_0)$. The integral is non-singular apart from the end-points since everywhere else $V(\phi)$ is positive.

Classical solitons of $\lambda\phi^4$ theory

Let us now demonstrate the general features of solitons discussed in the previous section with the prototype model of the a potential with a quartic interaction. Consider the potential,

$$V(\phi) = \frac{1}{4}\lambda \left(\phi^2 - \frac{m^2}{\lambda} \right)^2,$$

(5.19)

which has two minima at $\phi = \pm\frac{m}{\sqrt{\lambda}}$ and is obviously invariant under $\phi \to -\phi$. Substituting this potential into (5.18) and inverting it one finds when setting $\phi(x_0) = 0$ the following two possible solutions,

$$\phi(x) = \pm \frac{m}{\sqrt{\lambda}} \tanh \left[\frac{m}{\sqrt{2}}(x - x_0) \right],$$

(5.20)

which corresponds to either starting at $\phi = -\frac{m}{\sqrt{\lambda}}$ and ending at $\phi = \frac{m}{\sqrt{\lambda}}$, or vice versa. The former will be called a "kink" and the latter an "anti-kink". The invariance under $\phi \to -\phi$ and parity transformation are easily realized in the kink anti-kink system, namely, $\phi_{\text{kink}}(x) = -\phi_{\text{anti-kink}}(x)$, and for $x_0 = 0$ also $= -\phi_{\text{kink}}(-x)$ (otherwise reflect around x_0).

The energy density of the kink solution is given by

$$\epsilon(x) = \frac{1}{2}(\phi')^2 + V(\phi) = \frac{m^4}{2\lambda} \text{sech}^4 \left[\frac{m}{\sqrt{2}}(x - x_0) \right].$$

(5.21)

The total classical energy, which is refered to as the classical mass of the kink is (as our soliton is like a particle at rest),

$$M_{\text{cl}} = \int_{-\infty}^{\infty} dx \epsilon(x) = \frac{2\sqrt{2}}{3} \frac{m^3}{\lambda}.$$

(5.22)

Classical solitons of sine-Gordon theory

The sine-Gordon model, which will serve as a prototype model throughout this chapter is defined by the action given in (5.12) with the potential,

$$V(\phi) = -\frac{m^4}{\lambda} \left[\cos \left(\frac{\sqrt{\lambda}}{m}\phi \right) - 1 \right].$$

(5.23)

Later in Section 5.4 we will adopt a different convention where,

$$\frac{\sqrt{\lambda}}{m} = \beta \quad \mu^2 = \frac{m^3}{\sqrt{\lambda}}. \tag{5.24}$$

In terms of this parametrization the potential reads,

$$V(\phi) = -\frac{\mu^2}{\beta}[\cos(\beta\phi) - 1]. \tag{5.25}$$

The potential has an infinite set of discrete vacua at $\phi^k = 2\pi k \frac{m}{\sqrt{\lambda}}$ and again it is invariant under $\phi \to -\phi$. As for the ϕ^4 case, here as well the integral in (5.18) can be solved analytically. For the soliton that goes from 0 to $\frac{m}{\sqrt{\lambda}}2\pi$ and vice versa for the anti-soliton, and choosing $\phi(x_0) = \frac{m}{\sqrt{\lambda}}\pi$, we get,

$$\phi(x) = 4\frac{m}{\sqrt{\lambda}}\tan^{-1}[e^{\pm m(x-x_0)}]. \tag{5.26}$$

Adding $\frac{m}{\sqrt{\lambda}}2n\pi$ to this, gives a soliton that goes from $\frac{m}{\sqrt{\lambda}}2n\pi$ to $\frac{m}{\sqrt{\lambda}}2(n+1)\pi$, and vice versa for the anti-soliton. The soliton has a topological charge associated with (5.6) of $Q = 1$, the anti-soliton $Q = -1$.

Substituting the explicit expression of the soliton profile (5.26) into the expression for the energy one finds that the mass of the SG soliton is

$$M_{\text{SGsol}} = \frac{8m^3}{\lambda} = \frac{8m}{\beta^2}. \tag{5.27}$$

Classical stability of the solitons

So far we have shown that scalar field theories with degenerate vacua admit soliton solutions. Let us now address the question of whether these solutions are stable against small time-dependent perturbation. Consider the field configuration,

$$\phi(x,t) = \phi_{\text{sol}}(x) + \delta\phi(x,t), \tag{5.28}$$

where $\phi_{\text{sol}}(x)$ is a time-independent soliton solution and $\delta(x,t)$ is a small perturbation. Substituting this configuration into the equation of motion and retaining only linear terms in the perturbation we get,

$$\partial^\mu \partial_\mu \delta\phi + V''(\phi_{\text{sol}})\delta\phi = 0. \tag{5.29}$$

Since the equation is invariant under time translation, we express the perturbation as a superposition of normal modes in the following form,

$$\delta\phi(x,t) = \sum_n Re[a_n e^{iw_n t}\delta_n(x)]. \tag{5.30}$$

The normal modes obey the equation,

$$-\frac{d^2\delta_n}{dx^2} + V''(\phi_{\text{sol}})\delta_n = w_n^2\delta_n, \tag{5.31}$$

which is in fact a one-dimensional Schrodinger equation with $V''(\phi_{sol})$ as a potential. If this equation has eigenmodes with negative eigenvalues, the soliton is unstable.

It is easy to construct one eigenmode. Since the soliton is invariant under space translation $\phi_{sol}(x) \to \phi_{sol}(x+a)$, $\delta_0 = \frac{d\phi_{sol}(x)}{dx}$ is an eigenmode with a vanishing eigenvalue. Now since the soliton is a monotonic function of x, δ_o does not have nodes. A theorem about a one-dimensional Schrodinger equation tells us that the eigenmode with no nodes has the lowest eigenvalue and hence there are no negative modes and the soliton for any $V(\phi)$ is indeed stable.

The topological charge

Any two-dimensional scalar field theory in two dimensions admits the topological current (5.6), $J^{\mu}_{top} = e^{\mu\nu} \partial_\nu \phi$. Thus, the following difference is a conserved charge,

$$Q_{top} = \int dx \phi' = [\phi(t, +\infty) - \phi(t, -\infty)] \equiv \phi_+ - \phi_-. \qquad (5.32)$$

Often one refers to $\phi_\pm$ as the topological indices. In fact for theories with a potential that has a discrete number (finite or infinite) of vacua, non-singular field configurations of finite energy have both ϕ_+ and ϕ_- separately conserved. This results from the following argument. Finite energy implies that both ϕ_+ and ϕ_- are at absolute minima of the potential. Now since the non-singular configurations are continuous in time, and the potential has a set of discrete (finite or infinite) vacua, $\phi(t, \infty)$ must be stationary at ϕ_+, or $\partial_0 \phi(t, \pm\infty) = \partial_0 \phi_\pm = 0$, namely the indices are conserved.

In fact this conservation can be used to show the existence of non-dissipative solutions. For instance in the ϕ^4 theory we can show that a configuration with $\phi_+ = -\phi_-$ is non dissipative. By continuity in x there must be, for any t, some x for which $\phi = 0$. At this point $T_{00} \geq V(0)$ and since the definition of a dissipative solution is that the $\lim_{t\to\infty} \max_x T_{00} = 0$ it is clear that it is non-dissipative. Similar arguments hold for other cases of solitons.

Thus, one can divide the space of finite-energy non-singular solutions into topologically disconnected sub-spaces that are characterized by the two indices $\phi_\pm$. Such a sub-space cannot be continuously deformed into another one unless the finite energy condition is violated. For instance, in the ϕ^4 theory, the potential has two minima so that $\phi_+ = \frac{m}{\lambda}$ and $\phi_- = -\frac{m}{\lambda}$. Hence, there are four subspaces $(-,+), (+,-), (-,-), (+,+)$ associated with the soliton, the anti-soliton and the two trivial constant vacuum solutions. For the sine-Gordon the solitons belong to the subspaces characterized by $\phi_- = 2\pi n \frac{m}{\sqrt{\lambda}}$ and $\phi_+ = 2\pi(n+1)\frac{m}{\sqrt{\lambda}}$.

Obviously, non-trivial topological charges require multiple vacua. The latter situation occurs if and only if there is a spontaneous breaking of a symmetry. For instance in ϕ^4 and sine-Gordon it is the discrete $\phi \to -\phi$ symmetry which is broken.

Derrick's theorem

Consider a scalar field theory in $D + 1$ space-time dimensions described by the Lagrangian density,

$$\mathcal{L} = \frac{1}{2} \partial^\mu \phi \partial_\mu \phi - V(\phi), \tag{5.33}$$

where the potential $V(\phi)$ is non-negative and vanishes at its minima. The theorem states that *for $D \geq 2$ the only non-singular time-independent solutions of finite energy are the vacua.*

Let us denote by $\phi(\mathbf{x})$ a time-independent solution of the equation of motion. We now introduce a one-parameter family of field configurations defined as,

$$\phi(\lambda, \mathbf{x}) = \phi(\lambda \mathbf{x}), \tag{5.34}$$

where λ is a positive real number. The energy associated with the configuration $\phi(\lambda, \mathbf{x})$ is,

$$E(\lambda) = \lambda^{-D} \int d^D \mathbf{x} \left[\frac{1}{2} \lambda^2 (\nabla \phi)^2 + V(\phi) \right]. \tag{5.35}$$

By Hamilton's principle the energy as a function of λ is stationary at $\lambda = 1$ so that,

$$\int d^D \mathbf{x} \left[\frac{1}{2} (D - 2)(\nabla \phi)^2 + DV(\phi) \right] = 0. \tag{5.36}$$

For $D > 2$ the two terms in the integral have to vanish separately, which occurs only for the vacua. For $D = 2$, the potential term has to vanish, which again occurs only for the vacua. This proves the theorem.

The following remarks are in order:

(i) Derrick's theorem applies only to time-independent configurations.
(ii) It applies to field theories with only scalar fields. Once one introduces additional fields like gauge fields or fermions the theorem is not valid (see Section 20.3).

5.4 Breathers or "doublets"

So far we have discussed only time-independent solutions of the equations of motion. A natural question to ask is whether the equations also admit exact time-dependent solutions. Another question, seemingly unrelated, is that of the interactions between solitons and between solitons and anti-solitons. We will see shortly that these two puzzles are in fact related. We now proceed to examine these questions in the laboratory of the sine-Gordon model.

The following periodic configuration,

$$\phi(x, t) = \frac{4}{\beta} \tan^{-1} \left[\frac{\eta \sin(wt)}{\cosh(\eta w x)} \right], \tag{5.37}$$

where $\eta = \frac{\sqrt{(m^2 - w^2)}}{w}$, with $w \le m$, is a solution of the equation of motion (5.13). We will show now that this solution is related to a bound state of a soliton and anti-soliton.

Consider first the simple case of small w, namely $w \ll m$. For positive $\sin(wt)$ and finite x, the argument of $\tan^{-1}$ is very large, and thus $\phi \sim \frac{2\pi}{\beta}$. When x approaches $-\infty$ we can approximate $\phi(x,t)$ as,

$$\phi(x,t) \sim \frac{4}{\beta} \tan^{-1}\left[\exp\left(mx + \ln\left[\frac{2m}{w}\sin(wt)\right]\right)\right], \qquad (5.38)$$

which looks like a soliton to the left. Similarly, it looks like an anti-soliton to the right. The soliton and anti-soliton move further apart as $\sin(wt)$ increases to one, and then when $\sin(wt)$ decreases they approach each other. As $\sin(wt) \to 0$ the approximation that lead to (5.38) is no longer valid, in accordance with the fact that in this region the soliton and anti-soliton are on top of each other. A similar discussion applies also for negative $\sin(wt)$. It is thus clear that the solution (5.37) describes an oscillation of a soliton anti-soliton pair around their common center of mass.

Revealing a bound state solution implies that the system must be attractive at least in a certain region of the "coupling constant". Indeed if one uses the coupling constant,

$$\xi = \frac{\pi\beta^2}{8\pi - \beta^2}, \qquad (5.39)$$

then

$$\begin{aligned}
\infty > \xi > \pi & \quad \textit{repulsive interaction} \\
\xi = \pi & \quad \textit{free particle} \\
\pi > \xi > 0 & \quad \textit{attractive interaction}
\end{aligned} \qquad (5.40)$$

As will be clarified in the next chapter, the case of $\xi = \pi$ corresponds to a free massive Dirac fermion. This will be further discussed in Section 6.1 as a bosonization of the free massive Dirac fermion. Also, the attractive region corresponds to positive coupling of a four-fermion interaction, namely attraction, while the repulsive region above corresponds to a negative coupling four-fermion. The case of negative ξ leads to no ground state.

If indeed the breather describes a bound state, it has to have a mass which is smaller than twice the mass of the soliton. It is easy to compute the classical energy associated with (5.37) at $t = 0$, since both the potential and the ϕ' vanish. Thus,

$$E_{\text{breather}}^{(\text{clas})} = \int dx \left[\frac{1}{2}(\dot\phi)^2\right] = \frac{8(\eta w)^2}{\beta^2}\int dx \frac{1}{\cosh^2(\eta wx)} = 2M_{\text{sol}}\sqrt{1 - \left(\frac{w}{m}\right)^2}, \qquad (5.41)$$

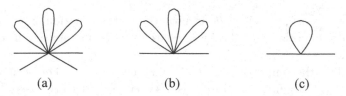

Fig. 5.1. Normal-order correction of the coupling (a), a correction to m^2 (b) and the lowest-order contribution to δm^2 (c).

where M_{sol} is the mass of the soliton. This verifies the existence of a binding energy since the mass of the breather is less than twice the mass of the soliton. In Section 5.5.1 the quantum description of the bound states will be addressed, and their scattering processes in Section 5.6.

5.5 Quantum solitons

The quantization of the soliton and breather was worked out in [74].

The classical mass of the soliton is $M = \frac{8m}{\beta^2}$, as in (5.27). Quantum mechanically this mass is corrected due to quantum fluctuations. We parametrize these space-time-dependent fluctuations as,

$$\phi_s \rightarrow \phi_s + \sum_n Re[e^{i w_n t} \delta_n(x)]. \tag{5.42}$$

The normal modes $\delta_n(x)$ obey the equation,

$$[-\partial_x^2 + m^2 \cos(\beta \phi_s)] \delta_n(x) = w_n^2 \delta_n(x). \tag{5.43}$$

The leading order of the quantum mass is then given by

$$M_{\text{quantum}} = M_s + \frac{1}{2} \sum_n w_n + V_{\text{ct}}(\phi_s), \tag{5.44}$$

where $V_{\text{ct}}(\phi_s)$ is a counter term that one has to add to the Lagrangian.

In two dimensions the only source of UV divergences in any order of perturbation theory are diagrams that contain a loop consisting of a single internal line. Stated differently, UV divergences are due to the fact that the action is not normal ordered. The corresponding diagrams are depicted in Figure 5.1. In fact the corrections (a) and (b) cancel and the only corrections follow from (c).

Let us first recall the normal ordering of $\phi^2(x)$, namely,

$$\phi^2(x) =: \phi^2(x) :_m + \frac{1}{4\pi} \int \frac{dk}{\sqrt{k^2 + m^2}} \tag{5.45}$$

where $: :_m$ indicates that the normal ordering is performed for a scalar of mass m. The last integral in obviously divergent so one introduces a cutoff $\Lambda \gg m$

such that

$$\frac{1}{4\pi} \int_{-\Lambda}^{\Lambda} \frac{dk}{\sqrt{k^2 + m^2}} = \frac{1}{4\pi} \ln \frac{4\Lambda^2}{m^2} + O\left(\frac{m^2}{\Lambda^2}\right). \tag{5.46}$$

For a general potential $V(\phi)$, Wick's theorem tells us that,

$$V(\phi) =: \left[e^{\frac{1}{8\pi} \ln \frac{4\Lambda^2}{m^2} \frac{d^2}{d\phi^2}} V(\phi)\right] :_m . \tag{5.47}$$

If one expands the exponent one finds a term with no contraction, next, with one contraction etc. We can pass from normal ordering at mass m to normal ordering at $\tilde{m}$. This transformation is independent of the cutoff, since

$$: \phi^2 :_m =: \phi^2 :_{\tilde{m}} + \frac{1}{4\pi} \ln \frac{m^2}{\tilde{m}^2}, \tag{5.48}$$

and hence

$$: V(\phi) :_m =: \left[e^{\frac{1}{8\pi} \ln \frac{m^2}{\tilde{m}^2} \frac{d^2}{d\phi^2}} V(\phi)\right] :_{\tilde{m}} . \tag{5.49}$$

When applied to the sine-Gordon case, the normal-ordered potential takes the form,

$$\frac{m^2}{\beta^2} : [\cos(\beta\phi) - 1] :_m \quad \frac{(m^2 - \delta m^2)}{\beta^2} [\cos(\beta\phi) - 1], \tag{5.50}$$

and where to the lowest order in β,

$$\delta m^2 = -\frac{m^2 \beta^2}{4\pi} \int^{\Lambda} \frac{dk}{\sqrt{k^2 + m^2}}. \tag{5.51}$$

Thus the counterterm potential reads,

$$V_{\text{counter}}(\phi) = -\frac{\delta m^2}{\beta^2} \int_{-\infty}^{\infty} dx[(1 - \cos(\beta\phi_s))] - E_{\text{vacuum}}, \tag{5.52}$$

where we further subtracted the energy of the vacuum. Finally the quantum mass takes the form

$$M_{\text{quantum}} = M_s + \frac{1}{2}\sum_n w_n - \frac{\delta m^2}{\beta^2} \int_{-\infty}^{\infty} dx[(1 - \cos(\beta\phi_s))] - \frac{1}{2}\sum_n \sqrt{k_n^2 + m^2}, \tag{5.53}$$

where $k_n = \frac{2\pi}{L}$, with L the size of the quantization length, to be sent to ∞ at the end of the calculation.

When substituting the set of all the frequencies w_n associated with solutions of (5.43) one finds that the quantum mass is finite and reads,

$$M_{\text{quantum}} = \frac{8m}{\beta^2} - \frac{m}{\pi} = \frac{m}{\xi} \tag{5.54}$$

up to corrections of order $m\beta^2$.

5.5.1 Quantization of the breather

Next we discuss the quantization of the classical time-dependent breather solution. First, as a warm up exercise, compute the spectrum approximately, using the Bohr–Sommerfeld "old quantization procedure". Adapting this recipe to field theory states that a one parameter family of periodic fields characterized by the period $\tau = \frac{2\pi}{w}$ has an energy eigenstate whenever

$$\int_0^\tau dt \int dx\pi(x,t)\partial_0\phi(x,t) = 2\pi N, \tag{5.55}$$

where N is an integer.

Using the relation between the Hamiltonian and Lagrangian densities, $\mathcal{H} = \pi\partial_0\phi - \mathcal{L}$, we find after integrating over one period that,

$$E\tau = 2\pi N - \int_0^\tau dt \int dx\mathcal{L}. \tag{5.56}$$

By differentiating with respect to τ (with N varying as a function of it, by analytic continuation), and using the equations of motion we find,

$$\frac{dN}{dE} = \frac{1}{w} = \frac{1}{m}\frac{1}{\sqrt{1 - \frac{E^2}{4M^2}}}, \tag{5.57}$$

where the expression for w in terms of E and M follows from the calculation of the energy which can be performed most conveniently at $t = 0$, as was done in (5.41).

Integrating this equation and using a natural boundary condition that $N = 0$ for $E = 0$ the Bohr–Sommerfeld procedure predicts the following spectrum,

$$M_N = 2M_{\text{sol}}\sin\left(\frac{N\beta^2}{16}\right) \quad N = 1, 2, \ldots, < \frac{8\pi}{\beta^2}. \tag{5.58}$$

Next we would like to describe the quantization procedure of Dashes, Hasslacher and Neveu (DHN). The classical action of the breather solution per period $\tau = \frac{2\pi}{w}$ is determined by substituting the breather solution (5.37) into the action and integrating,

$$S_{\text{cl}}(\phi_b) = \frac{32\pi}{\beta^2}\left[\cos^{-1}\left(\frac{w}{m}\right) - \eta\right]. \tag{5.59}$$

The stability of the breather solution is determined by the requirement that there are no negative eigenmodes to the stability equation,

$$[\partial^\mu\partial_\mu + m^2\cos(\beta\phi_b)]\delta_n(x,t) = 0, \tag{5.60}$$

where $\delta_n(x,t)$, which obeys $\delta_n(x, t + \tau) = e^{i\nu_n}\delta_n(x,t)$, is the fluctuation of the breather solution. The set of all the solutions of this equation was written down by DHN [74].

The corresponding spectrum of ν_n reads,

$$\nu_0 = 0 \quad \nu_1 = 0 \quad \nu_n = \frac{2\pi}{w}\sqrt{m^2 + q_n^2}, \tag{5.61}$$

where q_n obeys the equations

$$L q_n + f(q_n) = 2\pi n \quad f(q_n) = 4 \tan^{-1}\left(\frac{\eta w}{q_n}\right),$$ (5.62)

and where L is the size of the space direction. The two vanishing frequencies are associated with the invariance under space and time translations.

The WKB semi-classical quantization determines the energy level of the breather solution via the conditions,

$$E_{\rm cl}(\phi_b) + E_{\rm ct}(\phi_b) - 1/2 \sum_0^\infty \frac{w^2}{2\pi} \partial_w \nu_n = E$$

$$S_{\rm cl}(\phi_b) + S_{\rm ct}(\phi_b) + \frac{2\pi E}{w(\phi_b)} - 1/2 \sum_0^\infty \nu_n = 2\pi N,$$ (5.63)

where $E_{\rm ct}$ and $S_{\rm ct}$ are the energy and action associated with the counterterm. In the limit of $L \to \infty$ the sum over q_n turns into an integral. The integral has a quadratic as well as logarithmic divergences. These divergences will be cancelled out by the contribution of the counterterm such that $S_{\rm cl}(\phi_b) + S_{\rm ct}(\phi_b) - 1/2 \sum_0^\infty \nu_n$ is the same as the $S_{\rm cl}(\phi_b)$ given above with the renormalization of the coupling constant $\beta^2 \to \xi = \frac{\pi \beta^2}{8\pi - \beta^2}$. Using this result it is easy to determine the energy,

$$E = -\frac{d}{d\tau}\left(S_{\rm cl}(\phi_b) + S_{\rm ct}(\phi_b) - 1/2 \sum_0^\infty \nu_n\right)\frac{2w\eta}{\xi}$$ (5.64)

Substituting this into the second equation of (5.63) one finds that the energy levels take the form,

$$M_N = \frac{2m}{\xi} \sin\left(\frac{N\xi}{2}\right) \quad N = 1, 2, \ldots, < \frac{\pi}{\xi}.$$ (5.65)

Note that in spite of the fact that the quantization condition permits any N, only if it is smaller than $\frac{\pi}{\xi}$ the classical breather solution exists. Thus the interpretation of this result is that there is a finite number of quantum bound states corresponding to the classical breather solution. Even though the derivation of the mass spectrum of the bound states was based on a Wentzel, Krames and Brillonin (WKB) approximation, the final result turns out to be **exact**. This statement follows from the analysis of the physical poles of soliton anti-soliton scattering, and already indicated in perturbation theory from two loops. The latter, though, works for mass ratios only, in view of scale dependence for the normal ordering of each individual mass.

The spectrum (5.65) can be re-written in terms of the mass of the quantum soliton. Using (5.54) this takes the form,

$$M_N = 2M_{\rm sol} \sin\left(\frac{N\xi}{2}\right) \quad N = 1, 2, \ldots, < \frac{\pi}{\xi},$$ (5.66)

which indicates that the quantum breather states are indeed bound states of a quantum soliton anti-soliton pair.

At weak coupling $N\beta^2 \ll 1$, the mass spectrum reads,

$$M_N = Nm \left[1 - \frac{1}{6} \left(\frac{N\beta^2}{16} \right)^2 + O(N^2\beta^6) \right]. \tag{5.67}$$

Thus at weak coupling the lowest bound state has a mass of,

$$M_1 = m \left[1 - \frac{1}{6} \left(\frac{\beta^2}{16} \right)^2 + O(\beta^6) \right], \tag{5.68}$$

showing that first bound state is in fact the "elementary" boson of the theory. Moreover the higher bound states have a mass which is $NM_1 + O[(\beta^2)^2 N(1 - N^2)]$, namely bound states of N elementary bosons. These bound states are loosely bound with a binding energy of $\frac{m}{6}(\frac{\beta^2}{16})^2 N(N^2 - 1)$. Using perturbation theory one can show that each of these states is stable against decay to states with lower N. In fact the stability turns out to be an exact statement. The source of the stability of these states is the set of infinitely conserved charges as will be discussed in the following section.

5.6 Integrability and factorized S-matrix

One of the first papers that discusses integrability of the S-matrix is the seminal paper [235]. We follow this paper in describing the basic notions of integrability. The Yang–Baxter relations were derived in [230] and [29] and S-matrix results for solitons and breathers of the sine-Gordon model are analyzed in [198].

Consider an integrable theory with ∞ of conserved charges Q^n diagonalized in the single particle base such that

$$Q_n |p^{(a)}> = w_n^{(a)}(p)|p^{(a)}>, \tag{5.69}$$

where p is the momentum of the particle and (a) denotes its type. For the sine-Gordon case the eigenvalues $w_n^{(a)}(p)$ are given by,

$$w_{2n+1}^{(a)}(p) = p^{2n+1}, \quad w_{2n}^{(a)}(p) = p^{2n}\sqrt{p^2 + m_a^2}. \tag{5.70}$$

In general one assumes that the $w_n^{(a)}(p)$ form a set of independent functions. A multiple particle in or out state obeys

$$Q_n |p_1^{(a_1)} \ldots p_k^{(a_k)}, \text{in}> = \sum_{i=1}^{k} w_n^{(a_i)}(p_i)|p_1^{(a_1)} \ldots p_k^{(a_k)}, \text{in}>, \tag{5.71}$$

and since the charges Q_n are conserved one finds that,

$$\sum_{i \in \text{in}} w_n^{(a_i)}(p_i) = \sum_{i \in \text{out}} w_n^{(a_i)}(p_i). \tag{5.72}$$

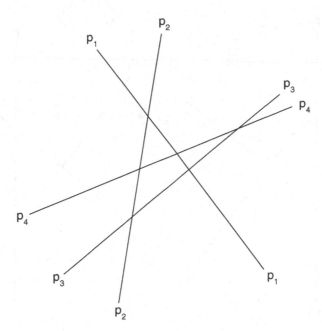

Fig. 5.2. Space-time picture of the multi-particle factorized scattering.

From this conservation one can deduce that,

- For any given mass m_a the number of initial and final particles of this mass is the same.
- The final set of momenta is the same as the initial one.

These two rules, that should apply also to intermediate states where particles are far enough from each other, together with the special kinematics of two dimensions, are behind the assertion that the multi-particle S-matrix of theories equipped with infinitely many conserved charges, can be expressed in terms of two-particle ones.

The factorized S-matrix corresponds to the following scattering process:

- In the infinite past a set of N particles with momenta $p_1 > p_2 > \ldots > p_N$ are spatially arranged in the opposite order, namely, $x_1 < x_2 < \ldots < x_N$.
- In the interaction region the particles collide in pairs. In each collision the momenta are conserved and in between collisions the particles move as free particles.
- The final state of the outgoing particles, achieved after $\frac{N(N-1)}{2}$ pair collisions, is built from the N particles arranged along the x coordinate in the order of increasing momenta.

The factorized scattering of the N particles is represented, for $N = 4$, by the space-time diagram Fig. 5.2, in which time is flowing up.

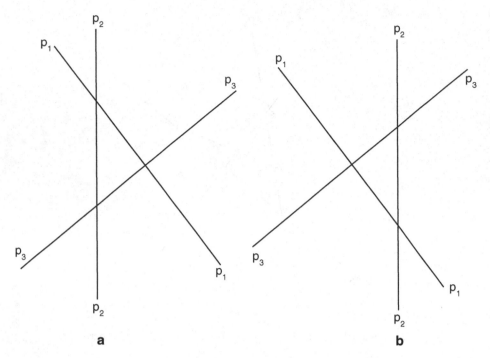

Fig. 5.3. Two possible ways of three-particle scattering.

- Each line corresponds to a given value of the momentum associated with the slope of the line.
- Each vertex corresponds to a two-particle collision. The two-particle amplitude $S_{ij}(p_i, p_j)$ has to be attached to each vertex.
- The total S-matrix element of the process is the product of all the $\frac{N(N-1)}{2}$ two-particle amplitudes $\prod_{ij} S_{ij}$, and then a sum over the different kinds of particles in the internal lines.

Take for example the case of $N = 3$. The same scattering can be represented in two ways, as is shown in Fig. 5.3. These two differ only by a parallel translation of a line, and thus represent the same process.

5.7 Yang–Baxter equations

The amplitudes and phases of the two diagrams should be the same. The requirement that they are indeed the same imposes cubic equations on the two-particle matrix elements which are called factorization equations or Yang–Baxter equations. We now proceed to analyze these conditions of factorization.

It is more convenient to discuss the S-matrix in terms of the rapidities of the massive particles. The rapidity β of a particle of mass m is defined via,[2]

$$p_\pm = m \exp(\pm\beta).$$ (5.73)

The scattering amplitude of a system of two particles $S(p_1, p_2)$ is a function of the rapidity difference $\beta = \beta_1 - \beta_2$ as can be seen from the fact that the s-channel invariant

$$s = (p_1 + p_2)^2,$$ (5.74)

is given by

$$s = m_1^2 + m_2^2 + 2m_1 m_2 \cosh(\beta_1 - \beta_2).$$ (5.75)

We now analytically continue s and define the amplitude $S(s)$ as an analytic function in the complex s-plane. This function has two cuts along the real axis for $s \geq (m_a + m_b)^2$ and $s \leq (m_a - m_b)^2$. The points $s = (m_a - m_b)^2$ and $s = (m_a + m_b)^2$ are square root branching points of $S(s)$. Using (5.75), $S(s)$ is mapped to $S(\beta)$, where $\beta = \beta_1 - \beta_2$. The physical sheet is mapped into the strip of $0 < \text{Im}\beta < \pi$, the branch cuts to the lines $\text{Im}\beta = 0$ and $\text{Im}\beta = \pi$, and the crossing transformation of $s \to 2m_1^2 + 2m_2^2 - s$ to $\beta \to i\pi - \beta$. If one assumes that there are no other cuts, then $S(\beta)$ is a meromorphic function in the above strip. In the non-relativistic limit, for $m_i \gg p_i^1$, the rapidity goes into the velocity $\beta_i \to v_i = \frac{p_i}{m_i}$.

5.8 The general solution of the S-matrix

Consider the two-particle S-matrix,

$$_{ik}S_{jl} = <A_j(\beta_1')A_l(\beta_2'), out|A_i(\beta_1)A_k(\beta_2), in>$$
$$= \delta(\beta_1' - \beta_1)\delta(\beta_2' - \beta_2)[\delta_{ik}\delta_{jl}S_1(s) + \delta_{ij}\delta_{kl}S_2(s) + \delta_{il}\delta_{kj}S_3(s)]$$
$$\pm [i \leftrightarrow k, \beta_1 \leftrightarrow \beta_2],$$ (5.76)

where $i, j, k, l = 1, 2$ so that the particles are in doublets of $O(2)$, and the $\pm$ refers to bosons $(+)$ and fermions $(-)$. This can be generalized to $O(N)$ in a straightforward way. Here we analyze only the case of the doublet. The amplitudes S_2, S_3 are the transition and reflection amplitudes, respectively, and S_1 corresponds to the process $A_i + A_i \to A_j + A_j$ for $(i \neq j)$. The $A_i(\beta)$, non-commutative

---

[2] Please note, that the β we had before, in the term $cos(\beta\phi)$ in the action, is not to be confused with the present one, which denotes the rapidity.

variables representing the particles, obey the relation,

$$A_i(\beta_1)A_j(\beta_2) = \delta_{ij} S_1(\beta) \sum_n A_n(\beta_2)A_n(\beta_1) + S_2(\beta)A_j(\beta_2)A_i(\beta_1)$$

$$+ S_3(\beta)A_i(\beta_2)A_j(\beta_1). \tag{5.77}$$

Incoming states are represented by products arranged by order of decreasing rapidities, while outgoing by increasing rapidities. The crossing symmetry relations are,

$$S_2(\beta) = S_2(i\pi - \beta) \quad S_1(\beta) = S_3(i\pi - \beta). \tag{5.78}$$

The unitarity conditions for the two-particle S-matrix are,

$$S_2(\beta)S_2(-\beta) + S_3(\beta)S_3(-\beta) = 1$$
$$S_2(\beta)S_3(-\beta) + S_2(-\beta)S_3(\beta) = 0$$
$$2S_1(\beta)S_1(-\beta) + S_1(\beta)S_2(-\beta) + S_1(\beta)S_3(-\beta) + S_2(\beta)S_1(-\beta)$$
$$+ S_3(\beta)S_1(-\beta) = 0. \tag{5.79}$$

Unitarity (5.79) and crossing symmetry (5.78) do not fix the S-matrix. The additional conditions one has to impose are those of the factorization or Yang–Baxter equations. The latter are obtained by considering all possible three-particle in-products $A_i(\beta_1)A_j(\beta_2)A_k(\beta_3)$, reordering them to get out-products using (5.77), and requiring that the results be independent of the order of successions of the two particle commutations, one then finds,

$$S_2 S_1 S_3 + S_2 S_3 S_3 + S_3 S_3 S_2 = S_3 S_2 S_3 + S_1 S_2 S_3 + S_1 S_1 S_2$$
$$S_3 S_1 S_3 + S_3 S_2 S_3 = S_3 S_3 S_1 + S_3 S_3 S_2 + S_2 S_3 S_1$$
$$+S_2 S_3 S_3 + 2S_1 S_3 S_1 + S_1 S_3 S_2 + S_1 S_3 S_3 + S_1 S_2 S_1 + S_1 S_1 S_1, \tag{5.80}$$

where, for each of the terms, the arguments for the three S factors are $\beta, \beta + \beta', \beta'$, respectively.

The general solution of the factorization equations is expressed in terms of one function which we take to be $S_2(\beta)$. The solution reads,

$$S_3(\beta) = ictg\left(\frac{4\pi\delta}{\gamma}\right) cth\left(\frac{4\pi\beta}{\gamma}\right) S_2(\beta)$$

$$S_1(\beta) = ictg\left(\frac{4\pi\delta}{\gamma}\right) cth\left(\frac{4\pi(i\delta - \beta)}{\gamma}\right) S_2(\beta) \tag{5.81}$$

with γ and δ real, but so far arbitrary. This solution as well as the restriction from unitarity (5.79) is valid also for a non-relativistic system. Crossing symmetry is a restriction that shows up only in the relativistic case. Imposing the latter (5.78) on the general solution fixes $\delta = \pi$. A "minimum" solution for $S_2(\beta)$ then takes the form,

$$S_2(\beta) = \frac{2}{\pi} \sin\left(\frac{4\pi^2}{\gamma}\right) sh\left(\frac{4\pi\beta}{\gamma}\right) \sin\left(\frac{4\pi(i\pi - \beta)}{\gamma}\right) U(\beta), \tag{5.82}$$

where,

$$U(\beta) = \Gamma\left(\frac{8\pi}{\gamma}\right)\Gamma\left(1 + i\frac{8\beta}{\gamma}\right)\Gamma\left(1 - \frac{8\pi}{\gamma} - i\frac{8\beta}{\gamma}\right)\prod_{n=1}^{\infty}\frac{R_n(\beta)R_n(i\pi - \beta)}{R_n(0)R_n(i\pi)},$$

$$R_n(\beta) = \frac{\Gamma\left(2n\frac{8\pi}{\gamma} + i\frac{8\beta}{\gamma}\right)\Gamma\left(1 + 2n\frac{8\pi}{\gamma} + i\frac{8\beta}{\gamma}\right)}{\Gamma\left((2n+1)\frac{8\pi}{\gamma} + i\frac{8\beta}{\gamma}\right)\Gamma\left(1 + (2n-1)\frac{8\pi}{\gamma} + i\frac{8\beta}{\gamma}\right)}. \tag{5.83}$$

It is a "minimal" in the number of singularities along the imaginary β axis, and more general solutions can be obtained from it by multiplying with a meromorphic function of the form $f(\beta) = \prod_{k=1}^{L}\frac{\mathrm{sh}\beta + i\sin\alpha_k}{\mathrm{sh}\beta - i\sin\alpha_k}$ for arbitrary real numbers α_k.

5.8.1 The S-matrix of the sine-Gordon model

The sine-Gordon model has a hidden $O(2)$ invariance, which is simplest to see via the soliton solutions, where the soliton and anti-soliton are incorporated in an $O(2)$ doublet. In terms of the $A_1(\beta)$ and $A_2(\beta)$ the soliton and anti-soliton amplitudes are,

$$A(\beta) = A_1(\beta) + iA_2(\beta) \quad \bar{A}(\beta) = A_1(\beta) - iA_2(\beta). \tag{5.84}$$

In terms of A and $\bar{A}$, (5.77) takes the form,

$$\begin{aligned} A(\beta_1)\bar{A}(\beta_2) &= S_T(\beta)\bar{A}(\beta_2)A(\beta_1) + S_R(\beta)A(\beta_2)\bar{A}(\beta_1) \\ A(\beta_1)A(\beta_2) &= S(\beta)A(\beta_2)A(\beta_1) \\ \bar{A}(\beta_1)\bar{A}(\beta_2) &= S(\beta)\bar{A}(\beta_2)\bar{A}(\beta_1), \end{aligned} \tag{5.85}$$

where $S(\beta), S_T(\beta)$ and $S_R(\beta)$ are the scattering amplitude of identical solitons, transition and reflection amplitude for soliton anti-soliton, which are related to $S_1(\beta), S_2(\beta)$ and $S_3(\beta)$ as,

$$\begin{aligned} S(\beta) &= S_3(\beta) + S_2(\beta) \\ S_T(\beta) &= S_1(\beta) + S_2(\beta) \\ S_R(\beta) &= S_1(\beta) + S_3(\beta). \end{aligned} \tag{5.86}$$

It follows from crossing symmetry (5.78) that,

$$S(\beta) = S_T(i\pi - \beta) \quad S_R(\beta) = S_R(i\pi - \beta). \tag{5.87}$$

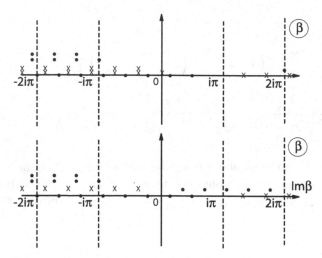

Fig. 5.4. The zeros (crosses) and poles (dots) of $S_T(\beta)$ (upper) and $S_R(\beta)$ (lower). All the singularities have pure imaginary values even though some of them are displaced from the imaginary axis for clarity [235].

Substituting these relations in the general solution derived in the previous section one finds for the SG model,

$$S_T(\beta) = -i\frac{\text{sh}\left(\frac{8\pi\beta}{\gamma}\right)}{\sin\left(\frac{8\pi^2}{\gamma}\right)}S_R(\beta)$$

$$S(\beta) = -i\frac{\text{sh}\left(\frac{8\pi(i\pi-\beta)}{\gamma}\right)}{\sin\left(\frac{8\pi^2}{\gamma}\right)}S_R(\beta), \tag{5.88}$$

where,

$$S_R(\beta) = \frac{1}{\pi}\sin\left(\frac{8\pi^2}{\gamma}\right)U(\beta). \tag{5.89}$$

The zeros and the poles of $S_T(\beta)$ and $S_R(\beta)$ are shown in Fig. 5.4.

The solution (5.88) is in fact the exact solution of the S-matrix of the SG model. This assertion is supported by the following properties:

- The poles of $S_T(\beta)$ are located at equidistance, and their values are in accordance with the semi-classical mass spectrum if one equates $\gamma = 8\xi$.
- Note that for the value of $\xi = \pi$ where the coupling of the associated Thirring model vanishes, and the SG model is a bosonized version of a free Dirac fermion,

$$S_T(\beta) \equiv S(\beta) = 1 \quad S_R(\beta) = 0. \tag{5.90}$$

- At $\xi \geq \pi$ all bound states, including the "elementary particle" associated with the field of the SG model, become unstable and the spectrum includes only soliton and anti-soliton. This situation follows from the fact that at this region the Thirring coupling is negative and there is a repulsion between the solition and anti-soliton.

- At $\xi = \frac{\pi}{n}$ the reflection amplitude vanishes identically.
- Expanding (5.88) in powers of $[(\frac{8\pi}{\gamma}) - 1]$, which means small coupling of the massive Thirring model, matches the perturbative expansion of the latter model.
- The limit $\beta^2 \to 0$ of the exact result (5.88) is equal to the semi-classical expression of the two-particle S-matrix.

The explicit expression for the two-particle S-matrix (5.88) enables one to also write down the S-matrix for any number of solitons and anti-solitons and the scattering of any number of bound states. This general approach to solving the S-matrix can be applied to the various integrable models. Here we demonstrate it on the sine-Gordon model. For soliton and anti-solitons we find the following S-matrix elements:

$$S_{\frac{1}{2},\frac{1}{2}}^{\frac{1}{2},\frac{1}{2}}(\beta) = S_{-\frac{1}{2},-\frac{1}{2}}^{-\frac{1}{2},-\frac{1}{2}}(\beta) = S(\beta)$$

$$S_{\frac{1}{2},-\frac{1}{2}}^{\frac{1}{2},-\frac{1}{2}}(\beta) = S_{-\frac{1}{2},\frac{1}{2}}^{-\frac{1}{2},\frac{1}{2}}(\beta) = -\frac{\mathrm{sh}\left(\frac{\pi}{\xi}\beta\right)}{\mathrm{sh}\left(\frac{\pi}{\xi}(\beta - \pi i)\right)}S(\beta)$$

$$S_{-\frac{1}{2},\frac{1}{2}}^{\frac{1}{2},-\frac{1}{2}}(\beta) = S_{\frac{1}{2},-\frac{1}{2}}^{-\frac{1}{2},\frac{1}{2}}(\beta) = \frac{\mathrm{sh}\left(\frac{\pi^2 i}{\xi}\right)}{\mathrm{sh}\left(\frac{\pi}{\xi}(\beta - \pi i)\right)}S(\beta)$$

$$S_{\epsilon_1,\epsilon_2}^{\epsilon_1',\epsilon_2'}(\beta) = 0 \quad \epsilon_1 + \epsilon_2 \neq \epsilon_1' + \epsilon_2'. \tag{5.91}$$

S can also be expressed as an exponential of an integral,

$$S(\beta) = -\exp\left[-i\int_0^\infty \frac{\sin(\kappa\beta)\,\mathrm{sh}\left(\frac{\pi-\xi}{2}\kappa\right)}{\kappa\,\mathrm{ch}\left(\frac{\pi\kappa}{2}\right)\mathrm{sh}\left(\frac{\xi\kappa}{2}\right)}\mathrm{d}\kappa\right]. \tag{5.92}$$

5.9 From conformal field theories to integrable models

So far we have analyzed integrable theories based on a scalar field theory with an integrable interacting potential. As was mentioned in Section 1 there is a more general scheme of constructing integrable models. This scheme is based on perturbing conformal field theories with relevant primary fields so that the action takes the form,

$$S = S_{\mathrm{CFT}} + \sum_i \lambda_i \int \mathrm{d}^2 z \Phi_i(z, \bar{z}). \tag{5.93}$$

Note that just as for the conformal field theories we use here complex two-dimensional coordinates. The $\Phi_i(z, \bar{z})$ are primary fields of conformal dimension $\Delta_i + \bar{\Delta}_i < 2$, namely relevant operators. Since these operators are

super-renormalizable, they do not affect the short distance behavior but do affect
the structure of the IR domain. In the analogy with statistical mechanical systems, where the CFT describes the behavior of the system at its fixed point, the
perturbation with the relevant primary fields describes the scaling region around
the fixed point.

A system described by an action of the form (5.93), is integrable provided that
one can identify a set of infinitely many conserved charges just as for the systems described previously. An important class of such theories are the integrable
minimal models, for example the tricritical Ising model $\mathcal{M}_{4,5}$ perturbed by $\Phi_{\frac{3}{5},\frac{3}{5}}$
and the tricritical Potts model $\mathcal{M}_{6,7}$ perturbed by $\Phi_{\frac{1}{7},\frac{1}{7}}$.

The renormalization group (RG) flow of the integrable systems follows a trajectory that starts at a fixed point and may end on another one in the IR, or
on a point that corresponds to a massive QFT. An important property of these
flows is the **c-theorem**[3] which states the following:

*Quantum field theories which possess rotational invariance, reflection pos-
itivity, and conservation of the energy momentum tensor admit a function
$c(\lambda_i)$ of the coupling constants λ_i which is non-increasing along the RG
trajectories and is stationary only at fixed points.*

The proof of the theorem is as follows. Consider the correlators of $T \equiv T_{zz}$
and $T_{z\bar{z}}$,

$$<T(z,\bar{z})T(0,0)> = \frac{F(z\bar{z})}{z^4}, \quad <T(z,\bar{z})T_{z\bar{z}}(0,0)> = \frac{G(z\bar{z})}{z^3\bar{z}},$$

$$<T_{z\bar{z}}(z,\bar{z})T_{z\bar{z}}(0,0)> = \frac{H(z\bar{z})}{z^2\bar{z}^2}. \tag{5.94}$$

We now use the conservation law,

$$\bar{\partial}T + \partial T_{z\bar{z}} = 0, \quad \partial\bar{T} + \bar{\partial}T_{z\bar{z}} = 0, \tag{5.95}$$

to deduce the following differential equations for F, G and H,

$$\dot{F} + (\dot{G} - 3G) = 0 \quad \dot{G} - G + \dot{H} - 2H = 0, \tag{5.96}$$

where $\dot{A} = \frac{\mathrm{d}A(x)}{\mathrm{d}\log x}$. Since the positivity condition implies that $H \geq 0$, the following
c function is non-increasing,

$$c = 2F - 4G - 6H, \quad \dot{c} = -12H. \tag{5.97}$$

At the fixed points $T_{z\bar{z}} = 0$, hence $G = H = 0$ and $c = 2F$, as indeed it should
be. Recall that the OPE in CFT is of the form $T(z\bar{z})T(0,0) = \frac{(\frac{c}{2})}{z^4} + \ldots$

One can further write down an expression for the integral of $\dot{c}$, namely for
the difference of c in the UV and IR regions. For the case of a perturbation with a single operator Φ, the trace of the energy-momentum tensor is

---

[3] Zamolodchikov c-theorem was derived in [232].

$T_{z\bar{z}} = \pi\lambda(1 - \Delta)\Phi$, where the total conformal dimension is 2Δ. Using (5.97) it was shown that,

$$c_{\text{UV}} - c_{\text{IR}} = 12\pi\lambda^2(1 - \Delta)^2 \int d^2x|x|^2 <\Phi(x)\Phi(0)> . \qquad (5.98)$$

This result has been applied to integrable minimal models yielding the correct difference in the Virasoro anomalies. Another example where this relation between the conformal data and the properties of the non-conformal theory can be tested is the sine-Gordon model. The model can be thought of as a perturbation on a free massless scalar field which has $c = 1$, and the massive model has $c = 0$. Let us check for this case the outcome of the relation (5.97). The perturbation now is,

$$\lambda\Phi = \frac{m^2}{\beta^2} : (\cos\beta\phi - 1): \quad 2\Delta = \frac{\beta^2}{4\pi}. \qquad (5.99)$$

If we expand $\lambda\Phi$ in β, then the leading order $\lambda\Phi$ is $\frac{1}{2}m^2\phi^2$ for which $\Delta = 0$ and,

$$<\lambda\Phi(x)\lambda\Phi(0)> = \frac{1}{4}m^2 <\phi(x)\phi(0)>^2 = \frac{m^4}{8\pi^2}K_0^2(m|x|), \qquad (5.100)$$

where K_0 is a Bessel function. Inserting this into (5.97) we get,

$$c_{\text{UV}} - c_{\text{IR}} = 3\pi\frac{m^4}{2\pi^2} \int d^2x|x|^2K_0^2(m|x|) = 3\int_0^\infty dr\, r^3 K_0^2(r) = 1, \qquad (5.101)$$

which verifies our data about the Virasoro anomalies at the two ends of the trajectory. One can further show that higher-order terms in β are in accordance with the fact that $c_{\text{UV}} - c_{\text{IR}}$ is β independent.

5.10 Conserved charges and classical integrability

We will now show that the classical sine-Gordon theory incorporates an infinite set of conserved charges. This will imply an exact determination of any S-matrix element of the theory. This property of having a set of an infinite number of conserved currents and charges is referred to as classical integrability. In the next section we will discuss the fate of the integrability in the quantum domain. We choose here to describe the sine-Gordon model, however this structure applies to a class of models.

There are several methods to determine these classically conserved charges. Here we will follow two of them: The Lax pair approach and a method based on a generating function.

The infinitely many charges of the integrable models were analyzed using various different techniques. The Lax pair method applied to the sine-Gordon is described in [61]. The method of multilocal charges from an integral equation was presented in [153]. The inductive method was introduced in [45].

Solitons and two-dimensional integrable models

5.10.1 The Lax pair method

In the Lax pair approach the idea is to rewrite the sine-Gordon equation of motion in terms of a commutator relation between two operators. Let us first rewrite the equation of motion of the sine-Gordon system in the light-cone coordinates,

$$\tilde{\partial}_+ \tilde{\partial}_- \tilde{\phi} = -\sin(\tilde{\phi}), \qquad (5.102)$$

where $\tilde{\phi} = \beta\phi$ and $\tilde{\partial}_\pm = m\partial_\pm$.

Next we define the following pair of 2×2 matrix Lax operators,

$$L = 2\sigma_3 \tilde{\partial}_- + \sigma_2(\tilde{\partial}_- \tilde{\phi})$$
$$B = \frac{1}{2}[\sigma_3 \cos(\tilde{\phi}) + \sigma_2 \sin(\tilde{\phi})]L^{-1}. \qquad (5.103)$$

In terms of these operators the sine Gordon equation takes the form,

$$\tilde{\partial}_+ L = [L, B]. \qquad (5.104)$$

The reason for rewriting the equation of motion in this form is the fact that the spectrum of L is conserved. To realize this property of the Lax pair, notice first that the solution of (5.104) can be parameterized as,

$$L(x^+) = S(x^+)L(0)S^{-1}(x^+) \quad \partial_+ S = -BS. \qquad (5.105)$$

Now consider the eigenvalue problem,

$$L(x^+)v(x^+) = \lambda v(x^+). \qquad (5.106)$$

It is easy to check that if $v(0)$ is an eigenfunction of $L(0)$ with eigenvalue λ, then $v(x^+)$ is an eigenfunction of $L(x^+)$ with the same eigenvalue λ, where,

$$v(x^+) = S(x^+)v(0). \qquad (5.107)$$

Since x^+ is the light-cone time direction, this implies the conservation of the eigenvalues. This conservation of the spectrum is the origin of the infinite set of conserved charges.

For the case that L can be represented by a finite matrix, it is obvious from (5.105), using the cyclicity of the trace, that $Q^n = Tr[L^n]$ are conserved charges. In general, and in particular in our case, L does not act only in a space of finite-dimensional matrices but also in the continuous space whose base vectors are $|x_->$. Thus the trace takes the form of the integral $\int_{-\infty}^{\infty} dx_-$. Due to the unitarity of $S(x^+)$ the cyclicity property of the trace is maintained and hence $Q^n = Tr[L^n]$ are indeed conserved charges.

It turns out that one can map the Lax pair of the sine-Gordon system to that of the Korteweg-deVries (KdV) equation $\partial_0 u(x,t) = 6u\partial_1 u - \partial_1^3 u$. This map is useful since it is more convenient to express the conserved charges of the sine-Gordon system in terms of the L operator of the KdV equation. In this format the first four charges of the set of infinite charges which are classically conserved are,

$$Q_1 = -\left(\frac{1}{4}\right) \int_{-\infty}^{\infty} (\partial_-\tilde{\phi})^2 \mathrm{d}x_-$$

$$Q_2 = +\left(\frac{1}{4}\right)^2 \int_{-\infty}^{\infty} [(\partial_-\tilde{\phi})^4 - 4(\partial_-^2\tilde{\phi})^2] \mathrm{d}x_-$$

$$Q_3 = -\left(\frac{1}{4}\right)^3 \int_{-\infty}^{\infty} [(\partial_-\tilde{\phi})^6 - 20(\partial_-\tilde{\phi})^2(\partial_-^2\tilde{\phi})^2 + 8(\partial_-^3\tilde{\phi})^2] \mathrm{d}x_-$$

$$Q_4 = +\left(\frac{1}{4}\right)^4 \int_{-\infty}^{\infty} [(\partial_-\tilde{\phi})^8 - \frac{112}{5}(\partial_-^2\tilde{\phi})^4 - 56(\partial_-\tilde{\phi})^4(\partial_-^2\tilde{\phi})^2$$

$$+ \frac{224}{5}(\partial_-\tilde{\phi})^2(\partial_-^3\tilde{\phi})^2 - \frac{64}{5}(\partial_-^4\tilde{\phi})^2] \mathrm{d}x_-. \tag{5.108}$$

5.10.2 The generating function method

A second method to determine the set of infinite conserved charges is based on a generating function. Define the generating function,

$$\psi = \phi + \frac{1}{\beta}\sin^{-1}(\epsilon\beta\partial_-\psi). \tag{5.109}$$

When ϕ obeys the sine-Gordon equation, ψ obeys the following equation,

$$\partial_+\left(\frac{1 - \sqrt{1 - \beta^2\epsilon^2(\partial_-\psi)^2}}{\epsilon^2}\right) - m^2\partial_-\left(\cos(\beta\psi) - 1\right). \tag{5.110}$$

Equation (5.109) determines ψ as a power series in ϕ and ϵ. Upon substituting this into (5.110), we get an infinite set of conserved charges, that are the coefficients of the even powers of ϵ. A dual sequence of charges can be obtained by interchanging ∂_+ and ∂_- in the equations above. The set of conserved currents is labeled as,

$$\partial_+ J_{2n}^+ + \partial_- J_{2n}^- = 0, \tag{5.111}$$

where $2n$ relates to the power of ϵ. The lowest-order current is the energy-momentum tensor,

$$J_0^+ \equiv T_{--} = \frac{1}{2}(\partial_-\phi)^2, \quad J_0^- \equiv T_{+-} = -\frac{m^2}{\beta^2}(\cos(\beta\phi) - 1). \tag{5.112}$$

The second-order currents are,

$$J_2^+ = \frac{1}{2}(\partial_-^2\phi)^2 - \frac{\beta^2}{8}(\partial_-\phi)^2, \quad J_2^- = \frac{m^2}{2}(\phi_-)^2\cos(\beta\phi), \tag{5.113}$$

and similarly one can write the expressions for the higher order currents J_{2n}. One can map the charges derived by the Lax pair procedure to those derived by the generating function method.

5.11 Multilocal conserved charges

In the previous section we analyzed the set of infinitely many conserved charges associated with local currents. It will be shown later that these conservation laws are responsible for the fact that the system is integrable, namely there is no particle production and the S-matrix is factorizable. We would like to show now that this type of structure may also follow from conservation laws associated with multilocal currents. The construction of the multi-local currents will be presented in two ways: (i) via an integral equation, (ii) by an inductive procedure.

We then show that the two types of charges are in fact equivalent.

5.11.1 *Multilocal charges from integral equation*

Consider first the $O(N)$ non-linear sigma model defined by the Lagrangian density,

$$S_{O(N)} = \frac{1}{2g_0^2} \int d^2x \left[(\partial_+ \vec{n}) \cdot (\partial_- \vec{n}) - u(\vec{n}^2 - 1) \right], \tag{5.114}$$

where $\vec{n}$ is an N-dimensional real vector and g_0 is the coupling constant. The fact that the field $\vec{n}$ is constrained $\vec{n}^2 = n_i n^i = 1$ $i = 1, \ldots, N$ is incorporated via the Lagrange multiplier u. The equations of motion that follow from this action are,

$$\partial_- \partial_+ \vec{n} + u\vec{n} = 0, \quad \vec{n}^2 = 1. \tag{5.115}$$

So u is actually the action density $u = (\partial_+ \vec{n}) \cdot (\partial_- \vec{n})$. Instead we can write the equation of motion as,

$$\partial_\mu \partial^\mu \vec{n} + \vec{n}(\partial_\mu \vec{n} \cdot \partial^\mu \vec{n}) = 0. \tag{5.116}$$

The action is classically invariant under $O(N)$ global symmetry, by construction. For the special case of $n = 3$, namely $O(3)$, the current takes the form,

$$j_\mu^k = \epsilon^{ijk} (n^i \partial_\mu n^j - \partial_\mu n^i n^j). \tag{5.117}$$

It is easy to check, using the equations of motion, that this current is indeed conserved. In addition the energy-momentum is conserved,

$$\partial_+ T_{--} = \partial_+ \left[\frac{1}{2} (\partial_- \vec{n})^2 \right] = \partial_- \vec{n} \partial_- \partial_+ \vec{n} = -u\partial_- \left[\frac{1}{2} (\vec{n})^2 \right] = 0. \tag{5.118}$$

Thus classically the trace of the energy-momentum tensor vanishes $T_{+-} = 0$.

In addition there is an infinite set of currents, the simplest of which takes the form,

$$J_- = \frac{1}{2|\partial_- \vec{n}|} \left[\partial_- \left(\frac{\partial_- \vec{n}}{|\partial_- \vec{n}|} \right) \right]^2, \quad J_+ = -\frac{u}{|\partial_- \vec{n}|}. \tag{5.119}$$

An alternative way to write down a set of non-local classically conserved charges is the following. We define at any given t the (2×2) $U(t, x)$ operator via the equation,

$$\partial_x U(t, x) = i \frac{w}{1 - w^2} \left[j_1^i - w j_0^i \right] \sigma^i U(t, x), \tag{5.120}$$

and the boundary condition $U(t, -\infty) = 1$, for any Cauchy data $n^i(t, x), \partial_t n^i(t, x)$, where w is the "spectral parameter" which is a complex parameter with $w \neq \pm 1$. Following this definition and using the equations of motion one can show that,

$$\frac{d}{dt} Q(w) \equiv \frac{d}{dt} U(t, \infty) = 0. \tag{5.121}$$

If one expands Q in terms of the spectral parameter one finds a set of infinitely many conserved charges,

$$Q(w) = \sum_{n=0}^{\infty} Q_n w^n \qquad \frac{d}{dt} Q_n = 0. \tag{5.122}$$

We can now rewrite the differential equation (5.120) in terms of the integral equation,

$$U(t, x) = 1 + \frac{w}{1 - w^2} \int_{-\infty}^{x} dy \left[j_0^i - w j_1^i \right] (t, y) \sigma^i U(t, y). \tag{5.123}$$

Inserting the expansion,

$$U(t, x) = \sum_{n=0}^{\infty} U_n w^n, \tag{5.124}$$

into the integral equation for $U(t, x)$, we find the following recurrence relation,

$$U_n(t, x) = i \int_{-\infty}^{x} dy \Big[j_0^i(t, y) \sigma^i \sum_{0 \leq k \leq \frac{(n-1)}{2}} U_{n-2k-1}(t, y)$$

$$- j_1^i(t, y) \sigma^i \sum_{1 \leq l \leq \frac{(n)}{2}} U_{n-2l}(t, y) \Big], \tag{5.125}$$

with $U_0(t, x) = 1$. Thus we can calculate Q_n recursively deriving explicit non-local expressions for the set of infinitely many conserved charges. The "lowest" charges are given by,

$$Q_1^0 = 0 \qquad Q_1^i = \int dy j_0^i(t, y)$$

$$Q_2^0 = -\frac{1}{2} Q_1^i Q_1^i$$

$$Q_2^i = \int dx \int dy dy' \epsilon_{ijk} j_0^j(t, y) j_0^k(t, y') \theta(y - y') - \int dy j_1^i(t, y). \tag{5.126}$$

The algebraic structure associated with these charges is the Yangian symmetry, the description of which is beyond the scope of this book. In the reference list we mention several that deal with these algebras. In Section 5.12 it will be further shown that both the charges of the form (5.119) as well as (5.126) are quantum mechanically conserved.

5.11.2 Charges by inductive procedure

The second method of obtaining non-local currents is as follows. Assume that the system admits a (non-abelian) conserved current which is also a pure gauge, namely,

$$J_\mu = g^{-1}\partial_\mu g, \quad \partial^\mu J_\mu = 0, \tag{5.127}$$

where g is a non-singular matrix (for instance $U(N)$ or $O(N)$ matrix). It follows that the current J is a flat gauge connection, since for $D_\mu = \partial_\mu + J_\mu$ we find that $[D_\mu, D_\nu] = 0$ and also $\partial_\mu D_\mu = D_\mu \partial_\mu$. Using the terminology of differential forms these properties can be rewritten as,

$$DJ \equiv dJ + J \wedge J = 0 \quad d * J = 0, \tag{5.128}$$

together with $[D, D] = 0$ and $D * d = d * D$. Now let us assume that there is an n-th conserved current $J_\mu^{(n)}$, then there is a function $\chi^{(n)}$ such that,

$$J_\mu^{(n)} = \epsilon_{\mu\nu}\partial^\nu \chi^{(n)}. \tag{5.129}$$

Then there is an (n+1)-th conserved current,

$$J_\mu^{(n+1)} = D_\mu \chi^{(n)}, \quad \partial^\mu J_\mu^{(n+1)} = 0. \tag{5.130}$$

The conservation follows easily,

$$d * J^{(n+1)} = d * D\chi^{(n)} = D * d\chi^{(n)} = DJ^{(n)} = DD\chi^{(n-1)} = 0, \tag{5.131}$$

or directly,

$$\partial^\mu J_\mu^{(n+1)} = D^\mu \partial_\mu \chi^{(n)} = -\epsilon^{\mu\nu} D_\mu J_\nu^{(n)} = -\epsilon^{\mu\nu} D_\mu D_\nu \chi^{(n-1)} = 0. \tag{5.132}$$

The sequence of $\chi^{(n)}$ starts with $\chi^{(0)} = 1$ and $J_\mu^{(1)} = J_\mu$.

Associated with the set of infinitely many conserved currents, there is obviously also a set of infinite number of conserved charges,

$$Q^{(n)} = \int dx\, J_0^{(n)}(t, x). \tag{5.133}$$

Consider the special case $n = 2$,

$$Q^{(2)}(t) = \int dx\, J_0^{(2)} = \int dx(\partial_0 + j_0^{(1)})\chi^{(1)}$$

$$= -\int dx\, j_1^{(1)}(t, x) + \int dx\, j_0^{(1)}(t, x)\chi^{(1)}(t, x). \tag{5.134}$$

We can now re-express $\chi^{(1)}$ as $\chi^{(1)}(t,x) = \int dx' j_0^{(1)}(t,x')$. When substituting this to the previous equation we discover the structure of multi-local charges of (5.126).

5.12 Quantum integrable charges in the $O(N)$ model

In Section 5.11 it was demonstrated that certain two-dimensional interacting models have an infinite set of conserved classical charges. This property constitutes the classical integrability of a given system. A natural question to ask is whether this integrability persists also in the quantum regime. We will analyze this question in the context of two models, the $O(N)$ sigma model in this section and the sine-Gordon model in the following one. In fact we have already seen in Section 5.8 how the quantum integrability of the sine-Gordon model fully determines the S-matrix of the theory.[4]

In Section 5.11.1 it was shown that classically the $O(N)$ model is scale invariant as the trace of the energy-momentum vanishes. Following our discussion of the Virasoro anomaly in Chapter 2, it is clear that quantum mechanically the classical conformal invariance of the $O(N)$ model is broken by an anomaly. The right-hand side of (5.118) should not vanish any more but rather be equal to some local terms. These terms can be determined by Lorentz $x^+ \to ax^+$, $x^- \to a^{-1}x^-$ and scale invariance $x^+ \to ax^+$, $x^- \to ax^-$. It turns out that up to a constant the quantum relation is,

$$\partial_+\left[\frac{1}{2}(\partial_-\vec{n})^2\right] = \hat{\beta}\partial_- u = \hat{\beta}\partial_-(\partial_-\vec{n}\partial_+\vec{n}). \qquad (5.135)$$

In fact one can show that the constant $\hat{\beta}$ is the one-loop beta function.

Consider now the next conservation law. Classically it reads,

$$\partial_-\left[\frac{1}{2}u(\partial_-\vec{n})^2\right] + \partial_+\left[\frac{1}{2}(\partial_-^2\vec{n})^2\right] = \frac{3}{2}(\partial_-\vec{n})^2(\partial_- u). \qquad (5.136)$$

One can show that this classical conservation is directly related to the conservation of the current in (5.119). For this make use of the classical scale invariance $x_- \to f(x_-)$ and the classical conservation $\partial_+(\partial_-\vec{n})^2 = 0$, to choose a gauge where $(\partial_-\vec{n})^2 = 1$. Now the classical conservation law takes the form,

$$\partial_+\left[\frac{1}{2}(\partial_-^2\vec{n})^2\right] = (\partial_- u). \qquad (5.137)$$

This is exactly the conservation of the current (5.119), when inserting the gauge $(\partial_-\vec{n})^2 = 1$. To get to the form in general coordinates, namely the form as in

---

[4] Quantum integrability was discussed in [176].

(5.119), make the substitution,

$$\partial_- \to \frac{1}{|\partial_- \vec{n}|} \partial_-, \tag{5.138}$$

noting that it also implies $u \to \frac{u}{|\partial_- \vec{n}|}$.

Quantum mechanically again the right-hand side of equation (5.136) can be corrected by several terms. To eliminate possible terms that are total derivatives it is convenient to analyze the integrated form of this conservation law,

$$\partial_+ \int dx^- \left[\frac{1}{2} (\partial_-^2 \vec{n})^2 \right] = (3+\gamma) \int dx^- \frac{1}{2} (\partial_- \vec{n})^2 (\partial_- u), \tag{5.139}$$

and finally using (5.135) to eliminate $\partial_- u$ we get the quantum conservation law,

$$I = \int dx^- \left[\frac{1}{2} (\partial_-^2 \vec{n})^2 - \frac{(3+\gamma)}{4\hat{\beta}} (\partial_- \vec{n})^4 \right], \qquad \partial_+ I = 0. \tag{5.140}$$

One can now show that this conservation law implies the conservation of $\sum_i P_{-i}^3$. Since I commutes with the S matrix $[I, S] = 0$ one has,

$$<b \text{ out}|I|b \text{ out}> <b \text{ out}|a \text{ in}> = <b \text{ out}|a \text{ in}> <a \text{ in}|I|a \text{ in}> . \tag{5.141}$$

For asymptotic states with N particles and momenta $P_1, \ldots, P_N$ we have

$$<N|I|N> = \text{Constant} \sum_i P_{-i}^3. \tag{5.142}$$

The reason that the conserved charge on an asymptotic state has to be proportional to $\sum_i P_{-i}^3$ is that its tensorial structure is of the form $---$, the only conserved quantity with $-$ Lorentz index is P_- and there are no higher tensorial charges that are not products of P_-.

In a similar manner one has a similar conservation law for P_+ so that,

$$\sum_{in} P_-^3 = \sum_{out} P_-^3 \qquad \sum_{in} P_+^3 = \sum_{out} P_+^3. \tag{5.143}$$

If we now write these conservation laws for a $2 \to N$ process combined with the ordinary conservation of momenta we get four equations for two quantities which, combined with the analyticity of the S-matrix, implies that there cannot be any multiple production, and the only allowed process is $2 \to 2$.

It can also be shown that the classically conserved charges constructed by an integral equation (5.126) are also quantum mechanically conserved.

5.13 Non-local charges and quantum groups

The discussion of quantum groups and non-local charges follows the paper [39]. For a review see, for instance, [114].

In the previous sections we have derived in various forms sets of infinitely many conserved charges and argued that they constitute the integrability of the corresponding models. In particular we described in Section 5.11 non-local charges. Here we will establish the algebraic structure of these charges. It will be shown that they involve non-trivial braiding and that they obey the algebra associated with "quantum groups".

Rather than discussing the generalities of this algebraic structure we analyze it in the context of our laboratory model, the sine-Gordon model. We now rewrite the Lagrangian density of the model as,

$$S = \frac{1}{4\pi} \int d^2 z \partial \phi \bar{\partial} \phi + \frac{\hat{\lambda}}{\pi} \int d^2 z : \cos(\hat{\beta}\phi) : . \tag{5.144}$$

It is straightforward to relate $\hat{\lambda}$ of this formulation to m, β of (5.23), and $\hat{\beta}$ here equals β there.

Recall (Section 5.9) that this action can be considered as a conformal field theory plus a relevant perturbation of the form (5.93). In such a case one can identify a conserved current that obeys the relation,

$$\bar{\partial} J(z, \bar{z}) = \partial H(z, \bar{z}) \quad \partial \bar{J}(z, \bar{z}) = \bar{\partial} \bar{H}(z, \bar{z}), \tag{5.145}$$

where for the conformal limit one has $\bar{\partial} J = \partial \bar{J} = 0$, and $H, \bar{H}$ are defined via,

$$\text{Res}_{z=w} (\phi_{\text{pert}}(w) J^a(z)) = \partial h^a(z) \quad H^a(z, \bar{z}) = 2\hat{\lambda} h^a(z) \bar{\phi}_{\text{pert}}(\bar{z}), \tag{5.146}$$

where the perturbation term is written, in the conformal limit, as $\phi_{\text{pert}}(z) \bar{\phi}_{\text{pert}}(\bar{z})$, and all this under the condition that the Res above are indeed total derivatives. A similar construction applies also for the anti-holomorphic current.

We can now identify a pair of non-local fields

$$\tilde{\phi}_\pm(t, x) = \frac{1}{2} \left[\phi(t, x) \pm \int_{-\infty}^x dy \partial_0 \phi(t, y) \right], \tag{5.147}$$

with which we can write a pair of conserved currents $J_\pm$ of the form (5.145) as,

$$J_\pm = e^{\pm \frac{2i}{\beta} \tilde{\phi}_+ (t, x)}$$

$$H_\pm = \lambda \frac{\hat{\beta}^2}{\hat{\beta}^2 - 2} e^{\pm i \left(\frac{2}{\hat{\beta}} - \hat{\beta} \right) \tilde{\phi}_+ (t, x) \mp \frac{i}{\hat{\beta}} \tilde{\phi}_- (t, x)}. \tag{5.148}$$

The conserved charges associated with the pair of currents are

$$Q_\pm = \frac{1}{2\pi i} \left(\int dz J_\pm + \int d\bar{z} H_\pm \right) \quad \bar{Q}_\pm = \frac{1}{2\pi i} \left(\int d\bar{z} \bar{J}_\pm + \int dz \bar{H}_\pm \right). \tag{5.149}$$

The charges $Q_\pm$ are non-local, as a consequence of being built from the non-local field $\tilde{\phi}$.

Using the basic canonical commutation relation $[\phi(t, x), \partial_0 \phi(t, y)] = 4\pi i \delta(x - y)$ and $e^A e^B = e^{[A,B]} e^B e^A$ (when [A,B] commutes with A and B) one finds the following braiding relations,

$$J_\pm(t, x) \bar{J}_\mp(t, y) = \frac{1}{q^2} \bar{J}_\mp(t, y) J_\pm(t, x)$$

$$J_\pm(t, x) \bar{J}_\pm(t, y) = q^2 \bar{J}_\pm(t, y) J_\pm(t, x), \tag{5.150}$$

where,

$$q = e^{-\frac{2\pi i}{\beta^2}}. \tag{5.151}$$

These non-trivial braiding relations of the currents imply similar relations for the conserved charges,

$$Q_+ \bar{Q}_+ - q^2 \bar{Q}_+ Q_+ = 0$$
$$Q_- \bar{Q}_- - q^2 \bar{Q}_- Q_- = 0$$
$$Q_+ \bar{Q}_- - \frac{1}{q^2} \bar{Q}_- Q_+ = a(1 - q^{2Q_{\text{top}}})$$
$$Q_- \bar{Q}_+ - \frac{1}{q^2} \bar{Q}_+ Q_- = a(1 - q^{-2Q_{\text{top}}})$$
$$[Q_{\text{top}}, Q_\pm] = \pm 2Q_\pm \quad [Q_{\text{top}}, \bar{Q}_\pm] = \pm 2\bar{Q}_\pm, \tag{5.152}$$

where $a = \frac{\lambda}{2\pi i}\gamma^2$, $\gamma^{-1} = \Delta = -\bar{\Delta}(Q_\pm)$ and the topological charge $Q_{\text{top}} = \frac{\hat{\beta}}{2\pi}(\phi(x = \infty) - \phi(x = -\infty))$ (compare with (5.6)).

This algebra of the charges is referred to as "q-deformation" $\hat{S}L_q(2)$ of the $SL(2)$ affine Lie algebra with zero center. Recall the basic $SL(2)$ algebra in the Chevaley basis (3.5),

$$[H, E_\pm] = \pm 2E_\pm \quad [E_+, E_-] = H. \tag{5.153}$$

Introducing the spectral parameter w the infinitely many generators of the $SL(2)$ affine Lie algebra are defined via $J^a = \sum_n J_n^a w^n$ with $J^a = H, E_\pm$. We then define the Chevaley basis of the affine algebra $\hat{SL}(2)$ as,

$$E_{+1} = wE_+ \quad E_{-1} = w^{-1}E_-$$
$$E_{+0} = wE_- \quad E_{-0} = w^{-1}E_+$$
$$H_1 = H \quad H_0 = -H. \tag{5.154}$$

In terms of these generators the $\hat{S}L_q(2)$ algebra reads,

$$[H_i, E_{+j}] = a_{ij} E_{+j}$$
$$[H_i, E_{-j}] = -a_{ij} E_{-j}$$
$$[E_{+i}, E_{-j}] = \delta_{ij} E_{-j} \frac{q^{H_i} - q^{-H_i}}{q - q^{-1}}, \tag{5.155}$$

and a_{ij} is the Cartan matrix of $SL(2)$.

The relations between the non-local charges $Q_\pm$ and $\bar{Q}_\pm$ and the generators of the $\hat{S}L_q(2)$ algebra are,

$$Q_+ = cE_{+1}q^{\frac{H_1}{2}} \quad Q_- = cE_{+0}q^{\frac{H_0}{2}}$$
$$\bar{Q}_- = cE_{-1}q^{\frac{H_1}{2}} \quad \bar{Q}_+ = cE_{-0}q^{\frac{H_0}{2}}$$
$$Q_{\text{top}} = H_1 = -H_0 \tag{5.156}$$

where $c^2 = \frac{\lambda}{2\pi i}\gamma^2(q^{-2} - 1)$.

5.14 Integrable spin chain models and the algebraic Bethe ansatz

The discussion of the algebraic Bethe ansatz follows closely the pedagogical paper of Faddeev [88] and also Beisert [31]. The use of the ansatz in a continuous system that we present follows that of Zamolodchikov [234].

A very useful class of two-dimensional integrable models are the spin chain models. In these models the space is divided into a discrete number of sites where spin variables are placed. So far, and in fact also in the rest of this book, we do not discuss discretized field theories. In this chapter we do since the spin chain models will be shown, in Section 18, to be intimately related to integrable sectors of gauge theories in four dimensions. We will demonstrate the techniques used to solve the spin chain models by applying them to a prototype model, the $XXX_{1/2}$ model. We will describe the model, write down the Bethe ansatz equations associated with it, solve them and extract the spectrum of the model. We then apply the technique to the discretized sine-Gordon model.

A discrete circle with N ordered points is taken to be the space direction. The "space" is periodic so that each site is identified with $i \equiv i + N$. The formal continuum limit can be taken by introducing a lattice spacing Δ such that $\Delta \to 0$, $N \to \infty$ while $x = N\Delta$ is kept finite.

At each site there is a dynamical variable X_i^α where i denotes the site and α is a set of finite number of values. One defines a quantum algebra of observables $\mathcal{A}$ by fixing a set of commutation relations between the X_i^α. When $[X_i^\alpha, X_j^\beta] = 0$ for any $i \neq j$ the algebra is called ultra local. Examples are canonical variables, and spin variables which will be used in the $XXX_{1/2}$ model we are about to describe.

The Hilbert space of the representations of the ultra local algebra has a natural tensor product,

$$\mathcal{H} = \prod_{i=1}^{N} \bigotimes h_i = h_1 \bigotimes h_2 \ldots \bigotimes h_i \ldots \bigotimes h_N, \qquad (5.157)$$

and the variables X_i^α act nontrivially only on h_i.

5.14.1 The $XXX_{1/2}$ model

The $XXX_{1/2}$ describes a spin chain model with N sites. At each site there is a spin variable $S_i^\alpha = \frac{\hbar}{2}\sigma^\alpha$ where σ^α are the Pauli matrices. The Hilbert space at each site is $\mathcal{C}^2$, the two-dimensional complex numbers. The Hamiltonian that defines the model is based on a nearest neighbor interaction of the form,

$$H = \sum_{i,\alpha} \left(S_i^\alpha S_{i+1}^\alpha - \frac{1}{4} \right). \qquad (5.158)$$

The spin of the system which is given by,

$$S^\alpha = \sum_i S_i^\alpha, \qquad (5.159)$$

and is conserved,

$$[H, S^\alpha] = 0. \tag{5.160}$$

The notion XXX associates with the fact that the coefficient of $(S_i^\alpha S_{i+1}^\alpha - \frac{1}{4})$ is a constant independent on i and α. In the case where the coefficient is α dependent, namely $J^\alpha (S_i^\alpha S_{i+1}^\alpha - \frac{1}{4})$, the model is referred as the XYZ model.

To extract the spectrum of the model we make use of the *Lax operator* defined by,

$$L_{k,a}(\lambda) = \lambda I_k \bigotimes I_a + i \sum_\alpha S_k^\alpha \bigotimes \sigma^\alpha, \tag{5.161}$$

where λ is a complex parameter referred to as the *spectral parameter*, I_k, S_k^α are the identity and spin operators acting on h_k, and I_a, σ^α act on an auxiliary space V which is also $\mathcal{L}^2$.

The Lax operator can also be written in the form,

$$L_{k,a}(\lambda) = \left(\lambda - \frac{i}{2}\right) I_{k,a} + i P_{k,a}, \tag{5.162}$$

where $P_{i,a}$ is the permutation operator in $\mathcal{L}^2 \bigotimes \mathcal{L}^2$, namely,

$$Pa \bigotimes b = b \bigotimes a, \tag{5.163}$$

and is given by,

$$P = \frac{1}{2}(I \bigotimes I + \sum_\alpha \sigma^\alpha \bigotimes \sigma^\alpha). \tag{5.164}$$

Equation (5.163) implies,

$$P_{a_1,a_2} = P_{a_2,a_1}$$
$$P_{n,a_1} P_{n,a_2} = P_{a_1,a_2} P_{n,a_1} = P_{n,a_2} P_{a_2,a_1}. \tag{5.165}$$

The Hamiltonian can also be expressed in terms of the permutation operator,

$$H = \frac{1}{2} \sum_i P_{i,i+1} - \frac{N}{2}. \tag{5.166}$$

The idea now is to relate the Hamiltonian and other conserved charges to the monodromy of a string of Lax operators along the whole chain. For that we have to analyze the commuting structure of the Lax operators. This structure is controlled by the *fundamental commutation relations* (FCR) which will be shown later to be part of the Yang–Baxter relations and read,

$$R_{a_1,a_2}(\lambda - \mu)L_{i,a_1}(\lambda)L_{i,a_2}(\mu) = L_{i,a_2}(\mu)L_{i,a_1}(\lambda)R_{a_1,a_2}(\lambda - \mu), \tag{5.167}$$

where,

$$R_{a_1,a_2}(\lambda) = \lambda I_{a_1,a_2} + i P_{a_1,a_2}. \tag{5.168}$$

To prove this relation one makes use of the relations of the permutation operator (5.165). The Lax operator can be interpreted as a connection along the chain in the sense,

$$\psi_{i+1} = L_i \psi_i. \tag{5.169}$$

This can be generalized to an ordered product which transports from i_1 to i_2,

$$T^{i_2}_{i_1,a}(\lambda) = L_{i_2-1,a}(\lambda) \ldots L_{i_1,a}(\lambda), \tag{5.170}$$

and to the full monodromy along the spin chain,

$$T_{N,a}(\lambda) = L_{N,a}(\lambda) \ldots L_{1,a}(\lambda). \tag{5.171}$$

We parameterize this monodromy operator in terms of a 2×2 matrix in the auxiliary space as,

$$T_{N,a}(\lambda) = \begin{pmatrix} A_N(\lambda) & a^\dagger_N(\lambda) \\ \tilde{a}_N(\lambda) & D_N(\lambda) \end{pmatrix}, \tag{5.172}$$

with entries in the full Hilbert space $\mathcal{H}$. In analogy to the FCR of the basic Lax operator it is straightforward to realize that there is a similar relation for the monodromy operator,

$$R_{a_1,a_2}(\lambda - \mu)T_{a_1}(\lambda))T_{a_2}(\mu) = T_{a_2}(\mu)T_{a_1}(\lambda)R_{a_1,a_2}(\lambda - \mu). \tag{5.173}$$

From this relation it follows that the trace of the monodromy operator,

$$F(\lambda) \equiv Tr[T(\lambda)] = A(\lambda) + D(\lambda), \tag{5.174}$$

is commuting, namely $[F(\lambda), F(\mu)] = 0$. We can now expand both T_N and $F(\lambda)$ as a polynomial of order N in λ as,

$$T_{a,N}(\lambda) = \lambda^N + i\lambda^{N-1} \sum_\alpha S^\alpha \otimes \sigma^\alpha + \ldots$$

$$F(\lambda) = 2\lambda^N + \sum_{l=0}^{N-2} Q_l \lambda^l. \tag{5.175}$$

We will see shortly that the set of $N - 1$ operators Q_l are commuting and constitute the set of conserved charges, including the Hamiltonian. Next we expand the monodromy at $\lambda = \frac{i}{2}$,

$$T_{a,N}\left(\frac{i}{2}\right) = i^N P_{N,a} P_{N-1,a} \ldots P_{1,a} = i^N P_{1,2} P_{2,3} \ldots P_{N-1,N} P_{N,a}. \tag{5.176}$$

This follows from $L_{i,a}(\frac{i}{2}) = iP_{i,a}$ and $\frac{d}{d\lambda}L_{i,a} = I_{i,a}$ and then taking the permutations one after the other of the term in the middle. The trace over the auxiliary space is $Tr[P_{N,a}] = I_N$ so that we can now define a shift operator in $\mathcal{H}$,

$$U = e^{i\mathcal{P}} \equiv i^{-N} Tr_a \left[T_N\left(\frac{i}{2}\right) \right] = P_{1,2} P_{2,3} \ldots P_{N-1,N}. \tag{5.177}$$

Using the properties of the permutation operator that $P^* = P$ and $P^2 = 1$, it follows that U is a unitary operator. Moreover, one can show that indeed it is a shift operator, namely,

$$U^{-1}X_iU = X_{i-1}. \tag{5.178}$$

To expand $F(\lambda)$ in the vicinity of $\lambda = \frac{i}{2}$ we first observe that,

$$\frac{\mathrm{d}}{\mathrm{d}\lambda}T_a(\lambda)|_{\lambda=1/2} = i^{N-1}\sum_i P_{N,a}\ldots\hat{P}_{i,a}\ldots P_{1,a}, \tag{5.179}$$

where $\hat{}$ means that the corresponding factor is absent. Using the same procedure as above we find that,

$$\frac{\mathrm{d}}{\mathrm{d}\lambda}F_a(\lambda)|_{\lambda=i/2} = i^{N-1}\sum_i P_{1,2}P_{2,3}\ldots P_{N-1,N}. \tag{5.180}$$

Most of the permutations can be cancelled by multiplying with U^{-1} so that

$$\left[\frac{\mathrm{d}}{\mathrm{d}\lambda}F_a(\lambda)\right]F_a(\lambda)^{-1}|_{\lambda=i/2} = \frac{\mathrm{d}}{\mathrm{d}\lambda}\ln(F_a(\lambda))|_{\lambda=i/2} = \frac{1}{i}\sum_i P_{i,i+1}. \tag{5.181}$$

Recalling the expression we found earlier for the Hamiltonian (5.166) we can now see that,

$$H = \frac{i}{2}\frac{\mathrm{d}}{\mathrm{d}\lambda}\ln(F_a(\lambda))|_{\lambda=i/2} - \frac{N}{2}. \tag{5.182}$$

We have just shown that the Hamiltonian is part of a set of $N-1$ commuting operators generated by $F(\lambda)$, the trace of the monodromy. In fact there are N such conserved charges if we add also one component, say S^3, of the spin. The model is characterized by its N degrees of freedom and is equipped with N conserved charges and hence is (at least classically) *integrable*.

5.14.2 Bethe ansatz equations

To diagonalize the family of operators $F(\lambda)$ one can generalize the procedure used in the quantum harmonic oscillator. In that case we have a non-trivial commutation relation $[a, a^\dagger] = 1$, a Hamiltonian which is $H = a^\dagger a + 1$ and a ground state which is annihilated by a, namely $a|0> = 0$. Let us start first with the commutation relations. These are determined from the FCR as,

$$[\tilde{a}(\lambda), \tilde{a}(\mu)] = 0$$

$$A(\lambda)\tilde{a}(\mu) = \frac{\lambda - \mu - i}{\lambda - \mu}\tilde{a}(\mu)A(\lambda) + \frac{i}{\lambda - \mu}\tilde{a}(\lambda)A(\mu)$$

$$D(\lambda)\tilde{a}(\mu) = \frac{\lambda - \mu + i}{\lambda - \mu}\tilde{a}(\mu)D(\lambda) - \frac{i}{\lambda - \mu}\tilde{a}(\lambda)D(\mu). \tag{5.183}$$

The last two relations generalize the relations,

$$aH = (H+1)a \quad a^\dagger H = (H-1)a^\dagger, \tag{5.184}$$

of the harmonic oscillator. To derive the above one uses an explicit 4×4 matrix formulation for the operators in $V \otimes V$. A natural basis for these matrices is

$$e_1 = e_+ \otimes e_+, \quad e_2 = e_+ \otimes e_-, \quad e_3 = e_- \otimes e_+, \quad e_4 = e_- \otimes e_-, \quad (5.185)$$

where,

$$e_+ = \begin{pmatrix} 1 \\ 0 \end{pmatrix}, e_- = \begin{pmatrix} 0 \\ 1 \end{pmatrix}. \tag{5.186}$$

In this basis the permutation operator and the R matrix read,

$$P = \begin{pmatrix} 1 & 0 & 0 & 0 \\ 0 & 1 & 0 & 0 \\ 0 & 0 & 1 & 0 \\ 0 & 0 & 0 & 1 \end{pmatrix} \tag{5.187}$$

$$R(\lambda) = \begin{pmatrix} \lambda+i & 0 & 0 & 0 \\ 0 & \lambda & i & 0 \\ 0 & i & \lambda & 0 \\ 0 & 0 & 0 & \lambda+i \end{pmatrix}. \tag{5.188}$$

The matrices $T_{a_1}(\lambda)$ and $T_{a_2}(\mu)$ read,

$$T_{a_1}(\lambda) = \begin{pmatrix} A(\lambda) & & a(\lambda) & \\ & A(\lambda) & & a(\lambda) \\ a^\dagger(\lambda) & & D(\lambda) & \\ & a^\dagger(\lambda) & & D(\lambda) \end{pmatrix} \tag{5.189}$$

$$T_{a_2}(\lambda) = \begin{pmatrix} A(\mu) & a(\mu) & & \\ a^\dagger(\mu) & D(\mu) & & \\ & & A(\mu) & a(\mu) \\ & & a^\dagger(\mu) & D(\mu) \end{pmatrix}. \tag{5.190}$$

Explicit multiplication of these matrices yields (5.183).

Similarly to the case of the harmonic oscillator we now define the *ground state*,

$$a(\lambda)|0> = a(\lambda) \prod_i \otimes |0>_i = 0. \tag{5.191}$$

We choose $|0>_i = e_+$ so that,

$$S^3|0> = \frac{N}{2}|0> \quad S^+|0> = 0. \tag{5.192}$$

Thus this state is the "highest weight state". It also follows that (the operator $*$ is left unspecified),

$$L_n(\lambda)|0>_i = \begin{pmatrix} \lambda + \frac{i}{2} & * \\ 0 & \lambda - \frac{i}{2} \end{pmatrix} |0>_i, \tag{5.193}$$

and hence,

$$T(\lambda)|0> = \begin{pmatrix} (\lambda+\frac{i}{2})^N & * \\ 0 & (\lambda-\frac{i}{2})^N \end{pmatrix} |0>, \tag{5.194}$$

which means that $|0>$ is an eigenstate of both $A(\lambda)$ and $D(\lambda)$ and thus also of $F(\lambda)$. Higher excited states are created from the ground state $|0>$ by a successive action with creation operators,

$$\Phi(\{\lambda\}) = a^\dagger(\lambda_1)\dots a^\dagger(\lambda_l)|0> . \tag{5.195}$$

Requiring that the state $\Phi(\{\lambda\})$ is an eigenstate of $F(\lambda)$ imposes a set of relations on the $\lambda_1, \dots, \lambda_l$. In particular using the FCR relation (5.183) we find,

$$A(\lambda)a^\dagger(\lambda_1)\dots a^\dagger(\lambda_l)|0>$$

$$= \prod_{k=1}^{l} \frac{\lambda - \lambda_k - i}{\lambda - \lambda_k}\left(\lambda + \frac{i}{2}\right)^N (\lambda)a^\dagger(\lambda_1)\dots a^\dagger(\lambda_l)|0>$$

$$+ \sum_{k=1}^{l} M_k(\lambda, \{\lambda\})a^\dagger(\lambda_1)\dots \hat{a}^\dagger(\lambda_k)\dots a^\dagger(\lambda_l)|0> . \tag{5.196}$$

The first term on the right-hand side of the equation has the form of an eigenstate equations but the rest of the terms do not. The idea is to choose the set $\{\lambda\}$ such that these terms will cancel out against similar terms in $D(\lambda)a^\dagger(\lambda_1)\dots a^\dagger(\lambda_l)|0>$. To get the value of the coefficient $M_1(\lambda, \{\lambda\})$ we use the second term on the right-hand side of the second equation in (5.183) when interchanging $A(\lambda)$ and $a^\dagger(\lambda_1)$ and in all other exchanges we use the first term. In this way we find that,

$$M_1(\lambda, \{\lambda\}) = \frac{i}{\lambda - \lambda_1} \prod_{k=2}^{l} \frac{\lambda_1 - \lambda_k - i}{\lambda_1 - \lambda_k}\left(\lambda_1 + \frac{i}{2}\right)^N . \tag{5.197}$$

Interchanging now $\lambda_1 \to \lambda_j$ we get similarly the expressions for all $M_j(\lambda, \{\lambda\})$. The same type of manipulations yield,

$$D(\lambda)a^\dagger(\lambda_1)\dots a^\dagger(\lambda_l)|0>$$

$$= \prod_{k=1}^{l} \frac{\lambda - \lambda_k - i}{\lambda - \lambda_k}\left(\lambda - \frac{i}{2}\right)^N (\lambda)a^\dagger(\lambda_1)\dots a^\dagger(\lambda_l)|0>$$

$$+ \sum_{k=1}^{l} N_k(\lambda, \{\lambda\})a^\dagger(\lambda_1)\dots \hat{a}^\dagger(\lambda_k)\dots a^\dagger(\lambda_l)|0>, \tag{5.198}$$

with,

$$N_j(\lambda, \{\lambda\}) = \frac{i}{\lambda - \lambda_1} \prod_{k=2}^{l} \frac{\lambda_1 - \lambda_k - i}{\lambda_1 - \lambda_k}\left(\lambda_1 - \frac{i}{2}\right)^N . \tag{5.199}$$

We now observe that if the set of $\{\lambda\}$ obey the condition,

$$\prod_{k\neq j}^{l} \frac{\lambda_1 - \lambda_k - i}{\lambda_1 - \lambda_k}\left(\lambda_1 + \frac{i}{2}\right)^N = \prod_{k\neq j}^{l} \frac{\lambda_1 - \lambda_k - i}{\lambda_1 - \lambda_k}\left(\lambda_1 - \frac{i}{2}\right)^N , \tag{5.200}$$

then the undesirable terms in (5.196) cancel out and we end with an eigenstate of $F(\lambda)$ and hence of the Hamiltonian. For the former it takes the form,

$$F(\lambda)\Phi(\{\lambda\}) = \left(\lambda + \frac{i}{2}\right)^N \prod_{j=1}^{l} \frac{\lambda - \lambda_j - i}{\lambda - \lambda_j} + \left(\lambda - \frac{i}{2}\right)^N \prod_{k\neq j}^{l} \frac{\lambda - \lambda_j - i}{\lambda - \lambda_j} \Phi(\{\lambda\}).$$

$$(5.201)$$

These conditions can be rewritten as,

$$\left(\frac{\lambda_j + i/2}{\lambda_j - i/2}\right)^N = \prod_{k\neq j}^{l} \frac{\lambda_j - \lambda_k + i}{\lambda_j - \lambda_k - i}. \qquad (5.202)$$

These conditions on the eigenvectors were derived originally by Bethe, though in a completely different way, and hence the names, the "Bethe ansatz equation" (BAE) for (5.202) and the "Bethe vector" for $\Phi(\{\lambda\})$. It is straightforward to observe that the poles in the eigenvalue of $F(\lambda)$ cancel out so that it is a polynomial in λ of degree N. One can further show that the full spectrum can be recast just with nonequal λ_j.

We now want to determine the eigenvalues of spin, momentum and Hamiltonian operators of the eigenstates just found.

Using the FCR relation (5.173) in the limit of $\mu \to \infty$ the $SL(2)$ invariance of the monodromy in $\mathcal{H} \otimes V$ is determined via,

$$\left[T_a(\lambda), \frac{1}{2}\sigma^\alpha + S^\alpha\right] = 0, \qquad (5.203)$$

which means in particular that,

$$[S^3, a^\dagger] = -a^\dagger \quad [S^+, a^\dagger] = A - D. \qquad (5.204)$$

The spin of the state is therefore given by,

$$S^3\Phi(\{\lambda\}) = \left(\frac{N}{2} - l\right)\Phi(\{\lambda\}). \qquad (5.205)$$

Furthermore it can be shown that the states $\Phi(\{\lambda\})$ are all highest weight states provided that the BAE is obeyed, namely,

$$S^+\Phi(\{\lambda\}) = 0. \qquad (5.206)$$

Since the S^3 eigenvalue of the highest weight states is non-negative it is obvious that $l \leq \frac{N}{2}$. When N is odd the spin of the state is half-integer, whereas when it is even the spin is even and in particular for $l = \frac{N}{2}$ there is an $SL(2)$ invariant state with vanishing spin.

Let us determine now the eigenvalue of the momentum operator. From the definition of the shift operator (5.177) it follows that,

$$U\Phi(\{\lambda\}) = i^N F\left(\frac{i}{2}\right)\Phi(\{\lambda\}) = \prod \frac{\lambda_j + \frac{i}{2}}{\lambda_j - \frac{i}{2}}\Phi(\{\lambda\}). \qquad (5.207)$$

The eigenvalue of the momentum operator is therefore given by,

$$P\Phi(\{\lambda\}) = \sum_j p(\lambda_j)\Phi(\{\lambda\}) \quad p(\lambda) = \frac{1}{i}\log\left[\frac{\lambda + \frac{i}{2}}{\lambda - \frac{i}{2}}\right]. \tag{5.208}$$

The energy eigenvalue is determined from (5.182),

$$H\Phi(\{\lambda\}) = \sum_j e(\lambda_j)\Phi(\{\lambda\}) \quad e(\lambda) = -\frac{1}{2}\frac{1}{\lambda^2 + \frac{1}{4}}. \tag{5.209}$$

The last expressions calls for a *"quasi particle"* interpretation. The operator $a^\dagger(\lambda)$ creates a quasi particle which reduces the spin S^3 by one unit and adds to the momentum and energy $p(\lambda)$ and $e(\lambda)$, respectively. We further observe that,

$$e(\lambda) = \frac{1}{2}\frac{d}{d\lambda}p(\lambda). \tag{5.210}$$

It is also evident that we can eliminate the dependence on the rapidity of the energy and momentum and read directly the dispersion relation,

$$e(p) = \cos(p) - 1. \tag{5.211}$$

Since this energy is always non-positive, the highest weight state $|0\rangle$ can be considered a ground state only if we take $-H$ as the Hamiltonian rather than H. In fact it will be shown shortly that the latter corresponds to a system of an antiferromagnet whereas the former corresponds to that of a ferromagnet.

5.14.3 The thermodynamic Bethe ansatz

The thermodynamic limit of the spin chain models is the limit of $N \to \infty$. Recall that the continuum limit of the model when $N \to \infty$ and the spacing $\Delta \to 0$. In the BAE N appears only in the left-hand side. If we take the log of the ansatz we find for real $\{\lambda\}$,

$$Np(\lambda_j) = 2\pi Q_j + \sum_{k=1}^{l} \varphi(\lambda_j - l_k), \tag{5.212}$$

where Q_j are integers $0 \le Q_j \le N - 1$ that define the branch of the log and $\varphi(\lambda)$ is a fixed branch of $log(\frac{\lambda+i}{\lambda-i})$. For large N and Q_j and fixed l one finds the usual expression for the momentum of a free particle on the chain,

$$p_j = 2\pi\frac{Q_j}{N}, \tag{5.213}$$

since the $\varphi(\lambda)$ is negligible.

The second term in (5.212) associates with the scattering of these particles. In fact by comparison with the quantum mechanics of a particle in a box we see that $\varphi(\lambda_i - \lambda_j)$ stands for the corresponding phase shift of the particles with rapidity λ_i and λ_j. Using this analogy we can now identify the S-matrix element with,

$$S(\lambda - \mu) = \frac{\lambda - \mu + i}{\lambda - \mu - i}. \tag{5.214}$$

The S-matrix also enters the large N commutation relations of the normalized creation operators $\tilde{a}^{\dagger}(\lambda) = a^{\dagger}(\lambda)A^{-1}(\lambda)$,

$$\tilde{a}^{\dagger}(\lambda)\tilde{a}^{\dagger}(\mu) = S(\lambda - \mu)\tilde{a}^{\dagger}(\mu)\tilde{a}^{\dagger}(\lambda). \tag{5.215}$$

In addition to the quasi-particle states in the Hilbert space, there are also bound states of the quasi-particles. These states correspond to complex solutions of the BAE. The simplest case is with two quasi-particles $l = 2$. In this case the two BAE read,

$$\left(\frac{\lambda_1 + i/2}{\lambda_1 - i/2}\right)^N = \frac{\lambda_1 - \lambda_2 + i}{\lambda_1 - \lambda_2 - i} \quad \left(\frac{\lambda_2 + i/2}{\lambda_2 - i/2}\right)^N \frac{\lambda_2 - \lambda_1 + i}{\lambda_2 - \lambda_1 - i}. \tag{5.216}$$

Using (5.208) it follows that $p(\lambda_1) + p(\lambda_2)$ is real. Furthermore, for $N \to \infty$ to compensate the exponential increase (decrease) of the left-hand side of the last equations, it is clear that the right-hand side must have $\mathrm{Im}(\lambda_1 - \lambda_2) = i$ (or $-i$) and thus the final form of λ_1 and λ_2 are,

$$\lambda_1 = \lambda_{1/2} + \frac{i}{2} \quad \lambda_2 = \lambda_{1/2} - \frac{i}{2}, \tag{5.217}$$

where $\lambda_{1/2}$ is real. The momentum and energy eigenvalues of the corresponding Bethe vector are,

$$p_{1/2} = \frac{1}{i} \ln \frac{\lambda + i}{\lambda - i} \quad e_{1/2} = \frac{1}{2} \frac{d}{d\lambda}[p_{1/2}(\lambda)] = \frac{1}{\lambda^2 + 1}. \tag{5.218}$$

The state is considered as a bound state since its energy is less than the sum of the energies of the two constituents,

$$e_{1/2} < [e_0(p - p_1) + e_0(p_1)], \tag{5.219}$$

for any $0 \le p, p_1 \le 2\pi$.

The bound state of two quasi-particles $l = 2$ can be generalized for $l > 2$. The roots λ_l are combined into complexes M, where M takes half integer values $M = 0, 1/2, 1, \ldots$ with $l = \sum_M \nu_M (2M + 1)$ where ν_M gives the number of complexes of type M. Each complex has roots of the type,

$$\lambda_{M,m} = \lambda_M + im \quad -M \le m \le M, \tag{5.220}$$

where λ_M is real and m are integers and half integers. The corresponding momentum and energy are given by,

$$p_M(\lambda) = \frac{1}{i} \ln \frac{\lambda + i(M + 1/2)}{\lambda - i(M + 1/2)} \quad e_M(\lambda) = \frac{1}{2} \frac{2M + 1}{\lambda^2 + (M + 1/2)^2}. \tag{5.221}$$

The S-matrix for scattering of complexes M and N are

$$S_{M,N} = \prod_{L=|M-N|}^{M+N} S_{0,L}(\lambda), \tag{5.222}$$

where,

$$S_{0,M}(\lambda) = \frac{\lambda + iM}{\lambda - iM} \frac{\lambda + i(M+1)}{\lambda - i(M+1)}. \tag{5.223}$$

To summarize, the ferromagnetic system with Hamiltonian $-H$ in the thermodynamics limit has a Hilbert space $\mathcal{H}_F$ with a ground state $|0\rangle = \prod(e_+)_i$. The excitations are quasi-particles characterized by M, $M = 0, 1/2, \ldots$ and the rapidity λ. The dispersion relation is given by $e_M = \frac{1}{2M+1}(1 - \cos(p_M))$ and the S-matrix is (5.222). Out of S^α, only S^3 makes sense as an operator on $\mathcal{H}_F$. The operators $S^\pm$ change the ground state at each site. This may be viewed as a symmetry-breaking phenomenon.

So far we have described the basic notions of the physics of the spin chain using the example of the $XXX_{1/2}$. One can further generalize these considerations in many directions such as the anti-ferromagnetic system, general spin states in the XXX model, namely, the $XXX_{s/2}$ model, the XXZ and many others.

The thermodynamic limit for the $XXX_{s/2}$ model

So far we have described the basic notions of the physics of the spin chain using the example of the $XXX_{1/2}$. One can further generalize these considerations also to the $XXX_{s/2}$ model. We state here the results without derivation. An eigenstate characterized by the set $\lambda_1, \ldots, \lambda_l$ which are determined by a straightforward generalization of the BAE (5.202),

$$\left(\frac{\lambda_j + (i/2)s}{\lambda_j - (i/2)s}\right)^N = \prod_{k \neq j}^{l} \frac{\lambda_j - \lambda_k + i}{\lambda_j - \lambda_k - i}. \tag{5.224}$$

The eigenvalues of the Hamiltonian and momentum are given by,

$$E = \sum_{k=1}^{l} \frac{s}{\lambda_k^2 + \frac{1}{4}s^2} \qquad P = \frac{1}{i} \sum_{k=1}^{l} \ln\left[\frac{\lambda_k + \frac{i}{2}s}{\lambda_k - \frac{i}{2}s}\right], \tag{5.225}$$

and the higher conserved charges that render the model integrable are,

$$Q_r = \frac{i}{r-1} \sum_{k=1}^{l} \left(\frac{1}{(\lambda_k + \frac{i}{2}s)^{r-1}} - \frac{1}{(\lambda_k - \frac{i}{2}s)^{r-1}}\right), \tag{5.226}$$

where $r \leq N$ and in particular $r = 2$ is the Hamiltonian.

In the thermodynamic limit, $N \to \infty$ states with low energy and zero momentum can be dealt with by introducing the scaling,

$$\tilde{E} = NE = \frac{1}{N}\sum_{k=1}^{l}\frac{s}{\tilde{\lambda}_k^2} \qquad 2\pi n = \frac{1}{N}\sum_{k=1}^{l}\frac{s}{\tilde{\lambda}_k} \qquad 2\pi n_k - \frac{s}{\tilde{\lambda}_k} = \frac{1}{N}\sum_{j=1,j\neq k}\frac{2}{\tilde{\lambda}-\tilde{\lambda}_k},$$

(5.227)

where $\lambda_k = N\tilde{\lambda}_k$ and where the second and third expressions were derived by taking the log of the zero momentum condition $U = 1$ and of the BAE, respectively. Using the same scaling one finds for the higher charges and the transfer matrix the results,

$$\tilde{Q}_r = \frac{Q_r}{N^{r-1}} = \frac{1}{N}\sum_{k=1}^{l}\frac{s}{\tilde{\lambda}_k^r} \qquad -i\log\tilde{T}(\tilde{\lambda}) = -i\log T(N\tilde{\lambda}) = \frac{1}{N}\sum_{j=1,j\neq k}^{l}\frac{s}{\tilde{\lambda}_j-\tilde{\lambda}_k}.$$

(5.228)

For $N \to \infty$ it is plausible to assume that the Bethe roots accumulate on smooth contours $(\mathcal{C}_1, \ldots \mathcal{C}_A) \equiv \mathcal{C}$ which are referred to as *"Bethe strings"*. Thus we replace the discrete $\tilde{\lambda}_k$ locations of the roots by a continuum variable $\tilde{\lambda}$ described by a density $\rho(\tilde{\lambda})$ so that the sum of the root translates into the integral,

$$\frac{1}{N}\sum_{k=1}^{l} \to \int_{\mathcal{C}}\mathrm{d}\tilde{\lambda}\rho(\tilde{\lambda}),$$

(5.229)

with the normalization that $\int_{\mathcal{C}}\mathrm{d}\tilde{\lambda}\rho(\tilde{\lambda}) = \frac{1}{N}$. Using the continuum formulation we can now rewrite the expressions for the Bethe ansatz, the energy and the higher charges as,

$$2\pi n = s\int_{\mathcal{C}}\frac{\mathrm{d}\tilde{\lambda}\rho(\tilde{\lambda})}{\tilde{\lambda}} \qquad 2\pi n_{\tilde{\lambda}} - \frac{s}{\tilde{u}} = 2\int_{\mathcal{C}}\frac{\mathrm{d}\tilde{\lambda}'\rho(\tilde{\lambda}')}{\tilde{\lambda}'-\tilde{\lambda}},$$

$$\tilde{E} = s\int_{\mathcal{C}}\frac{\mathrm{d}\tilde{\lambda}\rho(\tilde{\lambda})}{\tilde{\lambda}^2} \qquad \tilde{Q}_r = s\int_{\mathcal{C}}\frac{\mathrm{d}\tilde{\lambda}\rho(\tilde{\lambda})}{\tilde{\lambda}^r},$$

(5.230)

where $n_{\tilde{\lambda}}$ is the mode number n_k at the point $\tilde{\lambda} = \tilde{\lambda}_k$. It is expected to be a constant along each contour $\mathcal{C}_a$. In the second integral of the first line, a principal part prescription is implemented. An important result is that one can determine the set of conserved charges $\tilde{Q}_r$ using a resolvent, as,

$$G(\tilde{\lambda}) = |s|\int_{\mathcal{C}}\frac{\mathrm{d}\tilde{\lambda}'\rho(\tilde{\lambda}')}{\tilde{\lambda}'-\tilde{\lambda}} \qquad G(\tilde{\lambda}) = \sum_{r=1}^{\infty}\tilde{\lambda}^{r-1}\tilde{Q}_r.$$

(5.231)

The resolvent is related to the transfer matrix,

$$\tilde{T}(\lambda) = e^{iG(\tilde{\lambda})} + e^{-iG(\tilde{\lambda}-\frac{i}{\lambda})} = e^{-\frac{i}{2\tilde{u}}}2\cos\left(G(\tilde{\lambda}) + \frac{1}{2\tilde{\lambda}}\right).$$

(5.232)

We wish to conclude this chapter with a model that will enable us to connect the discussion of the spin chain models to continuum integrable models.

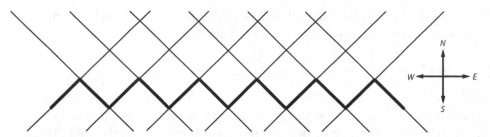

Fig. 5.5. Saw path on the two dimensional lattice.

5.14.4 Spin chain model in discrete time

So far we have discussed the two-dimensional integrable model with a discretized space dimension and a continuous time dimension. It turns out that to connect the spin chain models to continuum two-dimensional field theory it is useful to further also discretize the time direction. The shift operator (13.77) discussed above was determined from a trace of the monodromy at a particular value of the spectral parameter. To describe the system with both space and time discretized, one needs to distinguish values of the spectral parameter $\lambda \pm w$ for some fixed w. We define now an inhomogeneous monodromy built from the Lax operator $L_{i,f}$ acting on the quantum Hilbert space h_i and the auxiliary space V_f,

$$T_f(\lambda, w) = L_{2N,f}(\lambda + w)L_{2N-1,f}(\lambda - w) \ldots L_{2,f}(\lambda + w)L_{1,f}(\lambda - w). \quad (5.233)$$

The light-like shift operators U_+ and U_- are given by,

$$U_+ = tr_f[T_f(w,w)] \quad U_- = tr_f[T_f(-w,w)]. \quad (5.234)$$

The monodromy is along a saw path on the two-dimensional lattice as can be seen in Fig. 5.5.

$L_{2n,f}(\lambda + w)$ is a transport along the NW direction and $L_{2N-1,f}(\lambda - w)$ along the SW direction.

In analogy to the definition of the monodromy as a trace of $T(\lambda)$ (13.77), we now define the monodromy as,

$$F(\lambda, w|a, i, \mu) = tr_f[T_f(\lambda, w|a, i, \mu)], \quad (5.235)$$

where

$$T_f(\lambda, w|a, i, \mu) = L_{2N,f}(\lambda + w)L_{2N-1,f}(\lambda - w) \ldots$$
$$\ldots L_{2i,f}(\lambda + w)L_{f,a}^{-1}(\mu - \lambda)L_{2i-1,f}(\lambda - w) \ldots L_{2,f}(\lambda + w)L_{1,f}(\lambda - w).$$
$$(5.236)$$

It can be shown that $T_f(\lambda, w|a, i, \mu)$ is subjected to the FCR relations in a similar manner to $T(\lambda)$ (13.77). Due to the commutativity of F we get a zero curvature condition on the transport around an elementary plaquette of the space-time lattice,

$$L_{2i,a}(\lambda + w)L_{2i-1,a}(\lambda - w) = U_- L_{2i-1,a}(\lambda - w)U_-^{-1}U_+^{-1}L_{2i,a}(\lambda + w)U_+.$$
(5.237)

The light-like shifts $U_\pm$ are related to the shift in space and time in the usual form, namely,

$$U_+ = e^{-i(H-P)/2} \quad U_- = e^{-i(H+P)/2}.$$
(5.238)

As for the case of only space dimension discretized, here too one finds that the condition of having an eigenvector of the energy and momentum is the BAE which takes the form,

$$\left(\frac{\alpha(\lambda_j + w)\alpha(\lambda_j - w)}{\delta(\lambda_j + w)\delta(\lambda_j - w)} \right)^N = \prod_{j \neq k} S(\lambda_j - \lambda_k),$$
(5.239)

where $\alpha(\lambda), \delta(\lambda)$ are local eigenvalues and $S(\lambda)$ is the quasi particle phase factor.

5.14.5 The discretized version of the sine-Gordon model

As a particular example of an integrable lattice model, we consider a model with Weyl variables rather than spin variables on the lattice sites. These variables obey at each site the relations,

$$u_i v_i = q v_i u_i, \quad q = e^{i\gamma}.$$
(5.240)

The corresponding Lax operator is

$$L_{i,a}(x) = \begin{pmatrix} u_i & x v_i \\ -x v_i^{-1} & u_i^{-1} \end{pmatrix}.$$
(5.241)

To reduce the number of degrees of freedom per site from two to one we impose the constraint,

$$u_{2i}u_{2i-1}v_{2i}v_{2i-1}^{-1} = 1.$$
(5.242)

The BAE for this case take the form,

$$\left(\frac{\sinh(\lambda_j + w + i\gamma/2)\sinh(\lambda_j - w + i\gamma/2)}{\sinh(\lambda_j + w - i\gamma/2)\sinh(\lambda_j - w - i\gamma/2)} \right)^{N/2} = \prod_{k \neq j}^{l} \frac{\sinh(\lambda_j - \lambda_k + i\gamma)}{\sinh(\lambda_j - \lambda_k - i\gamma)},$$
(5.243)

where we have substituted $x = e^\lambda$ and $\kappa = e^w$.

The shift operator in the time direction related to the Hamiltonian can be determined in the following manner. Using the explicit form of the Lax operator we have,

$$L_{i_2,a}\left(\frac{1}{x}\right) L_{i_1,a}\left(\frac{1}{y}\right) r\left(\frac{x}{y}\right) = r\left(\frac{x}{y}\right) L_{i_1,a}\left(\frac{1}{y}\right) L_{i_2,a}\left(\frac{1}{x}\right).$$
(5.244)

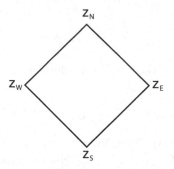

Fig. 5.6. Elementary plaquette.

It can be shown that this condition reduces to the functional equation,

$$\frac{r(x, qw)}{r(x, q^{-1}w)} = \frac{xw + 1}{x + w}. \tag{5.245}$$

Denoting

$$w_{2i} = u_{2i} u_{2i-1} v_{2i}^{-1} v_{2i-1} \qquad w_{2i+1} = u_{2i+1} u_{2i} v_{2i+1}^{-1} v_{2i}, \tag{5.246}$$

we find that the time shift operator is given by,

$$e^{-iH} = \prod r(\kappa^2, w_{2i}) \prod r(\kappa^2 w_{2i+1}). \tag{5.247}$$

With this operator at hand we can now determine the equation of motion of the model and show that it corresponds in the continuuum limit to the sine-Gordon equation. To accomplish this we define now z_i such that,

$$w_i = \frac{z_{i+1}}{z_{i-1}}. \tag{5.248}$$

It is easy to see that z_i does not commute only with one w, namely w_i for which we have,

$$z_i w_i = q^2 w_i z_i. \tag{5.249}$$

Now we apply the time evolution operator on z_i. It reads,

$$\hat{z}_i = e^{iH} z_i e^{-iH}. \tag{5.250}$$

When we substitute (5.247) we get the equation of motion,

$$\hat{z}_{2i+1} = z_{2i+1} \frac{\kappa^2 q^{-1} z_{2i+2} + z_{2i}}{q^{-1} z_{2i+2} + \kappa^2 z_{2i}}, \tag{5.251}$$

and similarly for z_{2i}. This equation connects z along an elementary plaquette (see Fig. 5.6). The equation can also be rewritten as,

$$(q^{-1} z_N z_W - z_S z_E) = \kappa^2 (q^{-1} z_S z_W - z_N z_E). \tag{5.252}$$

If we now define the variable,

$$\chi = e^{i\varphi} = \begin{pmatrix} z \\ z^{-1} \end{pmatrix}, \tag{5.253}$$

alternatively on each second SE characteristic line on our lattice, then the equations of motion take the form,

$$\chi_N = \chi - S^{-1} \frac{\kappa^2 \chi_W \chi_E + 1}{q^{-1} \chi_W \chi_E + \kappa^2}, \tag{5.254}$$

so that for large κ^2 and classical limit $q = 1$ we get,

$$\frac{\chi_N \chi_S}{\chi_E \chi_W} = 1 + \frac{1}{\kappa^2} \left(\frac{1}{\chi_E \chi_W} - \chi_E \chi_W \right) + \cdots, \tag{5.255}$$

and in terms of φ,

$$\frac{\chi_N \chi_S}{\chi_E \chi_W} = e^{i\frac{\Delta^2}{2}(\partial_t^2 \varphi - \partial_x^2 \varphi) + \cdots}, \tag{5.256}$$

and with the scaling $\frac{1}{\kappa^2} = m^2 \Delta^2$ we finally discover the sine-Gordon equation,

$$(\partial_t^2 \varphi - \partial_x^2 \varphi) + 2m^2 \sin(2\varphi) = 0. \tag{5.257}$$

In the quantum version one modifies the scaling to take into account the mass renormalization.

5.15 The continuum thermodynamic Bethe ansatz

Here we describe again the thermodynamic Bethe ansatz but now in the context of continuous models. This will not be a straightforward transition from a discretized to a continuous model via a certain limit, but rather a completely different derivation. Obviously, here as well the Bethe wave-function, the thermodynamic limit and the interplay between the spectrum and S-matrix elements will enter as essential ingredients.

Consider an integrable Euclidean field theory defined on a two-dimensional torus. We denote the two cycles of the torus as *cycle a* and *cycle b*, with corresponding circumferences of R and L and coordinates x and y, as shown in Fig. 5.7. Obviously, one can define in a twofold manner the states of the system and its Hamiltonian. We can consider the space of states on a denoted by $\mathcal{A}$ with the time direction along y, with the Hamiltonian,

$$H_a = \frac{1}{2\pi} \int_a T_{yy} \, dx, \tag{5.258}$$

and with the momentum,

$$P_a = \frac{1}{2\pi} \int_a T_{xy} \, dx, \tag{5.259}$$

which has quantized eigenvalues $\frac{2\pi n}{R}$.

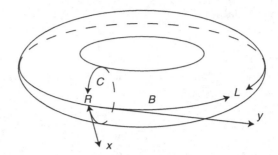

Fig. 5.7. Flat torus generated by two orthogonal geodesics C and B of circumference R and L, respectively.

Alternatively, one can consider the space of states $\mathcal{B}$ along the contour b with the time direction along $-x$, with the Hamiltonian,

$$H_b = \frac{1}{2\pi} \int_b T_{xx} \mathrm{d}y, \qquad (5.260)$$

and with the momentum,

$$P_b = -\frac{1}{2\pi} \int_b T_{yx} \mathrm{d}y, \qquad (5.261)$$

where now the quantization condition is that the eigenvalues of P_b are quantized in units of $\frac{2\pi n}{L}$.

Let us consider the cylindrical geometry via the limit $L \to \infty$ ($L \gg R$). For this case the partition function $Z(L, R)$ is dominated by the ground state of H_a with the ground state energy $E_0(R)$,

$$Z(R, L) \sim e^{-L E_0(R)}. \qquad (5.262)$$

On the other hand,

$$Z(R, L) = Tr_{\mathcal{B}}[e^{-R H_b}]. \qquad (5.263)$$

In $\mathcal{B}$, the thermodynamic limit, namely infinite space $L \to \infty$ which is the analog of the large N limit in Section 5.14.3, gives the free energy $f(R)$ at temperature $1/R$, via $\log Z(R, L) \sim -L\beta f(\beta)$, where $\beta = R$ is the inverse of the temperature. Hence,

$$E_0(R) = Rf(R). \qquad (5.264)$$

The ground state energy $E_0(R)$, which can be referred to also as the Casimir energy, can be related in the limit of conformal field theory, to the Virasoro anomaly. Define the scaling factor $\tilde{c}(r)$ via,

$$E_0(R) = -\frac{\pi \tilde{c}(r)}{6R}, \qquad (5.265)$$

where the dimensionless quantity $r = m_1 R$, with m_1 the lowest mass in the theory. The scaling factor will be determined by the TBA. On the other hand, in the case of conformal invariance, when $R \to 0$, using the relation between the Hamiltonian and L_0 and $\bar{L}_0$ we have,

$$E_0(R) = \frac{2\pi}{R}\left(\Delta_{\min} + \bar{\Delta}_{\min} - \frac{1}{12}c\right), \tag{5.266}$$

where $\Delta_{\min}$ denotes the conformal dimension of the lowest state. This means that the scaling factor should reduce to the effective Virasoro anomaly,

$$\lim_{r \to 0} \tilde{c}(r) = c - 24\Delta_{\min}. \tag{5.267}$$

Here we took the cases of $\Delta_{\min} = \bar{\Delta}_{\min}$.

The TBA method enables one to compute the spectrum of energies and momenta by combining the thermodynamic limit with the factorizable scattering amplitudes. We consider first for simplicity the case with only one type of particle of mass m with a pair scattering amplitude $S(\beta_1 - \beta_2)$ (see Section 5.7). Recall (5.73) that the energy and momentum are related to the rapidities β_i as

$$e_i(\beta) = m\cosh\beta_i \quad p_i(\beta) = m\sinh\beta_i, \tag{5.268}$$

and that the amplitudes obey the unitarity and crossing symmetry (5.78),

$$S(\beta)S(-\beta) = 1 \quad S(\beta) = S(i\pi - \beta). \tag{5.269}$$

For regions of configuration space where the particles are highly separated, namely where $|x_i - x_j| \gg R_c$, $i, j = 1, \ldots, N$ with R_c denoting the scale of the interaction, the particles can be treated as approximately free. In these regions it makes sense to introduce the wavefunction of the system $\Phi(x_1, \ldots, x_N)$ (in regions where the particles are not well separated, particle creation and annihilation prevent the use of a single wavefunction).

For integrable systems at any free region the number of particles will be the same, namely N, as well as the set of momenta p_i. The set of particles in a free region will be denoted by $(i_1, i_2, \ldots, i_N)$ where $x_{i_1} < x_{i_2} < \ldots < x_{i_N}$. A scattering process that yields a transition between $(i_1, i_2, \ldots, i_k, i_{k+1}, \ldots, i_N)$ and $(i_1, i_2, \ldots, i_{k+1}, i_k, \ldots, i_N)$ affects the wavefunctions, such that the latter wavefunction is given by the former multiplied by the scattering amplitude $S(\beta_k - \beta_{k+1})$. These matching conditions on the wavefunctions combined with the quantization of the momenta due to the fact that the "space" coordinate is compact, lead to the relation,

$$e^{ip_i L} \prod_{j \neq i} S(\beta_i - \beta_j) = 1, \quad \text{Or} \quad mL\sinh\beta_i + \sum_{j \neq i} \delta(\beta_i - \beta_j) = 2\pi n_i, \tag{5.270}$$

where for real β we rewrote the scattering amplitude in terms of a unimodular function $S(\beta)e^{i\delta(\beta)}$, with $\delta(\beta)$ a real phase.

This set of transcendental equations selects the admissible sets of rapidities and hence energies and momenta. The energy and momentum of the state

$(\beta_1, \ldots, \beta_N)$ are given by,

$$H_b = \sum_{i=1}^{N} m \cosh\beta_i, \quad P_b = \sum_{i=1}^{N} m \sinh\beta_i. \tag{5.271}$$

In the thermodynamic limit of $L \to \infty$ the set of equations (5.270) can be simplified. In this limit the number of particles becomes large since it grows $\sim L$ and the distance between adjacent levels behaves as $(\beta_i - \beta_{i+1}) \sim \frac{1}{mL}$. In that case one naturally defines the notion of a continuous rapidity density of particles $\rho_1(\beta)$. If there are n particles in a $\Delta\beta$ we take $\rho_1(\beta) = \frac{n}{\Delta\beta}$. We can now exchange the sum over β_i with an integral over β of the particle density, provided that the latter is independent on $\Delta\beta$ for $\frac{1}{mL} \ll \Delta\beta \ll 1$, so that (5.270) takes the form,

$$mL \sinh\beta_i + \int \delta(\beta_i - \beta')\rho_1(\beta')\mathrm{d}\beta' = 2\pi n_i. \tag{5.272}$$

If one further defines the level density $\rho(\beta)$ such that $n = \int \rho(\beta)\mathrm{d}\beta$ we get,

$$mL \cosh\beta + \int \varphi(\beta - \beta')\rho_1(\beta')\mathrm{d}\beta' = 2\pi\rho(\beta), \tag{5.273}$$

where $\varphi(\beta) = \frac{\partial \delta(\beta)}{\partial \beta}$.

In terms of the particle density the energy of the system is,

$$H_b = \int m \cosh\beta \rho_1(\beta)\mathrm{d}\beta. \tag{5.274}$$

At this point one has to distinguish between bosons and fermions. From the unitarity condition one can have either $S(0) = 1$ or $S(0) = -1$. In the former case bosons occupy each rapidity value in any number whereas for fermions the occupation number is at most one. In the latter case the situation is in some sense the opposite. $S(0) = -1$ implies that for two particles with the same rapidity the wavefunction is antisymmetric in their coordinates, and since this is incompatible with bose statistics, this state should be excluded. Hence the bosonic particles for $S(0) = -1$ behave like fermions, and indeed we will refer to this case as "fermionic". On the other hand identical particles that are fermion states of identical rapidity are allowed, and will be referred to as part of a "bosonic" system.

Now we would like to address the issue of the entropy for both the bosonic and fermionic cases. For small intervals of the rapidity $\Delta\beta_\alpha \ll 1$ but such that $\frac{1}{mL} \ll \Delta\beta_\alpha$ there is a large number of levels $N_\alpha \sim \rho(\beta_\alpha)\Delta\beta_\alpha$ and large number of particles $n_\alpha \sim \rho_1(\beta_\alpha)\Delta\beta_\alpha$ that are distributed among these levels. The number of different distributions for the two cases are,

$$\text{"fermionic"} = \frac{(N_\alpha)!}{(n_\alpha)!(N_\alpha - n_\alpha)!}, \quad \text{"bosonic"} = \frac{(N_\alpha + n_\alpha - 1)!}{(n_\alpha)!(N_\alpha - 1)!}. \tag{5.275}$$

The entropy $S(\rho, \rho_1) = \log \mathcal{N}(\rho, \rho_1)$, where $\mathcal{N}(\rho, \rho_1)$ is the total number of states, is given by,

$$S_{\text{Fermi}} = \int d\beta[\rho \log \rho - \rho_1 \log \rho_1 - (\rho - \rho_1) \log(\rho - \rho_1)],$$

$$S_{\text{Bose}} = \int d\beta[-\rho \log \rho - \rho_1 \log \rho_1 + (\rho + \rho_1) \log(\rho + \rho_1)]. \qquad (5.276)$$

Computing the partition function by performing the trace over all the states of the system translates to the minimization of the free energy,

$$-RLf(\rho, \rho_1) = -RH_b(\rho_1) + S(\rho, \rho_1), \qquad (5.277)$$

with respect to the densities ρ and ρ_1, subjected to the constraints (5.273). Using (5.273) the extremum equations take the form,

$$-Rm \cosh\beta + \epsilon(\beta) \pm \int \varphi(\beta - \beta') \log(1 \pm e^{-\epsilon(\beta')}) \frac{d\beta'}{2\pi} = 0, \qquad (5.278)$$

where the upper sign is for the fermionic case, the lower for the bosonic, and the "pseudoenergies" $\epsilon(\beta)$ are defined via,

$$\frac{\rho_1}{\rho} = \frac{e^{-\epsilon(\beta)}}{1 \pm e^{-\epsilon(\beta)}}. \qquad (5.279)$$

The extremal free energy is,

$$Rf(R) = \mp m \int (\cosh\beta) \log(1 \pm e^{-\epsilon(\beta)}) \frac{d\beta}{2\pi}. \qquad (5.280)$$

Comparing (5.273) with (5.280) determines a useful relation,

$$\rho(\beta) = \frac{L}{2\pi} \frac{\partial\epsilon(\beta)}{\partial R}. \qquad (5.281)$$

Finally we can also write down the TBA expressions for the expectation values of $T_{\mu,\nu}$. Using the last relation and (5.279) we get,

$$< T_{xx} > = 2\pi \frac{dE(R)}{dR} = \frac{2\pi}{L} m \int \rho_1(\beta) \cosh\beta d\beta,$$

$$< T_\mu^\mu > = \frac{2\pi}{R} \frac{d[RE(R)]}{dR}$$

$$= m \int \frac{2\pi}{L} [\rho_1(\beta) \cosh\beta - \frac{1}{R} \frac{\rho_1}{\rho} \frac{\partial\epsilon(\beta)}{\partial\beta} \sinh\beta] d\beta. \qquad (5.282)$$

These relations generalize in a straightforward manner to the more general case of N types of particles with masses m_a, $a = 1, \ldots, \hat{N}$ and scattering amplitudes S_{ab} which are now $\hat{N} \times \hat{N}$ matrices.

For integrable models where one can take the limit of $rRm_1 \to 0$, with m_1 being the lowest mass of the system, one can determine the scaling function $\tilde{c}(r)$. In this limit $\epsilon(\beta)$ become constant in the regime where $-\ln(2/r) \ll \beta \ll \ln(2/r)$. Their constant value is determined by the limit of equation (5.273) which now for the case of N particles reads,

$$\epsilon_a = -\sum_{b=1}^{\hat{N}} \int d\beta \phi_{ab}(\beta) \ln[1 + e^{-\epsilon_b}] = -\frac{1}{2\pi} \sum_{b=1}^{\hat{N}} [\delta_{ab}(+\infty) - \delta_{ab}(-\infty)] \ln[1 + e^{-\epsilon_b}],$$

(5.283)

recall that $\phi_{ab}(\beta) = \partial_\beta \delta_{ab}(\beta)$.

Interpolating between these values of $\epsilon(\beta)$ for $\beta \ll \ln(2/r)$ and the values for $\beta \to \infty$ where $\epsilon(\beta)$ go exponentially to infinity, and taking finally the $r \to 0$ limit one finds the effective Virasoro anomaly using the relation between the free energy and the scaling factor (5.265),

$$\tilde{c}(0) = \sum_{a=1}^{N} \tilde{c}_a(\epsilon_a) = \frac{6}{\pi^2} \sum_{a=1}^{N} L\left(\frac{1}{1 + e^{\epsilon_a}}\right),$$

(5.284)

where $\tilde{c}_a$ is the scaling factor associated with the particle of type a and $L(x)$ is the dilogarithm function,

$$L(x) = -\frac{1}{2} \int_0^x \left[\frac{\ln t}{1 - t} + \frac{\ln(1 - t)}{t}\right] dt.$$

(5.285)

Thus the determination of the Virasoro anomaly of the underlying CFT of our massive integrable theory with a purely elastic S-matrix follows from the solution for the pseudo-energies (5.283) and plugging it into the expression of the scaling factor. One can proceed with the continuous TBA and determine not only the ground state energy but also the full spectrum of energies as well as the spectrum of the eigenvalues of all the conserved charges, as was described for the discretized case discussed in Section 5.14. This is beyond the scope of this book. We refer the reader to the relevant papers in the reference list.

6

Bosonization

In one space dimension there are obviously no rotations and hence no angular momentum. This raises the possibility of equivalence relations between scalar fields and fields of higher tensorial structure like spinors, vectors etc. However, spinors and scalars seem to be distinct even in two dimensions due to their different statistics. An equivalence between these two types of fields should therefore incorporate the identification of operators, made out of scalars, that are anti-commuting and vice versa. It is well known that a bilinear of fermi fields is a commuting field, but it is less obvious how to construct a field that obeys the Fermi–Dirac statistics from scalars. This is precisely what the bosonization procedure does.

Coleman [63] and Mandelstam [159] introduced the concept of bosonization. Their construction is now referred to as the "abelian bosonization". An anti-commuting Fermi field, constructed from the exponential of a boson, was given explicitly by Mandelstam [159].

The fact that the theories of a free massless scalar and a free Dirac fermion are equivalent can be proven by showing that they fall into the same representation of the affine current algebra and the Virasoto algebra. The bosonic–fermionic duality can also be further elevated to the free massive theories and also to interacting ones.

It turns out that the original abelian bosonization is not convenient to accommodate color (or flavor) degrees of freedom and hence is inconvenient to address systems like QCD_2. A breakthrough in that direction was achieved by Witten, in his non-abelian bosonization [224].[1]

The equivalence enables one to use, as convenient, either the fermionization of scalar fields or the bosonization of fermions. The latter is useful in several cases. For instance in the case of duality between the Thirring model [205] and the sine-Gordon model,[2] which will be discussed in Section 6.2, the bosonization takes the form of a strong–weak duality. For strong fermionic interactions one finds a weak bosonic coupling. In applications to gauge theories (Section 9) it will be shown that the one loop anomaly behavior is encoded in classical bosonized theory. In QCD_2, as will be discussed in Section 9.3.2, the bosonic version of

---

[1] This paper discusses Majorana fermions. The construction for Dirac fermions was done in [7].
[2] This was proved by Coleman [63].

the theory admits a separation between the color and flavor degrees of freedom, which is very useful in describing the low energy color singlet states.

We start this chapter by introducing the set of rules that span abelian bosonization, including the rules for mass terms and the equivalence of the interacting Thirring model and the sine-Gordon model. We then describe Witten's non-abelian bosonization of Majorana fermions. This is further generalized to the case of massless Dirac fermions. We discuss the subtleties of the massive case and present two methods of handling the non-abelian bosonization of massive Dirac fields.[3] We then discuss in detail the action formulation of the chiral bosonization both abelian and non-abelian. We then depart from the applications that will be found to be relevant to QCD_2 and present topics in bosonization which are more relevant to conformal field and string theories like the bosonization of ghost fields and the Wakimoto bosonization [213]. We do not discuss bosonization on higher Riemann surfaces. The interested reader can consult for instance [211] and [84].

The topic of bosonization in two-dimensional field theories has been reviewed in several papers and books, like that of Stone [202]. Here we mainly follow the review of Frishman and Sonnenschein [101] for the basic ingredients, and update it to include more recent topics.

6.1 Abelian bosonization

6.1.1 Bosonization of a free massless Dirac fermion

Both the theory of a free massless real scalar field and the theory of a free massless Dirac field are conformal field theories invariant under affine Lie algebra. Recall from Chapter 2 that the former theory is defined by the action,

$$S = \int d^2x \mathcal{L} = \frac{1}{8\pi} \int d^2x \partial_\nu \hat{\phi} \bar{\partial}^\nu \hat{\phi}$$

$$= \frac{1}{4\pi} \int d^2\xi \partial_\xi \hat{\phi} \partial_\xi \hat{\phi} = \frac{1}{4\pi} \int d^2z \partial \hat{\phi} \bar{\partial} \hat{\phi}. \tag{6.1}$$

The solution of the equation of motion takes the form,

$$\hat{\phi}(z, \bar{z}) = \phi(z) + \bar{\phi}(\bar{z}). \tag{6.2}$$

The theory has holomorphically (and anti-holomorphically) conserved currents,

$$J(z) = i\partial\phi(z) \quad \bar{J}(z) = -i\bar{\partial}\bar{\phi}(\bar{z}), \tag{6.3}$$

and similarly holomorphic (and anti-holomorphic) energy-momentum tensors,

$$T(z) = -\frac{1}{2} : \partial\phi\partial\phi := -\frac{1}{2} : J(z)J(z) :$$

$$\bar{T}(\bar{z}) = -\frac{1}{2} : \bar{\partial}\bar{\phi}\bar{\partial}\bar{\phi} := -\frac{1}{2} : \bar{J}(\bar{z})\bar{J}(\bar{z}) :, \tag{6.4}$$

---

[3] A bosonization prescription for the mass term in the flavored case was suggested in [75] and [99].

which admit a Virasoro algebra with $c = 1$ and affine Lie algebra with level $k = 1$.

Recall also that the theory of a free massless Dirac field with the action,

$$S = \frac{1}{4\pi} \int d^2 z (\psi^\dagger \bar\partial \psi + \tilde\psi^\dagger \partial \tilde\psi) \tag{6.5}$$

admits conserved currents,

$$J(z) = \psi^\dagger \psi \quad \bar J(\bar z) = \tilde\psi^\dagger \tilde\psi, \tag{6.6}$$

and its energy-momentum tensor can be expressed as a bilinear of the currents using the Sugawara construction,

$$T(z) = -\frac{1}{2}[\psi^\dagger \partial \psi - \partial \psi^\dagger \psi] = -\frac{1}{2} : \psi^\dagger \psi \psi^\dagger \psi := -\frac{1}{2} : J(z)J(z) : . \tag{6.7}$$

The correponding level of the affine algebra and of the Virasoro anomaly are again $k = 1$, and $c = 1$, respectively.

Due to the uniqueness of the irreducible unitary $k = 1$ representation of the affine Lie algebras, and the fact that the infinite-dimensional algebraic structure fully determines the theories, we conclude that *in two space-time dimensions the theories of massless free scalar field and Dirac field are equivalent.*

The equivalence implies that every operator of one theory should have a partner in the other theory, in such a way that the OPEs of these dual operators should be identical. We have just realized such correspondence for the currents and energy-momentum tensor, namely,

$$J_b(z) = \partial\phi(z) \quad \leftrightarrow \quad J_f(z) =: \psi^\dagger \psi(z) :,$$
$$T_b(z) = -\frac{1}{2} : \partial\phi\partial\phi : \quad \leftrightarrow \quad T_f(z) = -\frac{1}{2}[\psi^\dagger \partial \psi - \partial \psi^\dagger \psi], \tag{6.8}$$

and similarly for the anti-holomorphic counterparts.

For completeness we now redescribe the currents using the "old" terminology of vector and axial currents. The vector current reads,

$$J_V^\mu =: \bar\psi \gamma^\mu \psi := -\frac{1}{\sqrt{\pi}} \epsilon^{\mu\nu} \partial_\nu \phi. \tag{6.9}$$

This identification of J^μ leads automatically to a conserved current,

$$\partial_\mu J_V^\mu = 0, \tag{6.10}$$

independent of the equations for ϕ. This is a "topological" conservation, connected with choosing the "vector conservation" scheme. In the applications to follow, we will demand more freedom in the scheme choice of interacting theories, in particular the possibility to have a vector current anomaly. The bosonization procedure will therefore be somewhat modified. The modification will correspond to a change of regularization scheme.

The overall coefficient of the current is such that the fermion number charge,

$$Q = \int_{-\infty}^{\infty} dx j_0(x) = 1, \tag{6.11}$$

for the ψ-field. In addition to the "topologically" conserved vector current, the bosonic theory has an axial current, which is equivalent to the fermionic axial current,

$$J_A^\mu =: \bar{\psi}\gamma^\mu\gamma^5\psi := \frac{1}{\sqrt{\pi}}\partial^\mu\phi. \tag{6.12}$$

The bosonic current is the Neother current associated with the invariance of the bosonic action under the global shift $\delta\phi = \epsilon$. The holomorphic and anti-holomorphic conserved currents discussed above are (in real coordinates) nothing but the left and right chiral currents $J_\pm^\mu = J_v^\mu \pm J_A^\mu$, which correspond to shifts with $\epsilon(x_+)$ and $\epsilon(x_-)$. Using the commutation relation (8.4) the ALA reads,

$$[J_\pm(x_\pm), J_\pm(x'_\pm)] = \frac{2i}{\pi}\delta'(x_\pm - x'_\pm). \tag{6.13}$$

This is the same algebra as that of the fermionic chiral currents. The Sugawara construction in this terminology reads,

$$T_{\pm\pm} = \pi : J_\pm J_\pm : . \tag{6.14}$$

These obey the Virasoro algebra,

$$[T_\pm(x_\pm), T_\pm(x'_\pm)] = 2i\left(T_\pm(x_\pm) + T_\pm(x'_\pm)\right)\delta'(x_\pm - x'_\pm) - \frac{i}{6\pi}\delta'''(x_\pm - x'_\pm), \tag{6.15}$$

which is identical to that of the fermionic energy-momentum tensor.

The equivalence of the bosons and the fermion bilinears is not only mathematical. The fermion Fock-space contains those bosons as physical states. The reason for this is that in one space dimension a massless field can move either to the left or to the right. A Dirac fermion and its anti-particle having together zero fermionic charge and moving in the same direction will never separate. They are therefore indistinguishable from a free massless boson. This picture changes when masses are introduced, and the above relations will be approached at momenta high compared to the mass scale (including high off-mass shell).

A natural question to ask is which operator of the bosonic picture corresponds to the basic Dirac ferion? Since the latter is in fact a combination of a left chiral spinor and a right one, we would like to determine the "bosonized" Weyl fermion $\psi(z)$. It is a holomorphic function of conformal dimension $1/2$, that transforms under the affine Lie transformation with a unit charge, namely, $\psi(z) \to e^{i\epsilon(z)}\psi(z)$. Due to the fact that under the same transformation the scalar field transforms as $\phi(z) \to \phi(z) + e(z)$ we are led to look for a candidate which is an exponential in the scalar field $e^{i\alpha\phi(z)}$. We now use $T(z)$ given in (6.4) to compute the confomal dimension of $: e^{i\alpha\phi(z)} :$ as follows,

$$T(z) : e^{i\alpha\phi(w)} := \frac{\frac{\alpha^2}{2} : e^{i\alpha\phi(w)} :}{(z-w)^2} + \cdots, \tag{6.16}$$

where ... stands for non-singular terms, Hence the conformal dimension is $\frac{\alpha^2}{2}$. Thus we conclude that the following equivalence should hold,

$$\psi(z) \leftrightarrow e^{i\phi(z)}, \quad \psi^\dagger(z) \leftrightarrow e^{-i\phi(z)}. \tag{6.17}$$

To confirm this bosonization rule we compute the OPEs in both descriptions and verify that they are indeed identical,

$$\psi(z)\psi^\dagger(-z) = \frac{1}{2z} + :\psi(0)\psi^\dagger(0): +2z : [\psi(0)\partial\psi^\dagger(0): -\partial\psi(0)\psi^\dagger(0))] + \mathcal{O}(z^2)$$

$$\psi(z)\psi^\dagger(-z) = \frac{1}{2z} + J(0) + 2zT(0) + \mathcal{O}(z^2)$$

$$: e^{i\phi(z)} :: e^{-i\phi(-z)} := \frac{1}{2z} + J(0) + 2zT(0) + \mathcal{O}(z^2)$$

$$: e^{i\phi(z)} :: e^{-i\phi(-z)} := \frac{1}{2z} + i\partial\phi(0) + 2z(\partial\phi\partial\phi(0)) + \mathcal{O}(z^2). \tag{6.18}$$

The bosonic version of the fermion ψ was originally proposed by Mandelstam. His formulation was done in terms of real coordinates and cannonical quantization. For completeness we now also present the "old" construction and proof of equivalence. The bosonized chiral fermion in the latter formulation takes the form,[4]

$$\psi_L = \sqrt{\frac{c\mu}{2\pi}} : \exp\left(-i\sqrt{\pi}\left(\int_{-\infty}^x d\xi[\pi(\xi) + \phi(x)]\right)\right) :$$

$$\psi_R = \sqrt{\frac{c\mu}{2\pi}} : \exp\left(-i\sqrt{\pi}\left(\int_{-\infty}^x d\xi[\pi(\xi) - \phi(x)]\right)\right) :, \tag{6.19}$$

where $\pi(x) = \dot\phi(x)$ is the conjugate momentum of $\phi(x)$, c is a constant. A computation yields $c = \frac{1}{2}e^\gamma \sim 0.891$, where γ is the Euler constant. The normal ordering denoted by : : is performed with respect to the scale μ.

The equal time commutation relations of the ϕ-field,

$$[\phi(x,t), \pi(y,t)] = i\delta(x - y),$$

imply, upon using the formula $e^A e^B = e^{[A,B]} e^B e^A$ (for $[A,B]$ a c-number) the canonical anti-commutation relations for the ψ field,

$$\{\psi_{L,R}^\dagger(x,t), \psi_{L,R}(y,t)\} = \delta(x - y). \tag{6.20}$$

The fermion field ψ is therefore, an inherently non-local functional of the scalar field. However fermion bilinears, such as the currents discussed above or the mass terms that will be described in the next section, are local functions.

So far we have addressed the map for massless theories. Let us now discuss the bosonization of a fermion mass bilinear operator. The mass term which mixes

---

[4] A generalization of this bosonization to a set of N fermions was done by Halpern [121].

the left and right chiral componenets of the Dirac fermion takes the following well-knows form,

$$m_f[\tilde{\psi}^\dagger(\bar{z})\psi(z) + \psi^\dagger(z)\tilde{\psi}(\bar{z})] = m_f(\psi_L^\dagger\psi_R + \psi_R^\dagger\psi_L). \tag{6.21}$$

Again for completeness we write down the expression both in the complex coordinates as well as in real coordinates.

Using the bosonization rules for chiral fermions (6.17) (to be justified below), we deduce the map of the fermion bilinear to the equivalent bosonic operator,

$$m_f[: e^{i\bar{\phi}(\bar{z})} :: e^{i\phi(z)} : + : e^{-i\phi(z)} :: e^{-i\bar{\phi}(\bar{z})} :] = m_f\mu : \cos(\hat{\phi}(z,\bar{z}) :, \tag{6.22}$$

where we have made use of $\hat{\phi}(z,\bar{z}) = \phi(z) + \bar{\phi}(\bar{z})$ and of the fact that there is no non-trivial OPE between $\phi(z)$ and $\bar{\phi}(\bar{z})$. Note that we write down the bosonic equivalent of the mass term operator in the context of the massless theory and hence the factorization to holomorphic and anti-holomorphic parts of the scalar field holds. Once we identify this operation relation we will then use it to add a fermion mass term to the bosonized action. The additional parameter which has a dimension of mass μ is the normal ordering scale.

The derivation of the bosonized mass term in the "old language" is somewhat more involved. We will return to this after we address the bosonization duality between the fermionic Thirring model and the bosonic sine-Gordon model.

We now summarize the equivalence relations between the bosonic and fermionic operators of the free theories, in both the "modern" complex coordinate formulation, as well as the "old" formulation in terms of real coordinates:

Operator	Fermionic	Bosonic
$J(z)$	$: \psi^\dagger\psi(z) :$	$i\partial\phi(z)$
$\bar{J}(\bar{z})$	$: \tilde{\psi}^\dagger\tilde{\psi}(\bar{z}) :$	$-i\bar{\partial}\phi(\bar{z})$
$T(z)$	$-\frac{1}{2} : [\psi^\dagger\partial\psi - \partial\psi^\dagger\psi] :$	$-\frac{1}{2} : \partial\phi\partial\phi(z) :$
$\bar{T}(\bar{z})$	$-\frac{1}{2} : [\tilde{\psi}^\dagger\partial\tilde{\psi} - \partial\tilde{\psi}^\dagger\tilde{\psi}]:$	$-\frac{1}{2} : \bar{\partial}\phi\bar{\partial}\phi(\bar{z}) :$
fermion$_L$	$\psi(z)$	$: e^{i\phi(z)} :$
fermion$_R$	$\tilde{\psi}(\bar{z})$	$: e^{i\phi(\bar{z})} :$
mass term	$\tilde{\psi}^\dagger(\bar{z})\psi(z) + \psi^\dagger(z)\tilde{\psi}(\bar{z})$	$\mu : \cos\hat{\phi}(z,\bar{z}) :$

Bosonization in "modern" complex formulation.

6.2 Duality between the Thirring model and the sine-Gordon model

The Thirring model is a fermionic theory with a current–current interaction, given by the Lagrangian density,

$$\mathcal{L} = i\bar{\psi}\,\partial\!\!\!/\psi - \frac{1}{2}gJ^\mu J_\mu,$$

Operator	Fermionic	Bosonic
$J_+(x^+)$	$: \psi_L^\dagger \psi_L :$	$\partial_+ \phi$
$J_-(x^-)$	$: \psi_R^\dagger \psi_R :$	$\partial_- \phi$
$T_{++}(x^+)$	$-\frac{1}{2} : [\psi_L^\dagger \partial \psi_L - \partial \psi_L^\dagger \psi_L] :$	$-\frac{1}{2} : \partial_+ \phi \partial_+ \phi(x^+) :$
$T_{--}(x^-)$	$-\frac{1}{2} : [\psi_R^\dagger \partial \psi_R - \partial \psi_R^\dagger \psi_R] :$	$-\frac{1}{2} : \partial_- \phi \partial_- \phi(x^+) :$
fermion$_L$	$\psi_L(x^+)$	$\sqrt{\frac{c\mu}{2\pi}} : \exp\left(-i\sqrt{\pi}\left(\int\limits_{-\infty}^{x} d\xi \pi(\xi) + \phi(x)\right)\right) :$
fermion$_R$	$\psi_R(x^-)$	$\sqrt{\frac{c\mu}{2\pi}} : \exp\left(-i\sqrt{\pi}\left(\int\limits_{-\infty}^{x} d\xi \pi(\xi) - \phi(x)\right)\right) :$
mass term	$\psi_L^\dagger(x^+)\psi_R(x^-)$ $+ \psi_R^\dagger(x^+)\psi_L(x^-)$	$\mu : \cos \hat{\phi}(x^+, x^-) :$

Bosonization in "old" formulation.

where $J_\mu = :\bar\psi\gamma_\mu\psi:$. The model is exactly solvable and meaningful for $g > -\pi$. The corresponding equation of motion reads,

$$i \, \partial\!\!\!/\psi(x) = g\gamma_\mu J^\mu(x)\psi(x). \tag{6.23}$$

The theory is invariant under vector and axial $U(1)$ global transformations. The corresponding conserved currents are,

$$J_\mu^V = J^\mu = :\bar\psi\gamma_\mu\psi: \quad J_\mu^A = \epsilon_{\mu\nu} J^{V\,\nu}. \tag{6.24}$$

The model can be studied by means of the operator product expansion on the light-cone [76]. The fermionic bilinears of the model are expressed as a function of the current, and the expressions obtained turn out to be very natural in the light of the bosonization procedure, which we now describe.

We start with the following generalization of the bosonization formula (6.19):

$$\psi_L = \sqrt{\frac{c\mu}{2\pi}} : \exp\left(-i\sqrt{\pi}\left(\frac{2\sqrt{\pi}}{\beta}\int\limits_{-\infty}^{x} d\xi \pi(\xi) + \frac{\beta}{2\sqrt{\pi}}\phi(x)\right)\right) :$$

$$\psi_R = \sqrt{\frac{c\mu}{2\pi}} : \exp\left(-i\sqrt{\pi}\left(\frac{2\sqrt{\pi}}{\beta}\int\limits_{-\infty}^{x} d\xi \pi(\xi) - \frac{\beta}{2\sqrt{\pi}}\phi(x)\right)\right) : \tag{6.25}$$

The meaning of the new parameter β will be clarified shortly. In a similar manner to the derivation of (6.20) we can verify that the equal-time anti-commutation relations are still obeyed. Furthermore one can show that the Dirac operator built from (6.25) obeys the equation of motion (6.23) provided that the bosonic field $\phi(x)$ obeys the equation of motion of the sine-Gordon model discussed in Section 5.3, namely,

$$\partial_\mu \partial^\mu \phi(x) + \frac{\mu^2}{\beta} : \sin(\beta\phi(x)) := 0, \tag{6.26}$$

and that the parameter β is related to the coupling constant g in the following way,

$$\frac{\beta^2}{4\pi} = \frac{1}{1 + \frac{g}{\pi}}. \tag{6.27}$$

From this last relation it follows that the special value $\beta^2 = 4\pi$ corresponds to $g = 0$ and hence a free Dirac fermion. Indeed as we shall see below for that value of β the sine-Gordon potential translates into the bosonized mass term. It is interesting to note a remarkable property of (6.27) which relates the coupling constants of the Thirring model and its bosonic equivalent, the sine-Gordon model. The weak coupling of one theory is the strong coupling of the other. This property often occurs in bosonized theories and hints at the usefulness of the method in dealing with theories for strong coupling, where perturbative methods fail.

The bosonization dictionary of the vector fermion number current is the following,

$$J_\mu^V = \, :\bar{\psi}\gamma_\mu\psi: \; \leftrightarrow \; J^\mu = -\frac{\beta}{2\pi}\epsilon^{\mu\nu}\partial_\nu\phi. \tag{6.28}$$

This expression differs from the bosonized current of the free Dirac theory (6.9) in its normalization factor. We immediately realize that for the special value $\beta^2 = 4\pi$ we precisely reproduce (6.9). The normalization factor can be determined from the assignment of the fermion number charge of a soliton that should be equal to the charge of the field ψ. Recall from Section 5.3 that the classical sine-Gordon model admits a finite energy soliton solution. It is time-independent and interpolates between adjacent wells of the scalar potential. In quantum theory this classical solution becomes a particle. The static soliton solution is given by (5.26),

$$\phi = \frac{4}{\beta}\tan^{-1}[\exp\mu(x - x_0)],$$

where x_0 is the "center" of the soliton. Substituting this into the integral of the current we find the fermion number of this solution to be,

$$Q = \frac{\beta}{2\pi}[\phi(\infty) - \phi(-\infty)] = 1. \tag{6.29}$$

Thus we see that indeed the normalization factor in the vector current (6.28) is the right one.

The relation (6.28) implies that the level in the affine Lie algebra will be $\frac{\beta^2}{4\pi}$, as compared to 1 for the free case.

Let us address again the issue of the bosonization of the fermion mass bilinear (6.22). The definition of the mass term in the "old formulation", as that of the current, requires some care due to the appearance of the products of operators at the same point. In fact when x approaches y one gets the following OPEs,

$$\psi_R^\dagger(x)\psi_L(y) = \frac{c\mu}{2\pi}|c\mu(x-y)|^\delta : e^{-i\beta\phi} :$$

$$\psi_L^\dagger(x)\psi_R(y) = \frac{c\mu}{2\pi}|c\mu(x-y)|^\delta : e^{i\beta\phi} : \tag{6.30}$$

with $\delta = -\frac{g}{2\pi}(1+\frac{\beta^2}{4\pi})$. The proper fermion mass term will therefore be defined by,

$$\lim_{y\to x} \int_{-\infty}^{\infty} dx \, |c\mu(x-y)|^{-\delta} m\bar{\psi}(x)\psi(y) = \frac{c\mu}{\pi}m \int_{-\infty}^{\infty} dx : \cos\beta\phi(x) : .$$

With μ chosen such that $m = \frac{\mu\pi}{c\beta^2}$, the mass term transforms in the bosonic language to,

$$\Delta\mathcal{L} = \frac{\mu^2}{\beta^2} : \cos\beta\phi : .$$

The normal ordering is with respect to μ.[5]

6.3 Witten's non-abelian bosonization

The non-abelian bosonization introduced by Witten is a set of rules assigning bosonic operators to fermionic ones, in a theory of free fermions invariant under a global non-abelian symmetry.[6] Originally the fermions considered were Majorana fermions and the corresponding global symmetry was $O(N)$. The bosonic operators are not expressed in terms of free bosonic fields as in abelian bosonization, but rather in terms of interacting group elements. In particular, bosonic expressions can be written for the energy-momentum tensor, various chiral currents, the mass term and the complete action.

The generalization to the case of N_f Dirac fermions was introduced in [112] and [7].

6.3.1 Bosonization of Majorana fermions

Let us start with N free Majorana fermions governed by the action,

$$S_\psi = \frac{i}{2} \int d^2x \sum_{k=1}^{N} (\psi_{Lk}\partial_+\psi_{Lk} + \psi_{Rk}\partial_-\psi_{Rk})$$

where ψ_L, ψ_R are left and right Weyl–Majorana spinor fields, $\partial_\pm = \frac{1}{\sqrt{2}}(\partial_0 \pm \partial_1)$ and $k = 1,\ldots,N$. The corresponding bosonic action is the Wess–Zumino–Witten

[5] The flavored Thirring model was studied in [73].
[6] A related approach is presented in [179].

(WZW) action discussed in Chapter 4:

$$S_b[u] = \frac{1}{16\pi} \int d^2 x Tr(\partial_\mu u \partial^\mu u^{-1})$$

$$+ \frac{1}{24\pi} \int_B d^3 y \varepsilon^{ijk} Tr(u^{-1}\partial_i u)(u^{-1}\partial_j u)(u^{-1}\partial_k u), \qquad (6.31)$$

where u is a matrix in $O(N)$ whose elements are bosonic fields. The second term, the Wess–Zumino (WZ) term, is defined on the ball B whose boundary Σ is taken to be the Euclidean two-dimensional space-time. Now, since $\pi_2[O(N)] = 0$, a mapping u from a two-dimensional sphere S into the $O(N)$ manifold can be extended to a mapping of the solid ball B into $O(N)$. The WZ term however is well defined only modulo a constant. It was normalized so that if u is a matrix in the vector representation of $O(N)$ the WZW term is well defined modulo $WZ \to WZ + 2\pi$. The source of the ambiguity is that $\pi_3[O(N)] \simeq Z$, namely there are topologically inequivalent ways to extend u into a mapping from B into $O(N)$.

Note that $O(2)$ is an exception, as $\pi_3[O(2)] = 0$.

Note that the equivalence is between a fermionic theory expressed in terms of an N-dimensional fermion in the vector representation of $O(N)$ and a group element which is an $N \times N$ matrix. Nevertheless, as will be shown below the two theories are fully equivalent.

Both the theory of N free Majorana fermions and the WZW model of (15.2) are invariant under ALA transformations of $O_L(N) \times O_R(N)$. The latter take the following forms for the bosonic and fermionic theories:

$$u \to g(z)u \quad \psi_i \to [g(z)]_i^j \psi(z)_j$$
$$u \to uh(\bar{z}) \quad \tilde{\psi}_i \to [h(\bar{z})]_i^j \tilde{\psi}(\bar{z})_j, \qquad (6.32)$$

where $g(z) \in O_L(N)$ and $h(\bar{z}) \in O_R(N)$. The corresponding currents in both pictures satisfy the ALA at level $k = 1$.

The two theories are also invariant under the conformal transformations,

$$z \to f(z) \quad \bar{z} \to \bar{f}(\bar{z}). \qquad (6.33)$$

The associated Virasoro central charges of the two descriptions are identical, as follows,

$$c_f = N \times \frac{1}{2} \quad c_b = \frac{k[\dim O(N)]}{k + N - 2} = \frac{1/2 N(N-1)}{1 + N - 2} = \frac{N}{2}. \qquad (6.34)$$

For the fermions it is just N times the central charge of a single Majorana fermion, whereas for the bosonized version we make use of the fact that the dual Coexter number of $O(N)$ is $N - 2$. The conformal invariance of the action (15.2) can be also shown by realizing that the corresponding β function vanishes. If one generalizes (15.2) by taking a coupling $\frac{1}{4\lambda^2}$ as a coefficient of the first term and $\frac{k}{24\pi}$ of the WZ term (k integer), the β function associated with λ is given at the one loop level (in the sense of expanding around $u = 1$), by,

$$\beta \equiv \frac{d\lambda^2}{d\ln\Lambda} = -\frac{(N-2)\lambda^2}{4\pi}\left[1 - \left(\frac{\lambda^2 k}{4\pi}\right)^2\right],$$

namely (15.2) is at a fixed point for $\lambda^2 = \frac{4\pi}{k}$ and hence exhibits conformal invariance there. By showing that the energy-momentum tensor obeys the Virasoro algebra, one can show that this property is in fact exact.

To summarize, the dictionary that translates the ALA currents and the energy momentum tensor of the fermionic theory, into the bosonic one and vice versa, is given by,

Operator	Fermionic	Bosonic
$J_{ij}(z)$	$: \psi_i \psi_j(z) :$	$\frac{iN}{4\pi}[u^{-1}\partial u]_{ij}(z)$
$\bar{J}_{ij}(\bar{z})$	$: \tilde{\psi}_i \tilde{\psi}_j(\bar{z}) :$	$\frac{iN}{4\pi}[u\bar{\partial}u^{-1}]_{ij}(\bar{z})$
$T(z)$	$-\frac{1}{2}\sum_{i=1}^{N}[\psi_i\partial\psi_i - \partial\psi_i\psi_i]:$	$-\frac{1}{2(N-1)} : J^a J^a(z) :$
$\bar{T}(\bar{z})$	$-\frac{1}{2}\sum_{i=1}^{N} : [\tilde{\psi}\partial\tilde{\psi}_i - \partial\tilde{\psi}_i\tilde{\psi}_i]:$	$-\frac{1}{2(N-1)} : \bar{J}^a \bar{J}^a :(\bar{z})$

6.3.2 Bosonization of Dirac fermions

The bosonic picture for the theory of N free massless Dirac fermions is built from a boson matrix $g \in SU(N)$ and a real boson ϕ. The bosonized action now has the form,

$$\begin{aligned}
S[g, \phi] = {} & \frac{1}{8\pi}\int d^2x Tr(\partial_\mu g \partial^\mu g^{-1}) \\
& + \frac{1}{12\pi}\int_B d^3y \varepsilon^{ijk} Tr(g^{-1}\partial_i g)(g^{-1}\partial_j g)(g^{-1}\partial_k g) \\
& + \frac{1}{2}\int d^2x \partial_\mu \phi \partial^\mu \phi.
\end{aligned} \tag{6.35}$$

Note the difference of factor two between the WZW action associated with the $SU(N)$ and the $O(N)$.

Here again both theories are conformal invariant with an identical Virasoro central charge,

$$c_f = N \times 1 \quad c_b = \frac{k \dim SU(N)}{k + N} + 1 = \frac{N^2 - 1}{1 + N} + 1 = N. \tag{6.36}$$

ALA transformation with respect to global $SU_L(N) \times SU_R(N) \times U(1)$,

$$\begin{aligned}
g \to h(z)u, \quad \phi \to \phi + a(z); \quad \psi_i \to [g(z)]_i^j \psi(z)_j \\
g \to u\bar{h}(\bar{z}), \quad \phi \to \phi + \bar{a}(\bar{z}); \quad \tilde{\psi}_i \to [h(\bar{z})]_i^j \psi(\bar{z})_j,
\end{aligned} \tag{6.37}$$

leave the actions of both pictures invariant.

One way to prove the equivalence of the fermionic and bosonic theories now, for N free massless Dirac fermions and the $k = 1$ WZW theory on $U(N)$ group manifold, is by showing that the generating functionals of the current Green functions of the two theories are the same. For the fermions we have,

$$e^{-iW_\psi(A_\mu)} = \int (d\psi_+ d\psi_- d\bar{\psi}_+ d\bar{\psi}_-) e^{i\int d^2x \bar{\psi} i \not{D} \psi}, \qquad (6.38)$$

where $D_\mu = \partial_\mu + iA_\mu$, $A_\mu = A_\mu^A(\frac{1}{2}T^A) + A_\mu^{(1)} \times 1$ and $(\frac{1}{2}T^A)$ generators of $SU(N)$. The term $W_\psi(A_\mu)$ was calculated by Polyakov and Wiegmann in a regularization scheme which preserves the global chiral $SU(N_L) \times SU(N_R)$ symmetry and the local $U(1)$ diagonal symmetry, leading to,

$$W_\psi(A_\mu) = S[\tilde{A}] + S[\tilde{B}] + \frac{1}{4\pi N} \int d^2x A_\mu^{(1)} A^{\mu(1)}, \qquad (6.39)$$

where $\tilde{A}, \tilde{B} \subset SU(N)$ are related to the gauge fields A_μ^A by $iA_+^A = (\tilde{A}^{-1}\partial_+ \tilde{A})^A$, $iA_-^A = (\tilde{B}^{-1}\partial_- \tilde{B})^A$.

In the bosonic theory one calculates,

$$e^{-iW_B(A_\mu^A)} = \int [du] e^{iS[u] + i\int d^2x(J_+^B A_+^B + J_+^B A_-^B)}$$

$$e^{-iW_B(A_\mu^{(1)})} = \int [d\phi] e^{\frac{i}{2}\int d^2x[(\partial\phi)^2 + (J_-A_+^{(1)} + J_+A_-^{(1)})]} \qquad (6.40)$$

where $J_+^B A_-^B$ and $J_+ A_-^{(1)}$ are the appropriate parts of $\frac{i}{4\pi}Tr[(g^{-1}\partial_+ g)A_-]$, and similarly for the $(-+)$ case and with $A_\pm^{(1)} = Tr(A_\pm)$. These functional integrals can be performed exactly, leading to,

$$W_B(A_\mu^A) = S[\tilde{A}] + S[\tilde{B}] \quad W_B(A^{(1)}) = \frac{1}{4\pi N} \int d^2x A_\mu^{(1)} A^{\mu(1)}.$$

Thus the bosonic current Green functions are identical to those of the fermionic theory, the latter regulated in the way mentioned above.

6.3.3 The bosonization of a mass bilinear of Dirac fermions

A further bosonization rule has to be invoked for the mass bilinear. For a theory with a $U(N)$ symmetry group the rule is,

$$\psi_+^{\dagger l} \psi_{-j} = \tilde{c}\mu N_\mu g_j^l e^{-i\sqrt{\frac{4\pi}{N}}\phi}, \qquad (6.41)$$

where N_μ denotes normal ordering at mass scale μ. The fermion mass term $m_q \bar{\psi}^i \psi_i$ is therefore,

$$m'^2 N_\mu \int d^2x Tr(g + g^\dagger),$$

where $m'^2 = m_q \tilde{c}\mu$, m_q is the quark mass, and c is the same constant as in (6.19). It is straightforward to show that the above bosonic operator transforms correctly under the $U(N)_L \times U(N)_R$ chiral transformations. On top of that it has the correct total dimension,

$$\Delta = \Delta_g + \Delta_\phi = \left(\frac{N-1}{N} + \frac{1}{N}\right) = 1, \qquad (6.42)$$

where $\Delta_g = \frac{N-1}{N}$ and $\Delta_\phi = \frac{1}{N}$ are the dimensions associated with the $SU(N)$ and $U(1)$ group factors, respectively. Moreover in Section 4.4 it was explicitly shown that the four-point function,

$$G(z_i, \bar{z}_i) = <g(z_1, \bar{z}_1)g^{-1}(z_2, \bar{z}_2)g^{-1}(z_3, \bar{z}_3)g(z_4, \bar{z}_4)>, \qquad (6.43)$$

is given by,

$$G(z_i, \bar{z}_i) = [(z_1 - z_4)(z_2 - z_3)(\bar{z}_1 - \bar{z}_4)(\bar{z}_2 - \bar{z}_3)]^{-\Delta_g} G(x, \bar{x}), \qquad (6.44)$$

where $G(x, \bar{x})$ is a function of the harmonic quotients,

$x = \frac{(z_1 - z_2)(z_3 - z_4)}{(z_1 - z_4)(z_3 - z_2)}$ and $\bar{x} = \frac{(\bar{z}_1 - \bar{z}_2)(\bar{z}_3 - \bar{z}_4)}{(\bar{z}_1 - \bar{z}_4)(\bar{z}_3 - \bar{z}_2)}$ only, and in the free case is,

$$G(x, \bar{x}) = [x\bar{x}(1 - x)(1 - \bar{x})]^{\frac{1}{N}} \times \left[I_1\frac{1}{x} + I_2\frac{1}{1-x}\right]\left[\bar{I}_1\frac{1}{\bar{x}} + \bar{I}_2\frac{1}{1-\bar{x}}\right], \qquad (6.45)$$

where $I_1, I_2, \bar{I}_1, \bar{I}_2$ are group invariant factors. This result for the correlation function, combined with the $U(1)$ part gives an expression identical to that for the fermionic bilinears. Moreover the result can be generalized to an n-point function.

6.3.4 Bosonization of Dirac fermions with color and flavor

In his pioneering work on non-abelian bosonization Witten also proposed a prescription for bosonizing Majorana fermions which carry both N_F "flavors" as well as N_C "colors", namely transform under the group $[O(N_F) \times O(N_C)]_L \times [O(N_F) \times O(N_C)]_R$. The action for free fermions is

$$S_\psi = \frac{i}{2}\int d^2x(\psi_{-ai}\partial_+\psi_{-ai} + \psi_{+ai}\partial_-\psi_{+ai}),$$

where now $a = 1, \ldots, N_C$ and $i = 1, \ldots, N_F$ are the color and flavor indices, respectively. The equivalent bosonic action is,

$$\tilde{S}[g, h] = \frac{1}{2}N_C S[g] + \frac{1}{2}N_F S[h]. \qquad (6.46)$$

The bosonic fields g and h take their values in $O(N_F)$ and $O(N_C)$, respectively and $S[u]$ is the WZW action given in (15.2).

The bosonization dictionary for the currents was shown to be,

$$J_{+ij} =: \psi_{+ai}\psi_{+aj} := \frac{iN_C}{2\pi}(g^{-1}\partial_+ g)_{ij} \quad J_{-ij} =: \psi_{-ai}\psi_{-aj} := \frac{iN_C}{2\pi}(g\partial_- g^{-1})_{ij}$$

$$(6.47)$$

$$J_{+ab} =: \psi_{+ai}\psi_{+bi} := \frac{iN_F}{2\pi}(h^{-1}\partial_+ h)_{ab} \quad J_{-ab} =: \psi_{-ai}\psi_{-bi} := \frac{iN_F}{2\pi}(h\partial_- h^{-1})_{ab},$$

$$(6.48)$$

where : : stands for normal ordering with respect to fermion creation and annihilation operators. As for the bosonic expressions for the currents, regularization
is obtained by subtracting the appropriate singular parts.

In terms of the complex coordinates $z = \xi_1 + i\xi_2$, $\bar{z} = \xi_1 - i\xi_2$ (where ξ_1
and ξ_2 are complex coordinates spanning C^2, and the Euclidian plane ($\xi_1 \to x$,
$\xi_2 \to -t$) and Minkowski space-time ($\xi_1 \to x, \xi_2 \to -it$) can be obtained as
appropriate real sections), one can express the currents as

$$J(z)_{ij} \equiv \pi J_{-ij} = \frac{iN_C}{2}(g\partial_z g^{-1})_{ij} \quad \bar{J}(\bar{z})ij \equiv \pi J_{+ij} = \frac{iN_C}{2}(g^{-1}\partial_{\bar{z}} g)_{ij},$$

and similarly for the flavored currents.

In a complete analogy the theory of $N_F \times N_C$ Dirac fermions can be expressed
in terms of the bosonic fields g, h, $e^{-i\sqrt{\frac{4\pi}{N_F N_C}}\phi}$ now in $SU(N_F), SU(N_C)$ and
$U(1)$ group manifolds respectively. The corresponding action is now,

$$S[g, h, \phi] = N_C S[g] + N_F S[h] + \frac{1}{2}\int d^2 x \partial_\mu \phi \partial^\mu \phi. \tag{6.49}$$

This action is derived simply by substituting $ghe^{-i\sqrt{\frac{4\pi}{N_C N_F}}\phi}$ instead of u in (6.31).

As for the equivalence between the bosonic and fermionic theories, we note
that in both theories the commutators of the various currents have the same
current algebra, and the energy-momentum tensor is the same when expressed in
terms of the currents. But the situation changes when mass terms are introduced
(see next section). The bosonization rules for the color and flavor currents are
obtained from (6.47) and (6.48) by replacing the Weyl–Majorana spinors with
Weyl ones, and in addition we have the $U(1)$ current,

$$J^{(1)}(z) \equiv \sqrt{\pi}J_-^{(1)} = : \psi^\dagger_{-ai}\psi_{-ai} := \sqrt{\frac{N_F N_C}{\pi}}\partial_{-\phi}$$

$$\bar{J}^{(1)}(\bar{z}) \equiv \sqrt{\pi}J_+^{(1)} = : \psi^\dagger_{+ai}\psi_{+ai} := \sqrt{\frac{N_F N_C}{\pi}}\partial_{+\phi}. \tag{6.50}$$

The affine Lie algebras are given by,

$$[J_n^A, J_m^B] = if^{ABC}J_{n+m}^C + \frac{i}{2}kn\delta^{AB}\delta_{n+m,0},$$

where $J^A = Tr(T^A J)$, T^A the matrices of $SU(N_C)$, $k = N_F$ for the colored currents and $J(z)$ is expanded in a Laurent series as $J(z) = \sum z^{-n-1}J_n$. A similar

expression will apply for the flavor currents with T^I the matrices of $SU(N_F)$, and the central charge $k = N_C$ instead of N_F. The commutation relation for $\bar{J}(\bar{z})$ will have the same form.

Generalizing the case of $SU(N) \times U(1)$ to our case, the Sugawara form for the energy-momentum tensor of the WZW action is given by,

$$T(z) = \frac{1}{2\kappa_C} \sum_A : J^A(z) J^A(z) : + \frac{1}{2\kappa_F} \sum_I : J^I(z) J^I(z) :$$
$$+ \frac{1}{2\kappa} : J^{(1)}(z) J^{(1)}(z) :, \tag{6.51}$$

where the dots denote normal ordering with respect to n ($n > 0$ meaning annihilation). The κs are constants yet to be determined. In terms of the affine Lie generators this can be written as,

$$L_n = \frac{1}{2\kappa_C} \sum_{m=-\infty}^{\infty} : J_m^A J_{n-m}^A : + \frac{1}{2\kappa_F} \sum_{m=-\infty}^{\infty} : J_m^I J_{n-m}^I :$$
$$+ \frac{1}{2\kappa} \sum_{m=-\infty}^{\infty} : J_m^{(1)} J_{n-m}^{(1)} : \tag{6.52}$$

Now, by applying the last expression on any primary field ϕ_l we can get a set of infinitely many "null vectors" of the form,

$$\chi_l^n = \left[L_n - \frac{1}{2\kappa_C} \sum_{m=n}^{0} : J_m^A J_{n-m}^A : \right.$$
$$\left. - \frac{1}{2\kappa_F} \sum_{m=n}^{0} : J_m^I J_{n-m}^I : - \frac{1}{2\kappa} \sum_{m=n}^{0} : J_m^{(1)} J_{n-m}^{(1)} : \right] \phi_l,$$

for any $n \leq 0$ (for $n > 0$ holds immediately). Since each of these vectors must certainly be a primary field, $L_m \chi^n = J_m^A \chi^n = J_m^I \chi^n = J_m \chi^n = O$, which holds trivially for $m > 0$. When checking for $m \leq 0$, it leads to expressions for the various κ, for the central charge c of the Virasoro Algebra, and for the dimensions of the primary fields $\Delta_l = \Delta_{l+} + \Delta_{l-}$, in terms of N_C, N_F and the group properties of the primary fields,

$$\kappa_C = \kappa_F = \frac{1}{2}(N_C + N_F), \quad \kappa = N_F N_C$$

$$c = \frac{N_C(N_F^2 - 1)}{(N_C + N_F)} + \frac{N_F(N_C^2 - 1)}{(N_C + N_F)} + 1 = N_F N_C$$

$$\Delta_{l\pm} = \frac{(c_{l\pm}^2)^F}{(N_F + N_C)} + \frac{(c_{l\pm}^2)^C}{(N_F + N_C)} + \frac{(c_{l\pm}^2)^{(1)}}{N_C N_F} \tag{6.53}$$

where $(c_{l\pm}^2)^C$ is the eigenvalue of the $SU_{R,L}(N_C)$ second Casimir operator in the representation of the primary field ϕ_l, namely $(\frac{1}{2}T^A)(\frac{1}{2}T^A) = (c_l^2)^C I$, and similarly for the flavor group. In the cases of $SU(N_C)$ and $SU(N_F)$ the discussion applies to Δ_{l+} or Δ_{l-} separately, with C_{l+}^2 and C_{l-}^2, respectively. Note that the expressions for κ_F and κ_C of equation (6.53) are an immediate generalization of the case of the group $SU(N)$ with the central term equal to one. There the factor was $N + 1$, the N being the second Casimir of the adjoint representation, and the 1 being the central term.

The equivalence of the bosonic and fermionic Hilbert spaces was demonstrated by showing that the two theories have the same current algebra (affine Lie algebra), and that the energy-momentum tensor can be constructed from the currents in a Sugawara form. Goddard *et al.* [110] showed that a necessary and sufficient condition for such a construction of the fermionic $T(z)$, in a theory with a symmetry group G, is the existence of a larger group $G \subset G'$ such that G'/G is a symmetric space with the fermions transforming under G just as the tangent space to G'/G does. Based on this theorem they found all the fermionic theories for which an equivalent WZW bosonic action can be constructed. The cases stated above fit in this category. Note in passing that this does not hold for cases where the symmetry group includes more non-abelian group factors, like for example $SU(N_A) \times SU(N_F) \times SU(N_C) \times U(1)$.

The prescription equation (6.49) described above , for the bosonic action that is equivalent to that of colored and flavored Dirac fermions, is by no means unique. In fact it will be shown that this prescription will turn out to be inconvenient once mass terms are introduced. Another scheme, based on the WZW theory of $U(N_F N_C)$ will be recommended.

6.3.5 Bosonization of mass bilinears in the product scheme

A natural question here is how to generalize the rule (6.41) to Majorana fermions with action (6.46), and its analog for the case of $SU(N_F) \times SU(N_C) \times U(1)$ given in (6.49). We call the latter the **product scheme**. The bosonization rule for the latter case is,

$$\psi_+^{\dagger ai} \psi_{-bj} = \tilde{c}\mu N_\mu g_j^i h_b^a e^{-i\sqrt{\frac{4\pi}{N_F N_C}}\phi}. \tag{6.54}$$

Consequently, the bosonic form of the fermion mass term $m_q \bar{\psi}^{ia} \psi_{ia}$ is,

$$m'^2 N_\mu \int d^2x (TrgTrh + Trh^\dagger Trg^\dagger) e^{-i\sqrt{\frac{4\pi}{N_F N_C}}\phi}, \tag{6.55}$$

with $m'^2 = m_q \tilde{c}\mu$. Once again the bosonic operator (6.54) has the correct chiral transformations and the proper dimension,

$$\Delta = \Delta_g + \Delta_h + \Delta_\phi = \frac{N_F^2 - 1}{N_F(N_F + N_C)} + \frac{N_C^2 - 1}{N_C(N_C + N_F)} + \frac{1}{N_C N_F} = 1.$$

Unfortunately, the explicit calculation of the four-point function reveals a discrepancy between the fermionic and bosonic terms in (6.54). This can actually be understood directly. Since g and h are fields defined on entirely independent group manifolds, then (ignoring for a moment the $U(1)$ factor) the four-point function of the mass term can be written as,

$$<g(z_1, \bar{z}_1)g^{-1}(z_2, \bar{z}_2)g^{-1}(z_3, \bar{z}_3)g(z_4, \bar{z}_4)>$$
$$<h(z_1, \bar{z}_1)h^{-1}(z_2, \bar{z}_2)h^{-1}(z_3, \bar{z}_3)h(z_4, \bar{z}_4)> .$$

This expression differs from the corresponding fermionic Green's function, as it includes independent "contractions" for the g and h factors, whereas in the fermionic correlation function the flavor and color contractions are correlated. Moreover, the expressions for the bosonic Green's function involve hypergeometric functions, and do not resemble the case of free fermions, which is a product of poles.

6.3.6 Bosonization of the $U(N_F \times N_C)$ WZW action

It is clear from the previous discussion that the bosonization prescription for our case needs alteration. A priori there can be two ways out, either modifying the rule for the bosonization of the mass bilinear or using a different bosonic scheme altogether. As for the first approach, (6.54) preserves the proper chiral transformation laws under the product group $SU(N_F) \times SU(N_C) \times U(1)$ as well as the correct dimension, and therefore the number of possible modifications is very limited. For example, one might think of multiplying the expression in (6.54) by an operator which is a chiral singlet under the above group, with zero dimension. We do not know of such a modification. Therefore we are going to try a different bosonic theory than (6.49). The symmetry of the free fermionic theory can actually be taken as $U_L(N_F \times N_C) \times U_R(N_F \times N_C)$ rather than $[SU(N_F) \times SU(N_C) \times U(1)]_L \times [SU(N_F) \times SU(N_C) \times U(1)]_R$. The natural bosonic action is hence a WZW theory of $u \subset U(N_F N_C)$ and with $k = 1$. The currents are now,

$$J(z)_{\alpha\beta} = \frac{i}{2}(u\partial_z u^{-1})_{\alpha\beta} \quad \bar{J}(\bar{z})_{\alpha\beta} = \frac{i}{2}(u^{-1}\partial_{\bar{z}}u)_{\alpha\beta},$$

with α, β running from 1 to $N_F \times N_C$. The expressions for the flavor and color currents can be obtained by appropriate traces, over color for the flavor currents and over flavor for the color currents.

The mass bilinear is now,

$$\psi^\dagger_{+\alpha}\psi_{-\beta} = \tilde{c}\mu N_\mu u_{\alpha\beta},$$

where now the $U(1)$ term is absorbed into u.

Clearly the requirement for Sugawara construction of T, for proper chiral transformations of all the operators and for a correct dimension for the mass bilinear are fulfilled. Since now the flavor and color degrees of freedom are attached to the same bosonic field, the previous "contraction problem" in the n point functions is automatically resolved. Moreover as stated above the four-point function and in fact any Green's function will now reproduce the results of the fermionic calculation.

The currents constructed from u obey the Affine Lie algebra with $k = 1$. The color currents, for instance, are $J^A = Tr(T^A J)$, where T^A are expressed as $(N_C N_F) \times (N_C N_F)$ matrices defined by $\lambda^A \otimes 1$, with λ^A the Gell–Mann matrices in color space and 1 stands for a unit $N_F \times N_F$ matrix. The central charge is $k = N_F$. The same arguments will apply for the flavor currents, now with $k = N_C$. The central charge for the $U(1)$ current is $N_C N_F$.

To see the difference between the present theory and the previous one let us express u in terms of $(N_F N_C) \times (N_F N_C)$ matrices $\tilde{g}, \tilde{h}$ and $\tilde{l}$ in $SU(N_F), SU(N_C)$ and the coset-space,

$$SU(N_F \times N_C)/\{SU(N_F) \times SU(N_C) \times U(1)\}$$

respectively, through,

$$u = \tilde{g}\tilde{h}\tilde{l}e^{-i\sqrt{\frac{4\pi}{N_C N_F}}\phi}.$$

Using the formula for expressing an action of the form $S[AgB^{-1}]$ we get

$$S[u] = S[\tilde{g}\tilde{h}\tilde{l}] + \frac{1}{2}\int d^2x \partial_\mu \phi \partial^\mu \phi$$

$$S[\tilde{g}\tilde{h}\tilde{l}] = S[\tilde{g}] + S[\tilde{l}] + S[\tilde{h}] + \frac{1}{2\pi}\int d^2x Tr(\tilde{g}^\dagger \partial_+ \tilde{g}\tilde{l}\partial_- \tilde{l}^\dagger + \tilde{h}^\dagger \partial_+ \tilde{h}\tilde{l}\partial_- \tilde{l}^\dagger).$$

We can now choose $\tilde{l} = l$ so that $l\partial_- l^\dagger$ will be spanned by the generators that are only in $SU(N_F \times N_C)/\{SU(N_F) \times SU(N_C) \times U(1)\}$. This can be achieved by taking $\tilde{u} = \tilde{g}\tilde{h}\tilde{l}$, which is u but without the $U(1)$ part, and then taking for $\tilde{h} = h \otimes 1$ a solution of the equation $\partial_- hh^\dagger = \frac{1}{N_F}Tr_F[(\partial_- \tilde{u})\tilde{u}^\dagger]$, and similarly for g with $\frac{1}{N_C}Tr_C$. These are also the conditions that the flavor currents should be expressed in terms of $\tilde{g}$ and the color currents in terms of $\tilde{h}$. For this choice, the mixed term in the above action, the term involving products of $\tilde{l}s$ with $\tilde{g}s$ or $\tilde{h}s$, is zero, and so the new action is,

$$S[u] = N_C S[g] + N_F S[h] + \frac{1}{2}\int d^2x (\partial_\mu \phi \partial^\mu \phi + S[l]).$$

Note that l is still an $SU(N_C N_F)$ matrix, while g and h are expressed as $SU(N_F)$ and $SU(N_C)$ matrices respectively, but the matrix l involves only products of color and flavor matrices (not any of them separately).

6.4 Chiral bosons

So far we have discussed the bosonization of a Dirac fermion via abelian bosonization, and N Dirac fermions using the $U(N)$ WZW model, or N Majorana fermions with an $SO(N)$ WZW model. What about the bosonization of chiral left or right Weyl fermions? In the fermionic language it is trivial to write an action of a fermion with one given chirality. It is also easy to factorize a scalar field into its left $\phi_L(x^-)(\phi(z))$ and right moving $\phi_R(x^+)(\bar{\phi}(\bar{z}))$ parts since the solution of equation of motion of a scalar field in real and complex coordinates takes the form

$$\phi(x_+, x_-) = \phi_L(x^-) + \phi_R(x^+) \quad \text{or} \quad \phi(z, \bar{z}) = \phi(z) + \bar{\phi}(\bar{z}). \tag{6.56}$$

However, as will be discussed shortly it turns out that it is quite subtle to write down an action of a chiral boson which is equivalent to that of a left or a right chiral fermion. Once we establish an action for a chiral boson, the question is how can one couple it to abelian and non-abelian gauge fields?

In this section we will construct two seemingly independent constructions of the action of a chiral boson. In fact, it will turn out that one formulation is a special case of the other. We start with Siegel's formulation which is based on the coupling of a scalar field to fictitious gravity in a light-cone gauge [194] and then we describe a manifestly non-Lorentz invariant action [92] which is a special case of the former.[7] In [36], [199], and in [100] the two formulations were related and further generalizations were discussed. We follow the latter paper here.

Chiral bosons play an important role in string theories and in particular chiral bosons on Riemann surfaces of any genus. Here we will not enter into discussions on these constructions and describe chiral bosons only on a two-dimensional flat Minkowsky space-time.

6.4.1 Chiral boson via coupling to fictitious "light-cone gravity"

A scalar field in two space-time dimensions couples to two-dimensional gravity via the following well-known action,

$$\mathcal{L} = \frac{1}{2}\sqrt{g}g_{\alpha\beta}\partial^\alpha\phi\partial^\beta\phi, \tag{6.57}$$

where $g_{\alpha\beta}$ is the two-dimensional metric. In the " light-cone gravitational gauge" the metric has the form,

$$g^{++} = 0 \quad g^{+-} = 1 \quad g^{--} = 2\lambda. \tag{6.58}$$

In this gauge the action (6.57) reads,

$$\mathcal{L} = \partial_+\phi\partial_-\phi + \lambda(\partial_-\phi)^2. \tag{6.59}$$

Since we have fixed only part of the local symmetries of (6.57) it is straightforward to realize that the last action is still invariant under the following local transformation

$$\delta\phi = \epsilon^-\partial_-\phi \quad \delta\lambda = -\partial_+\epsilon^- + \epsilon^-\partial_-\lambda - \partial_-(\epsilon^-\lambda) \tag{6.60}$$

---

[7] The anomalies of the system were analyzed in [129].

which is the transformation of the scalar field under a combination of x^- coordinate transformation and a Weyl rescaling,

$$\delta x^- = \epsilon^-(x) \quad \delta g_{\alpha\beta} = -g_{\alpha\beta}\partial_-\epsilon^-. \tag{6.61}$$

It is important to emphasize that we are in fact considering a flat two-dimensional Minkowski space-time and 2λ is the $--$ component of a fictitious metric. The action (6.59) is further invariant under the global shift symmetry $\phi \to \phi + a$. We denote the corresponding current as the axial current defined in (6.12) which takes the following form,

$$J^+_{(ax)} = \partial_-\phi \quad J^-_{(ax)} = \partial_+\phi + 2\lambda\partial_-\phi. \tag{6.62}$$

One can also define $J_{-(ax)} = J^+_{(ax)}$ and $J_{+(ax)} = J^-_{(ax)} - 2\lambda J_{-(ax)}$, namely using the fictitious metric to raise and lower indices. In addition we have the topological vector conserved current $J_{(v)\mu} = \epsilon_{\mu\nu}\partial^\nu\phi$. The vector and axial currents defined here are those defined in (6.9), times a factor of $-\sqrt{\pi}$. Following from these two currents we can obviously also define the left and right conserved currents $J_{l/r} = \frac{1}{2}(J_{(v)} \pm J_{(ax)})$.

The equations of motion derived by variations with respect to λ and to ϕ are,

$$\delta\lambda: \ (\partial_-\phi)^2 = 0 \quad \delta\phi: \partial_+\partial_-\phi + \partial_-(\lambda\partial_-\phi) = 0. \tag{6.63}$$

These equations imply classically the chiral nature of the boson, namely a left moving boson $\phi(x^+)$ and the conservation of the axial current. Note that unlike an ordinary scalar field the chiral scalar action (6.59), admits only the "holomorphic" conservation of the left current but not the "anti-holomorphic" conservation of the right one namely,

$$\partial_- J_{(l)}{}^- = 0 \quad \partial_+ J_{(r)}{}^- = -\partial_- J_{(r)}{}^- \neq 0, \tag{6.64}$$

which implies that there is only left affine symmetry but not a right one.

So far we have discussed the classical system. Quantum mechanically it turns out that the symmetry (6.60) is anomalous. There are several ways to verify this anomaly. Probably the easiest way is to realize the resemblance of the action (6.57) to that of the bosonic string. It is well known that the latter is consistent only in 26 dimensions and not, as our case seems to be, in one dimension. Technically this follows from the fact that the ghost system associated with the fixing of the fictitious diffeomorphism and Weyl invariance have a Virasoro anomaly (see Section 6.5.1) equal to -26. In fact following the discussion in Section 6.5.1 we know that there is another way to cancel this anomaly and that is to add to the action a background charge term of the following form,

$$\mathcal{L} = \mathcal{L}_{cl} + qR^{(2)}\phi = \mathcal{L}_{cl} + q\partial_-^2\lambda\phi = \mathcal{L}_{cl} - q\lambda\partial_-^2\phi, \tag{6.65}$$

where q is the background charge and $R^{(2)}$ is the fictitious scalar curvature. This modification of the action yields a modified energy-momentum tensor as will be discussed in Section 6.5.1, of the form,

$$T^{\phi}_{--} = -\frac{1}{2}(\partial_-\phi)^2 - q\partial_-^2\phi. \tag{6.66}$$

To fix the gauge associated with the symmetry (6.60) one introduces a (b, c) ghost system (see Section 6.5) that contributes $c = -26$ to the Virasoro anomaly. The modified energy-momentum tensor (6.66) contributes $c = 1 + 6\pi q^2$ so that to cancel the anomaly one has to take the background charge $q = \sqrt{\frac{25}{6\pi}}$ and hence the quantum mechanical action of the chiral boson is (6.65) with the value of q just quoted.

There are several possible methods to quantize this action. One approach is to follow the BRST quantization of the bosonic string. In this procedure one uses the Nielpotent operator associated with the Noether charge that corresponds to the BRST symmetry

$$Q_{\mathrm{BRST}} = \int dxc \left(T^{\phi}_{--} + \frac{1}{2}T^{\mathrm{ghost}}_{--} \right), \tag{6.67}$$

to construct the space of physical states. The latter furnish the cohomology of Q_{BRST}, namely,

$$Q_{\mathrm{BRST}}|\mathrm{phys}\rangle = 0 \quad |\mathrm{phys}\rangle \neq Q_{\mathrm{BRST}}|\mathrm{state}\rangle. \tag{6.68}$$

For the vanishing ghost sector this implies that,

$$T^{\phi}_{--}{}^{(+)}|\mathrm{phys}\rangle = 0 \quad \rightarrow \quad a(k)|\mathrm{phys}\rangle = 0 \; for \; k > 0, \tag{6.69}$$

where $T^{\phi}_{--}{}^{(+)}$ is the positive frequency part of T^{ϕ}_{--} and $a(k)$ is the annihilation operator of momentum k. Thus the space of physical states is made of only left-moving $(k > 0)$ states.

Once the local symmetry (6.60) has been made non-anomalous, one can safely choose any gauge fixing. In particular we can take $\lambda = 0$ while keeping the construction of the physical states discussed above, or fixing $\lambda = -1$. As will be seen in the following section this gauge will turn out to be convenient and rather than addressing the issues of coupling to abelian and non-abelian gauge fields of the action (6.59) we will do it instead with the gauge fixed action.

Another approach of quantizing the system is based on the implementation of Dirac brackets. Starting from (6.59) and its corresponding Hamiltonian density,

$$\mathcal{H} = \frac{1}{2}\left[\frac{(\pi + \lambda\phi')^2}{1+\lambda} + (1-\lambda)(\phi')^2\right], \tag{6.70}$$

we realize that the conjugate momentum of λ, π_λ vanishes. This is a primary constraint $\chi_1 = \pi_\lambda = 0$. Requiring that this constraint is preserved in time, namely, $\dot{\chi}_1 = \{\pi_\lambda(x), H\} = 0$ we find a secondary constraint $\tilde{\chi}_2 = \frac{(\pi-\phi')^2}{1+\lambda} = 0$. The Poisson bracket of χ_1 and $\tilde{\chi}_2$ vanish and hence they are a first-class constraint. However, if we replace $\tilde{\chi}_2$ with its classical equivalent constraint $\chi_2 = \pi - \phi'$ then the latter is a second-class constraint. If we add the additional constraint in the form of gauge fixing $\chi_3 = \lambda(x) - \lambda_0(x)$ with $\lambda_0(x)$ a given function, than all the constraints are second order with the constraint algebra,

$$c_{ij}(x,y) = \{\chi_i(x), \chi_j(y)\} \quad c_{22} = -2\delta'(x-y) \quad c_{13} = -\delta(x-y). \tag{6.71}$$

The next step is to define the Dirac bracket,

$$\{F(x), G(y)\}_D = \{F(x), G(y)\} - \int dz dw \{F(x), \chi_i(z)\} c_{ij}^{-1}(z,w) \{\chi_j(w), G(y)\}, \tag{6.72}$$

where $c_{ij}(x,z) c_{jk}^{-1}(z,y) = \delta_{ik}(x-y)$. The Dirac bracket rather than the Poisson bracket is then elevated to the commutator in the quantum theory $[\ ] = i\{\ \}$. Using this prescription one finds the desired result,

$$[\phi(x), \phi(y)] = \frac{1}{4i}\epsilon(x-y). \tag{6.73}$$

Implementing the constraint quantization in the path integral formulation, one has,

$$Z(J) = \int [d\phi][d\pi][d\lambda][d\pi_\lambda] \delta(\chi_1) \delta(\chi_2) \delta(\chi_3)$$

$$\times \sqrt{\mathrm{Det}[C_{ij}(x,y)]} e^{i \int d^2 x (\pi\dot\phi + \pi_\lambda \dot\lambda - \mathcal{H} - J\phi)}. \tag{6.74}$$

Using the δ functions and since $\mathrm{Det}[C]$ is field independent we find,

$$Z[J] = \int [d\phi] e^{i \int d^2 x \tilde{\mathcal{L}} - J\phi} \quad \tilde{\mathcal{L}} = \dot\phi\phi' - (\phi')^2. \tag{6.75}$$

Thus we see that this procedure yields the action (6.59) in the gauge $\lambda = -1$ which will be the topic of the following section.

6.4.2 Non manifestly Lorentz invariant classical action

Let us start with the following non-local action,

$$\mathcal{L} = \int dz dy \rho(z) \epsilon(z-y) \dot\rho(y) - \int dx \rho^2(x), \tag{6.76}$$

where $\rho(x)$ is a local bosonic field. The system can also be described in terms of the non-local bosonic field,

$$\phi(x) = \frac{1}{2} \int dy \epsilon(x-y) \rho(y), \quad \phi'(x) = \rho(x), \tag{6.77}$$

with a local Lagrangian density,

$$\mathcal{L} = \phi'\dot\phi - \phi'^2 = \partial_-\phi(\partial_+\phi - \partial_-\phi). \tag{6.78}$$

As we said the Lagrangian density is in fact (6.59) in the gauge $2\lambda = -1$. The classical equation of motion which corresponds to (6.76) is,

$$\partial_-\phi' = 0. \tag{6.79}$$

This equation has a general solution $\partial_-\phi = g(t)$, however by requiring $\partial_-\phi = 0$ at the spatial boundaries, we set $g(t) = 0$ and get the chiral solution $\partial_-\phi = 0$.

Even though the action (6.76) is not manifestly Lorentz invariant, it is easy to see that it is invariant under time translation $\delta\phi = \epsilon\dot\phi$, space translation $\delta\phi = \epsilon\phi'$, and the unconventional Lorentz transformation $\delta\phi = (t+x)\phi'$. For the above invariance transformations to exist we assume vanishing surface terms. The associated Noether charges are $H = \int \phi'^2$, $P = -H$ and $M = \int dx[(x+t)(\phi')^2]$. The system is, in fact, invariant under yet another unusual Lorentz transformation, $\delta(\partial_-\phi) = x^+\partial_+(\partial_-\phi) - x^-\partial_-(\partial_-\phi) - \partial_\phi$.

In addition, equation (6.76) is invariant under the global axial transformation $\phi \to \phi + \alpha$. The associated current $j^\mu_{(ax)}$ and the vector current which as was discussed in Section 6.1 is "topologically conserved" are given by

$$j_{(ax)_+} = \partial_+\phi - 2\partial_-\phi \quad j_{(ax)_-} = \partial_-\phi$$
$$j_{(v)_+} = \partial_+\phi \quad j_{(v)_-} = -\partial_-\phi. \tag{6.80}$$

As usual from the vector and the axial currents we can write down the left and right currents $J(lr) = \frac{1}{2}[j_{(v)} \pm j_{(ax)}]$, respectively. They have the following expressions,

$$J_{(l)_-} = 0, \quad J_{(l)_+} = \phi; \quad J_{(r)_-} = -\partial_-\phi, \quad J_{(r)_+} = \partial_-\phi. \tag{6.81}$$

Note however that only the left current is holomorphically conserved namely $\partial_- J_{(l)_+} = 0$, while the right current is not antiholomorphically conserved. This property is related to the invariance of the Lagrangian under $\delta\phi = \alpha(x^+)$ and not under $\delta\phi = \alpha(x^-)$. As will be explained below only the left $U(1)$ affine Lie algebra current exists in the quantum theory. Similarly, the Lagrangian (6.76) is invariant under the conformal transformations $\delta\phi = \epsilon(x^+)\phi'$ and $\delta\phi = \epsilon(t)\partial_-\phi$. The associated Sugawara type Noether currents are,

$$T_{(l)_{++}} = (\phi')^2 = J_{(l)_+}^2 \quad \partial_- T_{(l)_{++}} = 0$$
$$T_{(r)_{11}} = (\partial_-\phi)^2 = J_{(r)_1}^2 \quad \partial_1 T_{(r)_{++}} = 0. \tag{6.82}$$

Quantization of the model

By treating the system as a constrained system, we now invoke its canonical and path-integral quantization. We repeat here the Dirac bracket procedure discussed above. The constraint $\chi = \pi - \phi' = 0$, is a second-class constraint since $\{\chi(x), \chi(y)\} = -2\delta(x-y)$. The Hamiltonian density of the system takes the following form $\mathcal{H} = \mathcal{H}_c + v(\phi, \pi)\chi$, where $\mathcal{H}_c = \phi'^2 = \pi^2 = \pi\phi'$. Using this form for $\mathcal{H}$ it is easy to verify that the condition $\dot\chi = \{\chi, \mathcal{H}\} = 0$, does not lead to further constraints but fixes $v(\phi, \pi)$. The passage to the quantum theory is performed by passing from the Dirac brackets rather than the usual Poisson bracket to the commutator via,

$$-i[F(x), G(y)] = \{F(x), G(y)\} - \int d\xi_1 d\xi_2 \{F(x), G(\xi_1)\}$$

$$\times \left[-\frac{1}{4}\epsilon(\xi_1 - \xi_2)\{F(\xi_2), G(y)\}.\right. \qquad (6.83)$$

Following this definition, the operator algebra for π and ϕ takes the form,

$$[\phi(x), \phi(y)] = \frac{1}{4i}\epsilon(x-y),$$

$$[\pi(x), \phi(y)] = \frac{1}{2i}\delta(x-y),$$

$$[\pi(x), \pi(y)] = \frac{1}{2i}\delta(x-y). \qquad (6.84)$$

One also finds that, for an arbitrary operator $F(\phi, \pi)$, $[\chi(x), F(y)] = 0$ so that the constraint $\pi(x) - \phi'(x) = 0$, is now realized at the operator level. For example the Hamiltonian density can now be expressed in the three forms of $\mathcal{H}_c$ mentioned above. The system is solved by Fourier transforming,

$$\phi(x) = \int_{-\infty}^{0} \frac{dk}{2\sqrt{\pi k}}[a_k e^{-ikx} + a_k^\dagger e^{ikx}]$$

$$\pi(x) = \phi'(x), \qquad (6.85)$$

with the usual algebra,

$$[a(k), a^\dagger(k')] = \delta(k - k'). \qquad (6.86)$$

Note that only $k \leq 0$ appears in the decomposition of $\phi(x)$, which expresses the chiral nature of the field. The single-particle Hilbert space is then a continuum of states with energy $E = |k|$, $k \leq 0$. Hence the Hamiltonian formalism has correctly implemented the chirality constraint $\partial_-\phi = 0$. Furthermore, this property can also be deduced from the Hamiltonian equation of motion $\dot{\phi} = i[H, \phi] = \phi'$. Note that to get the chiral solution $\partial_-\phi = 0$ as a solution of the equation of the motion, we had to assume chiral boundary conditions. Here it looks at first that the chirality property was derived with no assumptions, but in passing from (6.83) to (6.84) we assumed that $(\pi - \phi')(x = \infty) = -(\pi - \phi')(x = -\infty)$, so together with choosing zero surface terms we in fact assumed chiral boundary conditions.

For the path integral quantization of the system we use the method developed for Hamiltonian systems with constraints. The generating functional is given by,

$$Z[J] = \int d\phi d\pi \delta(\chi) e^{\int d^2 x (\pi \dot{\phi} - \mathcal{H}_c - J\phi)}$$

$$= \int d\phi e^{\int d^2 x (\mathcal{L} - J\phi)}, \qquad (6.87)$$

where a normalization by $Z[O]^{-1}$ is implied. The Lagrangian density that emerges in (6.87) is clearly in the original form of equation (6.78). The functional integral (6.87) is not specified completely until we include $\partial_-\phi = 0$ on the boundary; thus we are in the same situation as in the canonical quantization.

Using the commutation relations (6.84), it is straightforward to verify that the Noether charges H, P, M, respectively generate the transformations $\delta_T\phi$, $\delta_S\phi$ and $\delta_M\phi$ given above. It can easily be shown that they satisfy the Poincare algebra,

$$[H, P] = 0 \quad [M, H] = iP \quad [M, P] = iH. \tag{6.88}$$

Abelian bosonization of a chiral fermion

Recall (Section 3.8) that the action of one left-handed complex chiral fermion,

$$S_{f_+} = \int d^2x\, \psi^\dagger \partial_- \psi, \tag{6.89}$$

is classically invariant under both an affine chiral transformation $\delta\psi = i\epsilon(x^+)\psi$ and the conformal transformation $\delta\psi = i\epsilon^+(x^+)\partial_+\psi$. The associated Noether currents were shown to have the following quantum form,

$$J_+ =: \psi^\dagger\psi \quad T_{++} = i : \psi^\dagger\partial_+\psi := \pi : J_+J_+ :, \tag{6.90}$$

obey the left $U(1)$ affine Lie and Virasoro algebras, respectively with the well-known central charges $k = 1$ and $c = 1$.

It is now straightforward to realize that the $J_{(l)} = \phi'$ and $T_{(l)} = (\phi')^2$ have the same $k = 1$ and $c = 1$ central charges. Using the operator algebra we can now evaluate the commutators of the chiral current and of the energy-momentum tensor. The results are as follows

$$[J_{(l)}(x), J_{(l)}(y)] = [\phi'(x), \phi'(y)] = \frac{i}{2}\delta'(x - y),$$

$$[T_{(l)}(x), T_{(l)}(y)] = [: (\phi'(x))^2, (\phi'(y))^2] = i(T_{(l)}(x) + T_{(l)}(y))\delta'(x - y)$$

$$- \frac{i}{24\pi}\delta'''(x - y). \tag{6.91}$$

From the general discussion in Sections 2.4 and 3.3 it follows now that these commutation relations correspond to central charges of $c = k = 1$. Hence our bosonic theory furnishes the same irreducible representation of the affine Lie and Virasoro algebras as one free left-handed chiral fermion. Since for $k = 1$ the affine Lie algebra has a unique irreducible unitary representation the two theories on flat two-dimensional space-time are therefore equivalent. Below it will be shown that the anomalies of the bosonized theory in coupling to gauge and gravitational background are the same as for the fermionic theory.

Combining left and right chiral bosons

The canonical quantization procedure of the last section can be repeated for a right chiral boson. Let us rename the operators for the $\partial_-\phi = 0$ case as ϕ_L and π_L and call the corresponding operators for the $\partial_+\phi = 0$ case as ψ_R and π_R. The Lagrangian density for the right moving field has the form $\mathcal{L}_R = \partial_+\phi(\partial_+\phi - \partial_-\phi)$ Then the combined Lagrangian,

$$\mathcal{L} = \mathcal{L}_L + \mathcal{L}_R \tag{6.92}$$

and the corresponding Hamiltonian density,

$$\mathcal{H} = (\pi_L)^2 + (\pi_R)^2 \tag{6.93}$$

describe a single free massless scalar defined by,

$$\phi = \phi_L + \phi_R \quad \pi = \pi_L + \pi_R \tag{6.94}$$

with a Hamiltonian density,

$$\mathcal{H} = \frac{1}{2}\pi^2 + \frac{1}{2}\phi'^2. \tag{6.95}$$

Using the commutation relations of the chiral bosons (6.84) we find as expected that,

$$[\phi(x), \phi(y)] = [\pi(x), \pi(y)] = 0 \quad [\phi(x), \pi(y)] = i\delta(x - y). \tag{6.96}$$

Unsurprisingly the Noether charges associated with the space-time translation and Lorentz transformation, $H = H_R + H_L$, $P = P_R + P_L$ and $M = M_L + M_R$ obey the usual Poincare algebra. The (lr) $U(1)$ affine Lie algebra currents of the combined system are given by the left current of the left system and the right current of the right system, respectively. The central charges are $k = 1$ for the algebras in both sectors. Similarly, because of the Sugawara construction, $T_{++} = T_{++L}$ and $T_{--} = T_{--R}$ with $c = 1$ for the left and the right Virasoro algebras.

Partition function

We would now like to compare the one loop partition function of a chiral fermion and that of our chiral boson. We therefore pass to a two-dimensional space-time domain with $1 \geq x \geq 0$. The mode expansion for ϕ previously given by equation (6.85) now takes the form,

$$\phi(x, t) = \phi_0 + p(x + t) + \sum_{n > 0} \frac{1}{\sqrt{2n}}[a_n^\dagger e^{2\pi i n x^+} + a_n e^{-2\pi i n x^+}]. \tag{6.97}$$

The one loop partition function which corresponds to this mode expansion is

$$Z(\tau) = \text{Tr}[e^{-\tau_2 H + i\tau_1 P}] = \text{Tr}[q^{\frac{H}{\pi}}] \tag{6.98}$$

where $q = e^{i\tau\pi}$ and we use the fact that $H = -P$. Let us first calculate the contribution to the trace of the oscillation modes,

$$\text{Tr}[q^{\frac{H}{\pi}}] = \text{Tr}[q^{(\sum_n na_n^\dagger a_n - \frac{1}{12})}] = \prod_{n=1}^{\infty} q^{-\frac{1}{12}}[1 - q^{2n}]^{-1} = \eta^{-1}(\tau), \qquad (6.99)$$

where the factor $-\frac{1}{12}$ is the output of the normal ordering and η is the Dedekin η function (see eqn. (2.48)). The Hamiltonian of the zero modes is $H = \frac{1}{4\pi}p^2$. If we now take the bosonic momenta to lie on a shifted lattice such that the eigenvalues of p are $2\pi(m + \alpha)$ and introduce the twist operator $g = e^{i\beta p}$, we get the zero mode partition function to be equal to the Riemann theta function $\Theta[(\alpha\beta)](\tau|0)$ and so that altogether the full partition function is given by,

$$Z(\tau) = \frac{\Theta[(\alpha\beta)](\tau|0)}{\eta(\tau)}. \qquad (6.100)$$

This corresponds to the partition function of a Weyl fermion with the boundary conditions $\psi(x + 1, t) = -e^{2\pi i\alpha}\psi(x, t); \psi(x + \text{Re}\tau, t + \text{Im}\tau) = -e^{-2\pi i\beta}\psi(x, t)$.

6.4.3 Coupling to abelian gauge fields

There are several ways to couple an abelian gauge field to a chiral boson corresponding to the various regularization schemes in the fermionic theory. We start by analyzing the bosonization in the vector conserving regularization scheme. The vector current is coupled to an abelian gauge field via,

$$\mathcal{L}_{(V)} = \mathcal{L}_0 + (J_{(V)-}A_+ + J_{(V)+}A_-), \qquad (6.101)$$

where $\mathcal{L}_0$ is the uncoupled Lagrangian density and V stands for the vector conserving scheme. The vector current is still obviously conserved, but the divergence of the axial current,

$$\partial_- J_{(\text{ax})_+} + \partial_+ J_{(\text{ax})_-} = \partial_- A_+ - \partial_+ A_- = \epsilon^{\mu\nu} F_{\mu\nu}, \qquad (6.102)$$

is now equal to the anomaly deduced in the fermionic theory from the one loop diagram.

Next we discuss the bosonization in the left-right scheme. For that purpose a term bilinear in the gauge fields has to be added to the $J_{(l)+}A_{(l)-}$ term. The Lagrangian then takes the form:

$$\mathcal{L}_{(\text{LR})} = \mathcal{L}_0 + J_{(l)+}A_{(l)-} + \frac{\sqrt{2}}{4}A_{(l)-}A_{(l)1}, \qquad (6.103)$$

where (LR) indicates the left-right scheme. The divergence of the left current which is derived from $\frac{\partial\mathcal{L}_{(\text{L,R})}}{\partial A_\pm}$ now has the form,

$$\partial_-(J_{(l)}^-)_{(\text{L,R})} + \partial_+(J_{(l)}^+)_{(\text{L,R})} = \frac{1}{4}(\partial_- A_+ - \partial_+ A_-) = \frac{1}{4}\epsilon^{\mu\nu}F_{\mu\nu}. \qquad (6.104)$$

The Lagrangian (6.103) is therefore really the bosonized fermionic action regularized in the left-right scheme. Obviously, a similar prescription for the (V) and (LR) schemes can be applied to the right chiral boson. It is straightforward to show that the vector as well as the left-right actions are invariant under "curved space-time" Lorentz transformations as discussed above.

6.4.4 Chiral WZW and coupling to non-abelian gauge fields

The non-abelian bosonization of N Majorana or N Dirac fermions using WZW theories of $SO(N)$ and $U(N)$, respectively was described in Section 6.3. For the non-abelian bosonization of left chiral fermions we propose to generalize the action (6.78) into an action that describes a map to the group manifold of the form,

$$S_+[u] = \frac{\sqrt{2}}{4\pi} \int \mathrm{d}^2 x Tr(\partial_- u \partial_1 u^{-1}) + S_{WZ}. \tag{6.105}$$

In a similar manner to the abelian case this is a WZW action coupled to fictitious chiral gravity in the gauge $h_{++} = -1$. In fact we can consider a generalization of this action to the so-called k-level chiral WZW namely $S_{+k}[U] = kS_+[u]$. The equation of motion that follows from the variation of (6.105) with respect to the variation of u can be expressed as,

$$\partial_-(u^{-1}\partial_1 u) = 0 \quad or \quad \partial_1(u\partial_- u^{-1}) = 0, \tag{6.106}$$

where each form can be obtained from the other.

The global, chiral transformations $u \to Au$ and $u \to uB^{-1}$ where $A, B \in G$ leave the action (6.105) invariant. However, out of the invariance under the two affine Lie algebra transformation of the original WZW action $u \to uB^{-1}(x^+)$ and $u \to A(x^-)u$, only the first survives. As for the abelian case the invariance under the second transformation is lost. The Noether currents associated with the left-right transformations of (6.105) are,

$$J_{(l)_-} = 0 \quad J_{(\ell)_+} = \frac{i\sqrt{2}k}{2\pi} u^{-1}\partial_1 u$$

$$J_{(r)_-} = \frac{ik}{2\pi} u\partial_- u^{-1} \quad J_{(r)_+} = -\frac{ik}{2\pi} u\partial_- u^{-1}. \tag{6.107}$$

The conservation of the left and right currents follow here simply from the equations of motion. However, unlike the ordinary WZW action only the left current is holomorphically conserved a $\partial_- J_{(l)} = 0$; whereas $\partial_+ J_{(r)} \neq 0$. Obviously this is a manifestation of the invariance of the action only under the left affine Lie algebra transformation $\delta u = -iu\epsilon(x^+)$, discussed above. The left current transforms as follows $\delta J_{(l)} = [i\epsilon(x^+), J_{(l)}] + \frac{\sqrt{2}k}{2\pi}\epsilon$, leading to the $O(N)(U(N))$ affine Lie algebra with central charge equal to k.

The action $S_{+k}[u]$ is invariant under the left affine transformation $\delta u = \epsilon(x^+)u$. The corresponding energy momentum tensor has again a Sugawara form and Virasoro central charge which are,

$$T_{(l)} = \frac{2\pi}{(c_2 + k)} \sum_a : J_{(l)}^a J_{(l)}^a : \qquad c = \frac{k \dim G}{c_2 + k}, \tag{6.108}$$

where as usual $J_{(l)} = J_{(l)}^a T^a$ and T^a are hermitian matrices representing the algebra of the group, c_2 is the second Casimir operator in the adjoint representation and $\dim G$, is the dimension of the group.

The coupling of non-abelian gauge fields to a chiral WZW action, is a straightforward generalization of the coupling of the abelian gauge fields. Again there are several ways to couple gauge fields corresponding to the various regularization schemes in the fermionic theory. The bosonized action (for $k = 1$) related to the vector conserving regularization scheme is given by,

$$S_V[u, A_-, A_+] = S_+[u] + \int d^2x \operatorname{Tr}[J_{(v)}^- A_- + J_{(v)}^+ A_+]$$

$$- \frac{\sqrt{2}}{2\pi} \int d^2x \operatorname{Tr}[u^{-1} A_1 u A_- - A_- A_1], \tag{6.109}$$

where $J_{(v)}$ and $J_{(ax)}$ are constructed from the left and right currents of (6.107) in the usual way. Using the equation of motion one finds the conservation of the vector current and the anomalous divergence of the axial current,

$$D_\mu(J_{(v)}^\mu)_V = 0 \qquad D_\mu(J_{(ax)}^\mu)_V = \frac{1}{\pi}\epsilon^{\mu\nu} F_{\mu\nu} \tag{6.110}$$

The coupling of a left non-abelian gauge field that corresponds to the fermionic description in the left-right regularization scheme is given by,

$$S_{(LR)}[u, A_-, A_1] = S_+[u] + \frac{\sqrt{2}}{2\pi} \int d^2x \operatorname{Tr}[(u^{-1}\partial_1 u + \frac{1}{2\pi}A_1)A_-]. \tag{6.111}$$

The associated current divergence is,

$$D_\mu(J_{(l)}^\mu)_{LR} = D_- J_l^-{}_{(LR)} + D_+ J_-^+ l_{(LR)} = \frac{1}{\pi}\epsilon^{\mu\nu} F_{\mu\nu}. \tag{6.112}$$

This expression for the anomalous divergence of the left current is identical to the result of the loop calculation in the fermionic version regularized in the left-right scheme.

6.5 Bosonization of systems of operators of high conformal dimension

Can bosonization, the equivalence map between systems of dimension half fermions and dimension zero bosons, be extended also to systems made out of higher-dimensional operators? In particular can one map theories of odd and even fields of higher-dimensional operators to theories built from dimension zero scalar fields. In this section it will be shown that indeed such a map exists for two families of theories, one with anti-commuting fields and the other with commuting ones with arbitrary integer and half-integer dimensions.

The bosonization of the b, c and β, γ systems was introduced in [94]

6.5.1   The bosonization of the "b,c" free CFT

We first briefly describe the system. Consider a system built from a pair of anti-commuting fields b and c which is described by the action,

$$S_{b,c} = \frac{1}{2\pi} \int d^2 z b \bar{\partial} c. \tag{6.113}$$

It is easy to see that this action is invariant under conformal transformation $z \to a^{-1} z$ $b \to a^\lambda b$ and $c \to a^{1-\lambda} c$, namely the b and c fields have conformal dimensions,[8]

$$h_b = \lambda \quad h_c = (1 - \lambda). \tag{6.114}$$

The classical equations of motion,

$$\bar{\partial} b(z, \bar{z}) = 0 \quad \bar{\partial} c(z, \bar{z}) = 0, \tag{6.115}$$

imply that both fields are holomorphic, namely $b(z), c(z)$. Similar to the derivation of the operator equations of motion using the path integral (1.55) for the scalar field we find for the (b, c) system that,

$$\bar{\partial} b(z) c(0) = 2\pi \delta^2 (z, \bar{z}). \tag{6.116}$$

This is compatible with the OPE,

$$b(z)c(w) =: b(z)c(w) : + \frac{1}{z - w} \quad c(z)b(w) =: c(z)b(w) : + \frac{1}{z - w}. \tag{6.117}$$

The OPEs $b(z)b(w)$ and $c(z)c(w)$ do not have singular parts when w it brought to z. To compute the energy-momentum tensor we use the Noether procedure. We vary the b and c fields as follows,

$$\delta b = \bar{\epsilon} \partial b + \lambda (\partial \bar{\epsilon}) b,$$
$$\delta c = \bar{\epsilon} \partial c + (1 - \lambda)(\partial \bar{\epsilon}) b. \tag{6.118}$$

For holomorphic $\bar{\epsilon}(z)$ this is a symmetry transformation. Taking now $\bar{\epsilon}(z, \bar{z})$ we read the energy-momentum tensor from the variation of the action as follows,

$$\delta S_{b,c} = -\frac{1}{2\pi} \int d^2 z \bar{\partial} \bar{\epsilon} T = \frac{1}{2\pi} \int d^2 z \bar{\partial} \bar{\epsilon} [(\partial b)c - \lambda \partial (bc)]. \tag{6.119}$$

---

[8] We denote by λ the dimension of b as is common in the literature. Obviously it has nothing to do with λ used in the previous section on chiral bosons.

Thus the energy-momentum tensor is,

$$T = (\partial b)c - \lambda\partial(bc). \tag{6.120}$$

It is straightforward to verify that the OPE of T with b and with c indeed yields the variation (6.118). From the OPE $T(z)T(w)$ we read the Virasoro anomaly associated with the (b, c) system,

$$c = -3(2\lambda - 1)^2 + 1. \tag{6.121}$$

The action (6.113) is also invariant under the fermion number transformation,

$$b \to e^{i\alpha(z)}b \quad c \to e^{-i\alpha(z)}c. \tag{6.122}$$

Using again the Noether procedure we find that the corresponding conserved fermion number current is,

$$j =: bc : \quad \bar\partial j = 0. \tag{6.123}$$

From the basic OPE of $b(z)$ and $c(w)$ one finds that the OPE of the energy-momentum tensor and the fermion number current reads,

$$T(z)j(w) = \frac{1 - 2\lambda}{(z - w)^3} + \frac{j(w)}{(z - w)^2} + \frac{\partial j(w)}{(z - w)}. \tag{6.124}$$

The $(b(z), c(z))$ conformal field theory is fully holomorphic. Needless to say that one can similarly write down an anti-holomorphic system $\bar b(\bar z), \bar c(\bar z)$. In fact in Section 2.12 we have already discussed a special case of the b, c family. For $\lambda = \frac{1}{2}$ we get the Weyl spinor ψ of dimension $\frac{1}{2}$ such that the Virasoro anomaly of the system is $c = 1$. Another "famous" case is that of the b and c ghosts associated with the covariant fixing of two-dimensional diffeomorphism. In this case $\lambda = 2$ the dimensions of b and c are 2 and -1, respectively and the corresponding Virasoro anomaly is -26.

Now we raise again the question, can one describe the b, c system in terms of a scalar CFT? Since the $\psi, \psi^\dagger$ is a special case of the b, c system and since we have already developed the bosonization rules for Dirac and Weyl spinors (see Section 6.1.1) we start with a similar ansatz for the bosonic version of b and c, namely,

$$b(z) \leftrightarrow: e^{i\phi(z)} : \quad c(z) \leftrightarrow: e^{-i\phi(z)} :, \tag{6.125}$$

Comparing (6.18) and (6.117) it is evident that indeed this map reproduces the algebra of the (b, c) system. It is also easy to realize that the fermion number current has the following bosonic equivalent,

$$: b(z)c(z) :\leftrightarrow i\partial\phi(z). \tag{6.126}$$

What is left to determine is whether the energy-momentum tensors of the two theories and correspondingly the dimensions of the fields match. Obviously the free scalar action (6.1) which is the bosonic dual of the action of the Dirac operator cannot describe the (b, c) systems which are a family of CFTs. We also need to identify a family characterized by the parameter λ of scalar field theories. A simple way to achieve this is to realize that the energy-momentum of the general (b, c) system can be written in terms of the spin $1/2$ fermion as follows,

$$T^{b,c} = T^\psi - \left(\lambda - \frac{1}{2}\right)\partial(bc). \tag{6.127}$$

Thus following (6.126) the scalar energy momentum has to have the form,

$$T^\phi_\lambda = T^\phi - \left(\lambda - \frac{1}{2}\right)\partial^2\phi. \tag{6.128}$$

This is the energy-momentum of a scalar field with a background charge, or the linear dilaton theory with $q = -i(\lambda - \frac{1}{2})$. The central charge of these theories was shown to be,

$$c^\phi = 1 + 12q^2 = 1 - 3(2\lambda - 1)^2 = c^{b,c}. \tag{6.129}$$

Moreover using the fact that the dimension of an operator $: e^{ik\phi(z)} :$ was shown to be $\frac{k^2}{2} + ikq$ it is easy to check that,

$$h_{e^{i\phi}} = \lambda = h_b \quad h_{e^{-i\phi}} = 1 - \lambda = h_c, \tag{6.130}$$

which verifies that the operators mapped by the bosonization indeed have the same conformal dimensions.

The anti-commutative nature of the b, c system is obeyed also by their bosonic duals, just as for the case of spin $1/2$ fields, namely,

$$: e^{i\phi(z)} :: e^{i\phi(w)} := e^{-[\phi(z),\phi(w)]} : e^{i\phi(w)} :: e^{i\phi(z)} := - : e^{i\phi(w)} :: e^{i\phi(z)} :, \tag{6.131}$$

at equal times, namely $|z| = |w|$ since for that case $[\phi(z), \phi(w)] = \pm i\pi$.

6.5.2 The bosonization of the β, γ system

The b, c system is built from anti-commuting fields which as we have just seen are describable in terms of a scalar field. In a complete analogy one would suspect that a similar bosonization also holds for a system built from commuting fields with the same structure of action. As we shall see shortly this is indeed the case. The so-called β, γ system defined by the action (6.132),

$$S_{\beta,\gamma} = \frac{1}{2\pi} \int d^2z \beta \bar\partial \gamma. \tag{6.132}$$

The fact that now the building blocks are commuting introduces of course a sign change with respect to the b, c system when one interchanges the fields, namely,

$$\beta(z)\gamma(w) = -\frac{1}{z - w} \quad \gamma(z)\beta(w) = \frac{1}{z - w}. \tag{6.133}$$

The energy-momentum tensor is the same as (6.120) when replacing b and c with β and γ, respectively. For the Virasoro anomaly one has just to reverse the sign of that of the b, c system. A distinguished member of this family of conformal field theories is the case of $\lambda = 3/2$ which describes the ghost system associated with the gauge fixing of the superdiffeomorphism. In this case the $c = 11$. As is well known this combined with $c = -26$ requires a contribution to the Virasoro anomaly of 15 of the non-ghost fields which requires ten dimensions for superstring theories.

The bosonization of this commuting system is slightly more involved than that of the b, c system. It turns out that now one has to invoke two scalar fields ϕ and χ with the OPEs,

$$\phi(z)\phi(w) = -\ln(z - w) + \ldots \quad \chi(z)\chi(w) = \ln(z - w) + \ldots, \tag{6.134}$$

where $\ldots$ stands for non-singular terms. The corresponding scalar theories have a background charge of $(\lambda' - 1/2)$ for ϕ and $i/2$ for χ so that the energy-momentum tensor of the full bosonic reads,

$$T_{\phi,\chi} = -\frac{1}{2}\partial\phi\partial\phi + \frac{1}{2}\partial\chi\partial\chi + \frac{1}{2}(1 - 2\lambda')\partial^2\phi + \frac{1}{2}\partial^2\chi, \tag{6.135}$$

which yields the desired Virasoro anomaly $c = 1 + 3(2\lambda' - 1)^2 = -c_{b,c}$. The bosonic operators that correspond to β and γ are,

$$\beta(z) \leftrightarrow e^{-\phi+\chi}\partial\chi \quad \gamma \leftrightarrow e^{\phi-\chi}, \tag{6.136}$$

which have the conformal dimensions $h_\beta = \lambda'$ and $h_\gamma = 1 - \lambda'$.

6.5.3 The Wakimoto bosonization

One application of the β, γ system enables us to transform the WZW model into a theory expressed in terms of free fields. Consider a combined system that includes a $(\lambda' = 1, 0)$ β, γ system and an additional scalar field with a background charge $-\frac{i}{\sqrt{2(k+2)}}$. This system is described by the following Lagrangian,

$$\mathcal{L}_{\text{Wak}} = \beta\bar{\partial}\gamma + \bar{\beta}\partial\bar{\gamma} + \partial\varphi\bar{\partial}\varphi. \tag{6.137}$$

Alternatively, as was discussed in Section 6.4 one can add to this Lagrangian density also a term of the form $-\frac{i}{\sqrt{2(k+2)}}R^2\phi$. The corresponding energy-momentum tensor $T(z)$ is given by,

$$T(z) = -\beta\partial\gamma - \frac{1}{2}\partial\varphi\partial\varphi - \frac{i}{\sqrt{2(k+2)}}\partial^2\varphi, \tag{6.138}$$

and the non-trivial OPEs of these field are given by,

$$\gamma(z)\beta(w) = \frac{1}{z - w} + O(z - w) \quad \varphi(z)\varphi(w) = -\ln(z - w) + O(z - w). \tag{6.139}$$

We define now the following holomorphic currents in terms of the β, γ and φ fields:

$$J^+ = \beta(z),$$
$$J^0 = i\sqrt{2(k+2)}\partial\varphi(z) + 2 : \gamma\beta(z) :,$$
$$J^- = -i\sqrt{2(k+2)} : \partial\varphi\gamma : (z) - k\partial\varphi(z) - : \beta\gamma\gamma : (z). \tag{6.140}$$

Using the OPEs it is straightforward to determine the OPEs of the currents,

$$J^+(z)J^-(w) = \frac{k}{(z-w)^2} + \frac{J^0(z)}{(z-w)},$$

$$J^0(z)J^+(w) = \frac{2J^+}{(z-w)},$$

$$J^0(z)J^0(w) = \frac{2k}{(z-w)^2},$$

$$J^0(z)J^+(w) = \frac{-2J^-}{(z-w)}, \tag{6.141}$$

which are the OPEs of the $SU(2)$ affine current algebra, with level k. Furthermore the Sugawara energy-momentum tensor,

$$T(z) = \frac{1}{2(k+2)}\left[\frac{1}{2}J^0 J^0 + J^+ J^- + J^- J^+\right], \tag{6.142}$$

is identical to the energy-momentum tensor given in (6.138), and the associated Virasoro anomaly is,

$$c = 2 + 1 - 24\left(\frac{1}{4(k+2)}\right) = \frac{3k}{k+2}. \tag{6.143}$$

Since β has dimension one and γ dimension zero, their mode expansion takes the form,

$$\beta = \sum_n \beta_n z^{-n-1}, \quad \gamma = \sum_n \gamma_n z^{-n}. \tag{6.144}$$

Substituting this into the expressions of the currents one finds,

$$J_n^+ = \beta_n,$$
$$J_n^0 = i\sqrt{2(k+2)}n\varphi_n + 2\sum_m : \beta_m \gamma_{m-n} :,$$

$$J_n^- = -i\sqrt{2(k+2)}\sum_m : n\varphi_m \gamma_{n-m} : -kn\varphi_n - \sum_{lm} : \beta_l \gamma_m \gamma_{n-m-l} : .$$

$$\tag{6.145}$$

7
The large N limit of two-dimensional models

7.1 Introduction

The number of approximation techniques in quantum field theory is very limited. Perturbation expansion in small interaction coupling, like $\alpha_{\text{em}} = \frac{1}{137}$ of QED, is obviously the most important one. Other methods include dimensional expansion, high temperature expansion and large radius expansion. Quite surprisingly, one of the most useful approximation techniques is expansion in the number of degrees of freedom. A priori one would tend to think that the larger the number of degrees of freedom, the more complex the system. However, it turns out that theories with infinitely many degrees of freedom are much easier to solve than those with a finite number of degrees of freedom. Once the system with $N \to \infty$ is known, a systematic expansion in $\frac{1}{N}$ provides an approximation procedure for computing quantities that describe systems of finite N.

Large N methods have been applied in a very wide range of physical systems. Starting from non-critical phenomena in spin systems like the Heisenberg ferromagnet (discussed in Section 5.14), then $SU(N)$ QCD theories in various dimensions, and later matrix models associated with either string models or two-dimensional models.

The large N approximation in field theory or correspondingly the planar expansion of Feynman diagrams was introduced by 't Hooft in his seminal paper [122].

In this book we will focus on four arenas where large N approximations are being used:

(i) Two-dimensional quantum field theory models, which include the Gross–Neveu model and the CP^N models, will be addressed in this chapter.

(ii) Quantum chromodynamics with large N $SU(N)$ gauge theory in two dimensions. In Chapter 10 of Part 2 the solution of two-dimensional QCD, following 't Hooft [124], will be described,[1] together with a certain generalization of it.

(iii) The approach to four-dimensional QCD based on the $\frac{1}{N}$ expansion.

(iv) Baryons in large N QCD.

---

[1] Known as the 't Hooft model.

The last two topics will be described in the third part of the book in Chapters 19 and 20.

In Nature the number of colors is three, and thus one may wonder whether it makes sense to in expand a not-so-small parameter 1/3. Even though there is no general proof that this expansion is indeed reliable, a vast literature on the subject brings out a large amount of evidence that indeed this is the case. It turns out that for certain quantities the $\frac{1}{N}$ term vanishes and the correction starts as $\frac{1}{N^2}$, and hence puts the approximation on a more solid base.

As an example of the accuracy of the large N limit consider the Stirling formula for N, where the leading term is $\sqrt{2\pi N}(N/e)^N$ for large N. But the correction is actually $\frac{1}{12N}$ as compared to 1, making it only an 8 percent correction even for $N = 1$.

In the next sections we describe the $O(N)$ model, the Gross–Neveu model and the CP^N models.

There are several review articles on large N expansions, for instance [46], [160], [165]. In this chapter we make use of [160].

7.2 The Gross–Neveu model

The Gross–Neveu (GN) model, proposed in [117], describes a set of N Dirac fermions interacting via a four-fermi interaction. It turns out that one can solve the model, and prove that it is asymptotically free and admits a dynamical symmetry breaking, using $1/N$ expansion. The Lagrangian of the system can be written in the form,

$$\mathcal{L}_{\text{GN}} = i\bar{\psi}^a \ \slashed{\partial}\psi^a + \frac{\lambda_0^2}{2}(\bar{\psi}^a\psi^a)^2, \tag{7.1}$$

where $a = 1...N$ and λ_0 is a bare coupling of dimension zero. The corresponding action is invariant under a continuous $SU(N)$ global transformation and a discrete chiral transformation,

$$\psi^a \to g_b^a\psi^b \quad g\in SU(N), \quad \psi^a \to \gamma_5\psi^a \quad \bar{\psi}^a \to \gamma_5\bar{\psi}^a. \tag{7.2}$$

The discrete chiral symmetry forbids a mass term. In fact this is the most general action invariant under these symmetry transformations, with terms of dimension two or less, and hence it is a renormalizable action.

Let us now check whether in this formulation of the Lagrangian, where λ_0 is fixed, one can make sense of a large N limit. Consider the scattering process of two fermions with flavor index a that turn into a pair of fermions with a different flavor index b. The leading Feynman diagrams that contribute to this process are given in Fig. 7.1. The first diagram which is the basic interaction vertex is of order λ_0, the second one is of order λ_0^2, the third is of order λ_0^2N due to the N different flavors of the fermions that can run in the loop, and the last two diagrams are of order $\lambda_0^3N^2$. It is thus clear that the perturbation expansion

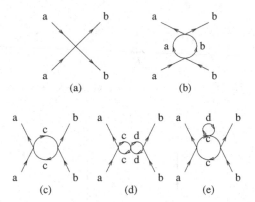

Fig. 7.1. Leading order diagrams for the $a + \bar{a} \rightarrow b + \bar{b}$ scattering.

expressed in terms of the fixed coupling λ_0 does not have a sensible large N expansion. However, one can easily cure this problem by defining the coupling $\lambda \equiv \lambda_0 N$, which is taken to be fixed when $N \rightarrow \infty$. The Lagrangian now reads,

$$\mathcal{L}_{GN} = i\bar{\psi}^a \not{\partial}\psi^a + \frac{\lambda}{2N}(\bar{\psi}^a \psi^a)^2. \tag{7.3}$$

It is now straightforward to see that the $\frac{1}{N}$ in front of the interaction term enables a well-defined large N limit. Consider again the diagrams in Fig. 7.1. The leading contribution is of order $\frac{\lambda}{N}$ and the loop corrections are now of order $\frac{\lambda^2}{N^2}$, $\frac{\lambda^2}{N}$ and $\frac{\lambda^3}{N}$, respectively. Hence it is obvious that the form of any scattering amplitude in perturbation theory is $\frac{1}{N}A(\lambda, 1/N)$, which becomes $\frac{1}{N}A(\lambda, 0)$ at the large N limit.

To further analyze the system it is convenient to introduce the auxiliary field σ, in terms of which the Lagrangian (7.3) can be written as,

$$\mathcal{L}_{GN} = i\bar{\psi}^a \not{\partial}\psi^a + \sigma\bar{\psi}^a \psi^a - \frac{N}{2\lambda}\sigma^2. \tag{7.4}$$

Integrating over σ or alternatively solving the classical equation of motion for σ and substituting it into the action yields the Lagrangian (7.1). The introduction of the auxiliary field, which will soon acquire a physical interpretation, is a standard step in the large N procedure which enables a simplified counting of powers of $\frac{1}{N}$.

The Feynman diagrams of the theory for this alternative formulation are then given in Fig. 7.2, with the full line representing the fermion propagator, and the dotted line that of σ.

Note that the only non-trivial interaction is the $\sigma\bar{\psi}^a \psi^a$ term and that each σ propagator contributes $\frac{i\lambda}{N}$. The diagrams Fig. 7.1 that contribute to the scattering process are converted to those in Fig. 7.3.

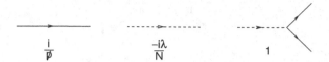

Fig. 7.2. Feynman rules of the Gross–Neveu model.

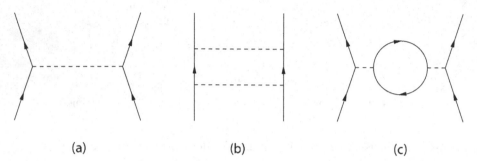

(a)   (b)   (c)

Fig. 7.3. Leading diagrams that contribute to the two-to-two scattering.

We would now like to integrate over the fermions and derive an effective Lagrangian, $\mathcal{L}_{\text{effective}}(\sigma)$. Since the Lagrangian is quadratic in the fermion fields, $\mathcal{L}_{\text{effective}}(\sigma)$ is given by the sum of terms (Fig. 7.4).

Note that all diagrams are with an even number of σ only, since σ is odd under the transformation (7.2). The first term is the tree level contribution $-\frac{N}{2\lambda}\sigma^2$ and the rest are the one loop contributions. Both are of order N, the latter due to the N fermions that can run in the loop. The N dependence of $\mathcal{L}_{\text{effective}}$ therefore has the form,

$$\mathcal{L}_{\text{effective}}(\sigma, \lambda, N) = N\hat{\mathcal{L}}_{\text{effective}}(\sigma, \lambda). \tag{7.5}$$

This makes the counting of powers of $\frac{1}{N}$ very easy. Consider a graph with E external σ lines, I internal σ lines, V vertices and L independent loops. The parameters (E, I, V, L) are not independent. For each internal line there is a momentum integration and hence a loop. However each vertex introduces a delta function in momenta that cancels one momentum apart from an overall delta function associated with the momentum conservation.

Thus one has,

$$L = I - V + 1. \tag{7.6}$$

Recall that each σ external or internal line carries a $\frac{1}{N}$ factor, while each vertex contributes a factor of N.

Thus the net power N of each graph is,

$$N^{-I+V-E} = N^{-E-L+1}. \tag{7.7}$$

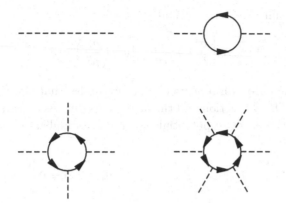

Fig. 7.4. Diagrams of $\mathcal{L}_{\text{effective}}(\sigma)$.

It is obvious from this expression that adding loops and external σ lines suppresses the corresponding contribution due to additional powers of $\frac{1}{N}$. Since the minimal number of σ external lines is two the leading behavior is of order $\frac{1}{N}$.

For the purpose of investigating the possibility of spontaneous breaking of the discrete symmetry of (7.2), it is enough to compute the effective potential $V(\sigma)$ rather than the effective action, namely, the limit where all the external lines carry zero momentum. The effective potential is given by the sum of the diagrams in Fig. 7.4:

$$-iV = -i\frac{N\sigma^2}{2\lambda} - N\sum_{n=1}^{\infty}\frac{1}{2n}\text{Tr}\int\frac{d^2p}{(2\pi)^2}\left[\frac{-\not{p}\sigma}{p^2+i\epsilon}\right]^{2n}, \qquad (7.8)$$

where $\frac{1}{2n}$ is the symmetry factor of the graph, N comes from summing over all possible flavors, -1 from the fermion loop and the expression in the bracket is the product of the propagator and $(i\frac{\not{p}}{p^2+i\epsilon})$ and the vertex $(i\sigma)$. Using the identity,

$$\sum_{n=1}^{\infty}\frac{x^{2n}}{2n} = -\frac{1}{2}\log(1-x^2), \qquad (7.9)$$

and analytically continuing to the Euclidean space gives,

$$V = N\left[\frac{\sigma^2}{2\lambda} - \int\frac{d^2p_E}{(2\pi)^2}\log\left(1+\frac{\sigma^2}{p_E^2}\right)\right]. \qquad (7.10)$$

The momentum integral is logarithmically divergent so by introducing a cutoff on the Euclidean momentum $p_E^2 \le \Lambda^2$ we find,

$$V = N\left[\frac{\sigma^2}{2\lambda} - \frac{1}{4\pi}\sigma^2\left[\log\left(\frac{\sigma^2}{\Lambda^2}\right)+1\right]\right]. \qquad (7.11)$$

The effective potential can be rewritten in terms of the coupling λ_r, renormalized at a scale μ, defined as,

$$\frac{1}{\lambda_r} \equiv \frac{1}{N}\frac{d^2V}{d\sigma^2}|_{\sigma=\mu} = \frac{1}{\lambda} + \frac{1}{2\pi}\log\left(\frac{\mu^2}{\Lambda^2}\right) + \frac{1}{\pi}, \qquad (7.12)$$

and substitute it into the $V(\sigma)$ to find,

$$V = N \left[\frac{\sigma^2}{2\lambda_r} + \frac{1}{4\pi}\sigma^2 \left[\log\left(\frac{\sigma^2}{\mu^2}\right) - 3 \right] \right] \tag{7.13}$$

The fact that the cutoff disappears obviously implies that the theory is indeed renormalizable. The β function and the anomalous dimension can be determined by substituting the effective potential into the renormalization group equation (see Section 17.6),

$$[\mu\partial_\mu + \beta(\lambda_r)\partial_{\lambda_r} - \gamma_\sigma(\lambda_r)\partial_\sigma] V(\sigma) = 0. \tag{7.14}$$

We thus find the exact (to all orders of λ_r) expression of $\beta(\lambda_r)$ and $\gamma_\sigma(\lambda_r)$, which take the form,

$$\beta(\lambda_r) = -\frac{\lambda_r^3}{2\pi} \quad \gamma_\sigma(\lambda_r) = 0. \tag{7.15}$$

Thus we have deduced that the Gross–Neveu model is asymptotically free, namely that the effective coupling goes to zero at high momenta. It turns out that the minus sign ensures this.

Let us now examine whether the chiral symmetry of this theory is spontaneously broken. For that we determine the extremum points of the potential. The vanishing points of the derivative,

$$\frac{dV}{d\sigma} = N \left[\frac{\sigma}{\lambda_r} + \frac{\sigma}{2\pi}\left(\log\frac{\sigma^2}{\mu^2} - 2\right) \right], \tag{7.16}$$

are at

$$\sigma = 0 \quad \text{and} \quad \sigma = \pm\sigma_0 = \pm\mu e^{1-\frac{\pi}{\lambda_r}}, \tag{7.17}$$

where

$$V(0) = 0 \quad \text{and} \quad V(\sigma_0) = -N\frac{\sigma_0^2}{4\pi} < 0. \tag{7.18}$$

Now since the potential vanishes at $\sigma = 0$ and it is negative at $\sigma = \pm\sigma_0$, its global minima are at $\pm\sigma_0$. Therefore the discrete chiral symmetry is broken and the massless fermions acquire mass which to the leading order is σ_0.

A further interesting property of the model is the dimensional transmutation. The bare theory depends on one continuous dimensionless parameter and the effective theory depends on one continuous parameter with dimensions, σ_0. Whereas one may anticipate that observables will depend on the dimensionless parameter in a complicated way, one finds a simple dependence on the parameter with dimensions which follows a dimensional analysis.

7.3 The CP^{N-1} model

Another model that can be solved using the large N expansion is the CP^{N-1} model. The model is an example of a non-linear sigma model where the fields live in a complex projective $N-1$ space,

$$CP^{N-1} = \frac{SU(N)}{SU(N-1) \times U(1)}. \tag{7.19}$$

The Lagrangian of the model can be written as,

$$\mathcal{L}_{CPN} = \partial_\mu Z^\dagger \partial^\mu Z - \frac{\lambda}{N} J_\mu J^\mu, \tag{7.20}$$

where $Z^\dagger \equiv (z_1, \ldots, z_N)$, namely, an N-dimensional "unit" vector of complex fields that obey the constraint,

$$Z^\dagger Z = \frac{N}{\lambda}, \tag{7.21}$$

and J_μ is given by,

$$J_\mu = -\frac{i}{2} \left[Z^\dagger \partial_\mu Z - (\partial_\mu Z^\dagger) Z \right]. \tag{7.22}$$

The Lagrangian describes a theory of massless particles with short range interaction which originates from both the explicit JJ interaction as well as from the constraint. The number of degrees of freedom of the CP^{N-1} model is $2N-2$. This is the dimension of the CP^{N-1} coset space,

$$\dim \left[CP^{N-1} \right] = \dim \left[\frac{SU(N)}{SU(N-1) \times U(1)} \right]$$

$$= (N^2 - 1) - ((N-1)^2 - 1 + 1) = 2N - 2. \tag{7.23}$$

Differently, we count N complex numbers Z_i, namely $2N$ real degrees of freedom, minus one degree of freedom due to the constraint (7.21), minus one degree of freedom due to the $U(1)$ local symmetry,

$$Z \rightarrow e^{i\alpha} Z. \tag{7.24}$$

It is easy to verify that (7.20) is indeed invariant under this transformation upon the use of the constraint. In fact one can also write the Lagrangian in the form,

$$\mathcal{L}_{CPN} = \partial_\mu \Phi^\dagger \partial^\mu \Phi, \tag{7.25}$$

where Φ is a traceless hermitian matrix built from Z and $Z^\dagger$ according to,

$$\Phi = \sqrt{\frac{\lambda}{N}} \left[ZZ^\dagger - \frac{1}{\lambda} \right], \tag{7.26}$$

It is clear that the local transformation of the above does not change Φ and hence the Lagrangian is invariant under this transformation.

The first step in the large N program is to eliminate the quartic interaction term, by introducing an auxiliary field in a similar manner to what was done in the Gross–Neveu model. However, since the interaction has the form of a vector times a vector, the auxiliary field should also be a vector. This shifts the Lagrangian according to,

$$\mathcal{L}_{CPN} \to \mathcal{L}_{CPN} + \frac{\lambda}{N}\left(J_\mu + \frac{N}{\lambda}A_\mu\right)^2$$

$$= \partial_\mu Z^\dagger \partial^\mu Z + 2J_\mu A^\mu + \frac{N}{\lambda}A_\mu A^\mu$$

$$= \left[(\partial_\mu - iA_\mu)Z^\dagger\right]\left[(\partial^\mu + iA_\mu)Z\right], \tag{7.27}$$

where in the third line we have used the constraint. It is clear from its last form that the Lagrangian is invariant under the $U(1)$ local transformation,

$$Z \to e^{i\alpha}Z \quad A_\mu \to A_\mu - \partial_\mu \alpha. \tag{7.28}$$

We now incorporate the fact that the Z are constraint variables by introducing another Lagrange multiplier into the Lagrangian,

$$\mathcal{L}_{CPN} = \left[(\partial_\mu - iA_\mu)Z^\dagger\right]\left[(\partial^\mu + iA_\mu)Z\right] + \sigma\left[Z^\dagger Z - \frac{N}{\lambda}\right]. \tag{7.29}$$

Obviously the path integral over σ, or equivalently using its equation of motion, implies that $Z^\dagger Z = \frac{N}{\lambda}$.

The action is now quadratic in Z, so we integrate out the Z fields similarly to what was done in the Gross–Neveu model. The Feynman diagrams that constitute the leading contributions to the effective action, which is now a functional of σ and A_μ, are drawn in Figure 7.5.

These diagrams, which include pure σ, pure A_μ and mixed diagrams, are all proportional to N. The computation of $V(\sigma)$ is similar to that in the GN model, leading to,

$$V(\sigma) = -N\left[\frac{\sigma}{\lambda} + \frac{\sigma}{4\pi}\left(\log\frac{\sigma}{\Lambda^2} - 1\right)\right], \tag{7.30}$$

where Λ is the cutoff. Again similar to the GN model the cutoff can be eliminated by performing a renormalization at a scale μ,

$$\frac{1}{\lambda_r} = -\frac{1}{N}\frac{dV}{d\sigma}\big|_{\mu^2}\frac{1}{\lambda} + \frac{1}{4\pi}\left(\log\frac{\mu^2}{\Lambda^2}\right), \tag{7.31}$$

so that the potential takes the form,

$$V(\sigma) = -N\left[\frac{\sigma}{\lambda_r} + \frac{\sigma}{4\pi}\left(\log\frac{\sigma}{\mu^2} - 1\right)\right]. \tag{7.32}$$

It is also evident that the model has a negative β function, or differently stated, for fixed λ_r and μ, when $\Lambda \to \infty$, λ vanishes, namely, the model is asymptotically free.

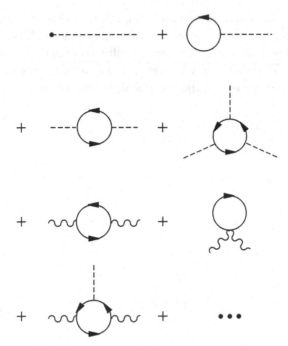

Fig. 7.5. Leading order contributions to the effective action.

Again the model admits a dimensional transmutation. The original dimensionless coupling is traded with a parameter σ_0 with dimensions, at which the potential has a minimum,

$$\frac{dV}{d\sigma} = -N\left[\frac{1}{\lambda_r} + \frac{1}{4\pi}\left(\log\frac{\sigma}{\mu^2}\right)\right] = 0, \quad \sigma_0 = \mu^2 e^{-\frac{4\pi}{\lambda_r}}. \tag{7.33}$$

The following remarks about the model are relevant:

(i) The model admits a dynamical generation of abelian gauge fields. From the diagrams on the third line of Fig. 7.5 we see that the contribution to $S_{\text{effective}}$ quadratic in A_μ is,

$$-\frac{iN}{4\pi}\left[g_{\mu\nu}p^2 - p_\mu p_\nu\right]\int_0^1 dx \frac{(1-2x)^2}{\sigma_0^2 - p^2 x(1-x) - i\epsilon}.$$

Now for long range interaction, namely, for small momenta we ignore the p^2 term to obtain,

$$-\frac{iN}{12\pi\sigma_0}\left[g_{\mu\nu}p^2 - p_\mu p_\nu\right],$$

which corresponds to the following term in the effective action,

$$S_{\text{effective}} = -\frac{N}{48\pi\sigma_0}\int d^2x F_{\mu\nu}F^{\mu\nu}, \tag{7.34}$$

namely an action of an abelian gauge field.

(ii) The Z fields can be interpreted as bosonic "quarks" in the fundamental representation of the group, though transforming in a non-linear way. These Z quarks are confined due to the dynamically generated abelian gauge interaction. As will be shown in Chapter 8, in two dimensions the abelian force between a quark anti-quark pair is linear in separation distance.

PART II

Two-dimensional non-perturbative gauge dynamics

In the first part of the book we have developed several non-perturbative tools for analyzing two-dimensional field theories. These include methods associated with conformal invariance, with affine Lie algebras and in particular the WZW model, techniques of integrable massive theories including solitons, S-matrix and the Yang–Baxter equation, sets of infinitely many conserved charges, the thermal Bethe ansatz, methods of bosonization and the large N limit approximation.

In this second part of the book the main idea is to implement those methods in the context of two-dimensional gauge dynamics aiming at extraction of the mesonic and baryonic spectra of two-dimensional QCD, decoding the confining behavior, versus a screening one, and analyzing models with other quark representations and models of generalized QCD.

The most important tool that will be used for this purpose will be bosonization. It will enable us to easily solve the Schwinger model, derive the baryonic spectrum of QCD_2 using a strong coupling limit, determine the screening nature of massless QCD models and compute the string tension for the massive ones. Using the large N approximation we will extract the mesonic spectrum of QCD.

In two respects this part is an intermediate stage on the way to four-dimensional gauge dynamics. Firstly, as was just mentioned, we will gain experience from applying non-perturbative methods on the simpler two-dimensional models, which will serve us when using them in the context of "real" physical systems. Secondly, two-dimensional gauge systems will serve as a toy model laboratory of four-dimensional ones. As will be shown in the third part of the book, certain aspects of two-dimensional physics will survive the transition to four dimensions. Obviously the challenge will be to identify those phenomena and devise some additional tools to handle the other cases.

8
Gauge theories in two dimensions – basics

As an introduction to the cast of characters of two-dimensional gauge theories, we briefly summarize here the basics of pure Maxwell theory, QED, pure YM theory and QCD. This includes the corresponding actions, symmetries, equations of motion and their solutions.

The basics of gauge theories in two dimensions is "standard material" which appears in many books and review articles, for instance [66], [178], [1] and [2]. For treatment in non-covariant gauges see [28].

8.1 Pure Maxwell theory

The simplest theory of gauge fields in two dimensions is obviously the abelian Maxwell theory defined by the classical action,

$$S = \int d^2 x \left[-\frac{1}{4} F_{\mu\nu} F^{\mu\nu} \right], \tag{8.1}$$

where the field strength

$$F_{\mu\nu} = \partial_\mu A_\nu - \partial_\nu A_\mu \tag{8.2}$$

has, in two dimensions, only one non-trivial component $E_1 \equiv F_{10} = -F_{01} = \partial_1 A_0 - \partial_0 A_1$. The action is invariant under the full global two-dimensional conformal symmetry $SO(2,2)$, discussed in Section 2.1, which includes in particular the $ISO(1,1)$, where I stands for inhomogeneous, namely adding the momenta, thus going over to the Poincare group from the Lorentz group. The action is by construction also invariant under the gauge transformation,

$$A_\mu(x,t) \to A_\mu(x,t) + \partial_\mu \Lambda(x,t). \tag{8.3}$$

The canonical dimension of A_μ is clearly zero. The corresponding equation of motion reads,

$$\partial^\mu F_{\mu\nu} = 0 \quad \partial_0 E_1 = \partial_1 E_1 = 0 \quad \to \quad E_1 = \text{constant}. \tag{8.4}$$

Thus we conclude that the two-dimensional Maxwell theory is an empty theory on an $R^{1,1}$ manifold. On such a space-time requiring finite energy implies that $E_1 = 0$. This is of course not surprising. In d-dimensional space-time the number of degrees of freedom of an abelian gauge field is $d - 2$ and hence there are no degrees of freedom in two dimensions.

8.2 QED_2 – Schwinger's model

Next we couple the two-dimensional abelian gauge fields to a Dirac fermion. The Lagrangian density of this model is given by,[1]

$$\mathcal{L} = -\frac{1}{4}F_{\mu\nu}F^{\mu\nu} + \bar{\Psi}(i\partial\!\!\!/ - e A\!\!\!/ - m)\Psi$$

$$= \frac{1}{2}(\bar{\partial}A - \partial\bar{A})^2 + \psi^\dagger\bar{\partial}\psi + \tilde{\psi}^\dagger\partial\tilde{\psi} + e\psi^\dagger\psi\bar{A} + e\tilde{\psi}^\dagger\tilde{\psi}A - m(\psi^\dagger\tilde{\psi} + \tilde{\psi}^\dagger\psi),$$

$$(8.5)$$

where in the second line the action is expressed in terms of the light-cone derivatives and components of the gauge fields, and the Dirac fermion is decomposed into its left and right chiral fermions, as discussed in Section 3.8. It is evident that with the gauge field having a vanishing dimension, the gauge coupling e has a dimension of mass. Thus the action is not invariant any more under the two-dimensional global conformal symmetry, but rather only under the $ISO(1,1)$ Poincare group. For the massless case, the action is classically invariant under the global transformations,

$$\Psi \to e^{i\alpha}\Psi \quad \Psi \to e^{i\tilde{\alpha}\gamma_5}\Psi$$
$$\psi \to e^{i(\alpha+\tilde{\alpha})}\psi \quad \tilde{\psi} \to e^{i(\alpha-\tilde{\alpha})}\tilde{\psi}.$$

$$(8.6)$$

In fact the left and right chiral transformations, for the massless case, can be lifted also into holomorphic and anti-holomorphic transformations, as was discussed in Section 3.7.1. The corresponding vector and axial currents

$$J^\mu = \bar{\Psi}\gamma^\mu\Psi \quad J_5^\mu = \bar{\Psi}\gamma^\mu\gamma_5\Psi$$
$$J = \psi^\dagger\psi \quad \bar{J} = \tilde{\psi}^\dagger\tilde{\psi}.$$

$$(8.7)$$

Again by construction the action is also invariant under the gauge transformation,

$$\Psi \to e^{-i\Lambda(x,t)}\Psi \quad A_\mu(x,t) \to A_\mu(x,t) + \frac{1}{e}\partial_\mu\Lambda(x,t).$$

$$(8.8)$$

Quantum mechanically the axial current is not conserved even for the massless case due to an anomaly,

$$\partial_\mu J_5^\mu = \frac{e}{2\pi}\epsilon_{\mu\nu}F^{\mu\nu}.$$

$$(8.9)$$

We will derive this result using the bosonized version, see Section 9.1. Unlike Maxwell's theory, this theory has non-trivial degrees of freedom. However, once again the gauge field is not dynamical. This phenomenon can be easily demonstrated in the axial gauge $A_1 = 0$, where the other component A_0 can be solved

---

[1] The Schwinger model was introduced in [190] and further analyzed in [68] and [64].

as a function of the electric current. The resulting electric field is,

$$E_1 = -F_{01} = -e\partial_1^{-1}J_0 - \frac{e\theta}{2\pi}, \quad J_0 =: \Psi^\dagger\Psi :, \tag{8.10}$$

θ is a new parameter in the theory, the vacuum angle.[2] In Part 3 of the book we will describe its four-dimensional analog which is the vacuum angle due to QCD_4 instanton tunneling.

The massless Schwinger model can easily be solved using the anomaly equation combined with the equation of motion of the system. This will be done in Section 9.1 using the bosonized version where we also address the massive case. In Chapter 15 we determine the spectrum of the massless case using a BRST quantization approach. In Chapter 14 we analyze the nature of the system and determine when it confines and when it admits a screening behavior.

8.3 Yang–Mills theory

It is straightforward to generalize the action of the Maxwell theory (8.1) to the non-abelian case.[3] The gauge fields are now in the adjoint representation of a non-abelian gauge group $\mathcal{G}$. We will mainly be interested in the groups $SO(N_c)$, $U(N_c)$ and $SU(N_c)$. Thus A_μ is an $N_c \times N_t$ either orthogonal, or hermitian or traceless hermitian matrix of the form $A_\mu = t^B A_\mu^B$ where t^B are the generators of the group, $B = 1, ..., \dim\mathcal{G}$ and $\dim\mathcal{G}$ is the dimension of the corresponding algebra $[\frac{1}{2}N_c(N_c - 1), N_c^2$ and $N_c^2 - 1$, respectively]. The field strength is now,

$$F_{\mu\nu} = \partial_\mu A_\nu - \partial_\nu A_\mu + i[A_\mu, A_\nu]$$
$$F_{\bar{z}z} = \bar{\partial}A - \partial\bar{A} + i[A, \bar{A}], \tag{8.11}$$

where again we write it out in light-cone coordinates. The action of two-dimensional Yang–Mills theory reads,

$$S_{YM2} = \int d^2x \left[-\frac{1}{2e_c^2}\text{Tr}(F_{\mu\nu}F^{\mu\nu}) \right] = \int d^2x \left[-\frac{1}{4e_c^2}F^a{}_{\mu\nu}F^{a\,\mu\nu} \right], \tag{8.12}$$

and the corresponding equations of motions are,

$$D_\mu F^{\mu\nu} = \partial_\mu F^{\mu\nu} + i[A_\mu, F^{\mu\nu}] = 0. \tag{8.13}$$

Note that we have rescaled the fields A by a factor of the gauge coupling, as compared with the abelian case. Note, however, that this does not affect the dynamical dimensions, namely the space-time behavior of Green's functions.

In this formulation A_μ has dimension one and so is the dimension of the color gauge coupling e_c. Again since the coupling constant has a dimension of mass the classical theory is not invariant under the full global conformal symmetry,

---

[2] The θ angle was introduced by Lowenstein and Swieca [152] and also by Coleman [64].
[3] The Yang–Mills non-abelian gauge theory was introduced in the seminal paper [229].

but only with respect to the $ISO(1,1)$ Poincare transformations. The action is invariant under a non-abelian gauge transformation, which in infinitesimal form is,

$$A_\mu \to A_\mu + D_\mu \Lambda = A_\mu + \partial_\mu \Lambda + i[A_\mu, \Lambda], \tag{8.14}$$

where $\Lambda = t^A \Lambda^A$. A priori this is not a free theory, but rather an interacting one. However, in a similar manner to Maxwell's theory, this model too on an $R^{1,1}$ manifold has no dynamical degrees of freedom just as the abelian model. This can easily be seen by fixing a gauge, for instance $A_0 = 0$. In this gauge the equations of motion read,

$$\partial_0 F^{01} = 0 \quad \partial_1 F^{10} + i[A_1, F^{01}] = 0. \tag{8.15}$$

From the first we get that $\partial_0^2 A_1 = 0$, and thus $A_1 = f_1(x_1) + x_0 f_2(x_1)$. Using the residual gauge invariance, of gauge transformations that depend only on x_1, we can go to $f_1 = 0$, and then the second equation implies that f_2 is a constant C, which yields $F_{01} = C$, and then again the requirement of finite energy results in $C = 0$. This will also be shown in a complicated way using a BRST approach in Chapter 15.

When the underlying manifold has a non-trivial topology like that of a torus then the theory is not totaly empty but instead has topological degrees of freedom. This will be described in Section 16.

Finally, the non-abelian case is different from the abelian in higher dimensions, as the former is not free there. While the abelian case represents free photons, the non-abelian case represents interacting gluons, which turn to interacting glue balls in the physical space.

8.4 Quantum chromodynamics

The theory of non-abelian gauge fields coupled to Dirac quarks in the fundamental representation of the gauge group, QCD_2, is described by the action,

$$S_{QCD_2} = \int d^2x \left\{ -\frac{1}{2e_c^2} \text{Tr}(F_{\mu\nu} F^{\mu\nu}) - \bar{\Psi}^{ai}[(i\slashed{\partial} - \slashed{A} + m)\Psi_i]_a \right\}. \tag{8.16}$$

The action is invariant under two-dimensional Poincare transformation and the non-abelian generalization of the gauge transformation of (8.8), which in infinitesimal form is,

$$\delta\Psi_a = -i[\Lambda(x,t)]_a^b \Psi_b \quad \delta A_\mu(x,t) = \partial_\mu \Lambda(x,t) + i[A_\mu, \Lambda], \tag{8.17}$$

with the non-abelian $\Lambda = \Lambda^A T_A$. Ψ is in the fundamental representation of the gauge group which we take to be $SU(N_c)$ where $a = 1, \ldots, N_c$ denote the color indices. As was discussed in Section 6.3.4 flavor degrees of freedom have been included by assigning a flavor index to the Dirac fermion Ψ_i, $i = 1, \ldots, N_f$. For this case the theory is obviously invariant classically under a global

$U_L(N_f) \times U_R(N_f)$ symmetry. Here there is no anomaly, as in 2d the anomaly occurs via the abelian gauge field only.

In a similar manner to the transition from the empty Maxwell theory to the dynamically viable Schwinger model, so is the transition from the two-dimensional pure Yang–Mills theory to QCD_2. The difference, however, is that in the non-abelian case, even for the massless case there is no simple way to solve the theory. Instead we will need to implement various different techniques developed in the first part of the book. In the next section we will describe both QED and QCD in two dimensions using the bosonization language. This will enable us to solve for the baryonic spectrum in the strong coupling limit. In Chapter 14 the string tension of several two-dimensional dynamical systems will be computed. An analysis of the spectrum of these theories will be derived using the BRST quantization approach in Chapter 15. In Chapter 10 we present the seminal 't Hooft solution of two-dimensional QCD in the large N limit. A current algebra generalization of the latter approach will enable us to solve the mesonic spectra of certain models. Finally in Chapter 12 we will implement a discrete light-cone quantization approach to solve QCD in two dimensions with quarks in the fundamental as well as the adjoint representation.

9

Bosonized gauge theories

Bosonization, the equivalence map between two-dimensional fermionic and bosonic operators, was developed in Chapter 6. In fact several such maps have been described. The simplest one has been the *abelian bosonization* that maps the free theory of a Dirac fermion into that of a single real scalar field. The map includes in particular an explicit bosonic expression for the left and right chiral fermions (6.19), the vector and axial abelian currents (6.3) and for a mass term (6.22). Using these transformations it is straightforward to write the bosonized Lagrangian or Hamiltonian that corresponds to two-dimensional QED and QCD. By its nature the abelian bosonization is more adequate to the abelian theory of QED. The bosonized version of QED will be discussed in the next section. We then apply this bosonization to QCD_2. Though it is possible to write QCD_2 in an abelian bosonization formulation, it will turn out not to be very useful. Instead, we will use the non-abelian bosonization discussed in Section 6.3. For that purpose we will need to gauge the WZW action. Once this is done the bosonized version of massless flavored QCD_2 follows easily. The massive case requires more care, as was explained in Section 6.3.3. Using the results of that section the full bosonized theory that corresponds to massive flavored QCD_2 will be written down.

References on bosonization were given in Chapter 6.

9.1 QED_2 – The massive Schwinger model

Recall that the fermionic Lagrangian of this model is given by,

$$\mathcal{L} = -\frac{1}{4}F_{\mu\nu}F^{\mu\nu} + \bar{\Psi}(i\not{\partial} - e\not{A} - m)\Psi. \tag{9.1}$$

The Hamiltonian density of the system in the $A_1 = 0$ gauge takes the form,

$$\mathcal{H} = \bar{\Psi}(i\gamma_1\partial_1 + m)\Psi + \frac{1}{2}(F_{01})^2.$$

In bosonic variables, using (8, 10), the Hamiltonian becomes[1]

$$\mathcal{H} =: \left[\frac{1}{2}\pi^2 + \frac{1}{2}(\partial_1\phi)^2 - \frac{cm^2}{\pi}\cos(2\sqrt{\pi}\phi) + \frac{e^2}{2\pi}\left(\frac{1}{2}\frac{\theta}{\sqrt{\pi}} - \phi\right)^2\right] :_m ,$$

---

[1] The treatment of the bosonized Schwinger model was done in [68] and [64].

where c is the constant of bosonization and the normal ordering is with respect to the mass m, as was explained in Section 6.1.1.

After a shift in the definition of ϕ,

$$\phi \to \phi + \frac{1}{2}\frac{\theta}{\sqrt{\pi}},$$

and normal ordering with respect to $\mu = e/\sqrt{\pi}$, one finds,

$$\mathcal{H} =: \left(\frac{1}{2}\pi^2 + \frac{1}{2}(\partial_1\phi)^2 + \frac{1}{2}\mu^2\phi^2 - \frac{cm\mu}{\pi}\cos(\theta + 2\sqrt{\pi}\phi)\right): . \tag{9.2}$$

From this expression the periodicity in θ is manifest. The angle θ is the conjugate to the winding number, appearing in two dimensions for the abelian case, since $\Pi_1[U(1)] = \mathcal{Z}$ (looking at a circle of large radius in the two-dimensional plane). Physics is invariant under $\theta \to \theta + 2\pi$. From (8.10) it is clear that $\frac{e\theta}{2\pi}$ corresponds to a background electric field. The periodicity is due to the ability to produce electron-positron pairs in the vacuum when $|\frac{e\theta}{2\pi}| > \frac{1}{2}e$, and these pairs create their own electric field which reduces the original one.

When we set $m = 0$ we discover that the massless Schwinger model is in fact a theory of one free bosonic field with a mass equal to μ.

In the strong coupling limit, the bosonized form of the Hamiltonian is very useful. The theory contains a meson of a mass that is approximately μ, and the number of bound states depends on the value of θ. It can be shown that there are no bound states for $|\theta| > \pi/2$. For $0 < |\theta| \leq \pi/2$ there is a stable two-body bound state, while for $\theta = 0$ there is also a three-body bound state.

Note that even though the Hamiltonian density (9.2) resembles that of a sine-Gordon model, it does not admit soliton solutions due to the mass term $\frac{1}{2}\mu^2\phi^2$. We will come back later to analyze this bosonized Hamiltonian, in the context of the question whether the system admits screening or confinement in Chapter 14.

Finally, let us show how the anomaly arises in the bosonized version. The equation of motion for the electromagnetic field is,

$$\partial^\mu F_{\mu\nu} = eJ_\nu. \tag{9.3}$$

The vector fermion current, in bosonic version, is (6.9),

$$J_\nu = \frac{1}{\sqrt{\pi}}\epsilon_{\nu\alpha}\partial^\alpha\phi. \tag{9.4}$$

Taking the time component, we get,

$$\partial^1\left(F_{10} - \frac{e}{\sqrt{\pi}}\phi\right) = 0. \tag{9.5}$$

From here, with the vanishing conditions at space infinity,

$$F_{10} = \frac{e}{\sqrt{\pi}}\phi. \tag{9.6}$$

Now, the axial current, in bosonic version, is (6.12),

$$J_5^\mu = \frac{1}{\sqrt{\pi}} \partial^\mu \phi. \tag{9.7}$$

For the case of a massless fermion, the scalar field is a free field of mass $\frac{e}{\sqrt{\pi}}$, and so,

$$\partial_\mu J_5^\mu = -\frac{1}{\sqrt{\pi}} \frac{e^2}{\pi} \phi = \frac{e}{\pi} F_{01}, \tag{9.8}$$

which is the anomaly equation.

Note that in the bosonic version, the anomaly is a result of the equations of motion, while in the fermionic one it is the result of one loop.

9.2 Abelian bosonization of flavored QCD_2

Let us now apply the prescription of flavored Dirac fermions for the analysis of QCD_2. It is convenient to start with the Hamiltonian of the theory in its fermionic formulation which we derive from (8.16),

$$H = (e_c)^2 \sum_{a,b=1}^{N_C} \left(E_b^a\right)^2 + \sum_{a,b=1}^{N_C} \sum_{i=1}^{N_F} \bar{\Psi}^{ai} \gamma_1 \left(i\delta_a^b \partial_1 - A_a^b\right) \Psi_{bi} + m \sum_{a=1}^{N_C} \sum_{i=1}^{N_F} \bar{\Psi}^{ai} \Psi_{ai}, \tag{9.9}$$

in the gauge,

$$A_0 = 0; \quad A_b^a = 0 \ for \ a = b; \quad E_b^a = 0 \quad for \ a \neq b. \tag{9.10}$$

The Gauss law of the system is given by,

$$\partial_1 E_b^a = i[A, E]_b^a + \frac{1}{2} \sum_{i=1}^{N_F} \Psi^{\dagger ai} \Psi_{bi} - \frac{\delta_b^a}{2N_C} \sum_{i=1}^{N_F} \sum_{d=1}^{N_C} \Psi^{\dagger di} \Psi_{di} \tag{9.11}$$

Bosonizing now the various parts of the Hamiltonian one then gets,[2]

$$H = H_\Psi^0 + H_E - H^I$$

$$H_\Psi^0 = \Sigma_{ai} \left[\frac{1}{2}[\pi_{ai}^2 + (\partial_1 \phi_{ai})^2] + \frac{cm\mu}{\pi} : (1 - \cos(2\sqrt{\pi}\phi_{ai})) : \right]$$

$$H_E = \frac{e_c^2}{8\pi N_c} \sum_{ab} \left[\sum_i (\phi_{ai} - \phi_{bi}) \right]^2$$

$$H^I = \frac{2c^2\mu^2}{\pi^{\frac{3}{2}}} \Sigma_{a\neq b} \Sigma_{ij} K_{ij,ab} N_\mu \left[\cos\sqrt{\pi} \int_{-\infty}^x (\pi_{ai} - \pi_{aj} + \pi_{bj} - \pi_{bi})(\xi)d\xi \right]$$

$$\left[\sin(\sqrt{\pi}(\phi_{ai} + \phi_{aj} - \phi_{bj} - \phi_{bi})(\xi)) \right] \left[\sum_{ab} (\phi_{ak} - \phi_{bk}) \right]^{-1}, \tag{9.12}$$

---

[2] Abelian bosonization of two-dimensional QCD was discussed in [24] and [201] and was further elaborated in [62].

H^0_Ψ is the free "fermionic" part, after bosonization, thus in terms of bosonic variables; H_E is the first term of the Hamiltonian (9.9) rewritten in terms of the boson variables corresponding to the fermions, by eliminating the electric fields through the Gauss law. Thus although originally coming from the kinetic part of the gauge potentials, it actually involves the interactions. This is a result of the fact that there are no transverse vectors in $1+1$ dimensions. K^{ab}_{ij} is a properly generalized ordering operator.[3]

In the case of one flavor, $i = j = 1$, H^I does not involve the π variables.

The interaction involves non-local terms which relate to color non-singlets. For static and $e_c \to \infty$ approximations one finds that for $N_F = 2$ the interaction is field independent. For $N_F \geq 3$, on the other hand, the limit is singular. This singularity should not be there in the predictions of physical quantities, but it renders any further treatment very complicated.

It is thus clear that a different method of bosonization is required for the treatment of flavored QCD_2. In the following it will be shown that the "non-abelian bosonization", based on the WZW model discussed in Section 6.3, is an adequate tool for this purpose.

Before proceeding to non-abelian bosonization and in particular to gauge the color symmetry group of the colored-flavored WZW model, we describe briefly another approach, in which the flavor sector appears in the form of a WZW model, but for the color degrees of freedom the gauged abelian bosonization is invoked. As we have seen above one can use the Gauss law to express the gauge fields in terms of the appropriate fermionic bilinear, which translate into bosonic group elements as,

$$2\partial_1 e_a = \sum_i \Psi^{\dagger ia} \Psi_{ia} = \frac{i}{\pi} \partial_1 \mathrm{Tr}_F (\log g_a), \qquad (9.13)$$

where $g_a \in U(N_F)$ is one out of N_C such matrices, and $e_a = 2\sqrt{\pi} E^a_a$, the diagonal element. One can also express A^b_a for $a \neq b$ in terms of fermion densities. Inserting these into the QCD_2 Hamiltonian one gets,

$$\mathcal{H} = \mathcal{H}^0 + \mathcal{H}^I$$

$$\mathcal{H}^I = -\sum_{a,b} \frac{(e_c)^2}{32\pi^2 N_C} \left[\mathrm{Tr} \log \left(g_a g_b^{-1} \right) \right]^2 - \sum_{a,b} \pi \mu^2 \frac{\mathrm{Tr}(g_a g_b^{-1})}{\mathrm{Tr} \log(g_a g_b^{-1})}$$

$$+ \sum_a mc\mu \sqrt{N_F} \, \mathrm{Tr}(g_a), \qquad (9.14)$$

$\mathcal{H}^0$ includes the fermion kinetic term. For $N_F = 2$ the potential is free from singularities, for $N_F \geq 3$ it is not. In the case of $N_F = 2$ the low lying baryonic spectrum can be extracted. Here we will not follow this approach further and instead will move on to the fully non-abelian bosonization.

---

[3] See Cohen *et al.* [62].

9.3 Non-abelian bosonization of QCD_2

Whereas abelian bosonization has been very useful to address various abelian systems, we have seen in the last section that the implementation of this approach to QCD_2 is quite limited. Instead the natural approach is to make use non-abelian bosonization, namely, the WZW action.[4] Recall from Section 6.3 that the bosonized action of massless free colored flavored fermions can be expressed either using an $SU(N_C) \times SU(N_F) \times U(1)$ scheme where it reads,

$$S = N_C S[g] + N_F S[h] + \frac{1}{2} \int d^2x \partial_\mu \phi \partial^\mu \phi, \tag{9.15}$$

or a $U(N_F \times N_C)$ where the action takes the form,

$$S[u] = N_C S[g] + N_F S[h] + \frac{1}{2} \int d^2x (\partial_\mu \phi \partial^\mu \phi + S[l]). \tag{9.16}$$

Note that l is still an $SU(N_C N_F)$ matrix while g and h are expressed now as $SU(N_F)$ and $SU(N_C)$ matrices, respectively, but the matrix l involves only products of color and flavor matrices (not any of them separately). For massive Dirac fermions we can use only the latter frame in which the mass term action reads,

$$S_m[u] = m'^2 N_{\tilde{m}} \int d^2x Tr(u + u^\dagger). \tag{9.17}$$

To determine the bosonized action of two-dimensional QCD_2 one needs to couple the colored degrees of freedom to the gauge fields. Thus, we first have to gauge the WZW model.

9.3.1 Gauging the WZW action

Since there are two possible bosonization schemes (for the massless case) we need to invoke a gauging procedure for both of them. We start first by gauging an $SU(N_C)$ WZW model which is what is needed in the product scheme, we later adopt it also to the $U(N_F \times N_C)$. Gauging the colored WZW is achieved by gauging the vector subgroup $SU_V(N_C)$ of $SU_L(N_C) \times SU_R(N_C)$. There are various methods to gauge the model. Here we present two of them. One is a trial and error method, and the other is by gauging via covariantizing the current. Those methods are applicable also in the $U(N_F N_C)$ bosonization scheme.

The gauging of the WZW model and the full non-abelian bosonization of QCD in two dimensions was analyzed in [75] and [99]. Bosonization of QCD in two dimensions was reviewed in [101].

Trial and error Noether method

The WZW action on the $SU(N_C)$ group manifold is, as stated above, invariant under the global vector transformation $h \to UhU^{-1}$, where $U \subset SU(N_C)$. Now

---

[4] The hybrid of abelian and non-abelian bosonizations was implemented in [107].

we want to vary h with respect to the associated local infinitesimal transformation $U = 1 + i\epsilon(x) = 1 + iT^A\epsilon^A(x)$,

$$\delta_\epsilon h = i[\epsilon, h], \quad \delta_\epsilon h^{-1} = i[\epsilon, h^{-1}]. \tag{9.18}$$

The variation of the action $S^{(0)}[h] \equiv S[h]$ under such a transformation is,

$$\delta_\epsilon S^{(0)}[h] = -\int d^2 x \mathrm{Tr}(\partial_\mu \epsilon J^\mu), \tag{9.19}$$

where the Noether vector current is given by,

$$J_\mu = \frac{i}{4\pi}\{[h^\dagger \partial_\mu h + h\partial_\mu h^\dagger] - \varepsilon_{\mu\nu}[h^\dagger \partial^\nu h - h\partial^\nu h^\dagger]\}. \tag{9.20}$$

We introduce now the first correction term $S^{(1)}$ given by,

$$S^{(1)} = \int d^2 x \mathrm{Tr}(A_\mu J^\mu) \quad \delta_\epsilon S^{(1)}[h] = -\int d^2 x \mathrm{Tr}[\partial_\mu \epsilon(J^\mu + J'^\mu)]. \tag{9.21}$$

The variation of $S^{(1)}$ is derived using the infinitesimal variation of the gauge field $\delta A_\mu = -D_\mu \epsilon = -(\partial_\mu \epsilon + i[A_\mu, \epsilon])$. J'^μ is found to be,

$$J'_\mu = \frac{-1}{4\pi}\{[h^\dagger A_\mu h + h A_\mu h^\dagger - 2A_\mu] - \varepsilon_{\mu\nu}[h^\dagger A^\nu h - h A^\nu h^\dagger]\}. \tag{9.22}$$

The second iteration will be given by adding $S^{(2)}$, where now J'^μ is replacing J^μ,

$$S^{(2)} = \int d^2 x \mathrm{Tr}(A_\mu J'^\mu), \quad \delta_\epsilon S^{(2)}[h] = -2\int d^2 x \mathrm{Tr}(\partial_\mu \epsilon J'^\mu). \tag{9.23}$$

It is therefore obvious that,

$$\delta_\epsilon\left[S^{(0)} + S^{(1)} - \frac{1}{2}S^{(2)}\right] = 0. \tag{9.24}$$

Hence the action we are looking for is $S[h, A_\mu] \equiv [S^{(0)} + S^{(1)} - \frac{1}{2}S^{(2)}]$, given by,

$$S[h, A_\mu] = \frac{1}{8\pi}\int d^2 x \mathrm{Tr}(D_\mu h D^\mu h^\dagger)$$

$$+ \frac{1}{12\pi}\int_B d^3 y \varepsilon^{ijk}\mathrm{Tr}(h^\dagger \partial_i h)(h^\dagger \partial_j h)(h^\dagger \partial_k h)$$

$$- \frac{1}{4\pi}\int d^2 x \varepsilon_{\mu\nu}\mathrm{Tr}[iA^\mu(h^\dagger \partial^\nu h - h\partial^\nu h^\dagger + ih^\dagger A^\nu h)], \tag{9.25}$$

which can also be written in light-cone coordinates,

$$S[h, A_+, A_-] = S[h] + \frac{i}{2\pi}\int d^2 x \mathrm{Tr}(A_+ h\partial_- h^\dagger + A_- h^\dagger \partial_+ h)$$

$$- \frac{1}{2\pi}\int d^2 x \mathrm{Tr}(A_+ h A_- h^\dagger - A_- A_+). \tag{9.26}$$

Gauging via covariantization of the Noether current

In four space-time dimensions the current, in terms of bosonic matrices, involves up to third power gauge potentials. In D space-time dimensions the bare current will contain (D-1) derivatives, and is gauged by replacing the ordinary derivatives with covariant derivatives and by adding terms which contain products of $F_{\mu\nu}$ with powers of h and $h^\dagger$ and also covariant derivatives $D_\mu h$ and $D_\mu h^\dagger$. In two dimensions, however, there is no room for such terms in the gauge covariant current, as these involve $\epsilon_{\mu_1...\mu_D}$ in D dimensions, with one free index and the others contracted with $F_{\mu\nu}$s and D_μs, and in two dimensions they cannot be constructed. Therefore the covariantized current is given by,

$$J_\mu(h, A_\mu) = \frac{i}{4\pi}\{[h^\dagger D_\mu h + h D_\mu h^\dagger] - \epsilon_{\mu\nu}[h^\dagger D^\nu h - h D^\nu h^\dagger]\}. \tag{9.27}$$

Knowing the current we deduce the action via, $J_\mu = \frac{\delta S}{\delta A_\mu}$, getting (9.26) directly.

Finally, we combine the gauged WZW action of the color group manifold, the WZW of the flavor group manifold and the action term for the gauge fields, to get the bosonic form of the action of massless QCD_2. The well known fermionic form of the action is (a mass term will be added later),

$$S_F[\Psi, A_\mu] = \int d^2x \left\{ -\frac{1}{2e_c^2}\text{Tr}(F_{\mu\nu}F^{\mu\nu}) - \bar{\Psi}^{ai}[(i\partial\!\!\!/ + A\!\!\!/\,)\Psi_i]_a \right\}, \tag{9.28}$$

where e_c is the coupling constant to the color potentials (note it has mass dimensions in 1+1 space-time), and,

$$F_{\mu\nu} = \partial_\mu A_\nu - \partial_\nu A_\mu + i[A_\mu, A_\nu]. \tag{9.29}$$

The bosonized action is,

$$S[g, h, A_+, A_-] = N_C S[g] + N_F S[h] + \frac{1}{2}\int d^2 \times \partial_\mu\phi\partial^\mu\phi$$

$$+ \frac{N_F}{2\pi}\int d^2x \text{Tr}[i(A_+ h\partial_- h^\dagger + A_- h^\dagger\partial_+ h)$$

$$- (A_+ h A_- h^\dagger - A_- A_+)]$$

$$- \frac{1}{2e_c^2}\int d^2x \text{Tr}F_{\mu\nu}F^{\mu\nu}. \tag{9.30}$$

9.3.2 Multiflavor QCD_2 using the $U(N_F \times N_C)$ scheme

Let us now repeat the gauging of the $SU_V(N_C)$ subgroup in the framework of the $U(N_F \times N_C)$ bosonization procedure.

Using the gauging prescription discussed in Section 9.3.1 we first get the action in which the whole $SU(N_C N_F)$ is gauged, namely,

$$S[u, A_+, A_-] = S[u] + \frac{i}{2\pi}\int d^2x \text{Tr}(A_+ u\partial_- u^\dagger + A_- u^\dagger\partial_+ u)$$

$$- \frac{1}{2\pi}\int d^2x \text{Tr}(A_+ u A_- u^\dagger - A_- A_+)$$

$$+ m'^2 N_{\tilde{m}}\int d^2x \text{Tr}(u + u^\dagger), \tag{9.31}$$

where we have also added a mass term with $m'^2 = m_q c\tilde{m}$. Now since we are interested in gauging only the $SU(N_C)$ subgroup of $U(N_F N_C)$, we take A_μ to be spanned by the generator $T^D \subset SU(N_C)$ via $A_\mu = e_c A_\mu^D T^D$. We then add to this action the kinetic term for the gauge fields $-\frac{1}{2e_c'^2} \int d^2 x \text{Tr}(F_{\mu\nu} F^{\mu\nu})$. The coupling e_c' is related to the color gauge coupling e_c by $e_c' = \sqrt{N_F} e_c$, so that after taking the trace over flavor we get the expected kinetic term with coupling e_c. The resulting action is invariant under local color and global flavor,

$$u \to V(x) u V^{-1}(x), \quad A_\mu \to V(x)(A_\mu - i\partial_\mu) V^{-1}(x); \quad V(x) \subset SU_V(N_C)$$

$$u \to W u W^{-1}; \quad W \subset U(N_F).$$

The symmetry group is now $SU_V(N_C) \times U(N_F)$, just as for the gauged fermionic theory. We choose the gauge $A_- = 0$, so now the action takes the form,

$$S[u, A_+] = S[u] + \frac{1}{e_c'^2} \int d^2 x \text{Tr}(\partial_- A_+)^2 + \frac{i}{2\pi} \int d^2 x \text{Tr}(A_+ u \partial_- u^\dagger)$$
$$+ m'^2 N_{\tilde{m}} \int d^2 x \text{Tr}(u + u^\dagger). \tag{9.32}$$

Upon the decomposition $u = \tilde{g} \tilde{h} l e^{-i\sqrt{\frac{4\pi}{N_C N_F}}\phi}$, we see that the current that couples to A_+ is $\tilde{h}\partial_- \tilde{h}^\dagger$. In terms of u it is the color projection $(u\partial_- u^\dagger)_C = \frac{1}{N_F} \text{Tr}_F [u\partial_- u^\dagger - \frac{1}{N_C} \text{Tr}_C u\partial_- u^\dagger]$. Thus the coupling of the current to the gauge field $\frac{i}{2\pi} \int d^2 x \text{Tr}(A_+ \tilde{h}\partial_- \tilde{h}^\dagger)$.

We can further manipulate the action to a form which will be convenient for taking the strong coupling limit (see Chapter 13). We define $\tilde{H}(x)$ by $\partial_- \tilde{H} = i\tilde{h}\partial_- \tilde{h}^\dagger$. We take the boundary conditions to be $\tilde{H}(-\infty, x_-) = 0$ and then integrate out A_+ obtaining,

$$\tilde{S}[u] = S[u] - (\frac{e_c}{4\pi})^2 N_F \int d^2 x \text{Tr}(\tilde{H}^2)$$
$$+ m'^2 N_{\tilde{m}} \int d^2 x \text{Tr}(u + u^\dagger). \tag{9.33}$$

In Chapter 13 this form of action will constitute the starting point of determining the baryonic spectrum of QCD_2 in the strong coupling limit. In Chapter (14) we will use this action to analyze the string tension and the confining behavior of massive QCD_2.

10

The 't Hooft solution of 2d QCD

Two-dimensional quantum chromodynamics involves an $SU(N)$ symmetry group. We saw that models get simplified in the large N limit, and so we would like to examine the question of whether the large N limit QCD in two dimensions can be solved. This question was addressed by 't Hooft who showed that indeed QCD_2 in this limit is almost exactly soluble. The simplest Green's functions can be solved in closed forms and the meson spectrum can be extracted by a non-elaborate numerical computation.

This was derived by 't Hooft in his seminal paper [124], and it had many follow ups. In this chapter we consider only [56], which discusses the scattering properties of QCD in the large N model.

Recall the action of a two-dimensional QCD,

$$S_{QCD} = -\frac{1}{2}\mathrm{Tr}[F_{\mu\nu}F^{\mu\nu}] + \bar{\Psi}_i(i\slashed{D} - m_i)\Psi_i,\qquad(10.1)$$

where the gauge fields are spanned by $N \times N$ Hermitian matrices T^A such that $A_\mu = A_\mu^A T^A$, $F_{\mu\nu} = \partial_\mu A_\nu - \partial_\nu A_\mu + i\frac{g}{\sqrt{N}}[A_\mu, A_\nu]$, the covariant derivative $D_\mu = \partial_\mu + i\frac{g}{\sqrt{N}}A_\mu$, the fermions Ψ are in the fundamental representation of the color group and $i = 1, \ldots, N_f$ indicates the flavor degrees of freedom. There is a sum over the flavor indices. Note that the gauge coupling was chosen to be $\frac{g}{\sqrt{N}}$, obviously to accommodate a large N approximation with g fixed.

It is convenient to impose the algebraic light-cone gauge. This gauge is advantageous at least for the following two reasons:

(i) The field strength F_{+-} becomes linear in the gauge potential,

$$A^+ = A_- = 0 \quad \Rightarrow \quad F_{+-} = -\partial_- A_+.\qquad(10.2)$$

(ii) The theory after gauge fixing is still Lorentz invariant. This is obviously a property of two dimensions only.

In this gauge the Lagrangian of the system becomes,

$$\mathcal{L} = -\frac{1}{2}\mathrm{Tr}[(\partial_- A_+)^2] + \bar{\Psi}_k\left(i\slashed{\partial} - m_k - \frac{g}{\sqrt{N}}\gamma_- A_+\right)\Psi_k.\qquad(10.3)$$

Recall that in the light-cone gauge there are no ghost fields.

The Feynman rules associated with this action in the so-called double line notation follow from Fig. 10.1, as explained below.

In the following we shall be taking one flavor, for simplicity.

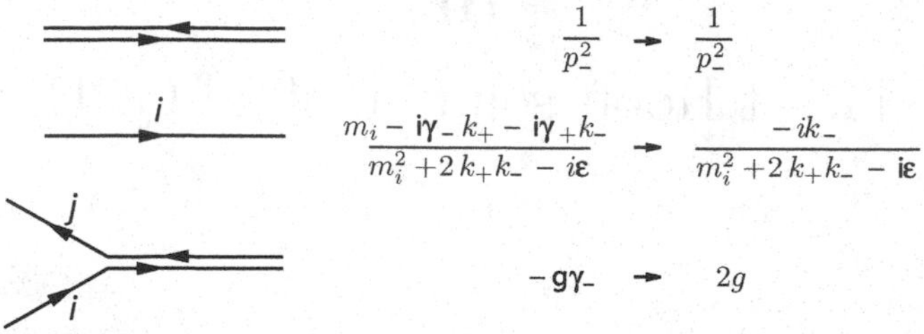

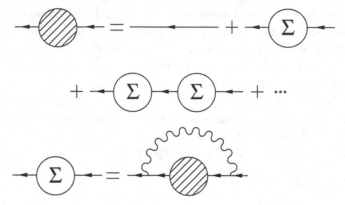

Fig. 10.1. The Feynman rules of QCD_2 in the light cone.

Fig. 10.2. The quark self-energy.

The light-cone gamma matrices obey the relations,

$$\gamma_-^2 = \gamma_+^2 = 0 \quad \{\gamma^+, \gamma^-\} = 2. \tag{10.4}$$

Since the vertex is proportional to γ_-, only that part of the propagator that is proportional to γ_+ can contribute. As a consequence we can eliminate all the γ dependence from the Feynman diagrams. Thus the double line, representing the gluon propagator, is $\frac{1}{p_-^2}$, the fermion line is $\frac{-ik_-}{m^2+2k_+k_--i\epsilon}$, and the coupling is $2g$.

Note that for the gauge field propagator one makes use of the principal value such that,

$$D_{++}(p) = \mathcal{P}\left(\frac{1}{p_-^2}\right) \equiv \frac{1}{2}\left[\frac{1}{(p_- + i\epsilon)^2} + \frac{1}{(p_- - i\epsilon)^2}\right]. \tag{10.5}$$

The dressed quark propagator and the quark self-energy, given in terms of the diagrams in Fig. 10.2, obey the coupled equations,

$$S(p) = \frac{ip_-}{2p_+p_- - m^2 - p_-\Sigma(p) + i\epsilon}$$

$$\Sigma(p) = 4g^2 \int \frac{dk_+ dk_-}{(2\pi)^2} S(p-k)\mathcal{P}\left(\frac{1}{(k_-)^2}\right), \tag{10.6}$$

where Σ is the γ_+ part, the only part that appears in the self-energy in our gauge. If we shift the integration variables $p_+ - k_+ \rightarrow -k_+$ we eliminate the dependence on p_+. Hence Σ is only a function of p_-. Due to its Lorentz structure it implies that Σ must be a constant times $\frac{1}{p_-}$, namely $m^2 + p_- \Sigma \equiv M^2$. Thus in the leading large N the sole effect of the interaction, for the propagator, is to replace the quark mass m by a renormalized quark mass M.

Integrating over k_+ we get,

$$\Sigma = \frac{g^2}{2\pi} \int dk_- \text{sgn}(p_- - k_-) \mathcal{P}\left(\frac{1}{(k_-)^2}\right) = -\frac{g^2}{\pi p_-}, \tag{10.7}$$

and hence,

$$M^2 = m^2 - \frac{g^2}{\pi}. \tag{10.8}$$

In the original treatment of 't Hooft, the regularization employed was not of principal value, but rather of a sharp cutoff, namely integrating over $|p_-| > \lambda$. This avoids the infrared divergence as well, but introduces a new scale, which is not gauge invariant. Obviously, one has to check that Green's functions of gauge invariant operators are independent of λ when $\lambda \rightarrow 0$. Thus we find that,

$$\Sigma(p) = \Sigma(p_-) = -\frac{g^2}{\pi}\left(\frac{\text{sgn}(p)}{\lambda} - \frac{1}{p_-}\right), \tag{10.9}$$

and correspondingly the dressed quark propagator is,

$$S(p) = \frac{ip_-}{2p_+ p_- - m^2 + \frac{g^2}{\pi} - \frac{g^2 |p_-|}{\pi \lambda} + i\epsilon}. \tag{10.10}$$

Now the pole of the quark propagator is shifted towards $k_+ \rightarrow \infty$ and hence there is no physical single quark state.

Let us consider now the spectrum of the mesonic bound states. The propagator of the meson is given by the sum of diagrams as is shown in Fig. 10.3.

This ladder sum is exact in the planar limit that follows from the large N approximation. If the propagator has a meson pole, then the ladder diagrams have to obey the Bethe–Salpeter equation as in Fig. 10.4.

The "blob" is the Fourier transform of the matrix element,

$$\tilde{\phi}(p, q) = \text{F.t.} <\text{meson}|T\bar{\psi}(x)\psi(0)|0>,$$

with external legs of a quark of mass m, momentum p, and an anti-quark of mass m and momentum $p - q$ (for simplicity, we take one flavor, and so the same mass for the quark and anti-quark). The Bethe–Salpeter equation reads,

$$\tilde{\phi}(p, q) = -4ig^2 S(p - q) S(p) \int \frac{d^2k}{(2\pi)^2} \mathcal{P}\left(\frac{1}{(k_- - p_-)^2}\right) \tilde{\phi}(k_-, q). \tag{10.11}$$

Defining

$$\phi(p_-, q) = \int dp_+ \tilde{\phi}(p, q),$$

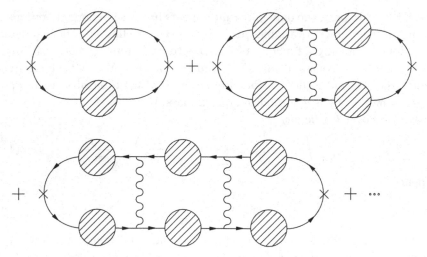

Fig. 10.3. The Green's function of the quark bilinear.

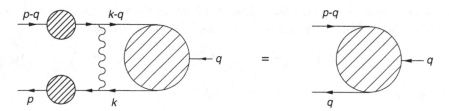

Fig. 10.4. The Bethe–Salpeter equation.

we get,

$$\phi(p_-, q) = -i\frac{g^2}{\pi^2} \int dp_+ S(p-q)S(p) \int dk_- \mathcal{P}\left(\frac{1}{(k_- - p_-)^2}\right)\phi(k_-, q). \quad (10.12)$$

The integral over p_+ can be done explicitly

$$I(p_-, q) \equiv \int dp_+ S(p-q)S(p)$$

$$= -\int dp_+ \frac{1}{\left[2(p_+ - q_+) - \frac{M^2 - i\epsilon}{p_- - q_-}\right]\left[2p_+ - \frac{M^2 - i\epsilon}{p_-}\right]}. \quad (10.13)$$

If p_- is outside the interval $[0, q_-]$ then the two poles are on same side of the real axis, and the integral vanishes. When p_- is inside the interval, the integral is (taking $q_- > 0$),

$$-i\pi\left[2q_+ - \frac{M^2}{p_-} - \frac{M^2}{(q_- - p_-)}\right],$$

so that,

$$\left[2q_+ - \frac{M^2}{p_-} - \frac{M^2}{q_- - p_-)}\right]\phi(p_-, q) = -\frac{g^2}{\pi}\int_0^{q_-} dk_- \mathcal{P}\left(\frac{1}{(k_- - p_-)^2}\right)\phi(k_-, q).$$

$$(10.14)$$

Defining x and y by,

$$p_- = xq_-, \quad k_- = yq_-,$$

and,

$$2q_+q_- = \mu^2,$$

one finally gets 't Hooft's equation,

$$\mu^2 \phi(x) = \left[\frac{M^2}{x} + \frac{M^2}{1-x}\right]\phi(x) - \frac{g^2}{\pi}\int_0^1 dy \frac{1}{(x-y)^2})\phi(y), \qquad (10.15)$$

with $\phi(x)$ defined on the interval $[0, 1]$.

The equation cannot be solved analytically, but one can compute the wave-functions that correspond to the various states numerically.

Before describing these solutions let us further discuss the equation. In fact one can derive the equation using a light-cone Schrödinger equation. In the light-cone coordinates a system is specified at x^+, and its dynamics is generated by P_+, the generator of translations of x^+. Since the latter commutes with P_-, the generator of translations of x^-, it is useful to use the eigenspace of P_-. For example, for a free single particle of mass M, $2P_+ = \frac{M^2}{P_-}$. Note however that unlike the ordinary Schrödinger formulation which is expressed in terms of a real line, the spectrum of P_1, in the light-cone case the spectrum of P_- is the positive half-line. For a system of two particles one can always choose to normalize the eigenvalue of P_- to be one, so that the eigenvalue of the operator on one of the two particles is x and the on the other it is $1 - x$, such that for two non-interacting particles $2P_+ = \frac{M^2}{x} + \frac{M^2}{1-x}$. This yields the first two terms in (10.15). The other term, the integral, is just a linear potential term. If we interpret temporarily x as a position operator, then the operator form of (10.15) is,

$$2P_+ = \frac{M^2}{x} + \frac{M^2}{1-x} + g^2|p|. \qquad (10.16)$$

This is the Hamiltonian of a massless particle moving in a potential and restricted to a box $[0, 1]$. This guarantees that the spectrum is discrete and there is no continuum of two free particles. Moreover we can go further with this interpretation and argue that at least for high-level states the eigenstates are like those of a free particle in a box namely,

$$\phi_n \approx \sin(\pi n x), \quad \mu_n^2 \approx g^2 \pi n, \qquad (10.17)$$

for $n = 1, 2, \ldots$ These states furnish a linear "Regge trajectory" with no continuum. We will verify shortly that for large n this is indeed the structure of the eigenstates and eigenvalues. Since the renormalized quark mass becomes tachyonic for large coupling constant g (eqn. 10.8) one may wonder whether the mesonic bound states can also be tachyonic. It turns out that this cannot occur.

From (10.15) it follows that,

$$\mu^2 \int_0^1 |\phi(x)|^2 dx = m^2 \int_0^1 |\phi(x)|^2 \left[\frac{1}{x} + \frac{1}{1-x}\right] dx$$

$$+ \frac{g^2}{2\pi} \int_0^1 dx \int_0^1 dy \frac{|\phi(x)||\phi(y)|}{(x-y)^2}. \qquad (10.18)$$

To solve the 't Hooft equation (10.15) we need to specify the boundary conditions.

At $x = 0$ ($x = 1$) the solution may behave like $x^{\pm\beta}$ $((1-x)^{\pm\beta})$ with,

$$\pi\beta \cot g(\pi\beta) + \frac{\pi M^2}{g^2} = 0. \qquad (10.19)$$

Let us define the "Hamiltonian" of the system as the right-hand side of equation (10.15), namely,

$$H\phi(x) \equiv \left[\frac{M^2}{x} + \frac{M^2}{1-x}\right] \phi(x) - \frac{g^2}{\pi} \int_0^1 dy P\left(\frac{1}{(x-y)^2}\right) \phi(y). \qquad (10.20)$$

This Hamiltonian is Hermitian only when acting on the space of functions that vanish on the boundary, as can be seen from (10.18). Using the latter one can show that ϕ_n with $\phi_n(0) = \phi_n(1) = 0$ constitute a complete orthonormal set,

$$\sum_n \phi_n(x)\phi_n(x') = \delta(x-x')$$

$$\int_0^1 \phi_n^*(x)\phi_m(x)dx = \delta_{nm}. \qquad (10.21)$$

Since the integral in (10.15) gets its main contribution from y close to x and since for a periodic function we have,

$$P\left(\int_0^1 \frac{e^{iwy}}{(x-y)^2} dy\right) \simeq P\left(\int_{-\infty}^\infty \frac{e^{iwy}}{(x-y)^2} dy\right) = -\pi|w|e^{iwx}, \qquad (10.22)$$

then the configurations given in (10.17) are a good approximation of the eigenstates of the system. The numerical solutions of eqn. (10.15) are drawn in Fig. 10.5.

In this figure the mass spectrum of mesons is shown for various values of quark mass. In cases when the mass of the quark and anti-quark are not equal, the term $[\frac{M^2}{x} + \frac{M^2}{1-x}]$ in (10.15) is replaced by $[\frac{M_1^2}{x} + \frac{M_2^2}{1-x}]$.

The masses and wavefunctions cannot be determined in general in an analytic form. However in certain limits one can write down approximate expressions. In [52] it was shown that the highly excited states $n \gg 1$, where n is the excitation number have masses given by,

$$(M_{\rm mes})_n^2 \sim \pi g^2 N \left(n + \frac{3}{4}\right) + (m_{q_1}^2 + m_{q_2}^2)\ln(n) + C(m_{q_2}^2) + C(m_{q_1}^2) + O\left(\frac{1}{n}\right), \qquad (10.23)$$

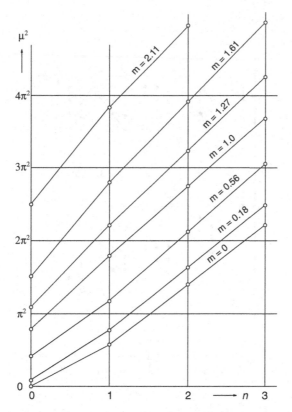

Fig. 10.5. The spectrum of mesons. The squared masses are in units of $\frac{g^2}{\pi}$ [124].

where m_{q_i} are the masses of the quark and anti-quark and where the functions $C(m_q^2)$ are given in [52].

The opposite limit of low-lying states and in particular the ground state can be deduced in the limit of large quark masses, namely $m_q \gg g$ and small quark masses $g \gg m_q$. For the ground state in the former limit one finds,

$$M^0_{\mathrm{mes}} \cong m_{q_1} + m_{q_2}. \tag{10.24}$$

In the opposite limit of $m_q \ll g$,

$$(M^0_{\mathrm{mes}})^2 \cong \frac{\pi}{3} \sqrt{\frac{g^2 N_c}{\pi}} (m_1 + m_2). \tag{10.25}$$

For the special case of massless quarks we find a massless meson.

In Fig. 10.6 the spectrum of meson nonets built from two triplets of flavor with masses

$$
\begin{aligned}
&(a) \; m_1 = 0 \quad m_2 = 0.2 \quad m_3 = 0.4 \\
&(b) \; m_1 = 0.8 \quad m_2 = 1.0 \quad m_3 = 1.2
\end{aligned}
\tag{10.26}
$$

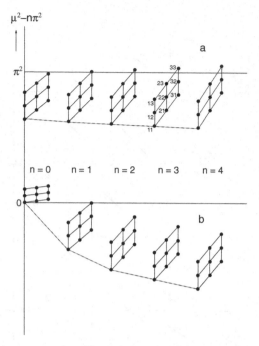

Fig. 10.6. Meson nonets for $N_c = 3$. In case (a) the masses of the triplet are $m_1 = 0.00, m_2 = 0.20, m_3 = 0.4$ and in (b) $m_1 = 0.80, m_2 = 1.00, m_3 = 1.2$ [124].

is shown, in units of $\frac{g}{\sqrt{\pi}}$. Then the ground state is at 2.7, the first excited state at 4.16, and level $n = 10$ is at 20.55. It is obvious from these cases that for larger n, the wavefunction gets more and more sharply picked around $x = 0.5$. For the case of unequal masses, the wavefunction ceases to be symmetric, as can be seen from Fig. 10.7 for $m_1 = 1, m_2 = 5$.

10.1 Scattering of mesons

In the previous section we have described the equation that governs the formation of mesonic bound states, and the corresponding meson spectrum follows from a homogeneous Bethe–Salpeter equation. This can be generalized to the equation for full quark anti-quark scattering amplitude, which takes the form of the non-homogeneous equation of Fig. 10.8.

The scattering amplitude has the following structure,

$$T_{\alpha\beta,\gamma\delta} = (\gamma_-)_{\alpha\gamma}(\gamma_-)_{\beta\delta}T(q,q',p). \tag{10.27}$$

The undressed amplitude $T(q, q', p)$ takes the form,

$$T(q, q', p) = \frac{ig^2}{(q_- - q'_-)^2} + \frac{ig^2 N}{\pi^2} \int \frac{dk_-\phi(k_-, q', p)}{(k_- - q_-)^2}, \tag{10.28}$$

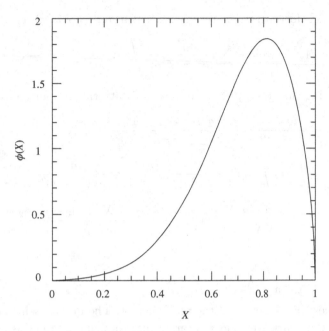

Fig. 10.7. Wavefunction for $m_1 = 1, m_2 = 5$.

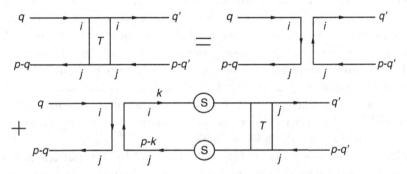

Fig. 10.8. The Bethe–Salpeter equation for quark anti-quark scattering.

where,

$$\phi(q_-, q'_-, p) = \int dq_+ \, S_E(q) S_E(q - p) T(q, q', p). \tag{10.29}$$

Similar to the equation for the "wave function" $\phi(x)$ we now get the generalization to $\phi(x, x', p)$ which reads,

$$\mu^2 \phi(x, x', p) = \left[\frac{M^2}{x} + \frac{M^2}{1 - x} \right] \phi(x, x', p)$$

$$+ \frac{\pi^2}{N p_-(x - x')^2} + \int_0^1 dy \frac{[\phi(x, x', p) - \phi(y, x', p)]}{(x - y)^2}. \tag{10.30}$$

It is now straightforward to express $\phi(x, x', p)$ in terms of $\phi(x)$ as,

$$\phi(x, x', p) = -\sum_n \frac{\pi g^2}{p^2 - p_n^2} \frac{1}{p_-} \int_0^1 dy \frac{\phi_n(x)\phi *_n (y)}{(x - y)^2},$$ (10.31)

and substituting this into (10.28) we find the scattering amplitude,

$$T(x, x', p) = \frac{ig^2}{p_-^2 (x' - x)^2} - \frac{ig^2 (g^2 N)}{\pi p_-^2} \sum_n \frac{1}{p^2 - p_n^2}$$

$$\times \int_0^1 dy \int_0^1 dy' \frac{\phi_n(y)\phi *_n (y')}{(x - y)^2 (x' - y')^2} = \frac{ig^2}{p_-^2 (x' - x)^2} - \sum_n \frac{1}{p^2 - p_n^2}$$

$$\left[\phi_n^*(x') \frac{2g}{\lambda} \sqrt{\frac{g^2 N}{\pi}} \left(\theta(x'(1 - x')) + \frac{\lambda}{2|p_-|} \left(\frac{\gamma_1 - 1}{x'} + \frac{\gamma_2 - 1}{1 - x'} - \mu_k^2 \right) \right) \right]$$

$$\times [(x' \leftrightarrow x)],$$ (10.32)

where γ_i for $i = 1, 2$ are $\frac{M_i}{(\frac{g}{\sqrt{\pi}})}$.

This clarifies the dynamics of the confinement. The infinite self-mass quark is cancelled by the quark anti-quark interaction producing finite mass color singlet bound states, whose mass squared, as we have seen above, increases linearly for high excited states. The infrared behavior is determined by the dependence on λ as in (10.9). The bound state wave function is of order $\frac{1}{\lambda}$ as $\lambda \to 0$. The fact that the amplitude for a bound state to decay into quarks is infinite as $\lambda \to 0$ compensates for the vanishing quark propagator in this limit to produce finite bound state amplitudes, which contain no multiquark discontinuities.

To test the consistency of the model one has to examine also the hadronic scattering processes. One has to check that these are finite in the limit of $\lambda \to 0$, unitary and Lorentz invariant. A consequence of the unitarity is the absence of long range forces among the color singlets.

In Fig. 10.9 the three-particle vertex function and the two-particle scattering are drawn. The three-particle vertex function, Fig. 10.9(a), is of order $g \sim \frac{1}{\sqrt{N}}$. Each quark propagator is of order λ. The k_+ loop momentum is of order $\frac{1}{\lambda}$ since it is dominated by the pole at $\frac{1}{\lambda}$. From the three bound state wave functions we get a factor of $(\frac{1}{\lambda})^2$ since at least one wave function must be of order unity to conserve momentum. So altogether the factors of λ cancel out and we get a finite result in the limit of $\lambda \to 0$.

The two-particle scattering is described in Fig. 10.9(b) and 10.9(c). The former describes a hadronic exchange and the latter a quark exchange. The quark exchange may seem to be infinite in the limit $\lambda \to 0$ since now the quark and anti-quark can move in the same direction with an amplitude that behaves like $\frac{1}{\lambda}$. The total dependence on λ is as follows: λ^4 from quark propagators, $\frac{1}{\lambda^4}$ from the wave functions and $\frac{1}{\lambda}$ from the loop momentum integration. However it can be shown that when one adds all diagrams that contribute to $\frac{1}{N}$ order, the terms of

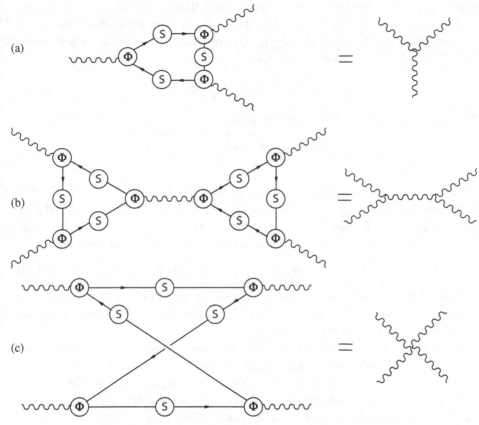

Fig. 10.9. (a) Three-particle vertex function. (b) Hadronic exchange contribution to two-particle scattering amplitude. (c) Quark exchange contribution to
two-particle scattering amplitude.

order $\frac{1}{\lambda}$ cancel, leaving a finite remainder. In this way we have verified unitarity
of the model to the first non-trivial order.

10.2 Higher $1/N$ corrections

At $N = \infty$ the mesons are stable since their decay rate, as will be shown shortly,
is proportional to $\frac{1}{N}$. Going to the $\frac{1}{N}$ corrections, a meson has the following
amplitude to decay into two mesons[1]

$$
\mathcal{A}(i, f_1, f_2; w) = \frac{4g^2\sqrt{N}}{\sqrt{\pi}} \left\{ \frac{1}{1-w} \int_0^w dx \phi_i(x) \phi_{f_1}\left(\frac{x}{w}\right) \Phi_{f_2}\left(\frac{x-w}{1-w}\right) \right.
$$

$$
\left. -\frac{1}{w} \int_w^1 dx \phi_i(x) \Phi_{f_1}\left(\frac{x}{w}\right) \phi_{f_2}\left(\frac{x-w}{1-w}\right) \right\}, \qquad (10.33)
$$

---

[1] The 1/N corrections were evaluated in [144].

where $\phi_i, \phi_{f_1}, \phi_{f_2}$ are the wave functions of the initial meson and first and second final mesons, respectively. The quark ends up being in the second final meson and the anti-quark in the first final meson. The vertex function $\Phi(x)$, with x *not* $\in$ $[0,1]$, is related to the wave function as,

$$\Phi(x) = \int_0^1 dy \frac{1}{(x-y)^2} \phi(y). \tag{10.34}$$

The kinematic parameter w takes the values,

$$w_\pm = \frac{\mu_i^2 + \mu_{f_1}^2 - \mu_{f_2}^2 \mp \sqrt{(\mu_i^2 + \mu_{f_1}^2 - \mu_{f_2}^2)^2 - 4\mu_i^2 \mu_{f_2}^2}}{2\mu_i^2}, \tag{10.35}$$

where w_+ and w_- correspond to the right and left moving final state f_1. The decay can take place only provided $\mu_i \geq \mu_{f_1} + \mu_{f_2}$. It is clear that for fixed $g^2 N$ the amplitude is of order $\mathcal{A} \sim O(\frac{1}{\sqrt{N}})$. The amplitude (10.33) is for a partial decay and for full-on shell amplitude one has to add the partial decays

$$\mathcal{A} = (1 - (-1)^{\sigma_i + \sigma_{f_1} + \sigma_{f_2}})(\mathcal{A}(i, f_1, f_2; w_+) + \mathcal{A}(i, f_1, f_2; w_-)), \tag{10.36}$$

with σ_+ for even parity state and σ_- for odd parity state. It was found that numerically these amplitudes for various excited states do not vanish. This also shows that the model is *not integrable*.

It was further found that the amplitudes for mesons made out of massless quark anti-quark pairs differ significantly from those of mesons made out of massive ones. An interesting result that follows from the computations of these amplitudes is that the amplitude for decay of an exited meson into a pion and another meson vanishes, in the case of massless quarks. This is actually to be expected, as for massless quarks the two-dimensional pion is massless and decoupled, since there is no chiral symmetry breaking in two dimensions.

11
Mesonic spectrum from current algebra

11.1 Introduction

In this chapter we study the mesonic spectrum of various QCD_2 theories. The main idea is to use the current algebra of the underlying ungauged theories. In addition we combine the bosonization techniques developed in Chapter 6 with that of a large N expansion of Chapter 7 and a light-front quantization as in Chapter 10. We will focus our attention on the massive mesonic spectrum of conformal field theories coupled to non-abelian gauge fields. In particular massless multi-flavor fundamental quarks and adjoint quarks that will be shown to correspond to the particular case of $N_f = N_c$.

First a universality theorem, that states that the massive mesonic spectrum does not depend on the representation of the matter field but rather only on its ALA level, will be derived, following Kutasov and Schwimmer [148].

We then present a detailed determination of the massive mesonic spectrum using a 't Hooft-like equation for the wave functions of "currentballs" states. We will discuss in particular the special cases of $N_f = 1, N_f = N_c$ and $N_f \gg N_c$. The last section is devoted to the spectrum of states built by the action of a single current creation operator on the adjoint vacuum. In both cases it will be shown that the bosonization approach leads to the introduction of current quanta as the basic degrees of freedom. Once the mass operator $P^+ P^- = M^2$ is expressed in terms of the current quanta, the bosonization has already left the scene.

The main content of this chapter, the mesonic spectrum from current algebra, is based on [17].[1] The spectrum based on the adjoint vacuum was introduced in [3].

11.2 Universality of conformal field theories coupled to YM_2

So far we have mainly discussed the coupling of matter in the fundamental representation to the two-dimensional YM fields. Obviously one can also couple other matter fields to these non-abelian gauge fields. A natural class of matter theories that one would like to gauge are the conformal field theories which admit on top of the Virasoro algebra also an affine Lie algebra structure. These theories which are characterized by the corresponding Lie algebra G and the level k of the

---

[1] This was previously also discussed in [18].

affine Lie algebra, are candidates for coupling to non-abelian gauge fields of the group G. A particular family of such theories are the WZW models, invariant under $G \times G$ of level k. We have discussed in Chapter 6 the gauging of such models. In this chapter we would like to address the issue of the spectrum of such gauged conformal field theories, and in particular the massive sector of the spectrum. In general the Lagrangian density of such a theory reads,

$$\mathcal{L} = \mathcal{L}_{CFT} - \frac{1}{2e^2} \text{Tr} \left[F_{\mu\nu}^2 \right] + \mathcal{L}_I$$

$$= \mathcal{L}_{CFT} - \frac{1}{2e^2} \text{Tr} \left[(\partial_- A_+)^2 \right] + \text{Tr} \left[A_+ J_- \right]$$

$$= \mathcal{L}_{CFT} - \frac{e^2}{2} \text{Tr} \left[J^+ \frac{1}{\partial_-^2} J^+ \right] = \mathcal{L}_{CFT} - \frac{e^2}{2} \text{Tr} \left[J \frac{1}{\partial^2} J \right], \qquad (11.1)$$

where we have used the light-cone gauge $A_- = 0$. We will be using the notation of J for J_- and $\bar{J}$ for J_+, and similar for other holomorphic and anti-holomorphic quantities.

A conformal field theory invariant under the symmetry generated by a G ALA has holomorphic currents J^a in the adjoint representation of G, as well as anti-holomorphic currents also in the adjoint representation of G. In general the holomorphic currents obey an ALA with level k and the anti-holomorphic currents an ALA of level $\bar{k}$. However, gauging the conformal theory requires vanishing of the chiral anomaly, namely it requires that,

$$k = \bar{k}. \qquad (11.2)$$

Next we quantize the system on the light-front. This framework is very convenient since both momenta P^- and P^+, or equivalently P and $\bar{P}$, can be expressed in terms of J only (with no reference to $\bar{J}$). This decoupling of one sector (the anti-holomorphic one) can be attributed to the fact that in a frame moving to the right with the speed of light there is no way to interact with massless left-moving particles. The light-cone Hamiltonian is given by,

$$P^+ = \frac{1}{[C(G) + k]} \int \mathrm{d}x^- : J^a(x^-) J^a(x^-)$$

$$= \sum_{n=1}^{\infty} \frac{1}{n^2} J^a_{-n} J^a_n, \qquad (11.3)$$

where in the last line we have assumed that the light-cone space direction $x^- = z$ has been put on a circle. Thus the Hamiltonian acts inside current blocks, and the problem of finding the massive spectrum splits into diagonalizing the decoupled blocks of P^+ on global G singlets. We want to emphasize again that the light-front dynamics is fully independent of the anti-holomorphic sector, apart from the constraint that $k = \bar{k}$. This clearly means that we can replace the anti-holomorphic sector with another anti-holomorphic sector, provided that the latter has a level that equals k. Obviously we could have fixed the opposite gauge $A_+ = 0$, leaving only the anti-holomorphic sector with currents $\bar{J}$. In that gauge we could have replaced the holomorphic sector with another one, again provided that it has level k. Thus we conclude that *the massive spectrum does not depend on the representations r and $\bar{r}$, but only on the gauge group G and the level k.*

We would like to demonstrate this universality in the context of a generalization of Schwinger's model, which contains n^R right moving fermions ψ_i^R $i = 1 \ldots n^R$ and n^L left-moving fermions ψ_i^L $i = 1 \ldots n^L$ [120]. Both the right- and left-moving fermions are charged with respect to an abelian $U(1)$ gauge symmetry with charges q_i^R and q_i^L respectively. The system is described by the Lagrangian density,

$$\mathcal{L} = \psi_i^{R\,\dagger} \bar{\partial} \psi_i^R + \psi_i^{L\,\dagger} \partial \psi_i^L + \bar{A}J - \frac{1}{4e^2}(\partial \bar{A})^2, \tag{11.4}$$

where $J = \sum_i^{n^R} q_i^R \psi_i^{R\,\dagger} \psi_i^R$ and we are using the gauge $A = 0$. Upon integrating $\bar{A}$ we get,

$$\mathcal{L} = \psi_i^{R\,\dagger} \bar{\partial} \psi_i^R + \psi_i^{L\,\dagger} \partial \psi_i^L - e^2 J \frac{1}{\partial^2} J. \tag{11.5}$$

We can now bosonize the system. Note that the fermions at hand are not Dirac fermions but rather n^R right and n^L left chiral fermions. The system is consistent in the sense that there is no chiral anomaly when,

$$k^R \equiv \sum_{i=1}^{N^R} q_i^R \quad k^L \equiv \sum_{i=1}^{N^L} q_i^L \quad k^R = k^L = k. \tag{11.6}$$

One can use the prescription for chiral bosonization described in Section 6.4. In fact it is enough to note that the interaction term takes the form,

$$\mathcal{L}_{\text{int}} = -e^2 J \frac{1}{\partial^2} J = e^2 (\phi)^2, \tag{11.7}$$

where $\phi = \sum_i^{N^R} q_i^R \phi_i^R = \sum_i^{N^L} q_i^L \phi_i^L$ and ϕ^L and ϕ^L are the right and left chiral bosons that corresponds to the right and left chiral fermions. Thus we conclude that the spectrum includes one massive mode corresponding to ϕ plus $n^R - 1$ and $n^L - 1$ massless right- and left-moving particles, respectively. It is now evident that indeed in accordance with the universality theorem, the massive sector does not depend on the explicit sequence of charges q_i^R and q_i^L but only on the combination expressed in ϕ.

Another example of the universality theorem is the case of adjoint fermions. The ALA associated with the currents built from the adjoint fermions $J^{ab} = \psi^{ac}\psi^{cb}$ is of level N_c. The CFT based on a WZW model of $SU(N_c)$ of level $k = N_c$ is another theory with the same ALA, and hence the massive sector of the spectrum of these theories should, according to the theorem, be the same.

In the next section we describe the massive spectrum of such models based on a 't Hooft-like equation for the currents.

11.3 Mesonic spectra of two-current states

In this section we derive the massive meson spectrum built from two current creation operators acting on the vacuum. In the next section we will discuss states constructed from a single current acting on the adjoint vacuum.

The first step in the determination of the spectrum is the derivation of a 't Hooft-like equation for the wave functions of the "currentball" states, at arbitrary level N_f. This equation should interpolate between the description of a single flavor ('t Hooft model), the model $N_f = N_c$ equivalent to adjoint fermions and the large N_f limit. We will argue that the equation obtained suggests that the underlying degrees of freedom in the problem are interacting "gluons" with mass $\frac{e^2 N_f}{\pi}$. Actually, these are related to the color currents, but are color singlets.

Then we will solve the equation for the lowest massive state. Whereas the 't Hooft model $N_f = 1$ is exactly solvable, the multi-flavor case with $N_f > 1$ is not solvable even in the Veneziano limit when both N_c and N_f are taken to infinity (with a fixed ratio), since pair creation and annihilation are not suppressed.

For the case of the adjoint quarks, the results derived using the current quanta will be shown to be compatible with those computed with fermions as the basic degrees of freedom discussed in Chapter 12. For large N_f it will be shown that the exact massive spectrum is a single particle with $M^2 = \frac{e^2 N_f}{\pi}$. This phenomenon is explained by the fact that this limit can be viewed as an "abelianization" of the model.

11.3.1 The basic setup

We now establish the basic setup. We start with the fermionic formulation of the various theories, impose the light-cone gauge, introduce the bosonized version and finally write down the mass operator.

In Section 8.4 the classical theory of QCD_2 with Dirac fermions in the fundamental representation was described. Here we will address this case as well as massless Majorana fermions in the adjoint representation. Recall that these theories are described by the following classical Lagrangian:

$$\mathcal{L} = -\text{Tr}\left[\frac{1}{2e^2}F_{\mu\nu}^2 + i\bar{\Psi}\,\slashed{D}\,\Psi\right], \tag{11.8}$$

where $F_{\mu\nu} = \partial_\mu A_\nu - \partial_\nu A_\mu + i[A_\mu, A_\nu]$ and the trace is over the color and flavor indices. For case (i) Ψ has the group structure Ψ_{ia} where $i = 1,\ldots,N_c$ and $a = 1,\ldots,N_f$ with $D_\mu = \partial_\mu - iA_\mu$, whereas for case (ii) $\Psi \equiv \Psi_j^i$ and $D_\mu = \partial_\mu - i[A_\mu, \ ]$. In both cases Ψ is two-spinor parametrized as $\Psi = \binom{\bar{\psi}}{\psi}$.

As we have seen in Chapter 10 it is useful to handle these models in the framework of light-front quantization, namely, to use light-cone space-time coordinates and to choose the chiral gauge $A_- = 0$. In this scheme the Lagrangian takes the form,

$$\mathcal{L} = -\frac{1}{2e^2}(\partial_- A_+)^2 + i\psi^\dagger \partial_+ \psi + i\bar{\psi}^\dagger \partial_- \bar{\psi} + A_+ J^+, \qquad (11.9)$$

where color and flavor indices were omitted and J^+ denotes the $+$ component of the color current $J^+ \equiv \psi^\dagger \psi$. This Lagrangian density is identical to (11.4) when one replaces the complex coordinates with light-cone ones.

By choosing x^+ to be the 'time' coordinate it is clear that A_+ and $\bar{\psi}$ are non-dynamical degrees of freedom. In fact, $\bar{\psi}$ are decoupled from the other fields, so in order to extract the physics of the dynamical degrees of freedom, one has to functionally integrate over A_+. The result of this integration is the following simplified Lagrangian,

$$\mathcal{L} = \mathcal{L}_0 + \mathcal{L}_I = i\psi^\dagger \partial_+ \psi + i\bar{\psi}^\dagger \partial_- \bar{\psi} - \frac{e^2}{2} J^+ \frac{1}{\partial_-^2} J^+. \qquad (11.10)$$

Since our basic idea is to solve the system in terms of the "quanta" of the colored currents, it is natural to introduce bosonization descriptions of the various fields.

(i) As was discussed in Section (9.3.2), the bosonized action of colored-flavored Dirac fermions in the fundamental representation is expressed in terms of a WZW action of a group element $u \in U(N_c \times N_f)$, with an additional mass term that couples the color, flavor and baryon number sectors. In the massless case when the latter term is missing, the action takes the form,

$$S_0^{\text{fund}} = S_{(N_f)}^{WZW}(g) + S_{(N_c)}^{WZW}(h) + \frac{1}{2}\int d^2x \partial_\mu \phi \partial^\mu \phi, \qquad (11.11)$$

where $g \in SU(N_c)$, $h \in SU(N_f)$ and $e^{i\sqrt{\frac{4\pi}{N_c N_f}}\phi} \in U_B(1)$, with $U_B(1)$ denoting the baryon number symmetry, and the WZW action was given in Section 4.1.

(ii) The current structure of free Majorana fermions in the adjoint representation can be recast in terms of a WZW action of level $k = N_c$, namely $S_0^{\text{adj}} = S_{(N_c)}^{WZW}(g)$, where now g is in the adjoint representation of $SU(N_c)$, so that it carries a conformal dimension of $\frac{1}{2}$. Multi-flavor adjoint fermions can be described as $S_{N_f}^{WZW}(g) + S_{N_c^2-1}^{WZW}(h)$ where $g \in SO(N_c^2 - 1)$ and $h \in SO(N_f)$. In the present work we discuss only gauging of $SU(N_c)$ WZW so the latter model would not be considered.

Substituting now S_0^{fund} or S_0^{adj} for S_0 the action that corresponds to 11.10 becomes,

$$S = S_0 - \frac{e^2}{2}\int d^2x J^+ \frac{1}{\partial_-^2} J^+, \qquad (11.12)$$

where the current J^+ now reads $J^+ = i\frac{k}{2\pi}g\partial_- g^\dagger$, and the level $k = N_f$ and $k = N_c$ for the multi-flavor fundamental and adjoint cases, respectively.

The light-front quantization scheme is very convenient because the corresponding momenta generators P^+ and P^- can be expressed only in terms of J^+. We would like to emphasize that this holds only for the massless case.

Using the Sugawara construction, the contribution of the colored currents to the momentum operator P^+ takes the simple form,

$$P^+ = \frac{1}{N_c + k} \int dx^- : J_j^i(x^-) J_i^j(x^-) :, \tag{11.13}$$

where $J \equiv \sqrt{\pi} J^+$, N_c in the denominator is the second Casimir operator of the adjoint representation and the level k takes the values mentioned above. Note that for future purposes we have added the color indices $i, j = 1 \ldots N_c$ to the currents. In the absence of the interaction with the gauge fields the second momentum operator P^- vanishes. For the various QCD_2 models it is given by,

$$P^- = -\frac{e^2}{2\pi} \int dx^- : J_j^i(x^-) \frac{1}{\partial_-^2} J_i^j(x^-) : . \tag{11.14}$$

In order to find the massive spectrum of the model we should diagonalize the mass operator $M^2 = 2P^+P^-$. Our task is therefore to solve the eigenvalue equation,

$$2P^+P^-|\psi\rangle = M^2|\psi\rangle. \tag{11.15}$$

We write P^+ and P^- in term of the Fourier transform of $J(x^-)$, defined by,

$$J(p^+) = \int \frac{dx^-}{\sqrt{2\pi}} e^{-ip^+ x^-} J(x^-).$$

Normal ordering in the expressions of P^+ and P^- are naturally with respect to p, where $p < 0$ denotes a creation operator, and to simplify the notation we will write from here on p instead of p^+. In terms of these variables the momenta generators are,

$$P^+ = \frac{2}{N+k} \int_0^\infty dp J_j^i(-p) J_i^j(p)$$

$$P^- = \frac{e^2}{\pi} \int_0^\infty dp \frac{1}{p^2} J_j^i(-p) J_i^j(p). \tag{11.16}$$

Recall that the light-cone currents $J_j^i(p)$ obey a level k, $SU(N_c)$ affine Lie algebra,

$$\left[J_i^k(p), J_l^n(p')\right] = \frac{1}{2}kp\left(\delta_i^n \delta_l^k - \frac{1}{N}\delta_i^k \delta_l^n\right)\delta(p+p') + \frac{1}{2}\left(J_i^n(p+p')\delta_l^k - J_l^k(p+p')\delta_i^n\right). \tag{11.17}$$

We can now construct the Hilbert space. The vacuum $|0, R\rangle$ is defined by the annihilation property,

$$\forall p > 0, \ J(p)|0, R\rangle = 0, \tag{11.18}$$

where R is an "allowed" representation depending on the level. Thus, a physical state in Hilbert space is,

$$\text{Tr } J(-p_1)\ldots J(-p_n)|0, R\rangle.$$

Note that this basis is not orthogonal.

11.3.2 't Hooft-like equation for the two-current wave function

We restrict ourselves to the simplest case of the two-current sector of the Hilbert space, (in Section 11.4 we will also mention the special case of one current on an adjoint vacuum),

$$|\Phi\rangle = \frac{1}{N_c N_f} \int_0^1 dk \ \Phi(k) J^a(-k) J^a(k-1)|0\rangle, \tag{11.19}$$

namely to states which are color singlets of two currents with total $P^+ = 1$ momentum and a distribution of P^- momentum $\Phi(k)$. Note that Φ is a symmetric function,

$$\Phi(k) = \Phi(1-k). \tag{11.20}$$

Our task now is to find the eigenvalue (Schrödinger) equation for the wave function $\Phi(k)$. Let us start by the action of the "Hamiltonian" P^- on the state $|\Phi\rangle$.

The commutator of P^- with a current $J^b(-k)$ yields the result,

$$\left[\int_0^\infty \frac{dp}{p^2} J^a(-p) J^a(p), J^b(-k)\right] \left(\left(\frac{1}{2}N_f - N_c\right)\frac{1}{k} + N_c\frac{1}{\epsilon}\right) J^b(-k)$$

$$+ \int_k^\infty dp \left(\frac{1}{p^2} - \frac{1}{(p-k)^2}\right) if^{abc} J^a(-p) J^c(p-k)$$

$$+ \int_0^k \frac{dp}{p^2} if^{abc} J^c(p-k) J^a(-p). \tag{11.21}$$

We introduced ϵ as an IR cutoff, namely, the lower limit of integration. This is the analog of λ in the derivation of the 't Hooft equation of Chapter 10. We take ϵ to go to zero at the end of the calculation.

The above expression (11.21) contains three terms on the right-hand side The first term contains a single creation operator. The second term contains an annihilation current and therefore should again be commuted with $J^b(k-1)$. The third term contains two creation currents and it would lead to a three-current state. This is a manifestation of the fact that pair creation is, generically, not suppressed in multi-flavor QCD$_2$.

Note that while deriving eqn (11.21) we get an "infinite" contribution $N_c \frac{1}{\epsilon} J^b(-k)$. This contribution will be cancelled by a counter contribution which comes from the regime $p \sim k$ in the first integral on the right-hand side of (11.21), as below.

The commutator of the second term on the right-hand side of (11.21) with $J^b(k-1)$ yields,

$$\left[\int_k^\infty dp \left(\frac{1}{p^2} - \frac{1}{(p-k)^2}\right) i f^{abc} J^a(-p) J^c(p-k), J^b(k-1)\right] \tag{11.22}$$

$$= N_c \int_k^\infty dp \left(\frac{1}{p^2} - \frac{1}{(p-k)^2}\right) (J^a(-p) J^a(p-1) - J^a(p-k) J^a(k-p-1)).$$

Our results can be summarized by the following set of equations,

$$M^2|\Phi\rangle = \frac{1}{N_c N_f} \int_0^1 dk \ \tilde{\Phi}(k) J^a(-k) J^a(k-1)|0\rangle + \frac{1}{(N_c N_f)^{\frac{3}{2}}} \tag{11.23}$$

$$\times \int_0^1 dk \ dp \ dl \ \delta(k+p+l-1) \Psi(k,p,l) i f^{abc} J^a(-k) J^b(-p) J^c(-l)|0\rangle,$$

with,

$$\Psi(k,p,l) = \frac{2e^2 (N_c N_f)^{\frac{1}{2}}}{\pi} \left(\frac{\Phi(l) - \Phi(k)}{p^2}\right), \tag{11.24}$$

and,

$$\tilde{\Phi}(k) = \frac{e^2}{\pi} \left((N_f - N_c)\left(\frac{1}{k} + \frac{1}{1-k}\right)\Phi(k) + \frac{2N_c}{\epsilon}\Phi(k)\right) \tag{11.25}$$

$$-N_c \int_0^{k-\epsilon} dp \frac{\Phi(p)}{(p-k)^2} - N_c \int_{k+\epsilon}^1 dp \frac{\Phi(p)}{(p-k)^2} + N_c \left(\frac{1}{k^2} - \frac{1}{(1-k)^2}\right) \int_0^k dp \Phi(p)$$

Ignoring the three-current term (see below), we get that $\Phi(k)$ obeys the eigenvalue equation,

$$\frac{M^2}{e^2/\pi}\Phi(k) = (N_f - N_c)\left(\frac{1}{k} + \frac{1}{1-k}\right)\Phi(k) \tag{11.26}$$

$$-N_c \mathcal{P}\int_0^1 dp \frac{\Phi(p)}{(p-k)^2} + N_c \left(\frac{1}{k^2} - \frac{1}{(1-k)^2}\right) \int_0^k dp \ \Phi(p).$$

We assumed that $\int_0^1 dp \ \Phi(p) = 0$, which we will justify shortly.

For general N_c and N_f, discarding the three-current term is unjustified. However, since the length of Ψ is $|\Psi(k,p,l)| \sim e^2 (N_c N_f)^{\frac{1}{2}}$, in the limit of large N_c with fixed $e^2 N_c$ and fixed N_f, or large N_f with fixed $e^2 N_f$ and fixed N_c, the three-current contribution is indeed negligible, as compared with the two-current term, the latter being of order 1.

The first integral in eqn. (11.26) should be calculated as a principal value integral (denoted by $\mathcal{P}$). The divergent part of this integral (arising from the regime $p \sim k$) cancels the previously mentioned infinity. In order to make contact with the ordinary 't Hooft equation, it is useful to integrate eqn. (11.26) with respect to k and rewrite it in terms of $\varphi(k) \equiv \int_0^k dp\, \Phi(p)$, to get,

$$\frac{M^2}{e^2/\pi}\varphi(k) = (N_f - N_c)\left(\frac{1}{k} + \frac{1}{1-k}\right)\varphi(k) - N_c \mathcal{P}\int_0^1 dp\frac{\varphi(p)}{(p-k)^2}$$
$$+ N_f \int_0^k dp\frac{\varphi(p)}{p^2} + N_f \int_k^1 dp\frac{\varphi(p)}{(1-p)^2}. \tag{11.27}$$

The derivation goes as follows. First, integrating eqn. (11.20) we get $\varphi(k) = -\varphi(1-k) + \text{const}$. Then taking $\varphi(1) = 0$ we get,

$$\varphi(k) = -\varphi(1-k). \tag{11.28}$$

Now $\varphi(1) = 0$ implies $\int_0^1 dk\Phi(k) = 0$, which was our assumption above. Then, differentiating (11.27) we do get (11.26), and by the last equation we also get that there is no extra integration constant.

We would like to comment on the issue of the Hermiticity of the "Hamiltonian" M^2. Naively, it seems that M^2 is not Hermitian with respect to the scalar product $<\psi|\varphi> = \int_0^1 dk\psi^\star(k)\varphi(k)$, since the Hermitian conjugate of (11.27) is,

$$\left(\frac{M^2}{e^2/\pi}\right)^\dagger \varphi(k) = (N_f - N_c)\left(\frac{1}{k} + \frac{1}{1-k}\right)\varphi(k)$$
$$- N_c \mathcal{P}\int_0^1 dp\frac{\varphi(p)}{(p-k)^2} - N_f\frac{1}{k^2}\int_0^k dp\varphi(p) - N_f\frac{1}{(1-k)^2}\int_k^1 dp\varphi(p). \tag{11.29}$$

However, as we shall see in the next subsection, the numerical solution yields real eigenvalues and eigenfunctions. Therefore, at least on the subspace which is spanned by the eigenfunctions, namely real functions that are zero at $k = 0, 1$ and anti-symmetric with respect to $k = \frac{1}{2}$, the operator M^2 is Hermitian. Note that (11.29) is "more regular" than (11.27), as in (11.27) it is $\varphi(p)/p^2$ that appears in the integration from zero.

Equation (11.27) is similar to the 't Hooft equation for a massive single flavor large N_c QCD$_2$, with $m^2 = \frac{e^2 N_f}{\pi}$. It differs from 't Hooft's equation by having two additional terms (the two last terms in (11.27)). It suggests that the dynamics that governs the lowest state of the multi-flavor model is given, approximately, by a model of a massive "glueball" with an $SU(N_c)$ gauge interaction and additional terms which are proportional to N_f.

Before we present our solution of (11.27) it is important to note that it is only an approximate solution. We neglected the three-current state with, a priori, no justification. We shall see, however, that the restriction to the truncated two-current sector is an excellent approximation for the lowest massive meson.

11.3.3 The two-current mesonic spectrum

The most convenient way to solve (11.27) is to expand $\varphi(k)$ in the basis,

$$\varphi(k) = \sum_{i=0}^{\infty} A_i \left(k - \frac{1}{2} \right) [k(1-k)]^{\beta+i} . \tag{11.30}$$

The value of β is chosen so that the Hamiltonian will not be singular near $k \to 0$ or $k \to 1$. This consideration leads to the equation,

$$\left(\frac{N_f}{N_c} - 1 \right) - \frac{N_f/N_c}{\beta+1} + \beta\pi \cot \beta\pi = 0, \tag{11.31}$$

as derived from eqn. (11.29). Had we started with (11.27), it would have been $-\beta$ replacing β in (11.31), and constrained to β larger than 1.

Upon truncating the infinite sum in (11.30) to a finite sum, the eigenvalue problem reduces to a diagonalization of a matrix. So, the problem can be reformulated as,

$$\lambda N_{ij} A_j = H_{ij} A_j, \tag{11.32}$$

with,

$$N_{ij} = \int_0^1 dk \left(k - \frac{1}{2} \right)^2 (k(1-k))^{2\beta+i+j} , \tag{11.33}$$

and,

$$H_{ij} = \left(\frac{N_f}{N_c} - 1 \right) \int_0^1 dk \left(k - \frac{1}{2} \right)^2 (k(1-k))^{2\beta+i+j-1}$$
$$- \frac{N_f}{N_c} \int_0^1 dk \left(k - \frac{1}{2} \right) (k(1-k))^{\beta+i} \frac{1}{k^2} \int_0^k \left(p - \frac{1}{2} \right) (p(1-p))^{\beta+j}$$
$$- \frac{N_f}{N_c} \int_0^1 dk \left(k - \frac{1}{2} \right) (k(1-k))^{\beta+i} \frac{1}{(1-k)^2} \int_k^1 \left(p - \frac{1}{2} \right) (p(1-p))^{\beta+j}$$
$$- \int_0^1 dk dp \frac{(k - \frac{1}{2}) (k(1-k))^{\beta+i} (p - \frac{1}{2}) (p(1-p))^{\beta+j}}{(k-p)^2} \tag{11.34}$$

Hence,

$$N_{ij} = \frac{B(2\beta+i+j+2, 2\beta+i+j+2)}{2(2\beta+i+j+1)}, \tag{11.35}$$

and,

$$H_{ij} = \left(\frac{N_f}{N_c} - 1 \right) \frac{B(2\beta+i+j+1, 2\beta+i+j+1)}{2(2\beta+i+j)}$$

$$- \frac{N_f}{N_c} \frac{B(2\beta+i+j+1, 2\beta+i+j+1)}{2(2\beta+i+j)(\beta+j+1)}$$

$$+ \frac{(\beta+i)(\beta+j)B(\beta+i, \beta+i)B(\beta+j, \beta+j)}{8(2\beta+i+j)(2\beta+i+j+1)}, \tag{11.36}$$

Table 11.1. *The mass of the lowest massive meson,*
in units of $\frac{e^2 N_c}{\pi}$, as a function of N_f/N_c and β.

β	N_f/N_c	M^2
0.0000	0	5.88
0.0573	0.2	6.91
0.1088	0.4	7.91
0.1552	0.6	8.91
0.1978	0.8	9.89
0.2366	1.0	10.86
0.2725	1.2	11.83
0.3050	1.4	12.77
0.3360	1.6	13.73
0.3645	1.8	14.67

where $B(x, y)$ is the beta function,

$$B(x, y) = \frac{\Gamma(x)\Gamma(y)}{\Gamma(x + y)}. \tag{11.37}$$

In practice, the process converges rapidly and a 5×5 matrix yields the 'continuum' results.

The lowest eigenvalues of (11.27) as a function of the ratio $\frac{N_f}{N_c}$ are listed in Table 11.1 (see also Fig. 11.1). Note that by $\beta = 0, N_f/N_c = 0$ we mean the limit $\beta \to 0, N_f/N_c \to 0$.

These values are in excellent agreement with recent DLCQ calculations, as will be given in the next chapter.

The typical error is less than 0.1 %.

An interesting observation is that the eigenvalues depend linearly on N_f, Fig. 11.1. The dependence is,

$$M^2 = \frac{e^2 N_c}{\pi} \left(5.88 + 5\frac{N_f}{N_c} \right). \tag{11.38}$$

We do not have a good understanding of this observation. It is not clear why the lowest eigenvalue sits on a straight line.

In the following sections we will consider some special cases.

11.3.4 Special cases: $N_f = 1$, $N_f = N_c$ and $N_f \gg N_c$

We now discuss three special cases, the massless 't Hooft model where the fermions are in the fundamental representation with $N_f = 1$, the case of adjoint fermions namely $N_f = N_c$ and the "abelianized" model of large $N_f \gg N_c$.

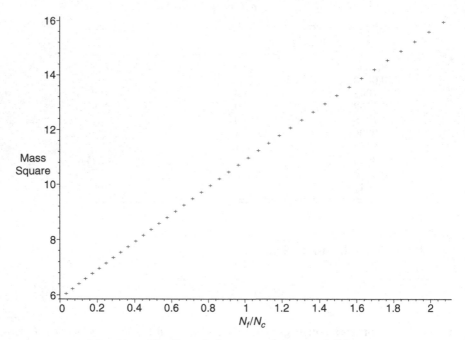

Fig. 11.1. The Green's function of the quark bilinear.

$N_f = 1$, *currentized massless 't Hooft model*

The limit $N_c \to \infty$ with $e^2 N_c$ fixed and $N_f \ll N_c$ corresponds to the well-known 't Hooft model. In this limit QCD_2 was solved exactly by 't Hooft [124] (see Chapter 10), using the fermionic basis. Let us see how our approach looks in the fermionic basis in this case. In the limit $N_f \ll N_c$ we can neglect terms which are proportional to N_f. Equation (11.27) takes the form,

$$\frac{M^2}{e^2/\pi} \varphi(k) - N_c \left(\frac{1}{k} + \frac{1}{1-k} \right) \varphi(k) - N_c \mathcal{P} \int_0^1 dp \frac{\varphi(p)}{(p-k)^2}, \qquad (11.39)$$

which is just the 't Hooft equation for the massless case. Note that (11.39) is *exact*, since in the small N_f limit the three-current state is suppressed by $N_c^{-\frac{1}{2}}$ with respect to the two-current state and therefore we can neglect it. Note also that in this eqn. (11.29) looks the same too.

Since the wave function $\varphi(k)$ is anti-symmetric, we will recover only the odd states in the spectrum of QCD_2 (the even states can be recovered by considering other sectors of the Hilbert space which decouple from the two-current state).

Though eqn. (11.39) is formally the same as the 't Hooft equation, the interpretation of $\varphi(k)$ should be different. It is the integral of the function $\Phi(k)$ which corresponds to the two-current state, namely to a mixture of 4-fermions and

2-fermions. What is the relation between the states that we find here and the mesons in 't Hooft's model?

In order to answer this question let us expand the currents in terms of fermions. It is useful to denote the current in double index notation

$$J^a(k) \to J^i_j(k) = \int_{-\infty}^{\infty} dq \left(\bar\Psi^i(q)\Psi_j(k-q) - \frac{1}{N_c}\delta^i_j\bar\Psi^k(q)\Psi_k(k-q) \right). \quad (11.40)$$

We do not bother about normal ordering, as no problem for k non zero, and we have to treat the $k=0$ part in a limiting way. The state $|\Phi\rangle$ can be written as,

$$|\Phi\rangle = \frac{1}{2N_c} \int_0^1 dk \ \Phi(k)J^i_j(-k)J^j_i(k-1)|0\rangle \quad (11.41)$$

$$= \frac{1}{2N_c} \int_0^1 dk \ \Phi(k) \int_{-\infty}^{\infty} dq \ dp \ \left(\bar\Psi^i(-q)\Psi_j(-k+q) - \frac{1}{N_c}\delta^i_j\bar\Psi^k(-q)\Psi_k(-k+q) \right)$$

$$\times \left(\bar\Psi^j(-p)\Psi_i(k+p-1) - \frac{1}{N_c}\delta^j_i\bar\Psi^k(-p)\Psi_k(k+p-1) \right)|0\rangle.$$

Note that the above expression (11.41) contains creation and annihilation fermionic operators. Written in terms of creation operators only; (11.41) reads,

$$|\Phi\rangle = \frac{1}{2N_c} \int_0^1 dk \int_0^k dq \int_0^{1-k} dp \ \Phi(k)\bar\Psi^i(-q)\Psi_j(-k+q)\bar\Psi^j(-p)\Psi_i(k+p-1)|0\rangle$$

$$- \frac{1}{2N_c^2} \int_0^1 dk \int_0^k dq \int_0^{1-k} dp \ \Phi(k)\bar\Psi^i(-q)\Psi_i(-k+q)\bar\Psi^j(-p)\Psi_j(k+p-1)|0\rangle$$

$$- \left(1-\frac{1}{N_c^2}\right) \int_0^1 dk \int_0^k dq \ \Phi(k)\bar\Psi^i(-q)\Psi_i(q-1)|0\rangle. \quad (11.42)$$

The last term in (11.42) corresponds to a meson. It can be written also as,

$$\int_0^1 dq \int_q^1 dk \ \Phi(k)\bar\Psi^i(-q)\Psi_i(q-1)|0\rangle = -\int_0^1 dq \ \varphi(q)\bar\Psi^i(-q)\Psi_i(q-1)|0\rangle, \quad (11.43)$$

which is exactly the 't Hooft meson. We conclude that the two-current state has an overlap with the 't Hooft meson and this is why (11.27) reproduces exactly the (odd part of the) spectrum of the 't Hooft model.

Large $N_f \gg N_c$ limit

In the limit $N_f \gg N_c$, with $e^2 N_f$ fixed, the truncation to two-current state should again predict exact results. The reason is that the three-current state is suppressed by $N_f^{-\frac{1}{2}}$ with respect to the two-current state.

In this limit eqn. (11.26) takes the form,

$$M^2 = \frac{e^2 N_f}{\pi} \left(\frac{1}{k} + \frac{1}{1-k} \right). \quad (11.44)$$

It describes a continuum of states with masses above $2m$, where $m^2 = \frac{e^2 N_f}{\pi}$. The interpretation is clear: in this limit the spectrum of the theory reduces to a single non-interacting meson (or "currentball") with mass m.

$N_f = N_c$, *The Adjoint Fermions Model*

The case $N_f = N_c$ is the most interesting one. It was shown that the massive spectrum of this model is equivalent to the massive spectrum of a model with a single adjoint fermion, due to 'universality' [148]. Since this model is not exactly solvable, it is interesting to see how our approach reproduces, almost accurately, previous numerical results.

The mass of the lowest massive meson, predicted by (11.27), is $M^2 = 10.86 \times \frac{e^2 N_c}{\pi}$. The values reported from DLCQ calculations are $M^2 = 10.8$ and $M^2 = 10.84$, in units of $\frac{e^2 N_c}{\pi}$, as will be detailed in the next chapter.

This agreement is very surprising. In the regime $N_f \sim N_c$, the three-current state is not suppressed by factors of color or flavor with respect to the two-current state. Why, therefore, is our approach so successful? The reason seems to be that as in the fermionic basis [38], the lowest massive state is an almost pure two-current state. However, the present approach is much more successful than the fermionic basis, where the prediction for the mass of the lowest massive boson of the adjoint model is twice as much as the lowest massive boson of the 't Hooft model. It seems that the "correct" underlying degrees of freedom are currents and not fermions, as predicted by the authors of [148].

To summarize, we have used a description of massless QCD$_2$ in terms of currents. With this basis we wrote down a 't Hooft-like equation (11.27) for the wave function of the two-current states.

The equation interpolates smoothly between the description of a single flavor model with large N_c ('t Hooft model), the adjoint fermions model $N_f = N_c$ and the large N_f model. The equation is derived by using an a-priori unjustified suppression of the three-current coupling. Nevertheless, we observe an excellent agreement with the DLCQ results for the first excited state. For higher excited states the agreement deteriorates and it is of the order of 20%.

The accuracy of the results for the first excited state, which implies that for this state the truncation of the "pair creation terms" is harmless, deserves further investigation.

11.4 The adjoint vacuum and its one-current state

Next we construct the spectrum of states, which is obtained by the action of a current on the "adjoint vacuum", in the color singlet combination. This way we get physical states, which are in a sense "one-current" states.

The "adjoint vacuum" is created from the singlet vacuum by applying the adjoint zero mode, which is taken as the limit $\epsilon \to 0$ of the product of quark and anti-quark creation operators, each one at momentum ϵ. Hence in our case,

$$|0, R\rangle = \lim_{\epsilon \to 0} \psi^i_{-1}(\epsilon)\psi^\dagger_{-1,j}(\epsilon)|0\rangle, \qquad (11.45)$$

where ψ^i_{-1} and $\psi^\dagger_{-1,j}$ are the creation operators of a quark and anti-quark respectively. We can represent the action of the above adjoint zero mode on the vacuum by the derivative of a creation current taken at zero momentum. Differentiating the current with respect to k, and acting on the vacuum we get,

$$J^{'i}_j(k)|0\rangle_{k=0-} = \sqrt{\frac{\pi}{2}}\frac{d}{dk}\int_0^\infty dp \int_0^\infty dq\, \delta(k+p+q)\psi^i_{-1}(p)\psi^\dagger_{-1,j}(q)|0\rangle_{k=0-}$$

$$= -\sqrt{\frac{\pi}{2}}\psi^i_{-1}(\epsilon)\psi^\dagger_j(\epsilon)|0\rangle_{\epsilon \to 0}. \qquad (11.46)$$

As the currents are traceless, we have to subtract the trace part for $i = j$. The latter can be neglected for large N_c. For any given N_c, results that follow are also the same after the trace is subtracted.

The adjoint vacuum we have is a bosonic one, constructed from fermion-antifermion zero modes, and as we show it can be written as the derivative of the current acting on the singlet vacuum. In the case of adjoint fermions there is another adjoint vacuum, a fermionic one, obtained by applying the adjoint fermion zero mode on the singlet vacuum.

As we showed already, $(J^a)'(0)|0\rangle$ represents the adjoint zero mode $b^\dagger(0)d^\dagger(0)|0\rangle$ (indices suppressed), for any N_f and N_c, so in particular also for $N_f = N_c$. But in the latter case the theory is equivalent to that of adjoint fermions, as follows from the equivalence theorem discussed in Section 11.2. As also stated there, states built on the adjoint vacuum above, cannot be distinguished from those built on the fermionic adjoint vacuum, the latter obtained by applying the adjoint fermions on the singlet vacuum.

The adjoint bosonic vacuum can also have flavor quantum numbers, when the fermion has flavor. This does not change our results about the mass of the new state we have. Our "currentball" will have flavor too in such a case. In our scheme of bosonization, which is the "product scheme", especially convenient when the quarks are massless, the flavor sector is decoupled, and so the flavor multiplets are given by the action of flavor zero modes, not changing the mass values.

Let us introduce the notation,

$$Z^a \equiv -\sqrt{\frac{2}{\pi}}(J^a)'(0).$$

The state we have in mind is,

$$|k\rangle = J^b(-k)Z^b|0\rangle.$$

This state is obviously a global color singlet, but in our light-cone gauge $A_- = 0$ it is also a local color singlet, as the appropriate line integral vanishes.

Now,

$$\sqrt{\frac{\pi}{2}} \left[J^a(p), Z^b \right] = \frac{1}{2} N_f \delta^{ab} \delta(p) - i f^{abc} (J^c)'(p), \qquad (11.47)$$

and thus, for $p > 0$,

$$J^a(p) Z^b |0\rangle = Z^b J^a(p) |0\rangle - i \sqrt{\frac{2}{\pi}} f^{abc} (J^c)'(p) |0\rangle = 0.$$

Hence the state $Z^b |0\rangle$ is annihilated by all the annihilation currents, and so it is indeed a colored vacuum.

Using,

$$\left[P^+, J^b(-k) \right] = k J^b(-k), \qquad (11.48)$$

we get that our state $|k\rangle$ is indeed of momentum k.

Note that when quantizing on a circle of radius R, the adjoint vacuum would be an eigenstate of P^+ with eigenvalue N_c/R. As we work in the continuum limit, we get zero.

11.4.1 The action of M^2 on the one-current states

First, we evaluate the commutator of P^- with a creation current,

$$\left[\int_0^\infty \mathrm{d}p \phi(p) J^a(-p) J^a(p), J^b(-k) \right]$$

$$= \frac{1}{2} N_f \frac{1}{k} J^b(-k) + i f^{abc} \int_0^k \mathrm{d}p \phi(p) J^a(-p) J^c(p-k)$$

$$+ i f^{abc} \int_k^\infty \mathrm{d}p \left(\phi(p) - \phi(p-k) \right) J^a(-p) J^c(p-k),$$

note that in P^- (and in P^+) we ignore contributions from zero-mode states, that is, we cut the integrals at ϵ, and then take the limit.

As P^+ and P^- act on a singlet state, and as $J^a(0)$, being the color charge, annihilates this state, the contribution from the zero modes in both P^+ and P^- is zero. Therefore it is legitimate to cut the integration limit above the zero mode and then take the cutoff to zero, as we have done. Note also that the integral of $\phi(p)$ around $p = 0$ is finite, and in fact zero when integrating over the whole line, therefore there are no divergences when we take the limit.

It is important, however, to remember that the zero mode does contribute when we act upon non singlet states, like the adjoint vacuum $Z^b|0>$ itself. When quantizing on a circle of radius R one gets that P^+ is of order $1/R$. And then, with P^- of order $e^2 R$, M^2 is R independent, and so remains finite in the continuum limit. However, this is subtle, as P^- becomes IR divergent in the continuum and needs to be regularized. This subtlety does not affect our calculation as we work in the singlet sector only.

Actually, the argument connected with P^- acting on singlets should be somewhat sharpened. Let us put the lower limit at ϵ, and let it go to zero at the end. This IR cutoff is similar to the one introduced in the derivation of 't Hooft's model discussed in Section 10. Then $J(\epsilon)$, when acting on a singlet, would go like ϵ. We have two currents in the integral, so we get ϵ^2. But then we have $1/\epsilon^2$ from the denominator, so a finite integrant. But the region for integration is of order ϵ, so indeed the total contribution goes to zero.

Now apply P^- on our state,

$$P^- J^b(-k) Z^b |0\rangle = [P^-, J^b(-k)] Z^b |0\rangle, \qquad (11.49)$$

as the Hamiltonian annihilates the color vacuum as well.

Using the commutator of the Hamiltonian with a current, we get,

$$\frac{\pi}{e^2} P^- J^b(-k) Z^b |0\rangle = \frac{1}{2} N_f \frac{1}{k} J^b(-k) Z^b |0\rangle$$
$$+ i f^{abc} \int_0^k dp \phi(p) J^a(-p) J^c(p-k) Z^b |0\rangle.$$

Note that we use the fact that annihilation currents do annihilate the colored vacuum also.

Let us apply the operator M^2 to our one-current state,

$$M^2 J^b(-k) Z^b |0\rangle = 2P^- P^+ J^b(-k) Z^b |0\rangle 2k P^- J^b(-k) Z^b |0\rangle$$

$$= \left(\frac{e^2 N_f}{\pi}\right) J^b(-k) Z^b |0\rangle + \left(\frac{2e^2}{\pi} k\right) i f^{abc} \int_0^k dp \phi(p) J^a(-p) J^c(p-k) Z^b |0\rangle. \qquad (11.50)$$

So it seems that, in the large N_f limit, the state $J^b(-k) Z^b |0\rangle$ is an (approximate) eigenstate, with eigenvalue $\frac{e^2 N_f}{\pi}$.

To see the exact dependence of the two terms in the equation above (the one- and two-current states) on N_f and N_c, we should normalize them. The normalization of $J^b(-k) Z^b |0\rangle$ is,

$$\langle 0| Z^a J^a(k) J^b(-k) Z^b |0\rangle = \langle 0| Z^a [J^a(k), J^b(-k)] Z^b |0\rangle$$
$$= \frac{1}{2} N_f k \delta(0) \langle 0| Z^b Z^b |0\rangle + i f^{abc} \langle 0| Z^a J^c(0) Z^b |0\rangle \qquad (11.51)$$
$$= \frac{1}{2} N_f k \delta(0) \langle 0| Z^b Z^b |0\rangle + N_c \langle 0| Z^b Z^b |0\rangle.$$

The second term in the last line can be neglected compared with the first, as it is a constant to be compared with $\delta(0)$ [the space volume divided by 2π].

Now,

$$\langle 0| Z^b Z^b |0\rangle = (N_c^2 - 1)\langle 0| Z^1 Z^1 |0\rangle,$$

and the factor $k\delta(0)$ is the normalization of a plane wave of momentum k. So the normalized state is, for $N_c \gg 1$,

$$\frac{1}{N_c\sqrt{\frac{1}{2}N_f}}J^b(-k)Z^b|0\rangle, \tag{11.52}$$

relative to $\langle 0|Z^1Z^1|0\rangle$.

The normalization of the second term is more complicated. A lengthy but straightforward calculation gives,

$$\left\|\left(if^{\text{def}}k\int_0^k dq\Phi(q)J^d(-q)J^f(q-k)\right)Z^e|0\rangle\right\|^2$$

$$= N_c\left(N_c^2 - 1\right)\left(\frac{1}{2}N_f\right)^2 k\delta(0)\langle 0|Z^1Z^1|0\rangle \tag{11.53}$$

$$\times k\left(\int_0^k dpp(k-p)\Phi(p)\left(\Phi(p) - \Phi(k-p)\right) - \frac{N_c}{N_f}\int_0^k dp\Phi(p)\int_0^{k-p} dqq\Phi(q)\right).$$

Using the following relations,

$$f^{abc}f^{abd} = \text{Tr}(T^cT^d) = N\delta^{cd}$$

$$f^{abc}f^{a'bc'}f^{aa'd}f^{cc'd} = \text{Tr}(T^bT^dT^bT^d)$$

$$= if^{bde}\text{Tr}(T^eT^bT^d) + \text{Tr}(T^bT^bT^dT^d) = \frac{1}{2}N^2(N^2-1),$$

we have evaluated only the terms proportional to $\delta(0)$ as they are the dominant ones.

The various momentum integrals (including the ones for the non dominant terms) are divergent for $\epsilon \to 0$, thus they should be regulated. We leave this problem for now, and assume henceforth that they are regulated and finite. For simplicity the integrals (including the factor k) appearing in the two dominant terms will be denoted R_1 and $-R_2$ in the following expressions. Note that we have $\frac{1}{\epsilon^2}$ and $\frac{1}{\epsilon}$ divergences and also $\ln(\frac{k^2}{\epsilon^2})$. It seems that these are cancelled in R_2.

Define now the normalized states,

$$|S_1\rangle = C_1\left(J^b(-k)Z^b|0\rangle\right) \tag{11.54}$$

$$|S_2\rangle = iC_2kf^{abc}\int_0^k dp\Phi(p)J^a(-p)J^c(p-k)Z^b|0\rangle, \tag{11.55}$$

where,

$$C_1\frac{1}{N_c\sqrt{\frac{1}{2}N_f}}, \quad C_2 = \frac{\frac{2}{N_f\sqrt{N_c^3}}}{\sqrt{R_1 + R_2\frac{N_c}{N_f}}}. \tag{11.56}$$

The mass eigenvalue equation of the normalized states is,

$$M^2|S_1\rangle = \frac{e^2}{\pi}N_f|S_1\rangle + \frac{e^2 N_c}{\pi}\sqrt{2}\sqrt{R_1\frac{N_f}{N_c} + R_2}|S_2\rangle,$$ (11.57)

thus, we see that in the large flavor limit, our state $|S_1\rangle$ is an eigenstate with mass

$$M = \sqrt{\frac{e^2 N_f}{2\pi}}.$$ (11.58)

In the large color limit, however, we actually get that the second term dominates by a factor of N_c. Moreover, while the first term goes to zero in the large N_c limit, due to the factor of e^2, the second term survives in that limit.

12

DLCQ and the spectra of QCD with fundamental and adjoint fermions

12.1 Discretized light-cone quantization

So far we have analyzed the mesonic spectra of QCD in low dimensions using the methods of the large N_c limit ('t Hooft model), and bosonization or currentization. In both these methods we have chosen a light-cone gauge and implemented a correlated light-front quantization. To get a broader perspective of the spectra in this framework, and in particular to extract the mesonic spectra using fermionic degrees of freedom with finite N_c, we now invoke another tool, the discrete light-cone quantization. We first describe the method and then apply it to both QCD with fundamental quarks and with adjoint quarks.

The discretized light-cone quantization (DLCQ) is a method devised to compute spectra and wave functions of physical states of quantum field theories.[1] It is based on the following ingredients:

- A Hamiltonian formulation of the theory.
- Calculations in momentum representation.
- Periodic boundary conditions and hence discretized momenta.
- Light-front quantization.

The Hamiltonian approach is used since it is more convenient for analyzing the structure of bound states. The periodic boundary conditions assure that charges associated with symmetries are strictly conserved. For a conserved current $\partial^+ J_+ + \partial^- J_- = 0$ the light-front charge is conserved,

$$Q(x^+) \equiv \int_{-L}^{L} dx^- J^+(x^-, x^+) \qquad \frac{dQ(x^+)}{dx^+} = 0, \qquad (12.1)$$

provided that,

$$J^+(x^+, +L) - J^+(x^+, -L) = 0, \qquad (12.2)$$

which is guaranteed by the periodic boundary conditions.

Note that, the light-front plane of constant x^+, serving as "time", is generally called "light-cone quantization", although the plane is only tangential to the light cone. In general d-dimensional space-time one may view the DLCQ

---

[1] For reviews see [48] and [47].

approach as a projection into non-relativistic dynamics since for a fixed light-cone momentum P^+ the Hamiltonian $H = P^-$ is quadratic in the transverse momenta $H = \frac{P_i^2}{2P^+} + \frac{M^2}{2P^+}$ where P^i are the transverse momenta.

In spite of a lot of progress in handling problems faced by the DLCQ there are still several unresolved issues such as:

- There is no proof that the light-front dynamics is fully equivalent to that of ordinary time evolution, in particular for massless chiral fermions.
- There are subtleties with the renormalization of a Hamiltonian matrix with a cutoff such that all physical results are independent of the cutoff.
- The implementation of a proper quantization for the system, with constraints which emerge from gauge fixing.

12.2 Application of DLCQ to QCD_2 with fundamental fermions

Instead of describing the method in general, we demonstrate the application of the DLCQ method to the case of QCD_2 with fundamental fermions. The light-front action of two-dimensional $SU(N)$ YM gauge fields coupled to Dirac fermions in the fundamental representation of the gauge group in the light-cone gauge $A^+ = 0$ reads,[2]

$$S_{QCD_2} = \int dx^+ dx^- \frac{1}{2} \text{Tr} \left[(\partial_- A_+)^2 + \bar{\Psi} \left(i\,\partial\!\!\!/ - m - \frac{g}{\sqrt{N}} \gamma_- A_+ \right) \Psi \right], \quad (12.3)$$

where $\Psi = \begin{pmatrix} \psi_L \\ \bar{\psi}_R \end{pmatrix}$, with ψ_L and $\bar{\psi}_R$ are Weyl fermions, the trace is over the color indices which are not written explicitly. To simplify the analysis we restrict ourselves to the case of a single flavor.

The corresponding equations of motion take the form,

$$i\partial_- \psi_L = m\psi_R, \quad (i\partial_+ + gA_+)\psi_R = m\psi_L, \quad \partial_-^2 A_+^a = g\psi_R^\dagger T^a \psi_R. \quad (12.4)$$

One can then express ψ_L and A_+ in terms of ψ_R only,

$$\psi_L(x^-, x^+) = \frac{-im}{2} \int_{-L}^{+L} dy^- \epsilon(x^- - y^-)\psi_R(x^+, y^-)$$

$$A_+^a(x^-, x^+) = \frac{g}{2} \int_{-L}^{+L} dy^- |x^- - y^-| \psi_R^\dagger T^a \psi_R(x^+, y^-), \quad (12.5)$$

where $\epsilon(x^- - y^-)$ is $+1$ for positive argument and -1 for negative.

---

[2] The application of the discrete light-front quantization to two-dimensional QCD was done in [128] and in [127].

The light-cone momentum and energy are given by,

$$P^+ = \int_{-L}^{+L} dx^- \, \psi_R^\dagger i \partial_- \psi_R (x^+, x^-)$$

$$P^- = \frac{-im^2}{2} \int_{-L}^{+L} dx^- \int_{-L}^{+L} dy^- \, \psi_R^\dagger (x^-) \epsilon(x^- - y^-) \psi_R (y^-)$$

$$- \frac{g^2}{2} \int_{-L}^{+L} dx^- \int_{-L}^{+L} dy^- \, \psi_R^\dagger (x^-) T^a \psi_R (x^-) |x^- - y^-| \psi_R^\dagger (y^-) T^a \psi_R (y^-).$$

$$(12.6)$$

When one substitutes into these expressions the expansion of the fields in anti-commuting modes, subjected to anti-periodic boundary conditions, one gets,

$$P^+ = \frac{2\pi}{L} \sum_{n=\frac{1}{2}, \frac{3}{2}, \ldots} n(b_n^\dagger b_n + d_n^\dagger d_n), \qquad (12.7)$$

where,

$$\psi_R (x^-) = \frac{1}{\sqrt{2L}} \sum_{n=\frac{1}{2}, \frac{3}{2}, \ldots} \left[b_n e^{-i\frac{n\pi x^-}{L}} + d_n^\dagger e^{i\frac{n\pi x^-}{L}} \right]. \qquad (12.8)$$

The creation and annihilation operators $b_n^\dagger, d_n^\dagger, b_n, d_n$ are all taken to be in the fundamental representation of $SU(N)$ and obey the usual algebra,

$$\{b_n^\dagger, b_m\} = \delta_{nm} \quad \{d_n^\dagger, d_m\} = \delta_{nm}. \qquad (12.9)$$

Since the eigenvalues of the momentum are proportional to $\frac{2\pi}{L}$ it is natural to define a dimensionless momentum,

$$K = \frac{L}{2\pi} P^+. \qquad (12.10)$$

Similarly one defines a dimensionless Hamiltonian,

$$H \equiv \frac{2\pi}{L} \frac{1 - \hat{\lambda}^2}{m^2} P^- \quad \hat{\lambda} \equiv \sqrt{\frac{1}{1 + \frac{\pi m^2}{g^2}}}, \qquad (12.11)$$

where we introduce the dimensionless coupling $\hat{\lambda}$. The rationale behind this parameterization is that the spectrum and wave function depend, apart from an overall mass scale, only on the ratio of $\frac{g}{m}$. The Hamiltonian H is decomposed into a free kinetic term H_0 and the potential V,

$$H = (1 - \hat{\lambda}^2)H_0 + \hat{\lambda}^2 V, \qquad (12.12)$$

where,

$$H_0 = \sum_{n=\frac{1}{2}, \frac{3}{2}, \ldots} \frac{1}{n} (b_n^\dagger b_n + d_n^\dagger d_n), \qquad (12.13)$$

and,

$$V = \frac{1}{\pi} \sum_{k=-\infty}^{\infty} J^a(k) \frac{1}{k^2} J^a(-k), \quad k \neq 0 \tag{12.14}$$

with the currents given by,

$$J^a(k) = \sum_{k=-\infty}^{\infty} [\theta(n)b_n^\dagger + \theta(-n)d_n]T^a[\theta(n-k)b_{n-k}^\dagger + \theta(k-n)d_{k-n}]. \tag{12.15}$$

We already have the expression for the potential (11.16).

Note that there is no contribution from $k = 0$, since we apply the Hamiltonian P^- only on singlet states, and $j^a(0)$ on these vanishes. Normal ordering of the potential gives,

$$V =: V : + \frac{\hat{\lambda}^2 C_F}{\pi} \sum_{n=\frac{1}{2},\frac{3}{2},\dots} \frac{I_n}{n} (b_n^\dagger b_n + d_n^\dagger d_n), \tag{12.16}$$

where $C_F = \frac{N^2-1}{2N}$ is the second Casimir operator in the fundamental representation, and I_n is the self-induced inertia $I_n = -\frac{1}{2n} + \sum_{m=1}^{n+1/2} \frac{1}{m^2}$. The self-induced inertia terms cancel the infrared singularity in the interaction term in the continuum limit. $: V :$ involves a sum of eight quartic terms in the fermionic creation and annihilation operators. For instance one such term is,

$$-\frac{1}{4N} \left(N\delta_{c_2}^{c_1}\delta_{c_4}^{c_3} - \delta_{c_4}^{c_1}\delta_{c_2}^{c_3} \right) \frac{1}{(n_4 - n_2)^2} \delta_{n_4-n_2+n_3-n_1,0} b_{n_4}^{\dagger c_4} b_{n_3}^{\dagger c_3} b_{n_2,c_2} b_{n_1,c_1}, \tag{12.17}$$

where c_i are the color indices that have been suppressed before and there is an implicit summation over the half integers n_i such that the momentum is conserved. Now since P^- and P^+ (or H and K) commute they can be diagonalized simultaneously. One fixes the value of $K = 1, 2, 3\dots$ and the corresponding Fock space is finite dimensional. One then diagonalizes H in the restricted subspace of gauge singlets such that the masses are given by,

$$M^2 = 2P^+P^- = \frac{2m^2}{1 - \hat{\lambda}^2} KH(K). \tag{12.18}$$

Notice that the dependence of the invariant masses on L the size of the space drops out.

In Fig. 12.1 the DLCQ spectrum of low-lying mesons is drawn as a function of m/g for $N = 2,3,4$ and compared with the t' Hooft large N calculation. A comparison with lattice calculation is presented in Fig. 12.2.

In performing these calculations it was found that, except for very small quark masses, there is a quick convergence of the numerics. This is a manifestation of the fact that the lowest Fock states dominate the hadronic state. It was found out that typically the momentum carried by sea quarks is less than one percent.

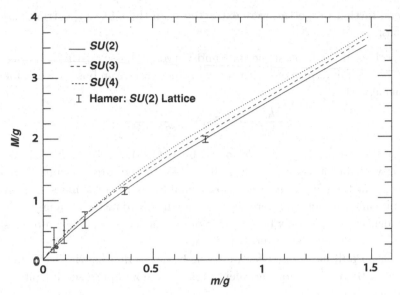

Fig. 12.1. Comparison of the DLCQ meson spectra for N = 2,3,4 and the spectrum derived from lattice calculations [127].

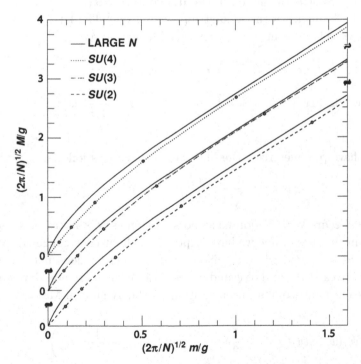

Fig. 12.2. Comparison of the DLCQ meson spectra for N = 2,3,4 and the 't Hooft large N spectrum [127].

Several further properties were extracted from the DLCQ spectrum and wave functions:

- The scaling of the lightest mesonic and baryonic masses with N. It was found that there is fair agreement with the result deduced from bosonization for small $\frac{m}{g}$, namely, that

$$\frac{M_{\text{meson}}}{M_{\text{baryon}}} = 2 \sin\left[\frac{\pi}{2(2N-1)}\right]. \tag{12.19}$$

  The results "measured" were found to be $1, .62(5), .46(4)$ for $N = 2, 3, 4$ comparing with bosonization result $1, .618, .445$. In the large N limit this result implies that the baryon mass is proportional to N times the mass of the meson.
- The mesonic form factors were shown to be in accordance with analytical work.
- The "deuteron", a loosely bound state of two nucleons, was shown to be stable, in QCD_2 with two colors and two flavors.
- The "anti Pauli-blocking" effect, for which the sea quarks with the same flavor as that of the majority of the valence ones, are not suppressed in spite of their fermionic nature.

12.3 The spectrum of QCD_2 with adjoint fermions

Our starting point is the action of two-dimensional $SU(N)$ YM theory coupled to Majorana fermions in the adjoint representation.[3] The latter is expressed in terms of a traceless Hermitian matrix ψ_{ij}. The action reads,

$$
\begin{aligned}
S_{\text{adj}} &= \int d^2x \text{Tr}\left[i\psi^T \gamma^0 \gamma^\mu D_\mu \psi - m\psi^T \gamma^0 \psi - \frac{1}{4g^2} F_{\mu\nu} F^{\mu\nu}\right] \\
&= \int dx^+ dx^- \text{Tr}\left[i(\psi \partial_+ \psi + \bar\psi \partial_- \bar\psi) - i\sqrt{2} m \bar\psi \psi + \frac{1}{2g^2}(\partial_- A_+)^2 + A_+ J^+\right],
\end{aligned}
\tag{12.20}
$$

where we have parameterized the Majorana fermions as follows,

$$\psi_{ij} = \frac{1}{\sqrt{\sqrt{2}}}\begin{pmatrix}\psi_{ij} \\ \bar\psi_{ij}\end{pmatrix} \qquad J^+_{ij} = 2\psi_{ik}\psi_{kj}, \tag{12.21}$$

where ψ and $\bar\psi$ are Weyl Majorana spinors written as $N \times N$ traceless Hermitian matrices. In the second line we have imposed the light-cone gauge $A_- = A^+ = 0$ and used $\gamma^0 = \sigma_2$ and $\gamma^1 = i\sigma_1$. Note that the action does not include time (x^+) derivatives of A_+ and of $\bar\psi$ and hence both of them are non-dynamical. The equal time (x^+) anti-commutation relation for the dynamical Majorana fermions is given by,

$$\{\psi_{ij}(x^-), \psi_{kl}(y^-)\} = \frac{1}{2}\delta(x^- - y^-)\left(\delta_{il}\delta_{jk} - \frac{1}{N}\delta_{ij}\delta_{kl}\right). \tag{12.22}$$

---

[3] Two-dimensional QCD with adjoint fermions was analyzed in several papers. Here we follow [38], [72], [147].

In analogy to the expression of P^+ and P^- given for the fundamental fermions in Section 12.2 and for the bosonized case in Section 11.3, we have now,

$$P^+ = \int dx^- \text{Tr}[i\psi\partial_-\psi]$$

$$P^- = \int dx^- \text{Tr}\left[\frac{-im^2}{2}\psi\frac{1}{\partial_-}\psi - \frac{1}{2}g^2 J^+\frac{1}{\partial_-^2}J^+\right]. \tag{12.23}$$

Denoting by Φ the physical states of the system, that obey the zero charge condition,

$$\int dx^- J^+|\Phi> = 0, \tag{12.24}$$

which is simultaneously an eigenstate of both P^+ and P^- since $[P^+, P^-] = 0$, the spectrum is then determined as usual in the light-cone quantization via,

$$2P^+P^-|\Phi> = M^2|\Phi> . \tag{12.25}$$

Next we introduce the mode expansion and transform the expressions from the configuration space to the space of momenta. In Section 12.2 we have done this directly in the discretized formalism. Here for completeness we first consider a continuous momentum and then perform the discretization.

The mode expansion reads,

$$\psi_{ij}(x^-) = \frac{1}{2\sqrt{\pi}}\int_0^\infty dk^+ [b_{ij}(k^+)e^{-ik^+x^-} + b_{ij}^\dagger(k^+)e^{ik^+x^-}], \tag{12.26}$$

and the non-trivial part of the algebra of the creation and annihilation operators is given by,

$$\{b_{ij}(k^+), b_{kl}(q^+)\} = \frac{1}{2}\delta(k^+ - q^+)\left(\delta_{il}\delta_{jk} - \frac{1}{N}\delta_{ij}\delta_{kl}\right). \tag{12.27}$$

From here on we will omit the $+$ of k^+ and denote it as k. Plugging the mode expansion into (12.23) we get,

$$P^+ = \int_0^\infty dk \, kb_{ij}^\dagger(k)b_{ij}(k), \tag{12.28}$$

and

$$P^- = \frac{m^2}{2}\int_0^\infty \frac{dk}{k}b_{ij}^\dagger(k)b_{ij}(k) + \frac{g^2 N}{\pi}\int_0^\infty dkC(k)b_{ij}^\dagger(k)b_{ij}(k)$$

$$+ \frac{g^2 N}{2\pi}\int_0^\infty dk_1 dk_2 dk_3 dk_4 [A(k_i)\delta(k_1 + k_2 - k_3 - k_4)b_{kj}^\dagger(k_3)b_{ji}^\dagger(k_4)b_{kl}(k_1)b_{li}(k_2)$$

$$+ B(k_i)\delta(k_1 + k_2 + k_3 - k_4)(b_{kj}^\dagger(k_4)b_{kl}(k_1)b_{li}(k_2)b_{ij}(k_3)$$

$$- b_{kj}^\dagger(k_1)b_{jl}^\dagger(k_2)b_{li}^\dagger(k_3)b_{ki}(k_4))], \tag{12.29}$$

where

$$A(k_i) = \frac{1}{(k_4 - k_2)^2} - \frac{1}{(k_1 + k_2)^2},$$

$$B(k_i) = \frac{1}{(k_2 - k_3)^2} - \frac{1}{(k_1 + k_2)^2},$$

$$C(k) = \int_0^k dp \frac{k}{(p-k)^2}. \tag{12.30}$$

From these expressions of P^+ and P^- it is obvious that the vacuum is annihilated by both P^+ and P^-,

$$P^+|0> = 0 \quad P^-|0> = 0. \tag{12.31}$$

The bosonic and fermionic states of the system take the following form,

$$|\Phi_b(p^+)> = \sum_{j=1}^{\infty} \int_0^{P^+} dk_1 \ldots dk_{2j} \delta \left(\sum_{i=1}^{2j} k_i - P^+ \right)$$

$$f_{2j}(k_1, k_2, \ldots k_{2j}) N^{-j} \mathrm{Tr}[b^\dagger(k_1) \ldots b^\dagger(k_{2j})]|0>,$$

$$|\Phi_f(p^+)> = \sum_{j=1}^{\infty} \int_0^{P^+} dk_1 \ldots dk_{2j+1} \delta \left(\sum_{i=1}^{2j} k_i - P^+ \right)$$

$$f_{2j}(k_1, k_2, \ldots k_{2j}) N^{-j} \mathrm{Tr}[b^\dagger(k_1) \ldots b^\dagger(k_{2j+1})]|0>, \tag{12.32}$$

where the wave functions obey the cyclicity relation due to the fermionic nature of the creation and annihilation operators,

$$f_i(k_2, k_3, \ldots, k_i, k_1) = (-1)^{i-1} f_i(k_1, k_2, \ldots k_i). \tag{12.33}$$

Unlike the case of fundamental fermions, pairs of adjoint fermions are not suppressed by additional factor of $\frac{1}{N}$ and hence the eigenstates are generated by applying operators on the vacuum with a mixture of different numbers of creation operators. This renders the extraction of the spectrum for adjoint fermions much harder to determine than that of the fundamental ones. These states are obviously eigenstates of P^+. We will have to ensure that they are also eigenstates of P^-. Following the same procedure as for 't Hooft's model of Chapter 10 and of Chapter 11 one derives a set of equations for the wavefunctions f_i by applying (12.25) on the bosonic and fermionic eigenstates which take the form,

$$M^2 f_i(x_1, x_2, \ldots x_i) = \frac{m^2}{x_1} f_i(x_1, x_2, \ldots x_i)$$

$$+ \frac{g^2 N}{\pi (x_1 + x_2)^2} \int_0^{x_1 + x_2} dy f_i(y, x_1 + x_2 - y, x_3, \ldots x_i)$$

$$\frac{g^2 N}{\pi} \int_0^{x_1 + x_2} \frac{dy}{(x_i - y)^2} [f_i(x_1, x_2, \ldots x_i) - f_i(y, x_1 + x_2 - y, x_3, \ldots x_i)]$$

$$\frac{g^2 N}{\pi} \int_0^{x_1} dy \int_0^{x_1-y} dz f_{i+2}(y, z, x_1 - y - z, x_2, \ldots x_i) \left[\frac{1}{(y+z)^2} - \frac{1}{(x_1-y)^2} \right]$$

$$\left[\frac{1}{(x_1+x_2)^2} - \frac{1}{(x_2+x_3)^2} \right] \pm \text{cyclic}, \tag{12.34}$$

where $x_i = \frac{k_i^+}{P^+}$ and the last term of the equation stands for cyclic permutations of $(x_1, x_2, \ldots, x_i)$ which for odd i comes with a $+$ sign and for even i with alternating signs. Similar to what happens in the 't Hooft model, the equation does not have an ambiguity once we incorporate a principal value prescription to the Coulomb double pole since at $x_1 = y$ the numerator also vanishes.

At this point we implement the idea of discretizing the light-cone momenta in the following way,

$$x \rightarrow \frac{n}{K} \qquad \int_0^1 dx \rightarrow \frac{2}{K} \sum_{\text{odd } n>0}^K, \tag{12.35}$$

where n is an odd positive integer and $K \rightarrow \infty$ is the continuum limit. The constraint $\sum_{j=1}^i x_j = 1$ eliminates all states with over K partons, where a parton is a state created from the vacuum by a single creation operator. In this way the discretized eigenvalue problem becomes finite dimensional. With this discretization the Fourier transform (12.26) translates into a sum,

$$\psi_{ij}(x^-) = \frac{1}{2\sqrt{\pi}} \sum_{\text{odd } n>0} [b_{ij}(n)e^{-ik^+ x^-} + b_{ij}^\dagger(n)e^{ik^+ x^-}]. \tag{12.36}$$

Similar to (12.8), the creation and annihilation operators of (12.27) also take discretized values, and obviously the Dirac delta function in (12.27) is replaced by a Kronecker delta function. The eigenvalue problem now reads,

$$2P^+ P^- = K \left[\frac{g^2 N}{\pi} T + m^2 V \right], \tag{12.37}$$

where the mass term is given by,

$$V = \sum_n \frac{1}{n} b_{ij}^\dagger(n) b_{ij}(n), \tag{12.38}$$

and,

$$T = 4 \sum_n b_{ij}^\dagger(n) b_{ij}(n) \sum_m^{n-2} \frac{1}{(n-m)^2} +$$

$$\frac{2}{N} \sum_m \{ \delta_{n_1+n_2, n_3+n_4} \left[\frac{1}{(n_4-n_2)^2} - \frac{1}{(n_1+n_2)^2} \right] b_{kj}^\dagger(n) b_{ji}^\dagger(n) b_{kl}(n) b_{li}(n)$$

$$+ \delta_{n_1+n_2+n_3, n_4} \left[\frac{1}{(n_3+n_2)^2} - \frac{1}{(n_1+n_2)^2} \right]$$

$$b_{kj}^\dagger(n_4) b_{kl}(n_1) b_{li}(n_2) b_{ij}(n_3) - b_{kj}^\dagger(n_1) b_{jl}^\dagger(n_2) b_{li}^\dagger(n_3) b_{ki}(n_4) \}, \tag{12.39}$$

where all the summations are over positive odd integers.

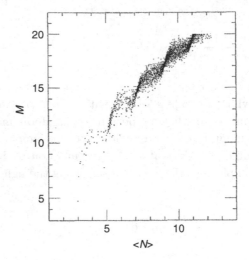

Fig. 12.3. The spectrum of fermionic states for K = 25, $m = 0$ [38].

One chooses a basis of states normalized to 1 in the large N limit,

$$\frac{1}{N^{i/2}\sqrt{s}}\text{Tr}[b^\dagger(n_1)\dots b^\dagger(n_i)]|0> \quad \sum_{j=1}^{i} n_j = K. \qquad (12.40)$$

The states are defined by ordered partitions of K into i positive odd integers, modulo cyclic permutations. If $(n_1, n_2, \dots, n_i)$ is taken into itself by s out of i possible cyclic permutations, then the corresponding state receives a normalization factor $\frac{1}{\sqrt{s}}$. Otherwise $s = 1$. For even i, however, all partitions of K where i/s is odd do not give rise to states.

Using the discretized Hamiltonian and the basis of states (12.40) one can diagonalize the Hamiltonian and compute the spectrum for a range of values of K and then extrapolate the results to infinite K, the continuum limit. One can extract certain properties of the spectrum also from the results at a fixed large K. In particular the dependence of the spectrum on the mass of the adjoint quark m is also of interest and the special cases of $m = 0$ and $m^2 = g^2 N/\pi$ where the model is supersymmetric.

The fermionic spectrum found by diagonalizing the system with $K = 25$ for the massless case and for $m^2 = \frac{g^2 N}{\pi}$ is described in Figs. 12.3 and 12.4 in the form of the mass of the bound state as a function of the expectation value of the parton number. The bosonic spectrum using $K = 24$ for the two masses is drawn in Figs. 12.5 and 12.6.

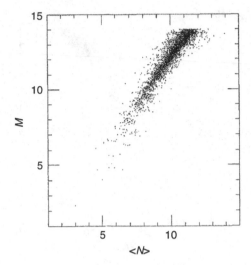

Fig. 12.4. The spectrum of fermionic states for K = 25, $m^2 = g^2/\pi$ [38].

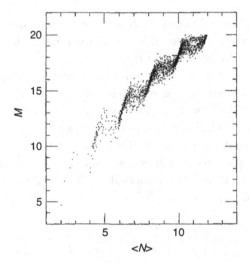

Fig. 12.5. The spectrum of bosonic states for K = 25, $m = 0$ [38].

The characteristic features of the spectra are the following:

- The density of states increases rapidly with the mass, and almost all the states lie within a band bounded by two $< N > \sim M$ lines. The system admits a Hagedorn behavior,

$$\rho(m) \sim m^\alpha e^{\beta m}, \qquad (12.41)$$

where $\rho(m)$ is the density and from the data it follows that $\beta \sim 0.7\sqrt{\frac{\pi}{g^2 N}}$.

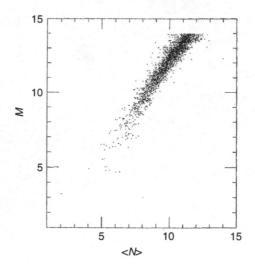

Fig. 12.6. The spectrum of bosonic states for K = 25, $m^2 = g^2/\pi$ [38].

- The mass increases roughly linearly with the average number of partons. Such a behavior characterizes a system of large N non-relativistic particles connected into a closed string by harmonic springs.
- For the low-lying states the wave function strongly peaks on states with a definite number of partons. For instance, for K = 25 the ground state has a probability of 0.9993 of consisting of 3 partons, and the first excited state has a probability of 0.99443 of consisting of 5 partons.
- Thus the low-lying states can be well approximated by truncating the diagonalization to a single parton number sector. For instance the bosonic ground state can be derived from a truncation of (12.34) to a two-parton sector which yields the following equation,

$$M^2\phi(x) = m^2\phi(x)\left(\frac{1}{x} + \frac{1}{1-x}\right) + \frac{2g^2 N}{\pi}\int_0^1 dy\,\frac{\phi(x) - \phi(y)}{(y-x)^2}, \qquad (12.42)$$

with $\phi(x) = f_2(x, 1-x)$. Note that this equation is the 't Hooft equation discussed in Chapter 10 with the replacement of $g^2 \to 2g^2$. This difference stems from the fact that unlike for mesons built from fundamental quarks, here there are two color flux tubes connecting two partons.

- Due to the fermionic statistics $\phi(x) = -\phi(1-x)$ half of the states of the 't Hooft model including the ground state are now excluded. In particular for $m = 0$ the state $\phi(x) = 1$ which associates with a massless bound state is missing. The absence of a massless ground state even in the limit of $m \to 0$ can be explained huristically as follows. For $m = 0$ the mass of the states is measured in units of the coupling constant g and hence the massless limit can be achieved in the strong coupling limit $g \to \infty$ for which the action takes the form,

$$S = \int d^2x \, \text{Tr}[i\Psi^T \gamma^0 \gamma^\mu \partial_\mu \Psi + A_\mu J^\mu]. \qquad (12.43)$$

Now the left and right currents $J_{ij}^{\pm}$ constitute two independent level N affine Lie algebras for which we have seen in Chapter 3 the corresponding Virasoro anomaly is,

$$c = c_0 - (N^2 - 1)\frac{k}{k + N} = \frac{N^2 - 1}{2} - (N^2 - 1)\frac{k}{k + N}, \qquad (12.44)$$

where c_0 is the central charge before gauging and k is the ALA level. Since $k = N$ it is obvious that $c = 0$ and hence there is no massless bound state. For fundamental quarks in the same limit we get, by taking $k = 1$ *and* $c_0 = N$, that $c = 1$, which means that for this case there is a massless bound state.

13

The baryonic spectrum of multiflavor QCD_2 in the strong coupling limit

We are now going to compute the baryonic spectrum of QCD_2, for which as it turns out the bosonic formulation is very convenient. The mesonic spectrum was found earlier, using large N with quark fields as variables in Chapter 10, as well as using currents as building blocks in Section 11.3. For the baryon spectrum, however, the large N limit, in terms of fermionic fields, is not the natural framework to use since in such a picture the baryon is a bound state of a large number N of constituents. Instead, it will be shown in this chapter that the bosonized version of QCD_2 in the strong coupling limit provides an effective description of the baryons.[1] We will start by deriving the effective action at the strong coupling limit. It will be argued that for the purpose of extracting the low-lying baryons, one can in fact use the product scheme instead of the $U(N_c \times N_c)$ scheme, with the former being more suitable for our purposes. Once the effective action is written down we will search for soliton solutions that carry a baryon number. It will be shown that for a static configuration the effective action reduces to a sum of sine-Gordon actions. Using the knowledge acquired on solitons, in Chapter 5, it will be easy to write down the classical baryonic configuration. We will then semi-classically quantize these solitons. This problem will be mapped into a quantum mechanical model on a $CP^{(N_f-1)}$ manifold. The energy and charges of the quantized soliton can be derived and thus the spectrum of the baryons is determined. We then analyze the quark flavor content of the baryons and discuss multi-baryon states. Finally, we include meson-baryon scattering, this time also for the case of any coupling.

13.1 The strong coupling limit

It turns out that the mass term plays an essential role in the determination of classical soliton solutions in 1+1 space-time dimensions. It is therefore required to switch on this term before deducing the low energy effective action. As was explained in Chapter 6, we know how to do this rigorously only in the scheme of $U(N_F N_C)$. It will turn out, however, that the product scheme can be used for the low mass states in the strong coupling limit.

---

[1] The spectrum of baryons of two-dimensional QCD extracted in the strong coupling limit was derived in [75].

Our starting point is the last equation of Chapter 9. In the strong coupling limit $\frac{e_c}{m_q} \to \infty$, the fields in $\tilde{h}$ which contribute to $\tilde{H}$ will become infinitely heavy. The sector $\tilde{g}l \subset \frac{SU(N_F N_C)}{SU(N_C)}$, however, will not acquire mass from the gauge interaction term. Since we are interested only in the light particles we can, in the strong coupling limit, ignore the heavy fields, if we first normal order the heavy fields at the mass scale $\tilde{\mu} = \frac{e_c \sqrt{N_F}}{\sqrt{2\pi}}$. Using the relation, for a given operator O,

$$\left(\frac{\tilde{\mu}}{\tilde{m}}\right)^\Delta N_{\tilde{\mu}} O = N_{\tilde{m}} O, \tag{13.1}$$

to perform the change in the scale of normal ordering, and then substituting $h_b^a = \delta_b^a$, we get for the low energy effective action,

$$S_{\text{eff}}[u] = S[\tilde{g}] + S[l] + \frac{1}{2}\int d^2x \partial_\mu \phi \partial^\mu \phi$$
$$+ cm_q \tilde{\mu} N_{\tilde{\mu}} \int d^2x \text{Tr}(e^{-i\sqrt{\frac{4\pi}{N_C N_F}}\phi} \tilde{g}l + e^{+i\sqrt{\frac{4\pi}{N_C N_F}}\phi} l^\dagger \tilde{g}^\dagger). \tag{13.2}$$

We can now replace the two mass scales m_q and $\tilde{\mu}$ by a single scale, by normal ordering at a certain m so the final form of the effective action becomes,

$$S_{\text{eff}}[u] = S[\tilde{g}] + S[l] + \frac{1}{2}\int d^2x \partial_\mu \phi \partial^\mu \phi$$
$$+ \frac{m^2}{N_C} N_m \int d^2x \text{Tr}(e^{-i\sqrt{\frac{4\pi}{N_C N_F}}\phi} \tilde{g}l + e^{+i\sqrt{\frac{4\pi}{N_C N_F}}\phi} l^\dagger \tilde{g}^\dagger), \tag{13.3}$$

with m given by,

$$m = \left[N_C cm_q \left(\frac{e_c \sqrt{N_F}}{\sqrt{2\pi}}\right)^{\Delta_C}\right]^{\frac{1}{1+\Delta_C}}, \tag{13.4}$$

here Δ_C, the dimension of $\tilde{h}$, is $\frac{N_C^2-1}{N_C(N_C+N_F)}$. For the $l = 1$ sector, defining $g' = \tilde{g}e^{-i\sqrt{\frac{4\pi}{N_C N_F}}\phi} \subset U(N_F)$ one gets the effective action,

$$S_{\text{eff}}[g'] = N_C S[g'] + m^2 N_m \int d^2x \text{Tr}_F(g' + g'^\dagger). \tag{13.5}$$

Thus, the low energy effective action in the $l = 1$ sector coincides with the result of the "naive" approach of the product scheme.

In the strong coupling limit $e_c/m_q \to \infty$ the low energy effective action reads,[2]

$$S[g] = N_C S[g] + m^2 N_m \int d^2x (\text{Tr}g + \text{Tr}g^\dagger), \tag{13.6}$$

with g in $U(N_F)$. Note that the analog of our strong coupling to the case of 3+1 space-time, would be that of light current quarks compared to the QCD scale Λ_{QCD}.

---

[2] From here on we omit the prime from g' so we denote $g \in U(N_F)$.

13.2 Classical soliton solutions

We now look for static solutions of the classical action. For a static field config-uration, the WZ term does not contribute. One way to see this is by noting that the variation of the WZ term can be written as,

$$\delta WZ \propto \int d^2x \varepsilon^{ij} \mathrm{Tr}(\delta g)g^\dagger(\partial_i g)(\partial_j g^\dagger), \tag{13.7}$$

and for g that has only spatial dependence $\delta WZ = 0$. Without loss of generality we may take, for the lowest energy, a diagonal $g(x)$,

$$g(x) = \left(e^{-i\sqrt{\frac{4\pi}{N_C}}\varphi_1}, \ldots, e^{-i\sqrt{\frac{4\pi}{N_C}}\varphi_{N_F}} \right). \tag{13.8}$$

For this ansatz and with a redefinition of the constant term, the action density reduces to,

$$\tilde{S}_d[g] = -\int dx \sum_{i=1}^{N_F} \left[\frac{1}{2}\left(\frac{d\varphi_i}{dx}\right)^2 - 2m^2\left(\cos\sqrt{\frac{4\pi}{N_C}}\varphi_i - 1\right)\right]. \tag{13.9}$$

This is a sum of decoupled standard sine-Gordon actions for each φ_i. The well-known solutions of the associated equations of motion are,

$$\varphi_i(x) = \sqrt{\frac{4N_C}{\pi}}\mathrm{arctg}\left[e^{\left(\sqrt{\frac{8\pi}{N_C}}mx\right)}\right], \tag{13.10}$$

with the corresponding classical energy,

$$E_i = 4m\sqrt{\frac{2N_C}{\pi}}, \quad i = 1, \ldots, N_F. \tag{13.11}$$

Clearly the minimum energy configuration for this class is when only one of the φ_i is nonzero, for example,

$$g_0(x) = \mathrm{Diag}\left(1, 1, \ldots, e^{-i\sqrt{\frac{4\pi}{N_C}}\varphi(x)}\right) \tag{13.12}$$

Conserved charges, corresponding to the vector current, can be computed using the definition,

$$Q^A[g(x)] = \frac{1}{2}\int dx \mathrm{Tr}(J_0 T^A), \tag{13.13}$$

where $\frac{1}{2}T^A$ are the $SU(N_F)$ generators and the $U(1)$ baryon number is generated by the unit matrix. This follows from $J_\mu = J_\mu^A T^A$, and in the fermionic basis $J_\mu^A = \bar{\psi}\gamma_\mu \frac{1}{2}T^A \psi$.

In particular, for eqn. (13.10), we get charges different from zero only for Q_B and Q_Y corresponding to baryon number and "hypercharge", respectively,

$$Q_B^\circ = N_C, \quad Q_Y^\circ = -\frac{1}{2}\sqrt{\frac{2(N_F-1)}{N_F}}N_C, \tag{13.14}$$

these charges are determined solely by the boundary values of $\varphi(x)$, which are,

$$\sqrt{\frac{4\pi}{N_C}}\varphi(\infty) = 2\pi, \quad \sqrt{\frac{4\pi}{N_C}}\varphi(-\infty) = 0. \tag{13.15}$$

Under a general $U_V(N_F)$ global transformation $g_o(x) \to \tilde{g}_o(x) = Ag_o(x)A^{-1}$ the energy of the soliton is obviously unchanged, but charges other than Q_B and Q_Y will be turned on. Let us introduce a parametrization of A that will be useful later,

$$A = \begin{pmatrix} & & & & z_1 \\ & A_{ij} & & & \vdots \\ & & & & \vdots \\ & & & & z_{(N_F-1)} \\ Y_1 & \cdots & \cdots & Y_{(N_F-1)} & z_{N_F} \end{pmatrix}. \tag{13.16}$$

Now,

$$\tilde{g}_o = 1 + (e^{-i\sqrt{\frac{4\pi}{N_C}}\varphi} - 1)z, \tag{13.17}$$

where $(z)_{\alpha\beta} = z_\alpha z_\beta^*$, and from unitarity $\sum_{\alpha=i}^{N_F} z_\alpha z_\alpha^* = 1$. The charges with $\tilde{g}_o(x)$ are,

$$(\tilde{Q}^\circ)^A = \frac{1}{2}N_C \operatorname{Tr}(T^A z). \tag{13.18}$$

Only the baryon number is unchanged. The discussion of the possible $U(N_F)$ representations cannot be done yet, since we are dealing so far with a classical system. We will return to the question of possible representations after quantizing the system.

13.3 Semi-classical quantization and the baryons

The next step in the semi-classical analysis is to consider configurations of the form,

$$g(x,t) = A(t)g_o(x)A^{-1}(t), \quad A(t) \in U(N_F), \tag{13.19}$$

and to derive the effective action for $A(t)$.[3] Quantization of this action corresponds to doing the functional integral over $g(x,t)$ of the above form. The effective action for $A(t)$ is derived by substituting $g(x,t) = A(t)g_o(x)A^{-1}(t)$ in the original action. Here we use the following property of the WZ action,

$$S\left[AgB^{-1}\right] = S\left[AB^{-1}\right] + S\left[g, \tilde{A}_\mu\right], \tag{13.20}$$

---

[3] The semi-classical quantization makes use of the Polyakov–Wiegmann formula [179].

where $S[g]$ is the WZW action and $S[g,\tilde{A}]$ is given by (9.25), respectively, with the gauge field $\tilde{A}_\mu$ given as,

$$i\tilde{A}_+ = A^{-1}\partial_+ A, \quad i\tilde{A}_- = B^{-1}\partial_- B; \quad A, B \in U(N_F). \tag{13.21}$$

Using the above formula for $A = B$, noting that $S(1) = 0$, and taking $A = A(t)$,

$$\partial_+ A = \partial_- A = \frac{\dot{A}}{\sqrt{2}}, \tag{13.22}$$

we get,

$$\tilde{S}\left[A(t)g_\circ(x)A^{-1}(t)\right] - \tilde{S}[g_\circ] = \frac{N_C}{8\pi} \int d^2x \mathrm{Tr}\left\{[A^{-1}\dot{A}, g_\circ][A^{-1}\dot{A}, g_\circ^\dagger]\right\}$$
$$+ \frac{N_C}{2\pi} \int d^2x \mathrm{Tr}\left\{(A^{-1}\dot{A})(g_\circ^\dagger \partial_1 g_\circ)\right\}. \tag{13.23}$$

This action is invariant under global $U(N_F)$ transformations $A \to UA$, where $U \in G = U(N_F)$. This corresponds to the invariance of the original action under $g \to UgU^{-1}$. On top of this it is also invariant under the local changes $A(t) \to A(t)V(t)$, where $V(t) \in H = SU(N_F - 1) \times U_B(1) \times U_Y(1)$, with the last two $U(1)$ factors corresponding to baryon number and hypercharge, respectively. This subgroup H of G is nothing but the invariance group of $g_\circ(x)$. In terms of $g_\circ(x)$ and $A(t)$ the charges associated with the global $U(N_F)$ symmetry, eqn. (13.13), have the form,

$$Q^B = i\frac{N_C}{8\pi} \int dx \mathrm{Tr}\left\{T^B A \left((g_\circ^\dagger \partial_1 g_\circ - g_\circ \partial_1 g_\circ^\dagger) + \left[g_\circ, \left[A^{-1}\dot{A}, g_\circ^\dagger\right]\right]\right) A^{-1}\right\}. \tag{13.24}$$

The effective action, eqn. (13.23), is an action for the coordinates describing the coset space,

$$G/H = SU(N_F) \times U_B(1)/SU(N_F - 1) \times U_Y(1) \times U_B(1)$$
$$= SU(N_F)/SU(N_F - 1) \times U_Y(1) = CP^N. \tag{13.25}$$

To see this explicitly we define the Lie algebra valued variables q^A through $A^{-1}\dot{A} = i\sum T^A \dot{q}^A$. In terms of these variables (13.23) takes the form (the part that depends on q^A),

$$S_q = \int dt \left[\frac{1}{2M}\sum_{A=1}^{2(N_F-1)}(\dot{q}^A)^2 - N_C\sqrt{\frac{2(N_F-1)}{N_F}}\dot{q}^Y\right]$$

$$\frac{1}{2M} = \frac{N_C}{2\pi} \int_{-\infty}^{\infty}(1 - \cos\sqrt{\frac{4\pi}{N_C}}\varphi)dx = \frac{\sqrt{2}}{m}(\frac{N_C}{\pi})^{3/2}. \tag{13.26}$$

The sum is over those q^A which correspond to the G/H generators and q^Y is associated with the hypercharge generator. Although the q^A seem to be a "natural" choice of variables for the action (13.23), which depends only on the combination $A^{-1}\dot{A}$, they are not a convenient choice of variables. The reason for that is the explicit dependence of the charges (13.24) on $A^{-1}(t)$ and $A(t)$ as well as on $A^{-1}\dot{A}(t)$.

Instead we found that a convenient parametrization is that of (13.16). One can rewrite the action (13.23), as well as the charges (13.24), in terms of the $z_1, \ldots, z_{N_F}$ variables, which however are subject to the constraint $\sum_{\alpha=1}^{N_F} z_\alpha z_\alpha^* = 1$. Thus,

$$\tilde{S}\left[A(t)g_\circ A^{-1}(t)\right] - \tilde{S}[g_\circ] = S[z_\alpha(t), \varphi(x)], \qquad (13.27)$$

where,

$$S\left[z_\alpha(t), \varphi(x)\right] = \tfrac{N_C}{2\pi} \int d^2x \{ (1 - \cos\sqrt{\tfrac{4\pi}{N_C}}\varphi)[\dot{z}_\alpha^* \dot{z}_\alpha$$
$$- (z_\gamma^* \dot{z}_\gamma)(\dot{z}_\beta^* z_\beta)] - i\sqrt{\tfrac{4\pi}{N_C}}\varphi' z_\alpha^* \dot{z}_\alpha \}. \qquad (13.28)$$

We can do the integral over x and rewrite (13.28) as,

$$S[z_\alpha(t)] = \frac{1}{2M} \int dt [\dot{z}_\alpha^* \dot{z}_\alpha - (z_\gamma^* \dot{z}_\gamma)(\dot{z}_\beta^* z_\beta)] - i\frac{N_C}{2} \int dt (z_\alpha^* \dot{z}_\alpha - \dot{z}_\alpha^* z_\alpha), \qquad (13.29)$$

where $1/M$ is defined in eqn. (13.26). The first term in (13.29) is the usual $CP^{(N_F - 1)}$ quantum mechanical action, while the second term is a modification due to the WZ term.

Similarly we express the $U(N_F)$ charges in terms of the z variables, using eqn. (13.24),

$$Q^C = \frac{1}{2} T_{\beta\alpha}^C Q_{\alpha\beta}$$
$$Q_{\alpha\beta} = N_C z_\alpha z_\beta^* + \frac{i}{2M}[z_\alpha z_\beta^*(z_\gamma^* \dot{z}_\gamma - \dot{z}_\gamma^* z_\gamma) + z_\alpha \dot{z}_\beta^* - z_\beta^* \dot{z}_\alpha]. \qquad (13.30)$$

Of course the symmetries of $S[z]$ are the global $U(N_F)$ group under which,

$$z_\alpha \rightarrow z_\alpha' = U_{\alpha\beta} z_\beta, \quad U \in U(N_F), \qquad (13.31)$$

and a local $U(1)$ subgroup of H under which,

$$z_\alpha \rightarrow z_\alpha' = e^{i\delta(t)} z_\alpha. \qquad (13.32)$$

As a consequence of the gauge invariance one can rewrite the action in a covariant form,

$$S[z_\alpha] = \frac{1}{2M} \int dt \mathrm{Tr}(Dz)^\dagger Dz + iN_C \int dt \mathrm{Tr}\dot{z}^\dagger z, \qquad (13.33)$$

where,

$$(Dz)_\alpha = \dot{z}_\alpha + z_\alpha(\dot{z}_\beta^* z_\beta). \qquad (13.34)$$

Constructing Noether charges of the $U(N_F)$ global invariance of (13.31) out of the action (13.33 leads to expressions identical with (13.30)). Note that in eqn. (13.34) we can view $\dot{z}_\beta^* z_\beta = ia(t)$ as a composite $U(1)$ gauge potential.

Now let us count the degrees of freedom. The local $U(1)$ symmetry allows us to take one of the zs to be real, and the constraint $\sum_\alpha z_\alpha z_\alpha^* = 1$ removes one more degree of freedom, so altogether we are left with $2N_F - 2 = 2(N_F - 1)$

physical degrees of freedom. This is exactly the dimension of the coset space $\frac{SU(N_F)}{SU(N_F-1)\times U(1)}$. The corresponding phase space should have a real dimension of $4(N_F - 1)$. Naively, however, we have a phase space of $4N_F$ dimensions and, therefore, we expect four constraints.

There are several methods of quantizing systems with constraints. Here we choose to eliminate the redundancy in the z variables and then invoke the canonical quantization procedure.[4]

But before following these lines let us briefly describe another method, through the use of Dirac's brackets. We outline the classical case. The quantum case is obtained by replacing $\{\,,\,\}$ with $i[\,,\,]$.

The first step in this prescription is to add to the Lagrangian a term of the form $\lambda(\sum_\alpha z_\alpha z_\alpha^* - 1)$, in which case the conjugate momentum π_λ of the Lagrange multiplier vanishes. By requiring that this condition be preserved in time one gets the secondary constraint $\Phi_1 = (\sum_\alpha z_\alpha z_\alpha^* - 1) = 0$. Further imposing $\dot\Phi_1 = \{\Phi_1, H\}_P = 0$, where $\{\,\}_P$ denotes a Poisson bracket, one finds another second-class constraint $\Phi_2 = \Pi \cdot z + z^\dagger \cdot \Pi^\dagger$. In addition there is a first-class constraint $\Phi_3 = \Pi \cdot z - z^\dagger \cdot \Pi^\dagger$, which corresponds to the local $U(1)$ invariance of the model. Fixing this symmetry one gets an additional constraint Φ_4. For instance one can choose the unitary gauge $\Phi_4 = z_{N_F} - z_{N_F}^*$. The next step is to compute the constraint matrix $\{\Phi_i, \Phi_j\}_P = c_{ij}$. In the constrained theory, the brackets between F and G are replaced by the Dirac brackets of those operators, given by

$$\{F, G\}_D = \{F, G\}_P - \{F, \Phi_i\}_P (c_{ij}^{-1})\{\Phi_j, G\}_P, \qquad (13.35)$$

where c_{ij}^{-1} is the inverse of the constraint matrix. Imposing the constraints as operator relations it is easy to see that z_{N_F}, Π_{N_F} and their complex conjugates can be eliminated. The brackets for the rest of the fields coincide with the results we derive below, when eliminating the constraints explicitly.

We now describe in some detail the quantization of the system using unconstrained variables. We want to choose a set of new variables so that the constraint $\sum_{\alpha=1}^{N_F} z_\alpha z_\alpha^* = 1$ is automatically fulfilled. There is a standard choice of such variables, namely (*for* $i = 1, \ldots, N_F - 1$),

$$z_i = \frac{k_i}{\sqrt{1+X}}, \quad z_i^* = \frac{k_i^*}{\sqrt{1+X}}, \quad z_{N_F} = \frac{e^{i\chi}}{\sqrt{1+X}}$$

$$X = \sum_{i=1}^{N_F-1} k_i^* k_i. \qquad (13.36)$$

The k_i, k_i^* and χ are $2N_F - 1$ real variables with no constraints on them. The phase space will now have dimension $2(2N_F - 1)$ and we still have two extra

---

[4] The quantization of the system including its constraint was done in [75]. For an alternative procedure of quantization in the presence of constraints see [181].

constraints. After some straightforward algebra we can write,

$$S[k, k^*, \chi] = \int dt L(k, k^*, \chi)$$

$$L(k, k^*, \chi) = \frac{1}{2M} k_i^* h_{ij} \dot{k}_j - i\frac{N_C}{2} \frac{k_i^* \dot{k}_i - \dot{k}_i^* k_i}{1 + X}$$

$$+ \frac{1}{2M} \frac{X}{(1 + X)^2} \dot{\chi}^2 + \dot{\chi} \left\{ \frac{i}{2M} \frac{k_i^* \dot{k}_i - \dot{k}_i^* k_i}{(1 + X)^2} + \frac{N_C}{1 + X} \right\}, \quad (13.37)$$

where,

$$h_{ij} = \frac{\delta_{ij}}{1 + X} - \frac{k_i k_j^*}{(1 + X)^2}. \quad (13.38)$$

The local $U(1)$ transformations of the z variables transcribe into the transformations,

$$\delta\chi = \epsilon(t), \quad \delta k_i = i\epsilon(t) k_i, \quad \delta k_i^* = -i\epsilon(t) k_i^*, \quad (13.39)$$

and $\delta L = -N_C \dot{\epsilon}$ just as in terms of the z variables. This local $U(1)$ symmetry can be made manifest by defining the covariant derivatives,

$$Dk_i = \dot{k}_i - i\dot{\chi} k_i \quad Dk_i^* = \dot{k}_i^* + i\dot{\chi} k_i^*. \quad (13.40)$$

The Lagrangian can then be recast in a manifestly gauge-invariant form,

$$L(k, k^*, x) = \frac{1}{2M} Dk_i^* h_{ij} Dk_j - i\frac{N_C}{2} \frac{k_i^* Dk_i - (Dk_i^*) k_i}{1 + X} + N_C \dot{\chi}. \quad (13.41)$$

Although one can now fix the gauge, for instance $\dot{\chi} = 0$, we will continue to work with (13.41). The conjugate momenta are given by,

$$\pi_i = \frac{\partial L}{\partial \dot{k}_i} = \frac{1}{2M} Dk_j^* h_{ji} - i\frac{N_C}{2} \frac{k_i^*}{1 + X}$$

$$\pi_i^* = \frac{\partial L}{\partial \dot{k}_i^*} = \frac{1}{2M} h_{ij} Dk_j + i\frac{N_C}{2} \frac{k_i}{1 + X}$$

$$\pi_\chi = \frac{\partial L}{\partial \dot{\chi}} = \frac{i}{2M} (k_i^* h_{ij} Dk_j - Dk_i^* h_{ij} k_j) + N_C \frac{1}{1 + X}. \quad (13.42)$$

Since h_{ij} is invertible we can solve for Dk_i^*, Dk_i in term of the phase space variables,

$$Dk_i^* = 2M \left[\pi_j + i\frac{N_C}{2} \frac{k_j^*}{1 + X} \right] h_{ji}^{-1}$$

$$Dk_i = 2M h_{ij}^{-1} \left[\pi_j^* - i\frac{N_C}{2} \frac{k_j}{1 + X} \right], \quad (13.43)$$

where,

$$h_{ij}^{-1} = (1 + X)(\delta_{ij} + k_i k_j^*). \quad (13.44)$$

Also,

$$\pi_\chi = i(k_i^* \pi_i^* - \pi_i k_i) + N_C, \quad (13.45)$$

giving the constraint equation,

$$\psi = \pi_\chi - i(k_i^* \pi_i^* - \pi_i k_i) - N_C = 0. \tag{13.46}$$

The canonical Hamiltonian is given by,

$$H_c = \pi_i \dot{k}_i + \pi_i^* \dot{k}_i^* + \pi_\chi \dot{\chi} - L$$

$$= 2M \left[\pi_i + i \frac{N_C k_i^*}{2(1+X)} \right] h_{ij}^{-1} \left[\pi_j^* - i \frac{N_C k_j}{2(1+X)} \right]$$

$$+ \dot{\chi}[\pi_\chi - i(\pi_i^* k_i^* - \pi_i k_i) - N_C], \tag{13.47}$$

and this can be further simplified to,

$$H_c = 2M(1+X) \left[\pi_i \pi_i^* + (\pi_i k_i)(\pi_i^* k_i^*) \right.$$

$$\left. - i \frac{N_C}{2} (\pi_i k_i - \pi_i^* k_i^*) + \frac{1}{4} \frac{N_C^2 X}{(1+X)} \right] + \dot{\chi}\psi. \tag{13.48}$$

Here H_c is obtained explicitly in terms of the canonical variables $k_i, k_i^*, \pi_i, \pi_i^*$. The $\dot{\chi}\psi$ term indicates that $\dot{\chi}$ also behaves as a Lagrange multiplier since, following the Dirac procedure, we should define,

$$H_T = H_c + \lambda(t)\psi, \tag{13.49}$$

where λ is a priori an arbitrary function of t. We could absorb the $\dot{\chi}$ in λ.

Quantization of this Hamiltonian is now essentially straightforward. Let us first consider the symmetry generators $Q_{\alpha\beta}$, which in terms of the new canonical variables take the form,

$$Q_{ij} = i(k_i \pi_j - \pi_i^* k_j^*)$$

$$Q_{i,N_F} = e^{-ix} \left[\frac{N_C k_i}{2} - i(\pi_i^* + k_i \pi_j k_j) \right]$$

$$Q_{N_F,i} = e^{ix} \left[\frac{N_C k_i^*}{2} + i(\pi_i + k_j^* \pi_j^* k_i^*) \right] = Q_{i,N_F}^*$$

$$Q_{N_F,N_F} = N_C - i(\pi_i k_i - \pi_i^* k_i^*). \tag{13.50}$$

We will now show that the H_T can be expressed in terms of the second Casimir operator of the $SU(N_F)$ group.

The second $U(N_F)$ Casimir operator is related to charge matrix elements $Q_{\alpha\beta}$ as,

$$Q_A Q^A = \frac{1}{2} Q_{\alpha\beta} Q_{\beta\alpha}. \tag{13.51}$$

A straightforward substitution gives,

$$\frac{1}{2} Q_{\alpha\beta} Q_{\beta\alpha} = (1+X)[\pi_i^* \pi_i + \pi_i k_i \pi_j^* k_j^*$$

$$- i \frac{N_C}{2} (\pi_i k_i - \pi_i^* k_i^*)] + \frac{1}{2} N_C^2 \left(1 + \frac{X}{2} \right). \tag{13.52}$$

Therefore, the Hamiltonian is,

$$H_T = 2M \left[Q^A Q^A - \frac{N_C^2}{2} \right] + \lambda(t)\psi. \tag{13.53}$$

Denoting the $SU(N_F)$ second Casimir operator by C_2, and using $Q_A Q^A = C_2 + \frac{1}{2N_F}(Q_B)^2$ we get (also applying the constraint $\psi = 0$),

$$H_T = 2M \left[C_2 - N_C^2 \frac{(N_F - 1)}{2N_F} \right]. \tag{13.54}$$

The fact that H_T is, up to a constant, the second Casimir operator, is another way to show that the charges $Q_{\alpha\beta}$ are conserved. These conserved charges will generate symmetry transformations via,

$$\delta k_i = i[\text{Tr}(\epsilon Q), k_i], \quad \delta k_i^* = i[Tr(\epsilon Q), k_i^*)]$$
$$\delta\chi = i[\text{Tr}(\epsilon Q), \chi], \tag{13.55}$$

and similar equations for the momenta π_i, π_i^*, π_χ. Here $\epsilon_{ij} = \frac{1}{2}\epsilon^A T_{ij}^A$ is the matrix of parameters. The transformation laws are derived using the constraint equation $\psi = 0$ after performing the commutator calculations. Notice that Q_{ij} and Q_{N_F, N_F} are linear in coordinates and momenta and therefore the $SU(N_F - 1) \times U_Y(1)$ transformations they generate are linear. The $Q_{N_F, i}$ and Q_{i, N_F} charges, on the other hand, have cubic terms as well (quadratic in coordinates), so that the coset-space transformations of $\frac{SU(N_F)}{SU(N_F-1)\times U(1)}$ are non-linear. This is a well-known property of CP^n models. Substitution of $Q_{\alpha\beta}$ in eqn (13.55) gives,

$$\delta k_l = i[\epsilon_{ji} k_i \delta_{jl} + e^{i\chi}\epsilon_{iN_F}\delta_{il} - e^{-i\chi}\epsilon_{N_F i} k_i k_l - \epsilon_{N_F N_F} k_l], \tag{13.56}$$

where we use $[k, \pi] = i$.

Inversely, starting with these transformation laws it is easy to verify the invariance of the action. The standard Noether procedure then gives the charges $Q_{\alpha\beta}$ in terms of the coordinates and velocities, which (not suprisingly) coincide with those given in eqn. (13.50). One could also deduce these transformation laws by making the change of variables $z_\alpha, z_\alpha^* \to k_i, k_i^*, \chi$ in (13.30) directly.

One can verify that,

$$[Q^A, Q^B] = if^{ABC} Q^C, \tag{13.57}$$

where f^{ABC} are the structure constants of the $U(N_F)$ group.

Do we have further restrictions on the physical states? We shall see now that in fact we do have. Remember that our Lagrangian (13.41) includes an auxiliary gauge field $A_\circ \equiv \dot\chi$ and thus has to obey the associated Gauss law,

$$\frac{\partial L}{\partial A_\circ} = \frac{\partial L}{\partial \dot\chi} = \pi_\chi = N_C - i(\pi_i k_i - \pi_i^* k_i^*) = 0. \tag{13.58}$$

Since π_χ is a linear combination of Q_B and Q_Y, and the first is constrained to be $Q_B = N_C$, the Q_Y is restricted as well. More specifically, $Q_Y = \bar Q_Y$, with,

$$\bar Q_Y = \frac{1}{2}\sqrt{\frac{2}{(N_F - 1)N_F}} N_C. \tag{13.59}$$

13.4 The baryonic spectrum

The masses of the baryons (13.11) and (13.54), and the two constraints on the multiplets of the physical states, namely $Q_B = N_C$ and that the multiplets contain $Q_Y = \bar{Q}_Y = \frac{1}{2}\sqrt{\frac{2}{(N_F-1)N_F}}N_C$, are the main results of the last section. All states of the multiplet with $Q_Y \neq \bar{Q}_Y$ will be generated from the state $Q_Y = \bar{Q}_Y$ by $SU(N_F)$ transformations as in (13.19). Using the above constraints we can investigate now what possible representations will appear in the low energy baryon sector. Considering states with quarks only (no anti-quarks), the requirement of $Q_B = N_C$ implies that only representations described by Young tableaux with N_C boxes appear. The extra constraint $Q_Y = \bar{Q}_Y$ implies that all N_C quarks are from $SU(N_F - 1)$, not involving the N_Fth. These are automatically obeyed in the totally symmetric representation of N_C boxes. In fact, this is the only representation possible for flavor space, since the states have to be constructed out of the components of one complex vector z as $\prod_{i=1}^{N_F} z_i^{n_i}$ with $\sum_i n_i = N_C$. See also more detailed discussion in the next section. For another way of deriving this result see Section 13.7.

Thus for $N_C = 3, N_F = 3$ we get only 10 of $SU(3)$. This is understandable, since there is no physical spin in two dimensions.

What about the masses of the baryons? The total mass of a baryons is given by the sum of (13.11) and (13.54), namely,

$$E = 4m\sqrt{\frac{2N_C}{\pi}} + m\sqrt{2}\sqrt{\left(\frac{\pi}{N_C}\right)^3\left[C_2 - N_C^2\frac{(N_F-1)}{2N_F}\right]}. \tag{13.60}$$

For large N_C, the classical term behaves like N_C, while the quantum correction like 1. This will be worked out in Section 13.7.

That the total mass goes like N_C for large N_C, and that the quantum fluctuations are $\frac{1}{N_C}$ of the classical result, is in accord with general considerations.

13.5 Quark flavor content of the baryons

A measure of the quark content of a given flavor q_i in a baryon state $|B\rangle$ is given by[5]

$$\langle\bar{q}_i q_i\rangle_B = \int dx \langle g_{ii}\rangle_B - \int dx \langle g_{ii}\rangle_0 \tag{13.61}$$

$$= \int dx z_i^* z_i \left\langle\left[e^{-i\sqrt{\frac{4\pi}{N_C}}\phi_c} - 1\right]\right\rangle_B \tag{13.62}$$

$$= \text{const.} \langle z_i^* z_i\rangle_B. \tag{13.63}$$

In order to make contact with the real world, we take here $N_C = 3$ and $N_F = 3$, getting the baryons in the 10 representation of flavor. Similarly, for $SU_F(2)$ there

---

[5] Quark solitons as constituents of hadrons were discussed in [86]. Following that, the flavor content of the baryons was discussed in [97].

is only the isospin $\frac{3}{2}$ representation. This is what we would expect from naïve quark model considerations. The total wave function must be antisymmetric. Baryon is a color singlet, so the wave function is antisymmetric in color and it must be symmetric in all other degrees of freedom. There is no spin, so the baryon must be in a totally symmetric representation of the flavor group, a 10 for three flavors. Therefore, strictly speaking there is no state analogous to the proton. On the other hand, there is a state which is the analog of the Δ^+, namely the charge 1 state in the 10 representation, $z_1^2 z_2$.

The 10 is the lowest baryon multiplet in QCD_2. In the following we shall be dealing with the relative weight of a given flavor in some baryon state. Thus, $\langle \bar{q}q \rangle_B$ will henceforth stand for the ratio,

$$\frac{\langle \bar{q}q \rangle_B}{\langle \bar{u}u + \bar{d}d + \bar{s}s \rangle_B}. \tag{13.64}$$

For $\Delta^+ \sim z_1^2 z_2$ we obtain

$$\langle \bar{s}s \rangle_{\Delta^+} = \frac{\int (d^2 z_1)(d^2 z_2)|z_3|^2 (z_1^2 z_2)(z_1^2 z_2)^*}{\int (d^2 z_1)(d^2 z_2)(z_1^2 z_2)(z_1^2 z_2)^*} = \frac{1}{6}, \tag{13.65}$$

as well as,

$$\langle \bar{u}u \rangle_{\Delta^+} = \frac{1}{2}, \qquad \langle \bar{d}d \rangle_{\Delta^+} = \frac{1}{3}. \tag{13.66}$$

In evaluating the integral in the numerator in eqn. (13.65) we have used $|z_3|^2 = 1 - |z_1|^2 - |z_2|^2$, which follows from the unitarity of the matrix A in (13.19). Similarly, for $\Delta^{++} \sim z_1^3$ we have,

$$\langle \bar{u}u \rangle_{\Delta^{++}} = \frac{2}{3}, \qquad \langle \bar{d}d \rangle_{\Delta^{++}} = \frac{1}{6}, \qquad \langle \bar{s}s \rangle_{\Delta^{++}} = \frac{1}{6}. \tag{13.67}$$

In the constituent quark picture Δ^{++} contains just three u quarks. Both the d-quark and the s-quark content of the Δ^{++} come only from virtual quark pairs. Therefore in the $SU(3)$-symmetric case $\langle \bar{s}s \rangle_{\Delta^{++}} = \langle \bar{d}d \rangle_{\Delta^{++}}$, and $\langle \bar{s}s \rangle_{\Delta^+} = \langle \bar{s}s \rangle_{\Delta^{++}}$, as expected.

From eqn. (13.67) one can also read the results for $\Omega^- \sim z_3^3$, by replacing $u \leftrightarrow s$. In the general case of N_F flavors and N_C colors, one obtains,

$$\langle (\bar{q}q)_{\text{sea}} \rangle_B = \frac{1}{N_C + N_F}, \tag{13.68}$$

where $(\bar{q}q)_{\text{sea}}$ refers to the non-valence quarks in the baryon B. Moreover, one can also compute flavor content of valence quarks. Consider a baryon B containing k quarks of flavor v. The v-flavor content of such a baryon is,

$$\langle \bar{v}v \rangle_B = \frac{k+1}{N_C + N_F}. \tag{13.69}$$

This implies an "equipartition" for valence and sea, each with a content of $1/(N_C + N_F)$. It also follows that the total sea content of N_F flavors is,

$$\sum_{q=1}^{N_F} \langle (\bar{q}q)_{\text{sea}} \rangle_B = \frac{N_F}{N_C + N_F}, \tag{13.70}$$

which goes to zero for fixed N_F and $N_C \to \infty$, as expected.

It is interesting to compare these results with the Skyrme model in 3+1 dimensions. For the proton,

$$\langle \bar{u}u \rangle_p^{3+1} = \frac{2}{5}, \qquad \langle \bar{d}d \rangle_p^{3+1} = \frac{11}{30}, \qquad \langle \bar{s}s \rangle_p^{3+1} = \frac{7}{30}, \tag{13.71}$$

and for the Δ,

$$\langle \bar{s}s \rangle_\Delta^{3+1} = \frac{7}{24}, \qquad \langle \bar{s}s \rangle_{\Omega^-}^{3+1} = \frac{5}{12}. \tag{13.72}$$

The qualitative picture is similar, although the $\bar{s}s$ content in the non-strange baryons is lower in $1+1$ dimensions. One may speculate that in $1+1$ dimensions the effects of loops are smaller than in $3+1$ dimensions, since the theory is super-renormalizable and there are only longitudinal gluons. In the $SU_F(3)$-symmetric limit the strange quark content of baryons with zero net strangeness is significant, albeit smaller than that of either of the other two flavors. The situation obviously is reversed for Ω^-.

In the real world the current mass of the strange quark is much larger than the current masses of u and d quarks. It is natural to expect that this will have the effect of decreasing the strange quark content from its value in the $SU_F(3)$ symmetry limit. We do not know the exact extent of this effect, but it is likely that the strange content decreases by a factor which is less than two. This estimate is based on both explicit model calculations and what we know from PCAC, namely that the analogous quark bilinear expectation values in the vacuum are not dramatically different from their $SU(3)$ symmetric values,

$$0.5 \leq \frac{\langle \bar{s}s \rangle_0}{\langle \bar{u}u \rangle_0} \leq 1. \tag{13.73}$$

13.6 Multibaryons

Let us now explore the possibility of having multi-baryons states.[6] The procedure follows similar lines to that of the baryonic spectrum, namely, we look for classical solution of the equation of motions with baryon number kN_C, and then we semiclassically quantize this. The ansatz for the classical solution of the

---

[6] Multibaryonic states were studied in [102] and [103].

low-lying k-baryon state is taken now to be,

$$
g_0(k) = \begin{pmatrix} \overbrace{1 \qquad}^{(N_F-k)} & & \\ & \ddots & \\ & & \underbrace{\exp[-i(\frac{4\pi}{N_C})^{\frac{1}{2}}\varphi_c]}_{k} \\ & & & \ddots \end{pmatrix} .
\tag{13.74}
$$

For the semi-classical quantization we generalize the parametrization given in (13.16) to,

$$
A = \begin{pmatrix} A_{ij} & z_{i\alpha} \end{pmatrix},
\tag{13.75}
$$

where i represents the rows $(1, \ldots, N_F)$ and α the columns $(N_F - k + 1, \ldots, N_F)$. The effective action in its covariant form (13.33) becomes,

$$
S[z_\alpha] = \frac{1}{2M} \int dt \, \mathrm{Tr}(Dz)^\dagger Dz + iN_C \int dt \, \mathrm{Tr}\dot{z}^\dagger z,
\tag{13.76}
$$

where now, instead of (13.34),

$$
(Dz)_{i\alpha} = \dot{z}_{i\alpha} + z_{i\beta}(\dot{z}_{j\beta}^* z_{j\alpha})eDz.
\tag{13.77}
$$

Using the same steps as those which led to (13.54) one finds now the Hamiltonian,

$$
H = 2M\left[C_2(N_F) - \frac{N_C^2}{2N_F}k(N_F - k)\right] + kE_c,
\tag{13.78}
$$

with E_c the classical contribution for one baryon, the first term in (13.60).

13.7 States, wave functions and binding energies

It was shown in [102] that the allowed k-baryon states contain (kN_C) boxes in the Young tableaux representation of the flavour group $SU(N_F)$. Let us recall that this result followed from the constraint implied by the local invariance,

$$
z_{i\alpha} \rightarrow e^{i\delta(t)} z_{i\alpha}.
\tag{13.79}
$$

Performing a variation corresponding to this invariance we find that the action S changes by

$$
\Delta S = (kN_C) \int \dot{\delta} \, dt.
\tag{13.80}
$$

This means that the N_z number is equal to (kN_C). Thus for any wave function, written as a polynomial in z and z^*, the number of zs minus the number of z^*s must equal (kN_C). Note that for $k = 1$ the transformation (13.79) represents also the N_F^{th} flavor number. Thus (13.80) entails that the representation contains a state with N_C boxes of the N_F flavor, and therefore must be the totally symmetric representation.

Now, the effective action (13.76) is invariant under a larger group of local transformations. In fact, we have extra $(k^2 - 1)$ generators, which correspond to $SU(k)$ under which (13.76) is locally invariant. This can be exhibited by defining "local gauge potentials",

$$\tilde{A}_{\beta\alpha}(t) = -(z^\dagger \dot{z})_{\beta\alpha}, \tag{13.81}$$

so that,

$$Dz = \dot{z} + z\tilde{A}. \tag{13.82}$$

Under the local gauge transformation corresponding to $\Lambda(t)$, $\tilde{A}$ transforms as,

$$\tilde{A}(t) \rightarrow e^{i\Lambda}\tilde{A}e^{-i\Lambda} + (\partial_t e^{i\Lambda})e^{-i\Lambda}, \tag{13.83}$$

which implies,

$$(Dz)_{i\alpha} \rightarrow (Dz)_{i\beta}(e^{-i\Lambda})_{\beta\alpha}, \tag{13.84}$$

and so $\Delta S = 0$. If we perform the $U(1)$ transformation (13.79) we obtain a contribution (13.80) from the Wess–Zumino term, which implies $N_z = (kN_C)$. But due to the larger local symmetry we have more restrictions; they imply that the allowed states have to be singlets under the above mentioned $SU(k)$ symmetry. This is analogous to the confinement property of QCD, which tells that, due to the non-abelian gauge invariance, the physical states have to be color singlets. Here we have an analogous singlet structure of the $SU(k)$ in the flavor space. Taking a wave function that has zs only (analogous to quarks only for QCD), it must be of the form,

$$\psi_k(z) = \prod_{i=1}^{N_C}\left(\epsilon_{\alpha_1...\alpha_k} z_{i_1 \alpha_1} ... z_{i_k \alpha_k}\right), \tag{13.85}$$

for a given set of $1 \leq i_1, ..i_k \leq N_F$.

The most general state will then be of the form,

$$\tilde{\psi}(z, z^*) = \psi_k(z)[\prod_{\{i,j\}}(z_{i\alpha}^* z_{j\alpha})^{n_{ij}}], \tag{13.86}$$

and the products are over given sets of indices.

Using the explicit formula from [103], we obtain the mass of the state represented by (13.85),

$$E[\psi_k] = Mk(N_F - k)N_C + kE_c. \tag{13.87}$$

To obtain binding energies, consider our k-baryon as built from constituents k_r, such that $k = \sum_r k_r$. Then,

$$B[k|k_r] = -(MN_C)[k^2 - \sum_i k_i^2]$$
$$= -(2MN_C)\sum_{r>s} k_r k_s. \tag{13.88}$$

When all $k_r = 1$, the sum gives us $\frac{1}{2}k(k-1)$, i.e. the number of one-baryon pairs in the k-baryon state. Note that the binding energy is always negative, thus the k-baryon is stable. The maximal binding corresponds to the case when all $k_r = 1$.

Note also that in the $N_C \to \infty$ limit, the binding tends to a finite value, since then,

$$\lim_{N_C \to \infty} (2MN_C) = (Cme_c)^{\frac{1}{2}} \left(\frac{2N_F}{\pi}\right)^{\frac{1}{4}} \pi^{\frac{3}{2}}. \tag{13.89}$$

Let us take as an example an analog of a deuteron, namely a di-baryon $k = 2$. Then for $N_C = 3$, $N_F = 2$ we find that its representation is a flavor singlet (this is the limiting case of $k = N_F$). The ratio of the binding to twice the baryon mass is given by,

$$\epsilon_2 = \frac{1}{1 + \frac{24}{\pi^2}} = 0.29. \tag{13.90}$$

For $k = 2$, $N_C = 3$ and $N_F = 3$ we find that the di-baryon is represented by, $\overline{10}$ and the ratio is given by,

$$\epsilon_3 = \frac{1}{2 + \frac{24}{\pi^2}} = 0.23. \tag{13.91}$$

For general N_F we obtain,

$$\epsilon_F = \frac{1}{(N_F - 1) + \frac{24}{\pi^2}} = \frac{1}{N_F + 1.43}. \tag{13.92}$$

Finally, let us make the following comment. The ratio of the quantum fluctuations term to the classical term, in the expression for the mass, eqn. (13.87), is given by,

$$\frac{\text{(Quantum corrections)}}{\text{(Classical term)}} = \left(\frac{\pi^2}{8}\right)\frac{N_F - k}{N_C}. \tag{13.93}$$

Thus, we do not expect our approximations to hold in the region $N_F \geq (N_C + 1)$. We expect it to start for $N_C \geq N_F$, and to be good in the region $N_C \gg N_F$.

13.8 Meson-baryon scattering

So far we have analyzed, using semiclassical quantization of the bosonized theory in the strong coupling limit, the spectrum of the baryons and their flavor content. Applying the same technique one can also study the scattering processes of mesons from baryons. The idea is to introduce perturbations around the classical soliton solutions and to compute the forward phase shifts. We start with the computation in the strong coupling limit [98] and then we discuss the general case of any coupling [87].

Our starting point is the soliton solution that describes the static classical baryon $g_c(x) = \exp[-i\Phi_c(x)]$ where,

$$\Phi_c(x) = \begin{pmatrix} \phi_c(x) & & & \\ & 0 & & \\ & & \ddots & \\ & & & 0 \end{pmatrix}, \tag{13.94}$$

and,

$$\phi_c(x) = 4\mathrm{arctg}\,(e^{\mu x}), \quad \mu = m\sqrt{\frac{8\pi}{N_C}}. \tag{13.95}$$

Note that we have shifted the non-trivial phase factor to the upper left-hand corner, whereas in (13.77) it was put in the lower right-hand one.

We introduce a fluctuation around it of the form,

$$g = \exp\{-i\,[\Phi_c(x) + \delta\phi(x,t)]\} \tag{13.96}$$

$$g \approx e^{-i\Phi_c(x)} - i \int_0^1 d\tau\, e^{-i\tau\Phi_c(x)}\delta\phi(x,t)e^{-i(1-\tau)\Phi_c(x)}. \tag{13.97}$$

Actually, to avoid integrals as in eqn. (13.97), which yield rather complicated expressions for fluctuations, we will adopt a different expansion, namely,

$$\begin{aligned} g &= e^{-i\Phi_c(x)}\, e^{-i\tilde\delta\phi(x,t)}, \\ &\approx e^{-i\Phi_c(x)} - ie^{-i\Phi_c(x)}\tilde\delta\phi(x,t), \end{aligned} \tag{13.98}$$

where we have denoted by $\tilde\delta\phi$ the new variation, different from the $\delta\phi$ of eqn. (13.97), but still a fluctuation about the classical solution. Now,

$$\begin{aligned} &\frac{N_c}{4\pi}\,\partial_+\left[e^{-i\Phi_c(x)}\left(\partial_-\tilde\delta\phi(x,t)\right)e^{i\Phi_c(x)}\right] \\ &+ m^2\left[e^{-i\Phi_c(x)}\tilde\delta\phi(x,t) + \tilde\delta\phi(x,t)e^{i\Phi_c(x)}\right] = 0. \end{aligned} \tag{13.99}$$

Obviously the two expressions coincide in the abelian case. In fact, the relation between $\delta\phi$ and $\tilde\delta\phi$ is

$$\tilde\delta\phi(x,t) = \int_0^1 d\tau\, e^{i\tau\Phi_c(x)}\,\delta\phi(x,t)\,e^{-i\tau\Phi_c(x)}. \tag{13.100}$$

Physical quantities should obviously come out to be the same for both types of fluctuation.

13.8.1 Abelian case

We start with the abelian fluctuation $\delta\phi$ that commutes with Φ_c. Denote this case by $\delta\phi_{ab}$, where the subscript "ab" stands for "abelian."

Then the fluctuation reads

$$\delta g = -i\delta\phi_{ab}(x)e^{-i\phi_c(x)}, \tag{13.101}$$

where,

$$\Box\, \delta\phi_{ab} + \mu^2(\cos\phi_c)\delta\phi_{ab} = 0, \tag{13.102}$$

and,

$$\cos\phi_c = \left[1 - \frac{2}{\cosh^2\mu x}\right]. \tag{13.103}$$

This equation of motion can be derived from the following effective action,

$$\mathcal{L}_{\text{eff}} = \frac{1}{2}\left(\partial_\mu\delta\phi_{ab}\right)^2 - \frac{1}{2}V(x)\left(\delta\phi_{ab}\right)^2 \tag{13.104}$$

$$V(x) = \mu^2\cos\phi_c(x) = \mu^2\left[1 - \frac{2}{\cosh^2\mu x}\right]. \tag{13.105}$$

For a solution with an harmonic time dependence of the form,

$$\delta\phi_{ab}(x,t) = e^{-i\omega t}\chi_{ab}(x), \tag{13.106}$$

the spatial part has to solve,

$$-\omega^2\chi_{ab} - \chi''_{ab} + V(x)\chi_{ab} = 0. \tag{13.107}$$

Note that asymptotically the potential approaches $x \to \pm\infty$, the potential $\to \mu^2$, and so asymptotically,

$$\chi''_{ab}(\pm\infty) + \omega^2\,\chi_{ab}(\pm\infty) = \mu^2\,\chi_{ab}(\pm\infty). \tag{13.108}$$

For asymptotic behavior of the form,

$$\chi_{ab}(x) \xrightarrow[|x|\to\infty]{} e^{ikx}, \tag{13.109}$$

with,

$$\omega^2 = k^2 + \mu^2, \tag{13.110}$$

the two asymptotic solutions are,

$$\chi_{ab}(x) \sim A(\omega)\sin kx + B(\omega)\cos kx, \tag{13.111}$$

and the S-matrix is,

$$S_{\text{forward}} = \frac{1}{2}(B - iA) \tag{13.112}$$

$$S_{\text{backward}} = \frac{1}{2}(B + iA), \tag{13.113}$$

for an incoming wave e^{ikx} from $x = -\infty$.

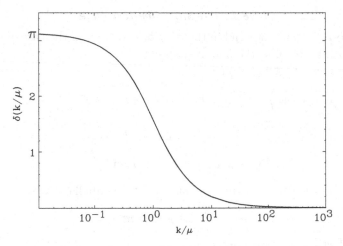

Fig. 13.1. The phase shift $\delta = 2\mathrm{ctg}^{-1}(k/\mu)$, eqn. (13.117), as a function of the normalized momentum k/μ, for the potential (13.105), governing the small fluctuations around the soliton in the abelian case. The phase shift is smooth and monotonically decreasing with momentum, indicating that no resonance is present. Note logarithmic momentum scale.

We can now proceed to derive the scattering matrix, using the standard procedure. The solution for $x \to \infty$ contains only the transmitted wave, $\psi(x \to \infty) \sim e^{ikx}$.[7]

It turns out that for the particular potential (13.105) there is no reflection at all, i.e. the wave function for $x \to -\infty$ contains only the incoming wave,

$$\psi(x \to -\infty) \sim e^{ikx - \delta} = \left(-\frac{1 + ik/\mu}{1 - ik/\mu}\right) e^{ikx}. \tag{13.114}$$

Thus,

$$\frac{1}{T} = -\frac{1 + ik/\mu}{1 - ik/\mu} \tag{13.115}$$

$$T = -\frac{1 - ik/\mu}{1 + ik/\mu} = e^{i\delta} \tag{13.116}$$

$$\mathrm{ctg}\,\frac{1}{2}\delta = \frac{k}{\mu}. \tag{13.117}$$

As shown in Fig. 13.1, δ varies smoothly and decreases monotonically from $\delta = \pi$ at $k = 0$ to $\delta = 0$ at $k = \infty$, indicating that there is no resonance.

The no-reflection potential we found is a special case of a well-known class of reflectionless in quantum mechanics.

---

[7] We take the convention where the scattering phase is taken to be zero at $x \to \infty$ and is therefore extracted from the wave function at $x \to -\infty$.

13.8.2 The non-abelian case

We got a no-reflection potential in the previous section, in the case of one flavor. We want to examine now the non-abelian case.

Following eqn. (13.99), we get,

$$\Box\, \tilde{\delta\phi} - i\left(\partial_+ \Phi_c\right)\left(\partial_- \tilde{\delta\phi}\right) + i\left(\partial_- \tilde{\delta\phi}\right)\left(\partial_+ \Phi_c\right) + \frac{1}{2}\mu^2\left[\tilde{\delta\phi}e^{-i\Phi_c(x)} + e^{i\Phi_c(x)}\tilde{\delta\phi}\right] = 0. \tag{13.118}$$

The equation for $\tilde{\delta\phi}_{ij}$ with $i, j \neq 1$ is as for the free case,

$$\Box\, \tilde{\delta\phi}_{ij} + \mu^2 \tilde{\delta\phi}_{ij} = 0, \qquad i \text{ and } j \neq 1, \tag{13.119}$$

whereas the $i = 1, j = 1$ matrix element is as in the abelian case,

$$\Box\, \tilde{\delta\phi}_{11} + \mu^2\left(\cos\phi_c(x)\right)\tilde{\delta\phi}_{11} = 0. \tag{13.120}$$

with no reflection and no resonance.

So in order to proceed beyond these results, we need to consider $\tilde{\delta\phi}_{1j}, \ j \neq 1$, or $\tilde{\delta\phi}_{i1}, \ i \neq 1$. As $\tilde{\delta\phi}$ is Hermitian, it is sufficient to discuss one of the above. Thus we take,

$$\tilde{\delta\phi}_{1j} = e^{-i\omega t}u_j(x) \qquad j \neq 1, \tag{13.121}$$

resulting in,

$$u_j''(x) - i\phi_c'(x)u_j'(x) + \left[\omega^2 + \omega\phi_c'(x) - \frac{1}{2}\mu^2\left(1 + e^{i\phi_c(x)}\right)\right]u_j(x) = 0. \tag{13.122}$$

Defining,

$$u_j \equiv e^{\frac{i}{2}\phi_c}\, v_j, \tag{13.123}$$

we find,

$$v_j'' + \left[\omega^2 + \omega\phi_c' - \frac{1}{2}\mu^2\left(1 + \cos\phi_c\right) + \frac{1}{4}\left(\phi_c'\right)^2\right]v_j = 0. \tag{13.124}$$

Using,

$$\frac{1}{2}\left(\phi_c'\right)^2 = \mu^2\left(1 - \cos\phi_c\right), \tag{13.125}$$

we get,

$$v_j'' + \left[\omega^2 + \omega\phi_c' - \mu^2\cos\phi_c\right]v_j = 0. \tag{13.126}$$

This can be rewritten as,

$$-v_j'' - \omega^2 v_j + V(x)v_j = 0, \tag{13.127}$$

where,

$$V(x) = -\omega\phi_c' + \mu^2\cos\phi_c =$$
$$= \mu^2 - 2\mu^2\left[\frac{(\omega/\mu)}{\cosh\mu x} + \frac{1}{\cosh^2\mu x}\right], \tag{13.128}$$

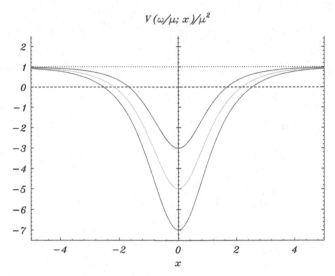

Fig. 13.2. The normalized potential $V(\omega/\mu; x)/\mu^2$ of eqn. (13.128), for $\omega/\mu = 1.01$ (upper), 2 (middle) and 3 (lower).

with $\omega = \sqrt{k^2 + \mu^2}$, as before. Note that the potential depends on the momentum of the incoming particle, as shown in Fig. 13.2.

Next we proceed to solve numerically for the reflection and transmission coefficient. It turns out that for numerical solution of the scattering problem it is more convenient to take the coefficient of the outgoing wave at $x \sim +\infty$ to be 1, instead of the T prefactor, and integrate eqn. (13.127) backward, reading off the T and R amplitudes from the solution at $x \sim -\infty$.

We thus use,

$$
\begin{aligned}
v_j(x) &= e^{ikx}, & x &\to +\infty \\
v_j(x) &= \tfrac{1}{T} e^{ikx} + \tfrac{R}{T} e^{-ikx}, & x &\to -\infty.
\end{aligned}
\tag{13.129}
$$

Since the potential is symmetric, the symmetric and anti-symmetric scattering amplitudes don't mix, yielding two independent phase shifts δ_S and δ_A, respectively. This leads to,

$$
\begin{aligned}
T &= \tfrac{1}{2} \left(e^{i\delta_S} + e^{i\delta_A} \right) \\
R &= \tfrac{1}{2} \left(e^{i\delta_S} - e^{i\delta_A} \right).
\end{aligned}
\tag{13.130}
$$

Defining,

$$
\delta_\pm = \frac{1}{2} \left(\delta_S \pm \delta_A \right),
\tag{13.131}
$$

we find that,

$$
\begin{aligned}
T &= e^{i\delta_+} \cos \delta_- \\
R &= i e^{i\delta_+} \sin \delta_-.
\end{aligned}
\tag{13.132}
$$

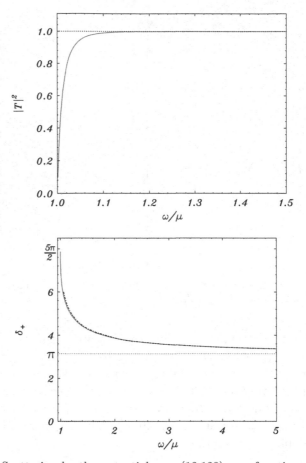

Fig. 13.3. Scattering by the potential eqn. (13.128) as a function of the normalized energy ω/μ. Upper plot: transmission probability $|T|^2$; lower plot: phase of T, δ_+ (continuous line). Also shown is the approximate result for δ_+ from WKB (dot-dashed line).

Note that R/T is purely imaginary. The transmission and reflections probabilities are,

$$|T|^2 = \cos^2 \delta_-$$
$$|R|^2 = \sin^2 \delta_-. \tag{13.133}$$

The numerical results for the transmission probability $|T|^2$ and for the phase of T, δ_+ are presented in Fig. 13.3. For comparison and as an extra check we also plot the WKB result for δ_+. Note that no resonance appears.

Note that the asymptotic value of the phase shift is π. This can also be obtained from a WKB calculation, which becomes exact at infinite energies.

13.8.3 Extension to arbitrary coupling

To analyze the system at any gauge coupling we go back to bosonized action prior to the implementation of the strong coupling limit, which we now rewrite in the form,

$$S_{\text{eff}}[u] = S_0[u] + \frac{e_c^2 N_f}{8\pi^2} \int d^2 x \, \text{Tr} \left[\partial_-^{-1} \left(u \, \partial_- u^\dagger \right)_c \right]^2 + m'^2 N_{\tilde{m}} \int d^2 x \, \text{Tr} \left(u + u^\dagger \right).$$

$$(13.134)$$

The strong coupling limit eliminates the second term of (13.134), for arbitrary coupling,

$$\frac{e_c^2 N_f}{8\pi^2} \int d^2 x \, \text{Tr} \left[\partial_-^{-1} \left(u \, \partial_- u^\dagger \right)_c \right]^2, \tag{13.135}$$

where $\left(u \, \partial_- u^\dagger \right)_c$ is the color part of $M \equiv u \partial_- u^\dagger$, to be computed as,

$$M_c = \text{Tr}_f M - 1/N_c \text{Tr}_{f \& c} M. \tag{13.136}$$

As already mentioned, this term represents the interactions, as it arises from integrating out the gauge potentials. However, we will see that, for the physical situation we discuss, this term does not contribute to meson-baryon scattering for any coupling. As a result, the latter is described by the effective action $\tilde{S}_{\text{eff}}[u]$, whereas in the strong coupling limit it is described by $\tilde{S}_{\text{eff}}[g]$.

In a similar manner to (13.95) we take u to be of the form,

$$u = \exp(-i\Phi_c) \exp(-i\delta\Phi), \tag{13.137}$$

corresponding to a classical soliton Φ_c, and a small fluctuation $\delta\Phi$ around it, representing the meson. The resulting action is then expanded to second order in $\delta\Phi$, yielding a linear equation of motion for $\delta\Phi$ in the soliton background. The latter serves as an external potential in which the meson is propagating.

We start by evaluating,

$$M \equiv u \partial_- u^\dagger =$$
$$= \exp(-i\Phi_c) \, \partial_- (\exp i\Phi_c) + \exp(-i\Phi_c) \exp(-i\delta\Phi) \left[\partial_- \exp(i\delta\Phi) \right] \exp(i\Phi_c),$$

$$(13.138)$$

and obtain the equations of motion for the meson field by varying with respect to $\delta\Phi$. The variation of (13.135) with respect to $\delta\Phi$ is proportional to,

$$\frac{\delta M_c}{\delta(\delta\Phi)} \partial^{-2} M_c. \tag{13.139}$$

To compute its variation with respect to $\delta\Phi$, we need only the second term M_2 of M, as the first term M_1 is independent of $\delta\Phi$.

We take for the soliton a diagonal ansatz (13.94) now in the form of a u matrix rather then a g one,

$$[\exp(-i\Phi_c)]_{aa'jj'} = \delta_{aa'}\delta_{jj'} \exp(-i\sqrt{4\pi}\chi_{aj}) :$$
$$a = 1,\ldots,N_c, \tag{13.140}$$
$$j = 1,\ldots,N_f,$$

so that,

$$\{\exp(-i\Phi_c)[\exp(-i\delta\Phi)\,\partial_-\,\exp(i\delta\Phi)]\exp(i\Phi_c)\}_{aj,a'j'}$$
$$= \exp(-i\sqrt{4\pi}\chi_{aj})\,[\exp(-i\delta\Phi)\,\partial_-\,\exp(i\delta\Phi)]_{aj,a'j'}\,\exp(i\sqrt{4\pi}\chi_{a'j'}). \tag{13.141}$$

The part of M that contributes to the effective action is its color projection (13.136). We note that $\mathrm{Tr}_{f\&c}M_2 = 0$, and thus,

$$[(M_2)_c]_{a,a'} = \sum_j \exp(-i\sqrt{4\pi}\chi_{aj})[\exp(-i\delta\Phi)\,\partial_-\,\exp(i\delta\Phi)]_{aj,a'j}\,\exp(i\sqrt{4\pi}\chi_{a'j}). \tag{13.142}$$

The mesons $\delta\Phi$ have to be diagonal in color, so,

$$[(M_2)_c]_{a,a'} = \sum_j [\exp(-i\delta\Phi)\,\partial_-\,\exp(i\delta\Phi)]_{aj,aj}\,\delta_{a,a'}. \tag{13.143}$$

We recall that the flavor structure of the mesons is independent of their color indices, and restrict our attention to mesons that have no $U(1)$ flavor part. In this way, we may be sure that classical solutions lead to stable particles, since their non-vanishing flavor quantum numbers put them in a different sector from the vacuum. We then have,

$$\sum_j [\exp(-i\delta\Phi)\,\partial_-\,\exp(i\delta\Phi)]_{aj,aj} = 0, \tag{13.144}$$

as shown earlier, and the effective meson-baryon action is,

$$\tilde{S}_{m\text{-}b}[\delta\Phi] = S_0[u] + m^2 N_m \int d^2x\,(\,\mathrm{Tr}\,u +\,\mathrm{Tr}\,u^\dagger\,), \tag{13.145}$$

with u depending on $\delta\Phi$ for fixed Φ_c as in (13.137).

Next we would like to evaluate the potential. The equation of motion for $\delta\Phi$ is obtained from (13.145), by first varying with respect to u and then varying u with respect to $\delta\Phi$. To first order in $\delta\Phi$, we find,

$$\delta u = -i[\exp(-i\Phi_c)]\delta\Phi. \tag{13.146}$$

The resulting equation of motion is then,

$$\frac{1}{4\pi}\partial_+\left[(\partial_-u)\,u^\dagger\right] + \left(um^2 - m^2u^\dagger\right) = 0, \tag{13.147}$$

where m is the diagonal mass matrix: $m = \delta_{ij}m_j$ with (possibly different) entries m_j corresponding to flavors j. We note that there is the possibility of an overall scale ambiguity in m, since, when the masses are different, there is a question of which normal-ordering scale to use. The resulting equation of motion for $\delta\Phi$ is,

$$\Box\,\delta\Phi - i\,(\partial_+\Phi_c)\,(\partial_-\delta\Phi) + i\,(\partial_-\delta\Phi)\,(\partial_+\Phi_c)$$
$$+ \frac{1}{2}\left[\delta\Phi\mu^2\,\exp(-i\Phi_c) + \exp(i\Phi_c)\mu^2\delta\Phi\right] = 0, \qquad (13.148)$$

where $\mu \equiv m\sqrt{8\pi}$.

As discussed before, both Φ_c and $\delta\Phi$ are diagonal in color. Moreover, Φ_c is diagonal in flavor too. So, taking the $aajj'$ matrix element of the equation of motion (13.148), we find,

$$\Box\,\delta\Phi_{ajj'} - i(\partial_+\Phi_c)_{aj}(\partial_-\delta\Phi)_{ajj'} + i\,(\partial_-\delta\Phi)_{ajj'}(\partial_+\Phi_c)_{aj'}$$
$$+ \frac{1}{2}\{\delta\Phi_{ajj'}\mu^2{}_{j'}[\exp(-i\Phi_c)]_{aj'} + [\exp(i\Phi_c)]_{aj}\mu^2{}_j\delta\Phi_{ajj'}\} = 0. \qquad (13.149)$$

Examining the classical solutions for the quark solitons inside the baryons, we see that, for a given color index a, there is only one flavor for which Φ_c is non-zero. We can now distinguish three cases:

- The first is when an index a and indices j and j' are chosen in such a way that both $(\Phi_c)_{aj}$ and $(\Phi_c)_{aj'}$ are zero. In such a case,

$$\Box\,\delta\Phi_{ajj'} + \frac{1}{2}[\mu^2{}_j + \mu^2{}_{j'}]\delta\Phi_{ajj'} = 0, \qquad (13.150)$$
$$\text{where } (\Phi_c)_{aj} = 0 \text{ and } (\Phi_c)_{aj'} = 0.$$

Thus $\delta\Phi_{ajj'}$ is a free field with squared mass given by the average of m_j^2 and $m_{j'}^2$ in this case, which we do not discuss further.
- The second case is that of $j = j'$, with a such that $(\Phi_c)_{aj}$ is a quark soliton inside the baryon. In this case,

$$\Box\,\delta\Phi_{ajj} + \mu^2{}_j\,\cos[(\Phi_c)_{aj}]\delta\Phi_{ajj} = 0. \qquad (13.151)$$

- The third case is when j is different from j', now with one of the Φ_c being a soliton and the other vanishing. Taking $(\Phi_c)_{aj}$ to be the soliton, we obtain,

$$\Box\,\delta\Phi_{ajj'} - i(\partial_+\Phi_c)_{aj}(\partial_-\delta\Phi)_{ajj'} + \frac{1}{2}\{\mu^2{}_{j'} + \mu^2{}_j[\exp(i\Phi_c)]_{aj}\}\delta\Phi_{ajj'} = 0, \qquad (13.152)$$

where $\qquad\qquad j' \neq j$ and $(\Phi_c)_{aj'} = 0$.

Next we want to proceed and evaluate the meson-baryon scattering. For that purpose we need to analyze the equations that determine the static solution $(\Phi_c)_{aj}$. First one defines,

$$(\Phi_c)_{aj} = \sqrt{4\pi}(\chi_c)_{aj}, \qquad (13.153)$$

where the $(\chi_c)_{aj}$ are canonical fields, whose equations of motion are,

$$\chi''_{\alpha j} - 4\alpha_c \left(\sum_l \chi_\alpha - \frac{1}{N_c} \sum_{\beta l} \chi_{\beta l} \right) - 2\sqrt{4\pi}m_j^2 \sin\sqrt{4\pi}\chi_{\alpha j} = 0.$$

Note the extra factor 2 in front of the mass term, as compared with eqn. (22) of [86], due to an error in this reference.

Choosing the boundary conditions $\chi_{aj}(-\infty) = 0$, we get as constraints for $\chi_{aj}(+\infty)$, denoted hereafter simply by χ_{aj},

$$\frac{1}{\sqrt{\pi}}\chi_{\alpha j} = n_{\alpha j} \quad \text{integers,} \tag{13.154}$$

and

$$\sum_l n_\alpha = n \quad \text{independent of } a. \tag{13.155}$$

The baryon number[8] associated with any given flavor l is given by,

$$B_l = \sum_a n_\alpha.$$

Combining the last two equations, we find,

$$B = \sum_l B_l = nN_c,$$

for the total baryon number.

We now continue in a similar manner to the discussion in the strong coupling limit, starting with the first non-trivial case (13.151) identified above. As the soliton solutions are such that there is a unique correspondence between the color index a and the flavour index j, we suppress a in what follows. Putting,

$$\delta\Phi_{jj} = e^{-i\omega_j t}u_j(x), \tag{13.156}$$

with,

$$u_j(x) \xrightarrow[x\to\infty]{} e^{ikx}, \tag{13.157}$$

we find,

$$\omega_j^2 = k^2 + \mu_j^2, \tag{13.158}$$

and the equation for $u_j(x)$ is,

$$u_j''(x) + \omega_j^2 u_j - \mu_j^2[\cos(\Phi_c)_j]u_j = 0. \tag{13.159}$$

We define the potential V_j for this scattering process via,

$$u_j''(x) + \omega_j^2 u_j - V_j u_j = 0. \tag{13.160}$$

---

[8] In our normalization, a single quark carries one unit of baryon number.

and find,

$$V_j = \mu^2{}_j[\cos{(\Phi_c)}_j]. \tag{13.161}$$

In our normalization the outgoing wave has coefficient 1, which is more convenient for numerical calculations, and the wave for $x \to -\infty$ is now,

$$u_j(x) = \frac{1}{T_j}\, e^{ikx} + \frac{R_j}{T_j}\, e^{-ikx}, \qquad x \to -\infty, \tag{13.162}$$

in this case.

In the second non-trivial case (13.152), we put,

$$\delta\Phi_{jj'} = e^{-i\omega_{jj'}t}u_{jj'}(x), \tag{13.163}$$

so that,

$$u''_{jj'}(x) - i(\Phi_c)'_j(x)u'_{jj'}(x) + \{\omega^2_{jj'} + \omega_{jj'}(\Phi_c)'_j(x)$$

$$-\frac{1}{2}\{\mu^2{}_{j'} + \mu^2{}_j[\exp(i\Phi_c)]_j\}\}u_{jj'} = 0. \tag{13.164}$$

To eliminate the first derivative term in u, we substitute,

$$u_{jj'} = \left[\exp\left(\frac{i}{2}\Phi_c\right)\right]_j v_{jj'}. \tag{13.165}$$

This results in,

$$v''_{jj'}(x) + \{\omega^2_{jj'} + \omega_{jj'}(\Phi_c)'_j(x) - \mu^2{}_j[\cos(\Phi_c)]_j\}v_{jj'} + \frac{1}{2}(\mu^2{}_j - \mu^2{}_{j'})v_{jj'}$$

$$+ \left\{\frac{1}{4}[(\Phi_c)'_j(x)]^2 - \frac{1}{2}\mu^2{}_j(1 - [\cos(\Phi_c)]_j)\right\}v_{jj'}$$

$$+ \frac{i}{2}\{(\Phi_c)_j''(x) - \mu^2{}_j[\sin(\Phi_c)_j]\}v_{jj'} = 0. \tag{13.166}$$

We note that the last three lines vanish when all the quark masses are equal, as then the soliton is a sine-Gordon one. Thus, the scattering would then only be elastic.

The potential of the scattering is defined here via,

$$v''_{jj'}(x) + \omega^2_{jj'}v_{jj'} - V_{jj'}v_{jj'} = 0, \tag{13.167}$$

so that,

$$V_{jj'} = -\omega_{jj'}(\Phi_c)'_j(x) + \mu^2{}_j[\cos(\Phi_c)]_j$$

$$-\frac{1}{2}(\mu^2{}_j - \mu^2{}_{j'})$$

$$-\left\{\frac{1}{4}[(\Phi_c)'_j(x)]^2 - \frac{1}{2}\mu^2{}_j(1 - [\cos(\Phi_c)]_j)\right\}$$

$$-\frac{i}{2}\{(\Phi_c)_j''(x) - \mu^2{}_j[\sin(\Phi_c)_j]\}. \tag{13.168}$$

Taking again,

$$v_{jj'}(x) \xrightarrow[x \to \infty]{} e^{ikx}, \tag{13.169}$$

we get,

$$\omega_{jj'} = \frac{1}{2}(\mu^2{}_j + \mu^2{}_{j'}), \tag{13.170}$$

and the wave for $x \to -\infty$ is,

$$v_{jj'}(x) = \frac{1}{T_{jj'}} e^{ikx} + \frac{R_{jj'}}{T_{jj'}} e^{-ikx}, \qquad x \to -\infty, \tag{13.171}$$

in this case.

To summarize we have shown that meson-baryon scattering in QCD_2 in the large-N_c limit is non-trivial for non-zero quark masses, and is described by two distinct effective potentials when the quark masses are unequal. These effective potentials are not of the sine-Gordon type found in previous cases, and we expect the scattering amplitudes also to be non-trivial. Their calculation will require numerical analysis.

14

Confinement versus screening

One of the most challenging problems of gauge dynamics in four dimensions is how to show that QCD admits confinement. One of the measures of confinement is the fact that the potential between an external quark and an external anti-quark placed at a separation distance L, as in Fig. 14.1, is dominated by a linear dependence, namely,

$$V = \sigma L. \tag{14.1}$$

The coefficient in this linear dependence is the *string tension*. Thus, a non-confining behavior, which will be referred to as a screening behavior, implies a vanishing string tension. Whereas in four dimensions the computation of the string tension is a formidable task, in two dimensions, as will be shown in this chapter, it is a fairly easy one. In this chapter we describe the extraction of the string tension in various two-dimensional gauge systems.

We start by calculating the string tension for the massive Schwinger model in both the fermionic and the bosonic languages. This is done in the small mass limit and then we discuss the corrections due to going beyond this limit. We then discuss the short range corrections to the confining potential. We focus on the abelian case, believing that the non-abelian case is very similar. Next we comment on the behavior of the string tension when finite temperature is introduced. Then we move to non-abelian generalization. We compute the string tension for the cases of matter in the fundamental and adjoint representations, followed by the symmetric and anti-symmetric representations.

Much of this chapter is based on [15] and [16].

The string tension of the massive Schwinger model was calculated using bosonizaton in [68]. The massless cases in gauge theories were analyzed in [116]. The next-to-leading order in small mass was computed by [4].

14.1 The string tension of the massive Schwinger model

We start with the derivation of the string tension in the massive Schwinger model, in the fermionic language. Consider the partition function of two dimensional massive QED_2,

$$Z = \tag{14.2}$$

$$\int DA_\mu \, D\bar\Psi \, D\Psi \exp\left(i \int d^2x \left(-\frac{1}{4e^2} F^2_{\mu\nu} + \bar\Psi i\partial\!\!\!/\Psi - m\bar\Psi\Psi - q_{\mathrm{dyn}} A_\mu \bar\Psi\gamma^\mu\Psi \right) \right),$$

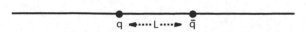

Fig. 14.1. Quark anti-quark separated at a distance L.

where q_{dyn} is the charge of the dynamical fermions. Gauge fixing terms were not written explicitly. Let us add an external pair with charges $\pm q_{\text{ext}}$ at $\pm L$, namely,

$$j_0^{\text{ext}} = q_{\text{ext}}\left(\delta(x+L) - \delta(x-L)\right), \tag{14.3}$$

so that the change of $\mathcal{L}$ is $-j_\mu^{\text{ext}} A^\mu(x)$. Note that by choosing j_μ^{ext} which is conserved, $\partial^\mu j_\mu^{\text{ext}} = 0$, the action including the coupling to the external current is also gauge invariant.

Now, one can eliminate this charge by performing a local, space-dependent left-handed rotation,

$$\Psi \to e^{i\alpha(x)\frac{1}{2}(1-\gamma_5)}\Psi \tag{14.4}$$
$$\bar{\Psi} \to \bar{\Psi}e^{-i\alpha(x)\frac{1}{2}(1+\gamma_5)}, \tag{14.5}$$

where $\gamma^5 = \gamma^0\gamma^1$. We choose a left-handed rotation (or equally well a right-handed one) rather than an axial one, since in the non-abelian case the former will be easier to implement.

The new action is,

$$S = \int d^2 x \left[-\frac{1}{4e^2}F_{\mu\nu}^2 + \bar{\Psi}i\slashed{\partial}\Psi - \bar{\Psi}\partial_\mu\alpha(x)\gamma^\mu\frac{1}{2}(1-\gamma_5)\Psi - m\bar{\Psi}e^{-i\alpha(x)\gamma_5}\Psi \right.$$
$$\left. - q_{\text{dyn}}A_\mu\bar{\Psi}\gamma^\mu\Psi - q_{\text{ext}}\left(\delta(x+L) - \delta(x-L)\right)A_0 + \frac{\alpha(x)q_{\text{dyn}}}{2\pi}F \right], \tag{14.6}$$

where the last term is induced by the chiral anomaly,

$$\delta S = \int d^2 x \frac{\alpha(x)q_{\text{dyn}}}{2\pi}F, \tag{14.7}$$

with F the dual of the electric field, $F = \frac{1}{2}\epsilon^{\mu\nu}F_{\mu\nu}$.

The external source and the anomaly term are similar, both being linear in the gauge potential. This is the reason that the θ-vacuum, to be discussed in Chapter 22, and electron-positron pair at the boundaries are the same in two dimensions.

In the following we assume $\theta = 0$, as otherwise we absorb it in to α. Choosing the $A_1 = 0$ gauge and integrating by parts, the anomaly term looks like an external source,

$$\frac{q_{\text{dyn}}}{2\pi}A_0\partial_1\alpha(x). \tag{14.8}$$

This term can cancel the external source by the choice,

$$a(x) = 2\pi \frac{q_{\text{ext}}}{q_{\text{dyn}}} (\theta(x + L) - \theta(x - L)). \tag{14.9}$$

Let us take the limit $L \to \infty$. The form of the action, in the region B of $-L < x < L$ is,

$$S_B = \int_B d^2x \left(-\frac{1}{4e^2} F_{\mu\nu}^2 + \bar{\Psi} i \partial \!\!\!/ \Psi - m \bar{\Psi} e^{-i2\pi \frac{q_{\text{ext}}}{q_{\text{dyn}}} \gamma_5} \Psi - q_{\text{dyn}} A_\mu \bar{\Psi} \gamma^\mu \Psi \right). \tag{14.10}$$

Thus the total impact of the external electron-positron pair is a chiral rotation of the mass term. This term can be written as,

$$\bar{\Psi} e^{-i2\pi \frac{q_{\text{ext}}}{q_{\text{dyn}}} \gamma_5} \Psi = \cos\left(2\pi \frac{q_{\text{ext}}}{q_{\text{dyn}}}\right) \bar{\Psi} \Psi - i \sin\left(2\pi \frac{q_{\text{ext}}}{q_{\text{dyn}}}\right) \bar{\Psi} \gamma_5 \Psi. \tag{14.11}$$

The string tension is the vacuum expectation value (v.e.v.) of the Hamiltonian density in the presence of the external source relative to the v.e.v. of the Hamiltonian density without the external source, in the $L \to \infty$ limit,

$$\sigma = <\mathcal{H}> - <\mathcal{H}_0>_0, \tag{14.12}$$

where $|0>_0$ is the vacuum state with no external sources. The change in the vacuum energy is due to the mass term. The change in the kinetic term which appears in (14.6) does not contribute to the vacuum energy.

Thus,

$$\sigma = m \cos\left(2\pi \frac{q_{\text{ext}}}{q_{\text{dyn}}}\right) <\bar{\Psi} \Psi> - m \sin\left(2\pi \frac{q_{\text{ext}}}{q_{\text{dyn}}}\right) <\bar{\Psi} i \gamma_5 \Psi> - m <\bar{\Psi} \Psi>_0 . \tag{14.13}$$

The values of the condensates $<\bar{\Psi} \Psi>$ and $<\bar{\Psi} \gamma_5 \Psi>$ are needed. The easiest way to compute these condensates is bosonization, but it can also be computed directly in the fermionic language. We state here the final result for the $m = 0$ case (the derivation can be found in the references of this chapter),

$$<\bar{\Psi} \Psi>_{m=0} = -e \frac{\exp(\gamma)}{2\pi^{3/2}} \tag{14.14}$$

$$<\bar{\Psi} \gamma_5 \Psi>_{m=0} = 0. \tag{14.15}$$

Equation (14.15) is due to parity invariance (with our choice $\theta = 0$). The resulting string tension, to first order in m,

$$\sigma = m e \frac{\exp(\gamma)}{2\pi^{3/2}} \left(1 - \cos\left(2\pi \frac{q_{\text{ext}}}{q_{\text{dyn}}}\right)\right). \tag{14.16}$$

Though this expression is only the leading term in a m/e expansion and might be corrected, when q_{ext} is an integer multiple of q_{dyn} the string tension is *exactly* zero, since in this case the rotated action (14.10) is not changed from the original one (14.2).

14.2 The Schwinger model in bosonic form

Next we derive the same result in the bosonized formulation.

The bosonized Lagrangian, in the gauge $A_1 = 0$, is given by,

$$\mathcal{L} = \frac{1}{2e^2}(\partial_1 A_0)^2 + \frac{1}{2}(\partial_\mu \phi)^2 + M^2 \cos(2\sqrt{\pi}\phi) + \frac{q_{\text{dyn}}}{\sqrt{\pi}} A_0 \partial_1 \phi - A_0 j_{\text{ext}}, \quad (14.17)$$

where $M^2 = m\mu$, $\mu = \frac{\exp(\gamma)}{2\pi}\mu_{(\phi)}$ with $\mu_{(\phi)} = \frac{e}{\sqrt{\pi}}q_{\text{dyn}}$ the mass of the photon, for $e \gg m$.

Chiral rotation corresponds to a shift in the field ϕ. Upon the transformation,

$$\phi = \tilde{\phi} + \sqrt{\pi}\frac{q_{\text{ext}}}{q_{\text{dyn}}}\left(\theta(x+L) - \theta(x-L)\right). \quad (14.18)$$

The Lagrangian (14.17) takes, in the region B, the form,

$$\mathcal{L}_B = \frac{1}{2e^2}(\partial_1 A_0)^2 + \frac{1}{2}(\partial_\mu \tilde{\phi})^2 + M^2 \cos\left(2\sqrt{\pi}\tilde{\phi} + 2\pi\frac{q_{\text{ext}}}{q_{\text{dyn}}}\right) + \frac{q_{\text{dyn}}}{\sqrt{\pi}} A_0 \partial_1 \tilde{\phi}. \quad (14.19)$$

Hence, similarly to the previous derivation, a local chiral rotation was used to eliminate the external source. The calculation of the string tension is exactly the same as in the previous section.

The relevant part of the Hamiltonian density is,

$$\mathcal{H} = -M^2 \cos\left(2\sqrt{\pi}\tilde{\phi} + 2\pi\frac{q_{\text{ext}}}{q_{\text{dyn}}}\right). \quad (14.20)$$

To zeroth order in $\left(\frac{M}{e}\right)^2$, the vacuum is $\tilde{\phi} = 0$. Setting this choice in (14.20) and subtracting the v.e.v. of the free Hamiltonian, we arrive at,

$$\sigma_{QED} = m\mu\left(1 - \cos\left(2\pi\frac{q_{\text{ext}}}{q_{\text{dyn}}}\right)\right), \quad (14.21)$$

where m is the electron mass, $\mu = e\frac{\exp(\gamma)}{2\pi^{3/2}}$, e the gauge coupling, γ the Euler number and q_{ext}, q_{dyn} are the external and dynamical charges, respectively (we measure charge in units of e, thus q_{ext} and q_{dyn} are dimensionless).

14.3 Beyond the small mass abelian string tension

The expression (14.21) contains only the leading $\frac{m}{e}$ contribution to the abelian string tension. This expression was computed in the previous section, using a classical average. However, as we used the normal ordering scale μ_ϕ which is the photon mass for $e \gg m$, taking $\tilde{\phi} = 0$ actually gives the full quantum answer, as is evident by comparing with the fermionic calculation in the section before that.

The full perturbative (in m) string tension can be written as,

$$\sigma_{QED} = m\mu \sum_{l=1}^{\infty} C_l \left(\frac{m}{eq_{\text{dyn}}}\right)^{l-1}\left(1 - \cos\left(2\pi l\frac{q_{\text{ext}}}{q_{\text{dyn}}}\right)\right). \quad (14.22)$$

The value of the first coefficient is $C_1 = 1$ and the next was found to be $C_2 = -8.9 \frac{\exp(\gamma)}{8\pi^{1/2}}$. Higher coefficients are not calculated yet.

Note that for finite $\frac{m}{e}$ we have to minimize the potential,

$$V = M^2 \left(1 - \cos \left(2\sqrt{\pi}\phi + 2\pi \frac{q_{ext}}{q_{dyn}} \right) \right) + \frac{1}{2}\mu_\phi^2 \phi^2 . \tag{14.23}$$

The minimum $\phi = \phi_m$ obeys,

$$2\sqrt{\pi}M^2 \sin \left(2\sqrt{\pi}\phi_m + 2\pi \frac{q_{ext}}{q_{dyn}} \right) + \mu_\phi^2 \phi_m = 0. \tag{14.24}$$

Thus, for the first-order $\left(\frac{m}{eq_{dyn}} \right)$ correction, we get a C_2 which is $-\left(\frac{1}{2}\right)\sqrt{\pi}(\exp \gamma)$. This has the same sign, but a factor 1.41 larger, than the instanton contribution.

Note that all above results for the string tension are symmetric under change of sign of the external charge, as expected on general grounds. However, when a θF term is introduced, we get odd terms as well, like $\sin(l\theta) \sin(2\pi l \frac{q_{ext}}{q_{dyn}})$. The even terms are multiplied by $\cos(l\theta)$.

Finally, let us remark that for very large $\frac{m}{e}$, the abelian case has a string tension which is $\frac{1}{2}e^2 q_{ext}^2$.

14.4 Correction to the leading long distance abelian potential

The potential (14.1) is the dominant long-range term. However, there are, of course, corrections. In this section we present these corrections.

The equations of motions which follow from the bosonized Lagrangian (14.17) are, in the static case,

$$-\frac{1}{e^2}\partial_1^2 A_0 + \frac{q_{dyn}}{\sqrt{\pi}}\partial_1 \phi - j_{ext} = 0 \tag{14.25}$$

$$-\partial_1^2 \phi + 2\sqrt{\pi}M^2 \sin 2\sqrt{\pi}\phi + \frac{q_{dyn}}{\sqrt{\pi}}\partial_1 A_0 = 0. \tag{14.26}$$

In order to solve these equation, it is useful to eliminate the bosonized matter field ϕ. Using the approximation $\sin 2\sqrt{\pi}\phi \sim 2\sqrt{\pi}\phi$, we arrive at (in momentum space),

$$A_0(k) = \frac{e^2 \left(k^2 + 4\pi M^2 \right)}{k^2 \left(k^2 + \left(4\pi M^2 + \frac{e^2}{\pi}q_{dyn}^2 \right) \right)} j_{ext}(k), \tag{14.27}$$

where k is the Fourier transform of the space coordinate. We will discuss the validity of our approximation for ϕ later in this section. The last equation can be rewritten as,

$$A_0(k) = \left(\frac{m_1^2}{m_2^2}\frac{1}{k^2} + \left(1 - \frac{m_1^2}{m_2^2} \right)\frac{1}{k^2 + m_2^2} \right) e^2 j_{ext}(k), \tag{14.28}$$

where,

$$m_1^2 = 4\pi M^2, \tag{14.29}$$

$$m_2^2 = 4\pi M^2 + \frac{e^2}{\pi} q_{\rm dyn}^2. \tag{14.30}$$

Note that the photon propagator has two poles, a massless pole that reproduces the string tension and a massive pole which adds a screening term to the potential. Note that there is no $\frac{\rm const.}{L}$ correction, which appears in higher dimensions, since in the present case the string cannot fluctuate in transverse directions.

Note also that in the massless case, when $M^2 = 0$, only the second term survives and the photon has only one pole with mass square $\frac{e^2}{\pi} q_{\rm dyn}^2$. This result is of course exact, independent of our approximation.

The resulting gauge field is,

$$A_0(x) = \frac{2\pi^2 M^2 q_{\rm ext}}{q_{\rm dyn}^2} (|x+L| - |x-L|)$$

$$- \frac{e\sqrt{\pi}}{2} \frac{q_{\rm ext}}{q_{\rm dyn}} \left(e^{-\frac{e}{\sqrt{\pi}} q_{\rm dyn} |x+L|} - e^{-\frac{e}{\sqrt{\pi}} q_{\rm dyn} |x-L|} \right), \tag{14.31}$$

where we took $M^2 \ll e^2$ for simplicity.

In order to calculate the potential we will use,

$$V = \frac{1}{2} \int A_0(x) j_{\rm ext}(x) {\rm d}x. \tag{14.32}$$

Hence the potential is,

$$V = 2\pi^2 M^2 \frac{q_{\rm ext}^2}{q_{\rm dyn}^2} \times 2L + \frac{e\sqrt{\pi}}{2} \frac{q_{\rm ext}^2}{q_{\rm dyn}} \left(1 - e^{-\frac{e}{\sqrt{\pi}} q_{\rm dyn} 2L} \right). \tag{14.33}$$

The first term is the confining potential which exists whenever the quark mass is non-zero. On top of this, there is always a screening potential.

The string tension which results from the above potential is,

$$\sigma = m\mu \times 2\pi^2 \frac{q_{\rm ext}^2}{q_{\rm dyn}^2}, \tag{14.34}$$

which is exactly (14.21) in the approximation $2\pi \frac{q_{\rm ext}}{q_{\rm dyn}} \ll 1$. This turns out to be also the condition for $\sin 2\sqrt{\pi}\phi \sim 2\sqrt{\pi}\phi$ that we assumed at the start of this section. To see that, we solve for ϕ from eqn. (14.25) as,

$$\phi(k) = -ik \frac{q_{\rm dyn}}{\sqrt{\pi}} \frac{e^2}{m_2^2} \left(\frac{1}{k^2} - \frac{1}{k^2 + m_2^2} \right) j_{\rm ext}(k). \tag{14.35}$$

Define $\phi = \phi_1 + \phi_2$, where ϕ_1 is the part with $\frac{1}{k^2}$, and ϕ_2 with $\frac{1}{k^2+m_2^2}$. The ϕ_2 part goes to zero at long distances, i.e. $k \to 0$. As for the ϕ_1 part, its x-space form is,

$$\phi_1(x) = \frac{e^2}{\sqrt{\pi}m_2^2} q_{\text{dyn}} q_{\text{ext}} \left(\theta(x+L) - \theta(x-L)\right), \tag{14.36}$$

which for small $\frac{m}{e}$ reduces to,

$$\phi_1(x) \sim \sqrt{\pi} \frac{q_{\text{ext}}}{q_{\text{dyn}}} \left(\theta(x+L) - \theta(x-L)\right). \tag{14.37}$$

Thus $2\sqrt{\pi}\phi$ small means,

$$(2\pi) \frac{q_{\text{ext}}}{q_{\text{dyn}}} \ll 1, \tag{14.38}$$

the condition mentioned before.

Note that we could generalize the argument to values of $2\pi\frac{q_{\text{ext}}}{q_{\text{dyn}}}$ that are close to $2\pi n$, with integer n.

14.5 Finite temperature

In this section we would like to comment on the behavior of the string tension in the presence of finite temperature. It is interesting to check whether the string is torn due to high temperature and whether the system undergoes a phase transition from confinement to deconfinement.

The prescription for calculating quantities at finite temperature T is to formulate the theory on a circle in Euclidean time with circumference $\beta = T^{-1}$.

For the purpose of calculating the string tension, we can follow the same steps which we employed previously, leading to a modification of eqn. (14.16) as,

$$\sigma = -m <\bar{\Psi}\Psi>_T \left(1 - \cos 2\pi \frac{q_{\text{ext}}}{q_{\text{dyn}}}\right). \tag{14.39}$$

It is enough to calculate $<\bar{\Psi}\Psi>_T$, the condensate at finite temperature, in the massless Schwinger model.

The chiral condensate behaves as,

$$<\bar{\Psi}\Psi>_{(T\to 0)} \to -\frac{e}{2\pi^{3/2}} e^\gamma, \tag{14.40}$$

and

$$<\bar{\Psi}\Psi>_{(T\to\infty)} \to -2Te^{-\frac{\pi^{3/2}T}{e}}. \tag{14.41}$$

This result indicates that the string is not torn even at very high temperatures. The explicit expression shows that $<\bar{\Psi}\Psi>_T$ is non-zero for all T. Thus, the system does not undergo a phase transition. It is just energetically favorable to have the electron-positron pair confined.

14.6 Two-dimensional QCD

The action of bosonized QCD_2 with massive quarks in the fundamental representation of $SU(N)$ (see Chapter 8) is

$$S_{\text{fundamental}} = \frac{1}{8\pi} \int_\Sigma d^2x \, \text{tr} \left(\partial_\mu g \partial^\mu g^\dagger \right) \tag{14.42}$$

$$+ \frac{1}{12\pi} \int_B d^3y \epsilon^{ijk} \, \text{tr}(g^\dagger \partial_i g)(g^\dagger \partial_j g)(g^\dagger \partial_k g)$$

$$+ \frac{1}{2} m\mu_{fund} \int d^2x \, \text{tr}(g + g^\dagger) - \int d^2x \frac{1}{4e^2} F^a_{\mu\nu} F^{a\mu\nu}$$

$$- \frac{1}{2\pi} \int d^2x \, \text{tr}(ig^\dagger \partial_+ g A_- + ig\partial_- g^\dagger A_+ + A_+ g A_- g^\dagger - A_+ A_-),$$

where e is the gauge coupling, m is the quark mass, $\mu = e^{\frac{\exp(\gamma)}{(2\pi)^{\frac{3}{2}}}}$, g is an $N \times N$ unitary matrix, A_μ is the gauge field and the trace is over $U(N)$ indices. Note, however, that only the $SU(N)$ part of the matter field g is gauged.

When the quarks transform in the adjoint representation, the expression for the action is,

$$S_{\text{adjoint}} = \frac{1}{16\pi} \int_\Sigma d^2x \, \text{tr} \left(\partial_\mu g \partial^\mu g^\dagger \right) \tag{14.43}$$

$$+ \frac{1}{24\pi} \int_B d^3y \epsilon^{ijk} \, \text{tr} \left(g^\dagger \partial_i g \right) \left(g^\dagger \partial_j g \right) \left(g^\dagger \partial_k g \right)$$

$$+ \frac{1}{2} m\mu_{\text{adj}} \int d^2x \, \text{tr} \left(g + g^\dagger \right) - \int d^2x \frac{1}{4e^2} F^a_{\mu\nu} F^{a\mu\nu}$$

$$- \frac{1}{4\pi} \int d^2x \, \text{tr} \left(ig^\dagger \partial_+ g A_- + ig\partial_- g^\dagger A_+ + A_+ g A_- g^\dagger - A_+ A_- \right).$$

The action (14.43) differs from (14.42) by a factor of one half in front of the WZW and interaction terms, because g is real and represents Majorana fermions. Another difference is that g now is an $(N^2 - 1) \times (N^2 - 1)$ orthogonal matrix. The two actions (14.42) and (14.43) can be schematically represented by one action,

$$S = S_0 + \frac{1}{2} m\mu_R \int d^2x \, \text{tr} \left(g + g^\dagger \right) \tag{14.44}$$

$$- \frac{ik_{\text{dyn}}}{4\pi} \int d^2x \left(g\partial_- g^\dagger \right)^a A^a_+,$$

where $A_- = 0$ gauge was used, S_0 stands for the WZW action and the kinetic action of the gauge field, k_{dyn} is the level (the chiral anomaly) of the dynamical charges ($k = 1$ for the fundamental representation of $SU(N)$ and $k = N$ for the adjoint representation).

Let us add an external charge to the action. We choose a static charge (with respect to the light-cone coordinate x^+) and therefore we can omit its kinetic term from the action. Thus an external charge coupled to the gauge field would be represented by,

$$-\frac{ik_{\text{ext}}}{4\pi}\int d^2x \, \left(u\partial_- u^\dagger\right)^a A_+^a.$$

Suppose that we want to put a quark and an anti-quark at a very large separation. A convenient choice of the charges would be a direction in the algebra in which the generator has a diagonal form. The simplest choice is a generator of an $SU(2)$ subalgebra. Since a rotation in the algebra is always possible, the results are insensitive to this specific choice. As an example we write down the generator in the case of fundamental and adjoint representations,

$$T_{\text{fund}}^3 = \text{diag}\left(\frac{1}{2}, -\frac{1}{2}, \underbrace{0, 0, \ldots, 0}_{N-2}\right)$$

$$T_{\text{adj}}^3 = \text{diag}\left(1, 0, -1, \underbrace{\frac{1}{2}, -\frac{1}{2}, \frac{1}{2}, -\frac{1}{2}, \ldots, \frac{1}{2}, -\frac{1}{2}}_{2(N-2)\ \text{doublets}}, \underbrace{0, 0, \ldots, 0}_{(N-2)^2}\right).$$

Generally T^3 can be written as

$$T^3 = \text{diag}(\lambda_1, \lambda_2, \ldots, \lambda_i, \ldots, 0, 0, \ldots),$$

where $\{\lambda_i\}$ are the 'isospin' components of the representation under the $SU(2)$ subgroup.

We take the $SU(N)$ part of u as,

$$u = \left[\exp -i4\pi \left(\theta(x^- + L) - \theta\left(x^- - L\right)\right)\right] T_{\text{ext}}^3, \tag{14.45}$$

for $N > 2$, and a similar expression but with a 2π factor for $N = 2$. T_{ext}^3 represents the '3' generator of the external charge and u is static with respect to the light-cone time coordinate x^+. The theta function is used as a limit of a smooth function which interpolates between 0 and 1 over a very short distance. In that limit $u = 1$ everywhere except at isolated points, where it is not well defined.

The form of the action (14.44) in the presence of the external source is,

$$S = S_0 + \frac{1}{2}m\mu R \int d^2x \left\{ \text{tr}\left(g + g^\dagger\right) \right.$$

$$\left. + \left[-\frac{ik_{\text{dyn}}}{4\pi}\left(g\partial_- g^\dagger\right)^a + k_{\text{ext}}\delta^{a3}\left(\delta\left(x^- + L\right) - \delta\left(x^- - L\right)\right)\right] A_+^a \right\}.$$

The external charge can be eliminated from the action by a transformation of the matter field. A new field $\tilde{g}$ can be defined as follows,

$$-\frac{ik_{\text{dyn}}}{4\pi}\left(\tilde{g}\partial_- \tilde{g}^\dagger\right)^a = -\frac{ik_{\text{dyn}}}{4\pi}\left(g\partial_- g^\dagger\right)^a + k_{\text{ext}}\delta^{a3}\left(\delta(x^- + L) - \delta(x^- - L)\right).$$

This definition leads to the following equation for $\tilde{g}^\dagger$,

$$\partial_-\tilde{g}^\dagger = \tilde{g}^\dagger \left(g\partial_-g^\dagger + i4\pi\frac{k_{\text{ext}}}{k_{\text{dyn}}}(\delta(x^- + L) - \delta(x^- - L))T_{\text{dyn}}^3 \right). \qquad (14.46)$$

The solution of 14.46 is,

$$\tilde{g}^\dagger = P\exp\left\{ \int dx^- \left(g\partial_-g^\dagger + i4\pi\frac{k_{\text{ext}}}{k_{\text{dyn}}}(\delta(x^- + L) - \delta(x^- - L))T_{\text{dyn}}^3 \right) \right\}$$

$$= e^{i4\pi\frac{k_{\text{ext}}}{k_{\text{dyn}}}\theta(x^- + L)T_{\text{dyn}}^3}\, g^\dagger e^{-i4\pi\frac{k_{\text{ext}}}{k_{\text{dyn}}}\theta(x^- - L)T_{\text{dyn}}^3}, \qquad (14.47)$$

where P denotes path ordering and we assume that T_{dyn}^3 commutes with $g\partial_-g^\dagger$ for $x^- \geq L$ and with $g^\dagger$ for $x^- = -L$ (as we shall see, this assumption is self consistent with the vacuum configuration).

Let us take the limit $L \to \infty$. For $-L < x^- < L$, the above relation simply means that,

$$g = \tilde{g}e^{i4\pi\frac{k_{\text{ext}}}{k_{\text{dyn}}}T_{\text{dyn}}^3}.$$

Since the Haar measure is invariant (and finite, unlike the fermionic case) with respect to unitary transformations, the form of the action in terms of the new variable $\tilde{g}$ reads,

$$S = S_{WZW}(\tilde{g}) + S_{\text{kinetic}}(A_\mu) - \frac{ik_{\text{dyn}}}{4\pi}\int d^2x \left(\tilde{g}\partial_-\tilde{g}^\dagger \right)^a A_+^a$$

$$+ \frac{1}{2}m\mu_R \int d^2x \, \text{tr}\left(\tilde{g}e^{i4\pi\frac{k_{\text{ext}}}{k_{\text{dyn}}}T_{\text{dyn}}^3} + e^{-i4\pi\frac{k_{\text{ext}}}{k_{\text{dyn}}}T_{\text{dyn}}^3}\tilde{g}^\dagger \right), \qquad (14.48)$$

which is QCD_2 with a chiraly rotated mass term.

The string tension can be calculated easily from (14.48). It is simply the vacuum expectation value (v.e.v.) of the Hamiltonian density, relative to the v.e.v. of the Hamiltonian density of the theory without an external source,

$$\sigma = <H> - <H_0> .$$

The vacuum of the theory is given by $\tilde{g} = 1$. In terms of the variable g, this configuration points in the '3' direction and hence satisfies our assumptions while solving eqn. (14.46). The v.e.v. is,

$$<H> = -\frac{1}{2}m\mu_R \, \text{tr}\left(e^{i4\pi\frac{k_{\text{ext}}}{k_{\text{dyn}}}T_{\text{dyn}}^3} + e^{-i4\pi\frac{k_{\text{ext}}}{k_{\text{dyn}}}T_{\text{dyn}}^3} \right)$$

$$= -m\mu_R \sum_i \cos\left(4\pi\lambda_i\frac{k_{\text{ext}}}{k_{\text{dyn}}} \right).$$

Therefore the string tension is,

$$\sigma = m\mu_R \sum_i \left(1 - \cos\left(4\pi\lambda_i\frac{k_{\text{ext}}}{k_{\text{dyn}}} \right) \right), \qquad (14.49)$$

which is the desired result.

We expect that similar corrections as those in eqn. (14.22) will occur also in non-abelian systems. For the fundamental/adjoint case, the following expression may correct the leading term,

$$\sigma_{QCD} = m\mu_R \sum_{l=1}^{\infty} \tilde{C}_l \left(\frac{m}{ek_{\text{dyn}}} \right)^{l-1} \sum_j \left(1 - \cos \left(4\pi\lambda_j l \frac{k_{\text{ext}}}{k_{\text{dyn}}} \right) \right). \qquad (14.50)$$

A few remarks should be made:

(i) The string tension (14.49) reduces to the abelian string tension (14.21) when abelian charges are considered. It follows that the non-abelian generalization is realized by replacing the charge q with the level k.

(ii) The string tension was calculated in the tree level of the bosonized action. Perturbation theory (with m as the coupling) may cause changes, eqn. (14.49), since the loop effects may add $O(m^2)$ contributions. However, we believe that it would not change its general character. In fact, one feature is that the string tension vanishes for any m when $\frac{k_{\text{ext}}}{k_{\text{dyn}}}$ is an integer, as follows from eqn. (14.48), since the action does not depend then on k_{ext} at all.

(iii) When no dynamical mass is present, the theory exhibits screening. This is simply because non-abelian charges at the end of the world interval can be eliminated from the action by a chiral transformation of the matter field.

(iv) When the test charges are in the adjoint representation $k_{\text{ext}} = N$, eqn. (14.49) predicts screening by the fundamental charges (with $k_{\text{dyn}} = 1$).

(v) String tension appears when the test charges are in the fundamental representation and the dynamical charges are in the adjoint. The value of the string tension is

$$\sigma = m\mu_{\text{adj}} \left(2 \left(1 - \cos \frac{4\pi}{N} \right) + 4(N-2) \left(1 - \cos \frac{2\pi}{N} \right) \right) \qquad (14.51)$$

as follows from eqn. (14.49) for this case.

The case of $SU(2)$ is special. The 4π which appears in eqn. (14.49) is replaced by 2π, since the bosonized form of the external $SU(2)$ fundamental matter differs by a factor of a half with respect to the other $SU(N)$ cases. Hence, the string tension in this case is $4m\mu_{\text{adj}}$.

(vi) We would like to add, that when computing the string tension in the pure YM case with external sources in representation R, the Wilson loop gives $\frac{1}{2}e^2 C_2(R)$, while our way of defining external source gives $\frac{1}{2}e^2 k_{\text{ext}}^2$. Thus we need a factor $\frac{C_2(R)}{k_{\text{ext}}^2}$ to bring our result to the Wilson loop case. Analogous factors should be computed for the other cases, when dynamical matter is also present.

14.7 Symmetric and antisymmetric representations

The generalization of (14.49) to arbitrary representations is not straightforward. However, we can comment about its nature (without rigorous proof).

Let us focus on the interesting case of the antisymmetric representation. One can show that the WZW action with g taken to be $\frac{1}{2}N(N-1) \times \frac{1}{2}N(N-1)$ unitary matrices, is a bosonized version of QCD_2 with fermions in the antisymmetric representation.

The antisymmetric representation is described in the Young-tableaux notation by two vertical boxes. Its dimension is $\frac{1}{2}N(N-1)$ and its diagonal $SU(2)$ generator is,

$$T_{as}^3 = \mathrm{diag}\Big(\underbrace{\frac{1}{2}, -\frac{1}{2}, \frac{1}{2}, -\frac{1}{2}, ..., \frac{1}{2}, -\frac{1}{2}}_{(N-2) \ \text{doublets}}, 0, 0, ..., 0 \Big), \qquad (14.52)$$

and consequently $k = N - 2$. When the dynamical charges are in the fundamental and the external in the antisymmetric the string tension should vanish because the tensor product of two fundamentals include the antisymmetric representation. Indeed, (14.49) predicts this result.

The more interesting case is when the dynamical charges are antisymmetric and the external are fundamentals. In this case the value of the string tension depends on whether N is odd or even.

When N is odd the string tension should vanish because the anti-fundamental representation can be built by tensoring the antisymmetric representation with itself $\frac{1}{2}(N-1)$ times. When N is even string tension must exist.

Note that (14.49) predicts,

$$\sigma = 2m\mu_{as}(N-2)\Big(1 - \cos\frac{2\pi}{N-2} \Big), \qquad (14.53)$$

which is not zero when N is odd, contrary to expectation.

The resolution of the puzzle seems to be the following. Non-abelian charge can be static with respect to its spatial location. However, its representation may change in time due to emission or absorption of soft gluons (without cost of energy). Our semi-classical description of the external charge as a c-number is insensitive to this scenario. We need an extension of (14.45) which takes into account the possibilities of all various representations. One possible extension is,

$$j_{ext}^a = \delta^{a3}k_{ext} \left(1 + lN \right) \left(\delta\left(x^- + L \right) - \delta\left(x^- - L \right) \right), \qquad (14.54)$$

where l is an arbitrary positive integer. This extension takes into account the cases which correspond to $1 + lN$ charges multiplied in a symmetric way. The resulting string tension is,

$$\sigma = m\mu_R \sum_i \left(1 - \cos\left(4\pi\lambda_i \frac{k_{ext}}{k_{dyn}} (1 + lN) \right) \right), \qquad (14.55)$$

which includes the arbitrary integer l. What is the value of l that we should pick?

The dynamical charges are attracted to the external charges in such a way that the total energy of the configuration is minimal. Therefore the value of l which is needed, is the one that guarantees minimal string tension.

Thus the extended expression for string tension is the following,

$$\sigma = \min_l \left\{ m\mu_R \sum_i \left(1 - \cos\left(4\pi\lambda_i \frac{k_{\text{ext}}}{k_{\text{dyn}}} (1 + lN) \right) \right) \right\}. \tag{14.56}$$

In the case of dynamical antisymmetric charges and external fundamentals and odd N, $l = \frac{1}{2}(N - 3)$ gives zero string tension. When N is even the string tension is given by (14.53).

The expression (14.56) yields the right answer in some other cases also, like the case of dynamical charges in the symmetric representation. The bosonization for this case can be derived in a similar way to that of the antisymmetric representation, and T^3 is given by

$$T^3_{\text{symm}} = \text{diag}\left(1, 0, -1, \underbrace{\frac{1}{2}, -\frac{1}{2}, \frac{1}{2}, -\frac{1}{2}, \dots, \frac{1}{2}, -\frac{1}{2}}_{(N-2) \text{ doublets}}, 0, 0, \dots, 0 \right). \tag{14.57}$$

Hence $k = N + 2$. When the external charges transform in the fundamental representation and N is odd, eqn (14.56) predicts zero string tension (as it should). When N is even the string tension is given by

$$\sigma = 2m\mu_{\text{symm}} \left(\left(1 - \cos\frac{4\pi}{N+2} \right) + (N-2)\left(1 - \cos\frac{2\pi}{N+2} \right) \right). \tag{14.58}$$

We have discussed only the cases of the fundamental, adjoint, anti-symmetric and symmetric representations, since we used bosonization techniques which are applicable to a limited class of representations.

15

QCD_2, coset models and BRST quantization

15.1 Introduction

In Chapter 9 we realized that the structure of the bosonized non-abelian massless QCD_2 is that of a gauged WZW model with an additional F^2 term of the gauge fields. Apart from the pure gauge term, this is therefore a special form of a two-dimensional coset model, discussed in Section 4.6. This naturally calls for a treatment of the system similar to that for a coset model. Using the form of the gauge fields in terms of scalars f and $\bar{f}$ as $A = if^{-1}\partial f$, $\bar{A} = i\bar{f}\bar{\partial}\bar{f}^{-1}$ with $f(z, \bar{z}), \bar{f}(z, \bar{z}) \in H^c$, the complexification of $H \equiv SU(N_C)$, leads to a convenient formulation of the model.[1] The main advantage of this approach is that one can then easily decouple the "matter" and the gauge degrees of freedom.

In this chapter we point out that the F^2 term requires a special treatment. The formulation of pure YM theory in terms of the f variables seems naively to contain unexpected "physical" massive color singlet states. This result is obviously neither in accordance with our ideas of the degrees of freedom of the model nor with the lattice and continuum solution of the theory. We show that similar "naive" manipulations in the case of QED_2 do reproduce the Schwinger model results. Using a coupling constant renormalization we show that in the limit of no matter degrees of freedom the coupling constant is renormalized to zero. In this case the unexpected states turn into unphysical massless "BRST" exact states. In the flavored QCD_2 case a similar analysis shows the existence of physical flavorless states of mass $m^2 = \frac{N_F}{2\pi}e_c^2$.

This chapter is based on [96].

15.2 The action

The bosonized version of QCD_2 was shown in Chapter 9 to be described by the action,

$$S_{QCD_2} = S_1(u) - \frac{1}{2\pi} \int d^2 z \operatorname{Tr}(iu^{-1}\partial u \bar{A} + iu\bar{\partial}u^{-1}A + \bar{A}u^{-1}Au - A\bar{A})$$
$$+ \frac{m^2}{2\pi} \int d^2 z : \operatorname{Tr}_G[u + u^{-1}] : + \frac{1}{e_c^2} \int d^2 z \operatorname{Tr}_H [F^2], \qquad (15.1)$$

---

[1] This parameterization of the gauge field was previously introduced in [9].

where $u \in U(N_F \times N_C)$, $S_k(u)$ is a level k WZW model,

$$S_k(u) = \frac{k}{8\pi} \int \mathrm{d}^2 x \mathrm{Tr}(\partial_\mu u \partial^\mu u^{-1}) + \frac{k}{12\pi} \int_B \mathrm{d}^3 y \varepsilon^{ijk} \mathrm{Tr}(u^{-1}\partial_i u)(u^{-1}\partial_j u)(u^{-1}\partial_k u),$$

(15.2)

A and $\bar{A}$ take their values in the algebra of $H \equiv SU(N_C)$, $F = \bar{\partial}A - \partial\bar{A} + i[A, \bar{A}]$, m^2 equals $m_q \mu C$, where μ is the normal ordering mass and $C = \frac{1}{2}e^\gamma$ with γ, Euler's constant. Apart from the last two terms which correspond to the quark mass term and the YM term, the rest of the action is a level-one $\frac{G}{H}$ coset model with $G = U(N_F \times N_C)$.

We now introduce the following parameterization for the gauge fields $A = if^{-1}\partial f, \bar{A} = i\bar{f}\bar{\partial}\bar{f}^{-1}$ with $f(z, \bar{z}), \bar{f}(z, \bar{z}) \in SU(N_C)^c$. These type of variables were used frequently in dealing with gauged WZW actions, for instance in computing the effective action of QCD_2 and in the $\frac{G}{G}$ models discussed in Section 4.7. They may be interpreted as Wilson lines along the z and $\bar{z}$ directions. The gauged WZW part of the action, first line of (15.1), takes the form,

$$S_1(u, A) = S_1(fu\bar{f}) - S_1(f\bar{f}).$$

The Jacobian of the change of variables from A to f introduces a dimension $(1, 0)$ system of anticommuting ghosts (ρ, χ) in the adjoint representation of H.[2] The WZW part of the action thus becomes

$$S_1(u, A) = S_1(fu\bar{f}) - S_1(f\bar{f}) + \frac{i}{2\pi} \int \mathrm{d}^2 z \mathrm{Tr}_H [\rho \bar{D}\chi + \bar{\rho}D\bar{\chi}],$$

(15.3)

where $D\chi = \partial\chi - i[A, \chi]$. Our integration variables in the functional integral are $if^{-1}df$ and $i\bar{f}d\bar{f}^{-1}$. This action involves an interaction term of the form $\mathrm{Tr}_H(\bar{\rho}[f^{-1}\partial f, \bar{\chi}])$ and a similar term for ρ, χ. By performing a chiral rotation, like those of Chapter 14, $\bar{\rho} \to f^{-1}\bar{\rho}f$ and $\bar{\chi} \to f^{-1}\bar{\chi}f$ with $\rho \to \bar{f}\rho\bar{f}^{-1}$ and $\chi \to \bar{f}\chi\bar{f}^{-1}$, one achieves a decoupling of the whole ghost system. The price of this is an additional $S_{-2N_C}^{(H)}(f\bar{f})$ term in the action (here trace over H only) resulting from the corresponding anomaly. This result can be derived by using a non-abelian bosonization of the ghost system. A different bosonization of a $(1,0)$ ghost system was described in Section 6.5.

In this language the ghost action takes the form,

$$S_{gh} = S_{N_c}(l_1, A, \bar{A}) + S_{N_c}(l_2, A, \bar{A}) + S_{\mathrm{twist}}(l_1) + S_{\mathrm{twist}}(l_2),$$

where l_1 and l_2 are in the adjoint representation and S_{twist} is a twist term given,

$$S_{\mathrm{twist}} = -\frac{N_c}{2\pi} \int \mathrm{d}^2 z \mathrm{Tr}[l\bar{\partial}l^{-1}f^{-1}\partial f].$$

(15.4)

---

[2] The ghost action was introduced in [8].

Now using the Polyakov–Wiegmann formula we get,

$$S_{gh}^b = S_{N_c}(fl_1\bar{f}) - S_{N_c}(f\bar{f}) + S_{\text{twist}}(l_1) + S_{N_c}(fl_2\bar{f}) - S_{N_c}(f\bar{f}) + S_{\text{twist}}(l_2)$$

$$= -S_{2N_c}(f\bar{f}) + \frac{i}{2\pi}\int d^2z\mathrm{Tr}_H[\bar{\rho}'\partial\bar{\chi}' + \rho'\bar{\partial}\chi'], \tag{15.5}$$

where the last line has been transferred back to the ghost language. Notice that unlike the ghost fields in (15.3) the new ghost fields ρ' and χ' are gauge invariant. It is interesting to note that the action given in (15.5) is non-local in terms of the local degrees of freedom A and $\bar{A}$. Note that had we done the right and left rotations separately, we would have got $S_{-2N_c}(f) + S_{-2N_c}(\bar{f})$, which however is not vector gauge invariant, but rather a left–right symmetric scheme.

The full gauge invariant action including the anomaly contribution of the anti-commuting part now reads,

$$S_{QCD_2} = S_1(u) + \frac{1}{e_c^2}\int d^2z\mathrm{Tr}_H[F^2] + \frac{m^2}{2\pi}\int d^2z : \mathrm{Tr}_G\left[f^{-1}u\bar{f}^{-1} + \bar{f}u^{-1}f\right] :$$

$$+ \left[S_{-(N_F+2N_C)}^{(H)}(f\bar{f}) + \frac{i}{2\pi}\int d^2z\mathrm{Tr}_H[\bar{\rho}'\partial\bar{\chi}' + \rho'\bar{\partial}\chi']\right]. \tag{15.6}$$

In deriving eqn. (15.6) we used a redefinition $fu\bar{f} \to u$. This does not require an extra determinant factor. Also, as $S^{(H)}(f\bar{f})$ involves Tr_H rather than the Tr_G in $S_1(f\bar{f})$ of eqn. (2.4), a factor of N_f appears. Note that had we introduced the special parameterization of the gauge fields in the fermionic formulation of QCD_2, we would have arrived at the same action after decoupling the fermionic currents from the gauge fields, by performing chiral rotation and then bosonizing the free fermions. Equation (15.6) was derived without paying attention to possible renormalizations. The latter will be treated in Section 15.7.

At this point one may choose a gauge. A convenient gauge choice is $\bar{A} = i\bar{f}\bar{\partial}\bar{f}^{-1} = 0$. Notice that since the underlying space-time is a plane this is a legitimate gauge. The gauge fixed action can be written down using the BRST procedure, namely,

$$S_{GF} = S_{QCD_2} + S^{(gf)} + S^{(gh)} = S_{QCD_2} + \delta_{BRST}(b\bar{A}) =$$

$$= S_{QCD_2} + \mathrm{Tr}_H[B\bar{A}] + \mathrm{Tr}_H[b\bar{D}c], \tag{15.7}$$

where S_{GF}, $S^{(gf)}$ and $S^{(gh)}$ are, respectively, the gauge fixed action, the gauge fixing term and the ghost action. The (b, c) fields are yet another $(1, 0)$ ghost system and B is a dimension-one auxiliary field, all in the adjoint representation of $SU(N_C)$. The integration over B introduces a delta function of the gauge choice to the measure of the functional integral. In addition we integrate over the ghosts b and c.

It is interesting to note that the QCD_2 action can be related to a "perturbed" topological $\frac{H}{H}$ coset model. To realize this face of QCD_2 we parameterize u as $ghle^{i\sqrt{\frac{4\pi}{N_C N_F}}\phi}$ and rewrite (15.7) accordingly. The Polyakov–Wiegmann relation implies,

$$S[u] = S[ghl] + \frac{1}{2} \int d^2x \partial_\mu \phi \partial^\mu \phi,$$

$$S[ghl] = S[g] + S[l] + S[h] + \frac{1}{2\pi} \int d^2x \mathrm{Tr}(g^\dagger \partial_+ gl\partial_- l^\dagger + h^\dagger \partial_+ hl\partial_- l^\dagger). \quad (15.8)$$

Since l is a dimension-zero field with an associated zero central charge we have $S[l] = 0$ and thus,

$$S_{GF} = S_{N_F}(h) + S^{(H)}_{-(N_F+2N_C)}(f) + \frac{i}{2\pi} \int d^2z \mathrm{Tr}_H[\bar\rho \partial \bar\chi + \rho \bar\partial \chi]$$

$$+ S_{N_C}(g) + \frac{1}{2\pi} \int d^2z [\partial\phi\bar\partial\phi]$$

$$+ \frac{m^2}{2\pi} \int d^2z \mathrm{Tr}_G : \left[f^{-1}ghle^{i\sqrt{\frac{4\pi}{N_C N_F}}\phi} + e^{-i\sqrt{\frac{4\pi}{N_C N_F}}\phi} l^{-1}h^{-1}g^{-1}f \right] :$$

$$+ \frac{1}{e_c^2} \int d^2z \mathrm{Tr}_H[(\bar\partial(f^{-1}\partial f))^2]. \quad (15.9)$$

It is now easy to recognize the first line in the action as the action of $\frac{SU(N_C)}{SU(N_C)}$ topological theory.

It is interesting to note that a WZW term $S_{-2N_C}(f)$ appears in the action even without the introduction of quarks. We therefore digress to an analysis of the pure YM theory in the formulation introduced above.

15.3 Two-dimensional Yang–Mills theory

Pure Yang–Mills theory has attracted much attention recently along the lines of an underlying string theory. Here we restrict our discussion to the 2D Minkowski or Euclidean space-time, where the rich structure of the model on a compact Riemann surface does not show up. In terms of the parameterization introduced in eqn. (15.6) the gauge invariant action of the pure YM theory is,

$$S_{YM_2} = S_{-(2N_C)}(f\bar{f}) + \frac{i}{2\pi} \int d^2z \mathrm{Tr}_H[\bar\rho \partial \bar\chi + \rho \bar\partial \chi]$$
$$+ \frac{1}{e_c^2} \int d^2z \mathrm{Tr}_H[F^2]. \quad (15.10)$$

Here again we remind the reader that the coupling constant undergoes a multiplicative renormalization. This will be discussed in Section 15.7. Let us first discuss the corresponding equations of motion for f and $\bar{f}$,

$$\delta f : \bar\partial A - D\bar{A} + \frac{2}{m_A^2}D\bar{D}F = \left(1 + \frac{2}{m_A^2}D\bar{D}\right)F = 0,$$

$$\delta \bar{f} : \partial \bar{A} - \bar{D}A - \frac{2}{m_A^2}\bar{D}DF = -\left(1 + \frac{2}{m_A^2}DD\right)F = 0, \quad (15.11)$$

where $D = \partial - i[A, \cdot]$, $m_A = e_c\sqrt{\frac{N_C}{\pi}}$ and $\square = 2\partial\bar\partial$. In fact these two equations are identical, as $[D, \bar{D}]F = 0$. The equation is that of a massive gauge field with self interaction. Note that in this approach, unlike the equations that follow from varying the action with respect to the gauge fields, one gets two derivatives of F. In deriving the above, it is convenient to remember that,

$$\delta S_{WZW}(f) = \frac{1}{2\pi} \text{Tr} \left\{ (f^{-1}\delta f)\bar{\partial}(f^{-1}\partial f) \right\}.$$

The YM action equation (15.10) is obviously invariant under the original gauge transformations,

$$f \to f v(z, \bar{z}) \quad \bar{f} \to v^{-1}(z, \bar{z})\bar{f},$$

with $v \in SU(N_C)$. In addition the action is invariant separately under the holomorphic and anti-holomorphic "color" transformations,

$$f \to u(\bar{z})f \quad \bar{f} \to \bar{f}w(z),$$

where $u, w \in SU(N_C)$. These are "spurious" transformations since they leave A and $\bar{A}$ invariant. The corresponding holomorphic and anti-holomorphic color currents are,

$$\bar{J}^s = -\frac{N_c}{\pi} [i(f\bar{f})\bar{\partial}(f\bar{f})^{-1} - \frac{2}{m_A^2} f\bar{D}Ff^{-1}],$$
$$J^s = -\frac{N_c}{\pi} [i(f\bar{f})^{-1}\partial(f\bar{f}) + \frac{2}{m_A^2}\bar{f}^{-1}DF\bar{f}]. \tag{15.12}$$

The gauge fixed ($\bar{f} = 1$) action takes the form,

$$S_{YM_2} = S_{-(2N_C)}(f) + \frac{1}{e_c^2} \int \mathrm{d}^2 z \text{Tr}_H [(\bar{\partial}(f^{-1}\partial f))^2]$$

$$+ \frac{i}{2\pi} \int \mathrm{d}^2 z \text{Tr}_H [\bar{\rho}\partial\bar{\chi} + \rho\bar{\partial}\chi]. \tag{15.13}$$

As is expected the equation of motion at present is just that of eqn. (15.11) after setting $\bar{A} = 0$. Naturally, the action now lacks gauge invariance, nevertheless, it is invariant under the following residual holomorphic transformations,

$$f \to u(\bar{z})f \quad f \to fw(z),$$

with the corresponding holomorphic and anti-holomorphic currents,

$$J_G = -\frac{N_C}{\pi} \left[A + \frac{2}{m_A^2} D(\bar{\partial}A) \right] \quad \bar{J}_G = -\frac{N_C}{\pi} \left[\tilde{A} + \frac{2}{m_A^2}\bar{D}(\partial\tilde{A}) \right],$$

and $\tilde{A} = if\bar{\partial}f^{-1}$. Notice that in spite of the similar structure, $\tilde{A}$ is not related to $\bar{A}$ which was set to zero. To better understand the physical picture behind these currents we defer temporarily to the abelian case.

15.4 Schwinger model revisited

Since in the pure Maxwell theory there is no analog to the $(-2N_C)$ level WZW term of eqn. (15.10), we study instead the Schwinger model in its bosonized form,

$$S_{(\text{Sch})} = \frac{1}{2\pi} \int \mathrm{d}^2 z \left[\partial X\bar{\partial}X - \sqrt{2}\partial X\bar{A} + \sqrt{2}\bar{\partial}XA + \frac{\pi}{e^2}(\partial\bar{A} - \bar{\partial}A)^2 \right].$$

In analogy to the change of variables in the non-abelian case, we now introduce the following parameterization of the gauge fields $A = \partial \varphi$, $\bar{A} = \bar{\partial} \bar{\varphi}$. In terms of these fields the action takes the form,

$$S_{(\text{Sch})} = \int \frac{d^2 z}{2\pi} \left\{ [\partial X \bar{\partial} X - \sqrt{2} X \partial \bar{\partial} (\varphi - \bar{\varphi}) + \frac{\pi}{e^2} [\partial \bar{\partial} (\varphi - \bar{\varphi})]^2 + i[\bar{\rho} \partial \bar{\chi} + \rho \bar{\partial} \chi] \right\}.$$

In the gauge $\bar{A} = 0$ and after the field redefinition $\tilde{X} = X + \frac{1}{\sqrt{2}} \varphi$, the action is decomposed into decoupled sectors,

$$S_{(\text{Sch})} = S(\tilde{X}) + S(\varphi) + S_{(\text{ghost})},$$

$$S(\tilde{X}) = \frac{1}{2\pi} \int d^2 z [\partial \tilde{X} \bar{\partial} \tilde{X}] \quad S(\varphi) = \frac{1}{4\pi} \int d^2 z \left\{ \frac{2}{\mu^2} [\partial \bar{\partial} \varphi]^2 - \partial \varphi \bar{\partial} \varphi \right\}, \quad (15.14)$$

where $\mu^2 = \frac{e^2}{\pi}$. The corresponding equations of motion are,

$$\partial \bar{\partial} \left[1 + \frac{2}{\mu^2} \partial \bar{\partial} \right] \varphi = 0, \quad \partial \bar{\partial} \tilde{X} = 0.$$

The invariance under the chiral shifts $\delta \varphi = \epsilon(\bar{z})$ and $\delta \varphi = \epsilon(z)$ are generated by the holomorphically conserved currents,

$$J_G = \partial \varphi + \frac{2}{\mu^2} \partial \bar{\partial} \partial \varphi, \quad \bar{J}_G = \bar{\partial} \varphi + \frac{2}{\mu^2} \partial \bar{\partial} \bar{\partial} \varphi.$$

To handle this type of "hybrid" current we suggest the following decomposition of the massless and massive modes $\varphi = \varphi_1 + \varphi_2$ with,

$$\partial \bar{\partial} \varphi_1 = 0 \quad [2 \partial \bar{\partial} + \mu^2] \varphi_2 = 0.$$

In the holomorphic quantization,

$$\Pi = \frac{\delta \mathcal{L}}{\delta(\partial \varphi)} = -\frac{1}{\pi \mu^2} \bar{\partial} \left(\partial \bar{\partial} + \frac{\mu^2}{4} \right) \varphi = \frac{1}{4\pi} \bar{\partial} (\varphi_2 - \varphi_1). \quad (15.15)$$

A unique solution to the commutation relations $[\varphi(z, \bar{z}), \Pi(w, \bar{w})]_{z=w} = i\delta(\bar{z} - \bar{w})$, $[\varphi, \varphi] = 0$ and $[\Pi, \Pi] = 0$ is,

$$[\varphi_1(z, \bar{z}), \varphi_1(w, \bar{w})]_{z=w} = \pi i \epsilon(\bar{z} - \bar{w})$$

$$[\varphi_2(z, \bar{z}), \varphi_2(w, \bar{w})]_{z=w} = -\pi i \epsilon(\bar{z} - \bar{w})$$

$$[\varphi_1(z, \bar{z}), \varphi_2(w, \bar{w})]_{z=w} = 0$$

$$\left[\tilde{X}(z, \bar{z}), \tilde{X}(w, \bar{w}) \right]_{z=w} = -\pi i \epsilon(\bar{z} - \bar{w}), \quad (15.16)$$

where ϵ is the standard antisymmetric step function. Notice that the massless degree of freedom has commutation relations which correspond to a negative metric on the phase space. These relations can also be translated to the following OPEs (choosing the part $\varphi_1(z)$ of φ_1),

$$\varphi_1(z) \varphi_1(w) = \log(z - w),$$

$$\varphi_2(z, \bar{z}) \varphi_2(w, \bar{w}) = -\log |(z - w)|^2 + O(\mu^2 |z - w|^2),$$

$$\varphi_1(z) \varphi_2(w, \bar{w}) = O(z - w). \quad (15.17)$$

It is thus clear that the model is invariant under a $U(1)$ affine Lie algebra of level $k = -1$ since $J_G(z)J_G(w) = \frac{1}{(z-w)^2}$, as $J_G = \partial\varphi_1$ with no contribution from φ_2.

The physical states of the model have to be in the cohomology of the BRST charge. Due to the fact that the current is holomorphically (and the other antiholomorphically) conserved, it follows that the same property holds for the BRST charge, and thus the space of physical states is an outer product of the cohomology of Q and $\bar{Q}$. The latter are given by,

$$Q = \chi J = \chi(i\partial\tilde{X} + \partial\varphi_1),$$
$$\bar{Q} = \bar{\chi}\bar{J} = \bar{\chi}(-i\bar{\partial}\tilde{X} + \bar{\partial}\varphi_1). \tag{15.18}$$

Expanding the fields $i\partial\tilde{X}$ and $\partial\varphi_1$ in terms of the Laurent modes $\tilde{X}_n$ and $(\tilde{\varphi}_1)_n$ with $[X_n, X_m] = n\delta_{n+m}$ and $[(\varphi_1)_n, (\varphi_1)_m] = n\delta_{n+m}$ we have

$$Q = \sum_n \tilde{\chi}_n \left[\tilde{X}_{-n} - i(\varphi_1)_{-n} \right].$$

Since $J_0 = \{Q, \rho_0\}$, physical states have to have a zero eigenvalue of J_0. The general structure of the states in the $\varphi_1, \tilde{X}, \rho, \chi$ Fock space is,

$$(\tilde{X}_n)^{n_X} (\varphi_{1m})^{n_f} (\chi_k)^{n_\chi} (\rho_l)^{n_\rho} |\text{vac}\,,$$

where obviously n_χ and n_ρ are either 0 or 1. It is straightforward to realize that only the vacuum state and states of the form $(\tilde{X}_0)^{n_X} (\varphi_{10})^{n_f}$ are in the BRST cohomology. Recall that being on the plane we exclude zero modes and thus only the vacuum state remains. Since there is no constraint on the modes of φ_2, the physical states are built solely of φ_2 which are massive modes. This result is identical to the well-known solution of the Schwinger model.

15.5 Back to the YM theory

Equipped with the lesson from the Schwinger model we return now to the YM case and introduce a decomposition of the group element f so that again the gauge currents obey an affine Lie algebra. Let us write $f = f_2 f_1$ which implies that,

$$A = if^{-1}\partial f = if_1^{-1}\partial f_1 + if_1^{-1}(f_2^{-1}\partial f_2)f_1 \equiv J_1 + J_2$$

With no loss of generality we take $\bar{\partial}f_1 = 0$ implying also $\bar{\partial}J_1 = 0$. Inserting these expressions into J_G of eqn (15.3) one finds,

$$J_G = -\frac{N_C}{\pi}\left[J_1 + J_2 + \frac{2}{m_A^2}(\partial\bar{\partial}J_2 + i[\bar{\partial}J_2, J_1 + J_2]) \right].$$

If one can consistently require that,

$$J_2 + \frac{2}{m_A^2}(\partial\bar{\partial}J_2 + i[\bar{\partial}J_2, J_1 + J_2]) = 0,$$

then, in a complete analogy with the abelian case, $J_G = -\frac{N_c}{\pi} J_1$. The latter is an affine current of level $k = -2N_C$. One can in fact show that (15.5) can be assumed without a loss of generality. $\bar{\partial} J_G = 0$ implies that $J_2 + \frac{2}{m_A^2}(\partial \bar{\partial} J_2 + i[\bar{\partial} J_2, J_1 + J_2]) = u(z)$, where $u(z)$ is some holomorphic function. We then introduce the shifted currents $\tilde{J}_2 = J_2 - u(z)$, $\tilde{J}_1 = J_1 + u(z)$. Now $\bar{\partial} \tilde{J}_1 = 0$ as does J_1, and $\tilde{J}_2$ obeys eqn (15.5) with $\tilde{J}_1$ replacing J_1. It is easy to check that the shifts in the currents correspond to $f_1 \to v(z) f_1$, $f_2 \to f_1 v(z)^{-1}$ with $u(z) = i f_1^{-1}(v(z)^{-1} \partial v(z)) f_1$.

Note that the equation for J_2 involves a coupling to J_1. This is related to the fact that, unlike the abelian case, one cannot write the action as a sum of decoupled terms which are functions of J_1 and J_2 separately.

Once the color current J_G is expressed in terms of the holomorphic current J_1, the analysis of the space of physical states is directly related to that of the topological $\frac{G}{G}$ model at $k = 0$. The physical states have to be in the cohomology of the BRST charge, which corresponds to the following holomorphically conserved BRST current,

$$Q(z) = \chi^a \left(J_G^a + \frac{1}{2} J_{gh}^a \right) = -\frac{N_C}{\pi} \chi^a \left(\left[A + \frac{2}{m_A^2} D(\bar{\partial} A) \right]^a + \frac{i}{2} f_{bc}^a \rho^b \chi^c \right).$$

An anti-holomorphic BRST current $\bar{Q}(z)$ determines the condition for physical states in the analogous manner to Q. From here on we restrict our description to the latter. We define now the zero level affine Lie algebra current,

$$J_{(\text{tot})}^a = J_G^a + J_{(gh)}^a = J_G^a + i, f_{bc}^a \chi_b \rho_c,$$

and the $c = 0$ Virasoro generator T,

$$T(z) = -\frac{1}{N_C} : J_G^a J_G^a : + \rho^a \partial \chi^a,$$

as well as dimension $(2,0)$ fermionic current,

$$G = -\frac{1}{2N_C} \rho_a J_G^a,$$

and realize the existence of the "topological coset algebra",

$$T(z) = \{Q, G(z)\}, \quad Q(z) = \{Q, j^\#(z)\}, \quad J_{(\text{tot})}^a = \{Q, \rho^a(z)\},$$
$$\{Q, Q(z)\} = 0, \quad\quad\quad\quad \{G, G(z)\} \equiv W(z),$$
$$W(z) = \{Q, U(z)\}, \quad\quad\quad\quad [W, W(z)] = 0, \quad\quad\quad (15.19)$$

where $J^\# = \chi_a \rho^a$ is "ghost number current",

$$W(z) = \frac{1}{4N_c} f_{abc} J_G^a \rho^b \rho^c + \partial \rho^a \rho_a,$$

and,

$$U = \frac{1}{12N_c} f_{abc} J_G^a \rho^b \rho^c.$$

A direct consequence is that any physical state has to obey,

$$J_{(tot)}{}^0_0|\text{phys}> = 0, \quad L_0|\text{phys}> = 0, \quad W_0|\text{phys}> = 0, \tag{15.20}$$

where $J_{(tot)}{}^i_n$, $\tilde{L}_n$ and W_n are the Laurent modes of $J_{(tot)}{}^i$ the Cartan sub-algebra currents, T and W, respectively. In fact the BRST cohomology of the present model is a special case of the set of G/G models.

We therefore refer the reader to those works [9], [200], [229] and present here only the result. On the plane where no ghost zero modes are allowed, the only state in the cohomology is the zero ghost number vacuum state of J_1.

This state can be a tensor-product with oscillators of the massive modes of J_2. Unlike the abelian case, J_G does not commute with J_2 so that in general the J_2 modes are not obviously in the BRST cohomology. However, there is no reason to believe that all the J_2 modes will be excluded by the BRST condition. Those J_2 modes that remain are by definition color singlets.

This result contradicts previous results on YM_2. Usually one believes that pure gluodynamics on the plane is an empty theory since all local degrees of freedom can be gauged away.

15.6 An alternative formulation

To get a better understanding of the subtleties of the Yang–Mills theory when expressed in terms of $A = if^{-1}\partial f$, $\bar{A} = if\bar{\partial}\bar{f}^{-1}$, and for future application, we compare now with another formulation of the theory. A similar approach will be used in the discussion of generalized YM theories in Chapter 16. Consider the following functional integral,

$$Z = \int DA D\bar{A} DB e^{iS(A,\bar{A},B)},$$
$$S = -\int d^2z Tr_H[\tfrac{1}{e_c}FB + \tfrac{1}{4}B^2], \tag{15.21}$$

where B is a pseudoscalar field in the adjoint representation. Obviously the integration over B produces the usual $Tr[F^2]$ action. It is also easy to realize that the action is invariant under the ordinary gauge symmetry provided that $\delta B = i[\epsilon, B]$. In terms of the f variables after imposing the gauge $\bar{f} = 1$ one finds,

$$S_{YM_2} = S_{-(2N_c)}(f) + \int d^2z Tr_H\left[\left(\frac{i}{e_c}(f^{-1}\partial f)\bar{\partial}B\right) - \frac{1}{4}B^2\right] + S^{(gh)},$$

where $S^{(gh)} = \frac{i}{2\pi}\int d^2z Tr_H[\bar{\rho}\partial\bar{\chi} + \rho\bar{\partial}\chi]$. One should again bear in mind that the coupling constant undergoes a multiplicative renormalization. This will be discussed in the next section. Using Polyakov–Wiegmann we get,

$$S_{YM_2}(B) = \Gamma_{2N_c}(B) - \tfrac{1}{4}Tr_H[B^2]$$
$$- S_{(2N_c)}(vf) + S^{(gh)}, \tag{15.22}$$

where $\Gamma_k(B) = S_k(v)$ with $\frac{1}{e_c}\bar\partial B = \frac{2N_C}{2\pi}(iv\bar\partial v^{-1})$. The second line in (15.22) is a $c = 0$ "topological system". Since the underlying Minkowski space-time does not admit zero modes we can safely integrate over the corresponding fields. We can further pass from functional integrating over B to $iv\bar\partial v^{-1}$. This involves the insertion of $\det\frac{\bar D}{\bar\partial}$ which will introduce a $\Gamma_{-2N_C}(B)$ term with no additional ghost terms. The functional integral (16.1) thus takes the final form,

$$Z = \int D[v]e^{-i\left(\frac{1}{4}\int d^2z\,\mathrm{Tr}_H\,[B^2]\right)}.$$

It is thus clear that in the present formulation there is no trace of the massive "physical modes" discussed in the previous section.

15.7 The resolution of the puzzle

Encouraged by the result of the last section, we proceed now to reexamine the steps that led to the unexpected massive modes in the pure YM_2 theory. In particular, we would like to check whether in addition to the implementation of proper determinants there is no coupling constant renormalization that has to be invoked when passing to the quantum theory expressed in the f variables. For this purpose we turn on again the matter degrees of freedom. We introduce N_F quarks in the fundamental color representation and explore the behavior of the system in the limit $N_F \to 0$. Recall that the action of this model is given in eqn. (15.9). Starting actually from eqn. (15.1), taking the massless limit, writing A in terms of f in the action but still with A as an integration variable, and using the formulation presented in the previous section, the path integral of the colored degrees of freedom now reads

$$Z^{(\mathrm{col})} = \int [DA][DB][Dh]e^{iS^{(\mathrm{col})}}$$

$$S^{(\mathrm{col})} = S_{N_F}(h) + \frac{N_F}{2\pi}\int d^2z\,\mathrm{Tr}_H\,[h\bar\partial h^{-1}f^{-1}\partial f]$$

$$+ \int d^2z\,\mathrm{Tr}_H\left[\left(\frac{i}{e_c}(f^{-1}\partial f)\bar\partial B\right) - \frac{1}{4}B^2\right], \qquad (15.23)$$

where we have also gone from u to h as in Section 15.2.

It was found out that quantum consistency imposes finite renormalization on the coupling constant of the current-gauge field interaction.[3] This renormalization is expressed in the following equality,

$$Z(\bar J) \equiv \int DAe^{i\left[S_k(f)+\frac{1}{2\pi}\int d^2z\,\mathrm{Tr}_H\,[i(f^{-1}\partial f)\bar J]\right]}$$

$$= \int Dfe^{i\left[S_{k-2N_C}(f)+\frac{e(-k)}{2\pi}\int d^2z\,\mathrm{Tr}_H\,[(f^{-1}\partial f)\bar J]\right]}\int D(gh)e^{iS^{(gh)}}$$

$$= e^{i\Gamma_{-k+2N_C}\left[\left(\frac{e(-k)}{-k+2N_C}\right)\bar J\right]}\int D(gh)e^{iS^{(gh)}}, \qquad (15.24)$$

---

[3] The finite renormalization of the coupling was introduced by D. Kutasov in [146].

where k is an arbitrary level and $\Gamma_k(L) = S_k(w)$ for $L = iw\bar{\partial}w^{-1}$. The renormalization factor $e(k)$ has to satisfy $\frac{e(-k-2N_C)}{e(k)} = \frac{k}{k+2N_C}$. In addition it is clear from eqn. (15.24) that it has to be singular at the origin. It can be shown that $e(k)$ takes the form $e(k) = \sqrt{\frac{k+2N_C}{k}}$. Implementing this renormalization in our case, eqn. (15.23) takes the form,

$$S^{(\mathrm{col})} = S_{N_F}(\tilde{h}) + S_{-(N_F+2N_C)}(f)$$
$$+ \int d^2z\,\mathrm{Tr}_H \left[\left(i\sqrt{\frac{N_F+2N_C}{N_F}} \frac{1}{e_c}(f^{-1}\partial f)\bar{\partial}B \right) - \frac{1}{4}B^2 \right]$$
$$+ S^{(gh)}, \tag{15.25}$$

where $\tilde{h} = fh$. After integrating the auxiliary field B the action becomes,

$$S^{(\mathrm{col})} = S_{N_F}(\tilde{h}) + S_{-(N_F+2N_C)}(f)$$
$$+ \int d^2z\,\mathrm{Tr}_H \left[\left(\frac{N_F+2N_C}{e_c^2 N_F} \right) \right] [\bar{\partial}(f^{-1}\partial f)]^2] + S^{(gh)}. \tag{15.26}$$

It is now straightforward to realize that the equation of motion which follows from the variation with respect to f is that of eqn. (15.11) where now $m_A = e_c\sqrt{\frac{N_F}{2\pi}}$. Thus, the coupling constant renormalization turns the massive modes into massless ones in the case of pure YM theory ($N_F = 0$). Notice that to reach this conclusion it is enough to use the fact that $e(k)$ has to be singular at $k = 0$ and the explicit expression of $e(k)$ is really not needed. Following the arguments presented in Chapter 5, it is clear that these states that became massless are not in the BRST cohomology and thus not in the physical spectrum.

A somewhat similar derivation of the triviality of the model in the $N_F = 0$ limit is the following. We integrate in eqn. (15.25) over the ghost fields and over f, using again the coupling constant renormalization, and find,

$$Z = \int D[v]e^{-\left\{ iS_{N_F}(v) + \frac{e_c^2 N_F^2}{4} \int d^2z\,\mathrm{Tr}_H[B^2(v)] \right\}}.$$

It is now clear that the action vanishes at $N_F = 0$ and hence again, on trivial topology, the theory is empty. Notice, however, that the implementation of renormalization modifies also the result of the previous section.

The final conclusion is that in both methods one finds that indeed the pure YM theory has an empty space of physical states as of course is implied by the original formulation in terms of A. We have demonstrated that in this formulation it follows only after taking subtleties of renormalization into account.

15.8 On bosonized QCD_2

To resolve the puzzle of the YM theory we were led to analyze the color and flavor sectors of QCD_2. The full bosonized QCD_2 includes in addition the baryon number degrees of freedom. The corresponding action is given by eqns. (15.6),

(15.7) or by eqn. (15.9). In the past the low-lying baryonic spectrum in the strong coupling limit $\frac{m_q}{e_c} \to 0$ was extracted using a semi-classical quantization. In this chapter our analysis was based on switching off the mass term, $m_q = 0$. This limit cannot be treated by the semi-classical approach, as the soliton solution is not there for $m_q = 0$. In our case here one finds a decoupled WZW action for the flavor degrees of freedom $S_{N_C}(g)$ and a decoupled free field action for the baryon degree of freedom, in addition to the action of the colored degrees of freedom which is given in eqn. (15.26) or eqn. (15.7). The general structure of a physical state in this case is that of a tensor product of g and ϕ with the colored degrees of freedom f, h and the ghosts. The structure of QCD_2 which emerged from the semi-classical quantization for $m_q \neq 0$ involves g and ϕ only. In our case here the f colored degrees of freedom acquire mass $m_A = e_c \sqrt{\frac{N_F}{2\pi}}$ while the h degrees of freedom remain massless. In the limit $e_c \to \infty$ the f degrees of freedom decouple. It is thus clear that one has to introduce the mass term which couples the three sectors. The massless limit of QCD_2 can then be derived by taking the limit $m_q \to 0$ after solving for the physical states. Indeed, it was shown in the limit of $e_c \to \infty$ that turning on $m_q \neq 0$ results in a hadronic spectrum where the flavor representation and the baryon number were correlated. The analysis of the spectrum of the massive multi-flavor QCD_2 in the approach of this work remains to be worked out.

15.9 Summary and discussion

In this chapter we have analyzed 2D YM and QCD theories using a special parametrization of the gauge fields in terms of group elements. In the $m_q = 0$ case it enabled us to decouple the matter and gauge degrees of freedom. However, this formulation led, in a naive treatment, to unexpected massive modes. Even though we did not present a full solution of the theory we had reasonable arguments to believe that the BRST projection would not exclude these modes. The fact that a similar approach to QED_2 reproduced the known results of the Schwinger model, enhanced the puzzling phenomenon. Eventually, we showed that a coupling constant renormalization, renders the unexpected massive modes into massless un-physical states. The benefit of this detective work is the appearance of "physical" massive states in massless QCD_2.

16
Generalized Yang–Mills theory on a Riemann surface

16.1 Introduction

Pure gauge theory in two dimensions is locally trivial and has no propagating degrees of freedom. This was discussed in the first chapter of this part of the book and now in the last chapter we will describe the global properties of gauge theories in two dimensions. For the latter to be non-trivial we will either take the underlying manifold to be a compact Riemann surface or introduce Wilson loops external sources. We will show that in those cases the gauge theory has a rich structure and in fact is almost a topological field theory, which is a theory with no propagating degrees of freedom (see also Section 4.7). Moreover, it will be shown that the theory has an interpretation in terms of a string theory.

It is easy to realize that in two dimensions the pure YM theory is in fact the simplest member of a wide class of renormalizable theories that incorporate only gauge fields. These will be referred to as the generalized gauge theories gYM. In Chapter 15 we introduced an alternative formulation of the YM theory using the action[1]

$$S = -\int d^2z \, \text{Tr}[iFB + g^2 B^2].$$
(16.1)

Now it is easy to realize that the B^2 term can be generalized to an arbitrary invariant function $\Phi(B)$. This will constitute the family of gYM.[2] The partition function of the generalized theories on Riemann surfaces and the computation of Wilson loops for these theories will also be described in this chapter.

Pure YM$_2$ theory defined on an arbitrary Riemann surface is known to be exactly solvable. In one approach the theory was regularized on the lattice and, using a heat kernel action, explicit expressions for the partition function and loop averages were derived.[3] Identical results were derived also in a continuum path-integral approach.

In the following sections we briefly review the former derivation of the partition function and determine in a similar way the results for the partition function and

---

[1] In fact in (16.1) we have used a slightly different formulation which is however equivalent to the one used in Chapter 15.

[2] The notion of the generalized QCD theory in two dimensions was introduced and analyzed in [82].

[3] A lattice version of two-dimensional Yang–Mills theory was shown to be exactly solvable by Migdal in [161]. Correlators of Wilson lines on this formulation were computed in [139].

Wilson loops in the gYM_2 case. Since this chapter is based on non-trivial two-dimensional topology which has not been dealt with intensively in this book, to fully understand its content the reader will need to consult the references to this chapter. The reader who is not interested in the topological aspects of two-dimensional gauge dynamics may skip this chapter and proceed directly to the third part of the book.

This chapter is based mainly on [115], [118], [119] and [105].

16.2 The partition function of the YM_2 theory

The partition function for the ordinary YM_2 theory defined on a compact Riemann surface $\mathcal{M}$ of genus H and area A is,

$$\mathcal{Z}(N, H, \lambda A) = \int [DA^\mu] \exp\left[-\frac{1}{4g^2} \int_\Sigma \mathrm{d}^2 x \sqrt{\det G_{\mu\nu}} \ \mathrm{tr} F_{\mu\nu} F^{\mu\nu}\right], \qquad (16.2)$$

where the gauge group G is taken to be either $SU(N)$ or $U(N)$, g is the gauge coupling constant, $\lambda = g^2 N$, $G_{\mu\nu}$ is the metric on $\mathcal{M}$, and tr stands for the trace in the fundamental representation.[4]

As was emphasized in Chapter 8 the pure YM theory defined on a flat Minkowski space-time with trivial topology is empty since one can gauge away the gauge fields. However this does not hold if the underlying manifold $\mathcal{M}$ is topologically non-trivial. If $\mathcal{M}$ contains a non-trivial cycle γ such that $\mathrm{tr}[Pe^{\oint_\gamma A_\mu \, \mathrm{d}x^\mu}] \neq 1$, where P stands for path-ordering, one cannot gauge A_μ away along γ. Thus, the partition function depends on the topology of $\mathcal{M}$ and in fact only on the topology and its area. This follows from the fact that the theory is invariant under *area preserving diffeomorphism*. The field strength can be written in the form $F_{\mu\nu} = \epsilon_{\mu\nu} F$ and hence the action takes the form,

$$S = -\frac{1}{4g^2} \int_\Sigma \mathrm{d}^2 x \sqrt{\det G_{\mu\nu}} F^2. \qquad (16.3)$$

Apart from the area form $\mathrm{d}^2 x \sqrt{\det G_{\mu\nu}}$ this action is independent of the metric and is therefore invariant under area preserving diffeomorphism.

The lattice partition function defined on an arbitrary triangulation of the surface, as described in Fig. 16.1, is given by,

$$\mathcal{Z}_\mathcal{M} = \int \prod_l \mathrm{d}U_l \prod_\triangle Z_\triangle[U_\triangle], \qquad (16.4)$$

where $\prod_l$ denotes a product over all links, $U_\triangle$ is the holonomy around a plaquette $U_\triangle = \prod_{l \in \triangle} U_l$, and $Z_\triangle$ is a plaquette action. For the latter one uses a heat kernel

---

[4] The partition function on any Riemann surface of the discretized theory was written down in [184]. An identical result was found also in the continuum formulation [228].

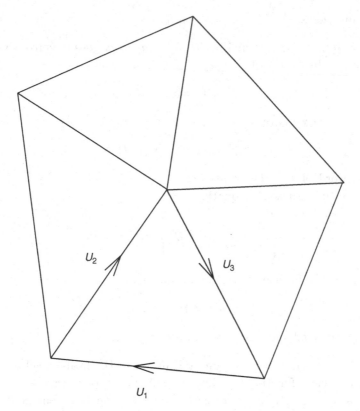

Fig. 16.1. A triangulation of the Riemann surface $\mathcal{M}$. A group matrix is placed on each link.

action rather than the Wilson action (which is $Z_\triangle[U_\triangle] = \mathrm{e}^{-\frac{1}{g^2}\mathrm{tr}(U+U^\dagger)}$), i.e.

$$Z_\triangle[U] = \sum_R d_R \chi_R(U)\mathrm{e}^{-tc_2(R)}, \qquad (16.5)$$

where the summation is over the irreducible representations R of the group; d_R, $\chi_R(U)$ and $c_2(R)$ denote the dimension, character of U and the second Casimir operator of R, respectively, and $t = g^2 a^2$ with a^2 being the plaquette area. The character, which was also discussed in Section 3.5 in relation to the ALA algebras, is defined here as $\chi_R(U) \equiv \mathrm{tr}_R[U]$. The holonomy U (its subscript $\triangle$ is omitted from here on) behaves as $U \approx 1 - iaF$ when a is small. Note that the region of validity of (16.5) is not only $a \to 0$ with F fixed, but actually also $a \to 0$ with F going to infinity as $a^{-1/2}$ because this is the region for which the exponential $-\frac{1}{4t}a^2 \,\mathrm{Tr}\, F^2$ is of order unity.

We will briefly review the derivation that singles out (16.5) as a convenient choice among the different lattice theories which belong to the same universality class. Let us look for a function $\Psi(U,t)$ that will replace the continuum $\mathrm{e}^{-\frac{1}{4t}a^2 \,\mathrm{Tr}\, F^2}$.

The requirements which we impose on Ψ are:

1. As t goes to zero (and, therefore, for finite g also a goes to zero) we want the holonomy to be close to 1,

$$\Psi(U, 0) = \delta(U - 1). \tag{16.6}$$

2. For any $V \in G$ we have,

$$\Psi(V^{-1} U V, t) = \Psi(U, t). \tag{16.7}$$

In other words Ψ is a *class function*.
3. Ψ satisfies the heat kernel equation,

$$\left\{ \frac{\partial}{\partial t} - \sum_{a,b} g_{ab} \partial^a \partial^b \right\} \Psi(U, t) = 0, \tag{16.8}$$

where g_{ab} is the inverse of the Cartan metric,

$$g^{ab} = \mathrm{tr}(t^a t^b), \tag{16.9}$$

which was defined and discussed in Section 3.2.1.

To see that (16.5) is an approximate solution to the heat-kernel equation we note that any class function is a linear combination of characters. The differentiation of a character in the direction of a Lie algebra element t^a is given by,

$$\partial^{a_1} \partial^{a_2} \cdots \partial^{a_k} \chi_R(U) = \frac{i^k}{k!} \chi_R(U t^{(a_1} t^{a_2} \cdots t^{a_k)}) + O(U - 1). \tag{16.10}$$

The notation $\chi_R(U t^{a_1} t^{a_2} \cdots t^{a_k})$ stands for the trace of the multiplication of the matrices which represent U and $t^{a_1}, \ldots, t^{a_k}$ in the representation R. The brackets $(\cdots)$ imply symmetrization with respect to the indices. The term $O(U - 1)$ means that the corrections are of the order of $U - 1 \sim aF \sim t^{1/2}$. Since,

$$\sum_{a,b} g_{ab} \partial^a \partial^b \chi_R(U) \approx -\frac{1}{2} \chi_R(U \sum_{a,b} g_{ab} t^{(a} t^{b)}) = -c_2(R) \chi_R(U), \tag{16.11}$$

we see that (16.5) is the correct answer up to terms of the order of $O(t^{3/2})$ which drop in the continuum limit. Using (16.5) as the starting point, we finally find the following form of the partition function,

$$\mathcal{Z}(N, H, \lambda A) = \sum_R d_R^{2-2H} e^{-\frac{\lambda A c_2(R)}{2N}}. \tag{16.12}$$

To get from (16.5) to (16.12) we take the following steps. First we make use of the additivity property of the heat-kernel action. Consider two triangles glued along U_1 as depicted in Fig. 16.2.

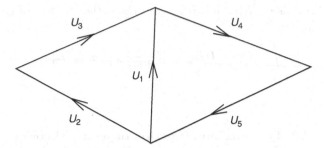

Fig. 16.2. Integrating over U_1 on a link which is common to the two triangles.

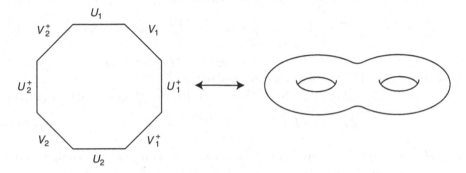

Fig. 16.3. Opening up of a genus-two surface.

Using the orthogonality of characters,

$$\int dU \chi_{R_1}(VU)\chi_{R_2}(U^\dagger W) = \delta_{R_1,R_2}\frac{\chi_{R_1}(VW)}{\dim R_1}, \tag{16.13}$$

we find,

$$\int dU_1 Z_{\triangle_1}(U_2U_3U_1)Z_{\triangle_2}(U_1^\dagger U_4U_5) = Z_{\triangle_1+\triangle_2}(U_2U_3U_4U_5). \tag{16.14}$$

This relation can be used to argue that the lattice representation is exact and independent of the triangulation since using this we can add as many triangles as desired, thus reaching the continuum limit. We can also use this relation to reduce the number of triangles to the minimum needed to capture the topology of $\mathcal{M}$. Describing a genus H manifold in term of a 4H-polygon with identified sides as described in Fig. 16.2 for a genus-two Reimann surface $a_1 b_1 a_1^{-1} b_1^{-1} \dots a_H b_H a_H^{-1} b_H^{-1}$. The partition function on such a manifold can be written as,

$$\mathcal{Z}_\mathcal{M} = \sum_R d_R e^{-\frac{\lambda A c_2(R)}{2N}} \int \prod DU_l DV_l \chi_R[U_1 V_1 U_1^\dagger V_1^\dagger \dots U_H V_H U_H^\dagger V_1^\dagger]. \tag{16.15}$$

We can simplify this expression using again the orthogonality of the characters and the relation,

$$\int DU \chi_R[AUBU^\dagger] = \frac{1}{\dim_R} \chi_R[A]\chi_R[B], \tag{16.16}$$

to arrive at (16.12).

16.3 The partition function of gYM_2 theories

Pure YM_2 theory is in fact a special representative of a wide class of 2D gauge theories which are invariant under area preserving diffeomorphisms. These generalized YM_2 theories are described by the following generalized partition function,

$$\mathcal{Z}(G, H, A, \Phi) = \int [DA^\mu][DB] \exp[\int_\Sigma d^2x \sqrt{\det G_{\mu\nu}} \, \text{tr}(iBF - \Phi(B))], \tag{16.17}$$

where $F = F^{\mu\nu}\epsilon_{\mu\nu}$ with ϵ_{ij} being the antisymmetric tensor $\epsilon_{12} = -\epsilon_{21} = 1$. B is an auxiliary Lie-algebra-valued pseudo-scalar field.[5]

We wish to generalize the substitution (16.5) for the plaquette action (16.17),

$$Z_\triangle[U] = \int DB e^{\text{tr}\{iaBF - t\Phi(B)\}} \xrightarrow{?} \Psi(U, t). \tag{16.18}$$

Here B is a Hermitian matrix and Φ is an invariant function (invariant under $B \to U^{-1}BU$ for $U \in G$). The quadratic case $\Phi(X) = g^2 \text{tr}(X^2)$ obviously corresponds to the YM_2 theory. We will take Φ to be of the form,

$$\Phi(X) = \sum_{\{k_i\}} a_{\{k_i\}} \prod_i \text{tr}(X^i)^{k_i}, \tag{16.19}$$

(e.g. $\text{tr}(X^3)^2 + \text{tr}(X^6)$) For $SU(N)$ ($U(N)$), $\text{tr}(X^i)$ can be expressed for $i \geq N$ ($i > N$) in terms of $\text{tr}(X^i)$ for smaller i. Thus the summands in (16.19) are not independent. This does not affect the following discussion. Moreover, in the large N limit that we will discuss in the following section, the terms do become independent.

We define the general structure constants $d_{abc...k}$ to be,

$$d_{abc...k} \stackrel{\text{def}}{=} g_{aa'}g_{bb'}g_{cc'}\cdots g_{kk'}\text{tr}(t^{a'}t^{b'}t^{c'}\cdots t^{k'}). \tag{16.20}$$

For every partition $r_1 + r_2 + \cdots + r_j$, we define the Casimir,

$$C_{\{r_1+r_2+\cdots+r_j\}} \stackrel{\text{def}}{=}$$

$$\frac{1}{(r_1 + r_2 + \cdots + r_j)!} d_{a_1^{(1)}...a_{r_1}^{(1)}} d_{a_1^{(2)}...a_{r_2}^{(2)}} \cdots d_{a_1^{(j)}...a_{r_j}^{(j)}} \, t^{(a_1^{(1)}} \cdots t^{a_{r_1}^{(1)}} t^{a_1^{(2)}} \cdots t^{a_{r_j}^{(j)})} \tag{16.21}$$

---

[5] In principle, we could perturb the ordinary YM_2 with operators of the form $\frac{1}{g^{2k-2}}\text{tr}(F^k)$, without the need for an auxiliary field.

Note that the index of $C_{\{.\}}$ will always pertain to a partition. Thus $C_{\{p\}} \neq C_{\{r_1+r_2+\cdots+r_j\}}$ even if $p = r_1 + r_2 + \cdots + r_j$. The brackets in the t-s mean a total symmetrization ($(r_1 + r_2 + \cdots + r_j)$! terms).

C_ρ can easily be seen to commute with all the group elements and so, by Schur's lemma, is a constant matrix in every irreducible representation.

We claim that the correct lattice generalization of (16.5) is,

$$\sum_R d_R \chi_R(U) e^{-t\Lambda(R)}, \tag{16.22}$$

where,

$$\Lambda(R) = \sum_{\{k_i\}} a_{\{k_i\}} C_{\{k_1 \cdot 1 + k_2 \cdot 2 + k_3 \cdot 3 + \cdots\}}(R). \tag{16.23}$$

This results from the requirements that $\Psi(U, t)$ must satisfy:

1. $\Psi(U, 0) = \delta(U - 1)$.
2. Ψ is a class-function.
3. Ψ satisfies the equation,

$$\left\{ \frac{\partial}{\partial t} - \sum_{\{k_j\}} a_{\{k_j\}} \prod_l \left((ia)^{-l} d_{a_1 a_2 \dots a_l} \frac{\partial^l}{\partial F_{a_1} \cdots \partial F_{a_l}} \right)^{k_l} \right\} \Psi(U, t) + O(U - 1) = 0 \tag{16.24}$$

For the U s that are important in the weight for a single plaquette, $U - 1$ is of the order of magnitude of aF which, in turn, is of the order of magnitude of $O(t^{1/\nu})$ where ν is the maximal degree of Φ. Thus, the corrections to Ψ are $O(a^{-(1+1/\nu)})$ and drop out in the continuum limit.

The partition function for the generalized YM_2 theory is therefore,

$$\mathcal{Z}(G, \Sigma_H, \Phi) = \sum_R (\dim R)^{2-2H} e^{-\frac{\mathcal{A}A}{2N} \Lambda(R)}, \tag{16.25}$$

where A is the area of the surface and $\Lambda(R)$ is defined in (16.23).

16.4 Loop averages in the generalized case

The full solution of the YM_2 theory includes, in addition to the partition function, closed expressions for the expectation values of products of any arbitrary number of Wilson loops,

$$W(R_1, \gamma_1, \dots R_n \gamma_n) = \left\langle \prod_{i=1}^n \mathrm{Tr}_{R_i} \mathcal{P} e^{i \oint_{\gamma_i} A \, dx} \right\rangle, \tag{16.26}$$

where the path-ordered product around the closed curve γ_i is taken in the representation R_i. Using loop equations, one can derive an algorithm to compute Wilson loops on the plane. This can be further generalized into a prescription for

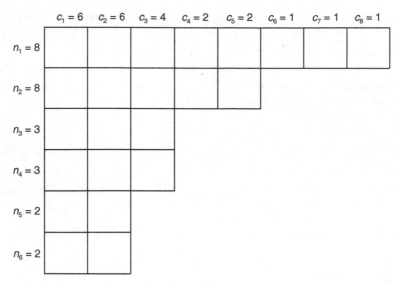

Fig. 16.4. Wilson loops on a torus.

computing those averages for non-intersecting loops on an arbitrary two manifold. Let us briefly summarize the latter. One cuts the 2D surface along the Wilson loop contours, see Fig. 16.4, forming several connected "windows". Each window contributes a sum over all irreducible representations of the form of (16.12). In addition, for each pair of neighbouring windows, a Wigner coefficient,

$$D_{R_1 R_2 f} = \int dU \chi_{R_1}(U) \chi_{R_2}(U^\dagger) \chi_f(U), \qquad (16.27)$$

is attached. Altogether, one finds,

$$W(R_1, \gamma_1, \ldots R_n \gamma_n) = \frac{1}{\mathcal{Z}} \frac{1}{N^n} \sum_{R_1} \cdots \sum_{R_n} D_{R_1 \ldots R_n} \prod_{i=1}^{N_w} d_{R_i}^{2 - 2G_i} e^{\frac{-\lambda A_i C_2(R_i)}{2N}}, \qquad (16.28)$$

where N_w is the number of windows, $2 - 2G_i$ is the Euler number associated with the window i and $D_{R_1 \ldots R_n}$ is the product of the Wigner coefficients for neighboring windows. For the case of intersecting loops a set of differential equations provides a recursion relation by relating the average of a loop with n intersections to those of loops with $m < n$ intersections.

Generalizing these results to the gYM_2 is straightforward. The only alteration that has to be invoked is to replace the $e^{\frac{-\lambda A C_2(R)}{2N}}$ factors that show up in those algorithms with similar factors where the second Casimir operator is replaced by the generalized Casimir operator (16.23). For instance the expectation value of a simple Wilson loop on the plane is given by,

$$\langle W(R, \gamma) \rangle = e^{\frac{-\lambda A_\gamma \Lambda(R)}{2N}}, \qquad (16.29)$$

where A_γ is the area enclosed by γ.

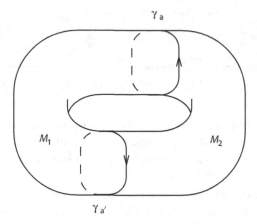

Fig. 16.5. Wilson loops on a torus.

It is interesting to note that for odd Casimir operators the expectation values of real representations (like the adjoint representation) equal unity due to the fact that the corresponding Casimirs vanish.

16.5 Stringy YM_2 theory

Before dwelling into the stringy description of the generalized YM_2 thoeries, we first review the proof of the stringy nature of pure YM_2 theory.[6] We then generalize this construction to the generalized YM_2 and present several examples to demonstrate the nature of the maps that contribute to the gYM_2 and their weights.

The partition function expressed as a sum over irreducible $SU(N)$ $(U(N))$ representations (16.12) can be expanded in terms of powers of $\frac{1}{N}$. This involves expanding the dimension and the second Casimir operator of the various representations. The representation of $U(N)$ or $SU(N)$ are described by Young tableau $Y(R)$, see for instance Fig. 16.5, composed of $r \leq N$ horizontal lines each with n_i boxes so that $n_1 \geq n_2 \geq \ldots \geq n_r$. The $U(N)$ and $SU(N)$ second Casimir operators of a representation R and its dimension are given by,

$$C_2^{U(N)}(R) = N \sum_{i=1}^{r} n_i + \sum_{i=1}^{r} n_i(n_i + 1 - 2i) = Nn + 2\hat{P}_{\{2\}}(R)^{U(N)}(R),$$

$$C_2^{SU(N)}(R) = N \sum_{i=1}^{r} n_i + \sum_{i=1}^{r} n_i(n_i + 1 - 2i) - \frac{(\sum_{i=1}^{r} n_i)^2}{N} = Nn + 2\hat{P}_{\{2\}}(R)^{SU(N)},$$

$$d_R = \frac{\prod_{i \leq j \leq N} (n_i - i - n_j + j)}{\prod_{i \leq j \leq N} (i - j)}, \qquad (16.30)$$

[6] The stringy description of Yang–Mills theory in two-dimensional Riemann surfaces was introduced by D. Gross and W. Taylor in [115], [118] and [119]. The formulation of the two-dimensional Yang–Mills theory in terms of topological string theories was done in [126] and [69].

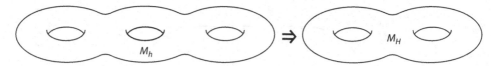

Fig. 16.6. A map from $\mathcal{M}_h$ to $\mathcal{M}_H$.

with $n = \sum_{i=1}^{r} n_i$. For every representation R there is a conjugate representation $\bar{R}$ whose Young tableau $Y(\bar{R})$ has its rows and columns interchanged. To determine the Casimir operator of the conjugate representation we use (16.30) with $2\hat{P}_{\{2\}}(R) = -2\hat{P}_{\{2\}}(R)^{U(N)}(R)$.

Using the Frobenius relations between representations of the symmetric group S_n and representations of $SU(N)$ $(U(N))$, the coefficients of this asymptotic expansion were written in terms of characters of S_n. The latter can be shown to correspond to permutations of the sheets covering the target space. The final result takes the form of,

$$
Z(A,H,N) \sim \sum_{n^{\pm},i^{\pm}=0}^{\infty} \sum_{p_1^{\pm},\ldots,p_{i^{\pm}}^{\pm} \in T_2 \subset S_{n^{\pm}}} \sum_{s_1^{\pm},t_1^{\pm},\ldots,s_H^{\pm},t_H^{\pm} \in S_{n^{\pm}}} \left(\frac{1}{N}\right)^{(n^+ + n^-)(2H-2)+(i^+ + i^-)}
$$

$$
\frac{(-1)^{(i^+ + i^-)}}{i^+! i^-! n^+! n^-!} (\lambda A)^{(i^+ + i^-)} e^{-\frac{1}{2}(n^+ + n^-)\lambda A} e^{\frac{1}{2}((n^+)^2 + (n^-)^2 - 2n^+ n^-)\lambda A/N^2}
$$

$$
\delta_{S_{n^+} \times S_{n^-}} \left(p_1^+ \cdots p_{i^+}^+ \, p_1^- \cdots p_{i^-}^- \, \Omega_{n^+,n^-}^{2-2H} \prod_{j=1}^{H} [s_j^+, t_j^+] \prod_{k=1}^{H} [s_k^-, t_k^-] \right), \tag{16.31}
$$

where $[s,t] = sts^{-1}t^{-1}$. Here δ is the delta function on the group algebra of the product of symmetric groups $S_{n^+} \times S_{n^-}$, T_2 is the class of elements of $S_{n^{\pm}}$ consisting of transpositions, and Ω_{n^+,n^-}^{-1} are certain elements of the group algebra of the symmetric group $S_{n^+} \times S_{n^-}$.

The formula (16.31) nearly factorizes, splitting into a sum over $n^+, i^+, \cdots$ and $n^-, i^-, \cdots$. The contributions of the $(+)$ and $(-)$ sums were interpreted as arising from two "sectors" of a hypothetical worldsheet theory. These sectors correspond to orientation reversing and preserving maps, respectively. One views the $n^+ = 0$ and $n^- = 0$ terms as leading order terms in a $1/N$ expansion. At higher orders the two sectors are coupled via the $n^+ n^-$ term in the exponential and via terms in $\Omega_{n^+ n^-}$.

Thus, the conventional YM_2 theory has an interpretation in terms of sums of covering maps of the target space, see Fig. 16.5. Those maps are weighted by the factor of $N^{2-2h} e^{-\frac{1}{2}n\lambda A}$ where h is the genus of the world-sheet and A is the area of the target space. The power of N^{2-2h} is obtained from the Riemann–Hurwitz formula,

$$
2h - 2 = (n^+ + n^-)(2H - 2) + (i^+ + i^-), \tag{16.32}
$$

where $B = i^+ + i^-$ is the total branching number. The number of sheets above each point in target space (the degree of the map) is n and $\lambda = g^2 N$ is the string tension. Maps that have branch points are weighted by a factor of λA. The dependence on the area A results from the fact that the branch point can be at any point in the target space.

16.6 Toward the stringy generalized YM_2

Note that the stringy description of the YM_2 theory does not attribute any special weight for maps that have branch points of a degree higher than one, nor is there a special weight for two (or more) branch points that are at the same point in the target space. The latter maps are counted with weight zero, at least for a toroidal target space, since they constitute the boundary of map space.

The main idea behind a stringy behavior of the gYM_2 is to associate nonzero weights to those boundary maps, once one considers the general $\Phi(B)$ case rather than the B^2 theory. In other words, we anticipate that we will have to add for the $\mathrm{tr}(B^3)$ theory, for example, maps that have a branch point of degree 2 and count them, as well, with a weight proportional to A. From the technical point of view the emergence of the YM_2 description in terms of maps followed from a large N expansion of the dimensions and the second Casimir operators of (16.12). Obviously a similar expansion of the former applies also for the generalized models and therefore what remains to be done is to properly treat the Casimirs appearing in the exponents of (16.25).

In [119], the expansion of the second Casimir operator $C_2(R)$ of a representation R introduced the branch points and the string tension contributions to the partition function. The $C_2(R)$ was expressed in terms of the eigenvalue of the sum of all the $\frac{n(n-1)}{2}$ transpositions of n elements (permutations containing a single cycle of length 2), where n is the number of boxes in R. This is the outcome of the following formula,

$$C_2(R) = nN + 2\hat{P}_{\{2\}}(R), \tag{16.33}$$

where $\hat{P}_2(R)$ is the value of the scalar matrix representing the sum of transpositions $\sum_{i<j\leq n}(ij)$ in the representation R of S_n (the matrix commutes with all permutations and thus is scalar). In the partition function, $C_2(R)$ was multiplied by $\frac{\lambda A}{2N}$. The resulting term $\frac{1}{2}n\lambda A$ arises from the action and is proportional to the string tension. The term $\lambda A \hat{P}_{\{2\}}$ arises from the measure and is interpreted as the contribution of branch points to the weight of a map.

Our task is, therefore, to express the generalized Casimirs C_ρ of (16.21) in terms of $\hat{P}_{\rho'}$, the generalizations of $\hat{P}_{\{2\}}(R)$. This is expressed as,

$$C_\rho(R) = \sum_{\rho'} \alpha_\rho^{\rho'} N^{h_\rho^{\rho'}} \hat{P}_{\rho'}(R), \tag{16.34}$$

where $\alpha_\rho^{\rho'}$ are coefficients that are independent of R and the power factors $h_\rho^{\rho'}$, are adjusted so that a string picture is achieved.

The $\hat{P}_{\{\rho'\}}$ factors are associated with ρ' which is an arbitrary partition of certain numbers, namely,

$$\rho' : \sum_i k_i \cdot i = \overbrace{1 + 1 + \cdots + 1}^{k_1} + \overbrace{2 + 2 + \cdots + 2}^{k_2} + \cdots , \qquad (16.35)$$

$\hat{P}_{\{\rho'\}}(R)$ is the product of two factors. The first is the sum of all the permutations in S_n (n is the number of boxes of R) which are in the equivalence class that is characterized by having k_i cycles of length i for $i \geq 2$. Just like the case of $\hat{P}_{\{2\}}(R)$, the matrix $\hat{P}_{\{\rho'\}}(R)$ commutes with all permutations and thus is a scalar. The sum is taken in the representation R of S_n. The second factor is,

$$\binom{n - \sum_{i=2} i k_i}{k_1} , \qquad (16.36)$$

which can be interpreted later as the number of ways to put k_1 marked points on the remaining sheets that do not participate in the branch points.

16.7 Examples

A complete diagrammatic expansion of the operators was determined in [105]. Using this expansion one can write down the stringy description of the partition function for any generalized YM theory. We end this chapter with a few examples of the Casimir factors for various choices of $\Phi(B)$ for both $U(N)$ and $SU(N)$ groups.

1. For $\frac{\lambda}{N}\mathrm{tr}(B^2)$ which is the conventional YM_2 theory we get (16.33),

$$\frac{2\lambda}{N}\hat{P}_{\{2\}} + \lambda\hat{P}_{\{1\}}. \qquad (16.37)$$

The first term means that we give a factor of $\frac{2\lambda A}{N}$ for each branch point, and the second term means that we have a factor of λ for each marked point (i.e. this is the string tension).

2. For $\alpha N^{-2}\mathrm{tr}(B^3)$ in $U(N)$ we get,

$$3\alpha N^{-2}\hat{P}_{\{3\}} + 3\alpha N^{-1}\hat{P}_{\{2\}} + 3\alpha N^{-2}\hat{P}_{\{1+1\}} + \frac{1}{2}\alpha\hat{P}_{\{1\}} + \frac{1}{2}\alpha N^{-2}\hat{P}_{\{1\}}. \quad (16.38)$$

The first term is the contribution from branch points of degree 2 (the simple branch points are of degree 1). The next term is a modification to the weight of the usual branch points. The third is the weight of two marked points at the same point (but different sheets), which will translate into $n^+(n^+ - 1)$ $+ n^-(n^- - 1)$ in the weight of a map for which (n^+, n^-) are the numbers of sheets of each orientability. The last two terms are modifications to the cosmological constant (or, in our terminology, to the weight of the single marked point).

Note that, because of the original power of N^{-2} we do not get the usual N^{2-2h} stringy behaviour in the partition function. We can overcome this problem by interpreting the last term not as the usual marked point, but as a *microscopic handle* that is attached to the point. By interpreting certain marked points as actually being microscopic handles (or higher Riemann surfaces) we can always adjust the power of N to be N^{2-2h}. Similarly, we should interpret the term $3\alpha N^{-2}\hat{P}_{\{1+1\}}$ as a connecting tube. We will come back to this interpretation towards the end of this section and investigate it further in the next subsection.

3. For $N^{-2}\mathrm{tr}(B^3)$ in $SU(N)$ we obtain the following corrections to (16.38),

$$-\frac{6}{N^3}\hat{P}_{\{2+1\}} - \frac{12}{N^3}\hat{P}_{\{2\}} + \frac{12}{N^4}\hat{P}_{\{1+1+1\}}$$
$$-\left(\frac{6}{N^2} - \frac{12}{N^4}\right)\hat{P}_{\{1+1\}} - \left(\frac{3}{N^2} - \frac{2}{N^4}\right)\hat{P}_{\{1\}}. \tag{16.39}$$

These terms and the terms in the previous example (16.38) do not mix chiralities (i.e. sheets of opposite orientations). In the full theory (chiral and anti-chiral sectors) there is the corresponding anti-chiral term:

$$+\frac{6}{N^3}\hat{P}_{\{\bar{2}+\bar{1}\}} + \frac{12}{N^3}\hat{P}_{\{\bar{2}\}} - \frac{12}{N^4}\hat{P}_{\{\bar{1}+\bar{1}+\bar{1}\}}$$
$$+\left(\frac{6}{N^2} - \frac{12}{N^4}\right)\hat{P}_{\{\bar{1}+\bar{1}\}} + \left(\frac{3}{N^2} - \frac{2}{N^4}\right)\hat{P}_{\{\bar{1}\}}. \tag{16.40}$$

For $SU(N)$ there are additional terms that do mix chiralities. They are,

$$-\frac{6}{N^3}(\hat{P}_{\{\bar{2}+1\}} - \hat{P}_{\{2+\bar{1}\}}) + \frac{12}{N^4}(\hat{P}_{\{\bar{1}+\bar{1}+1\}} - \hat{P}_{\{\bar{1}+1+1\}}). \tag{16.41}$$

The first term is the contribution of maps that have a branch point in one orientability and a marked point in the other (at the same target space point). The second term is the contribution of maps with three marked points – two for one orientability and one for the other.

To illustrate the content of these formulae in terms of representations, we will calculate the value of the third Casimir $\frac{1}{6}d_{abc}t^{(a}t^bt^{c)}$ for a totally anti-symmetric representation of $SU(N)$ with k boxes. The term $\hat{P}_{\{3\}}$ is translated into the sum of all the permutations of the k indices of a totally antisymmetric tensor that are 3-cycles, this gives $\frac{1}{3}k(k-1)(k-2)$. The term $\hat{P}_{\{2\}}$ gives the sum of all the permutations that are 2-cycles, that is $-\frac{1}{2}k(k-1)$ (a minus sign comes from antisymmetry). All in all we get,

$$C_{\{3\}}(k) \equiv \frac{1}{6}d_{abc}t^{(a}t^b t^{c)}$$

$$= k(k-1)(k-2) - \frac{3}{2}k(k-1) + \frac{3}{2}Nk(k-1)$$

$$+ \frac{1}{2}k + \frac{1}{2}N^2 k + \frac{3}{N}k(k-1)(k-2)$$

$$- 3k(k-1) + \frac{6}{N}k(k-1) - 3k + \frac{2}{N^2}k(k-1)(k-2) + \frac{6}{N^2}k(k-1) + \frac{2k}{N^2}$$

$$= \frac{k}{2N^2}(N+1)(N+2)(N-k)(N-2k). \tag{16.42}$$

4. For $N^{-3}\mathrm{tr}(B^4)$ in $U(N)$ we get,

$$4N^{-3}\hat{P}_{\{4\}} + 6N^{-2}\hat{P}_{\{3\}} + 6N^{-3}\hat{P}_{\{2+1\}} + \frac{8}{3}N^{-2}\hat{P}_{\{1+1\}}$$

$$+ \left(\frac{4}{3}N^{-1} + 6N^{-3}\right)\hat{P}_{\{2\}} + \left(\frac{1}{6} + \frac{5}{6}N^{-2}\right)\hat{P}_{\{1\}}.$$

The terms that have an extra microscopic handle are,

$$6N^{-3}\hat{P}_{\{2+1\}} + \frac{8}{3}N^{-2}\hat{P}_{\{1+1\}} + 6N^{-3}\hat{P}_{\{2\}} + \frac{5}{6}N^{-2}\hat{P}_{\{1\}}.$$

5. For $N^{-4}(\mathrm{tr}(B^2))^2$ in $U(N)$ we get,

$$24N^{-4}\hat{P}_{\{3\}} + 8N^{-4}\hat{P}_{\{2+2\}} + 4N^{-3}\hat{P}_{\{2+1\}}$$

$$+ \frac{16}{3}N^{-3}\hat{P}_{\{2\}} + \left(2 + \frac{8}{3}N^{-4}\right)\hat{P}_{\{1+1\}} + \left(\frac{2}{3}N^{-2} + \frac{1}{3}N^{-4}\right)\hat{P}_{\{1\}}. \tag{16.43}$$

and additional terms,

$$8N^{-4}\hat{P}_{\{2+\bar{2}\}} + 4N^{-3}(\hat{P}_{\{\bar{2}+1\}} + \hat{P}_{\{2+\bar{1}\}}) + 2N^{-2}\hat{P}_{\{1+\bar{1}\}},$$

that mix the two chiral sectors.

The meaning of the term $2N^{-2}\hat{P}_{\{1+\bar{1}\}}$ is a factor of $N^{-2}e^{-2\alpha n^+ n^- A}$, where α is the coefficient of the $N^{-4}(\mathrm{tr}(B^2))^2$ term in the action. n^+ is the number of sheets of positive orientability and n^- is the number of sheets of negative orientability for a given map.

16.8 Summary

In this chapter we studied the generalized two-dimensional Yang–Mills theory on Riemann surfaces. We reviewed the exact formulae for the partition function and Wilson loop averages of the conventional YM theory. We then presented the generalization of these results in the context of the generalized YM theories. These expressions are based on a replacement of the second Casimir operator with more general Casimir operators depending on the particular model. There is another method to obtain these results [228], i.e. by regarding the general Yang–Mills actions as perturbations of the topological theory at zero area.

Using the relations between $SU(N)$ representations and representations of the symmetric groups S_n, we wrote down the generalizations that have to be made in the Gross–Taylor string rules for 2D Yang–Mills theory, so as to make the generalized Yang–Mills theory for $SU(N)$ or $U(N)$ a local string theory as well. The extra terms are special weights for certain maps with branch points of a degree higher than one.

An obvious extension of the results presented in this chapter is to consider other gauge groups. The conventional YM_2 theory with gauge groups $O(N)$ or $Sp(N)$ which were shown to be related to maps from non-orientable world-sheets.

One can further couple the gYM_2 theories to fermionic matter in analogy to 't Hooft's analysis presented in Chapter 10. This domain of research is far from being fully explored. A particularly interesting question is to find out certain $\Phi(B)$s that lead to a special behaviour of the coupled system. For example, in the $U(1)$ case, the representations R are labeled by an integer n and for $\Phi(B) = -\alpha \log(1 + \lambda B^2)$ we get,

$$\mathcal{Z}(U(1), A) = \sum_n (1 + \lambda n^2)^{-\alpha A}$$

which has a singularity for $2\alpha A = 1$.

PART III

From two to four dimensions

17

Conformal invariance in four-dimensional field theories and in QCD

Conformal symmetry was shown to be extremely powerful in two-dimensional field theories and obviously also in string theories. This is due to the fact that the conformal algebra is infinite dimensional and hence supplies a set of infinitely many conserved charges. In four dimensions the conformal algebra is finite and therefore less powerful.[1] The purpose of this chapter is to explore the use in 4d conformal field theories of notions and tools that we encountered in two-dimensional CFT, such as primary fields, conformal operator expansion, conformal anomalies and Ward identities.

Free massless theories are obviously scale and conformal invariant. However, field theories that describe the elementary particles of nature and their interactions, are interacting field theories. The question is thus, to what extent can one apply the techniques of CFT to those theories and in particular QCD in four dimensions? QCD with massless quarks is a prototype model of theories which are classically conformal invariant. In fact even for theories with masses and other dimension-full parameters, in certain cases these can be neglected in the high energy and high momentum transfer regime of the theory. However, even in the massless case and with only dimensionless couplings, it is easy to realize that the quantum picture lacks conformal invariance. This follows from the fact that one has to introduce dimension-full parameters such as UV cutoff, which turns into a scale where the coupling is defined, after renormalization. Thus there is an anomaly in the conformal symmetry in the sense that it is a symmetry of the classical system but not of the quantum one.

We will investigate in this chapter conformal symmetry in four-dimensional field theories and in particular its applications in the context of QCD in four dimensions. We start with the description of conformal transformations, their corresponding generators and the $SO(2,4)$ conformal algebra. We then analyze the Noether currents that follow from the conformal transformation and their conservation laws. Next we present the $SL(2,\mathcal{R})$ collinear subgroup associated with light-cone conformal transformations. In a similar manner to the treatment of two-dimensional conformal symmetry, we define primary and descendant operators of the collinear group. We then define and study the conformal operator product expansion (COPE). We proceed and describe the conformal Ward identity, the Callan–Symanzik equation. We then make use of the conformal toolbox

---

[1] The full conformal algebra in four dimensions was introduced in [156]

in the study of four dimensional QCD which is conformal only at the classical level. We analyze the non-local operators built from a quark and an anti-quark and expand them in terms of Gegenbauer polynomials. We use the COPE to write down the operator product of two electromagnetic currents. Finally we determine, in the limit of large momentum exchange, the pion distribution amplitude.

Conformal invariance in four dimensions was described in several review papers and books. The original studies on conformal symmetry are summarized in [207] and [66]. A modern review that we follow in this chapter is [43].

17.1 Conformal symmetry algebra in four dimensions

In general in d space-time dimensions the conformal group is the subgroup of coordinate transformations that leaves the metric invariant up to a scale, namely,

$$g_{\mu\nu}(x) \to g'_{\mu\nu}(x') = \Omega(x)g_{\mu\nu}(x). \tag{17.1}$$

It is obvious from (2.2) that the 2d conformal transformations (2.1) indeed produce such a variation of the metric. An important property of conformal transformations in any dimension is that they preserve the angle $\frac{\vec{A}\cdot\vec{B}}{\sqrt{A^2 B^2}}$ between two vectors $\vec{A}$ and $\vec{B}$.

Starting from flat space-time, the general infinitesimal coordinate transformation $x^\mu \to x^\mu + \epsilon^\mu(x)$ induces a change of the metric,

$$ds^2 \to ds^2 + (\partial_\mu \epsilon_\nu + \partial_\nu \epsilon_\mu)dx^\mu dx^\nu, \tag{17.2}$$

so that the condition for conformal transformations reads,

$$\partial_\mu \epsilon_\nu + \partial_\nu \epsilon_\mu = \frac{2}{d}(\partial \cdot \epsilon)g_{\mu\nu}, \tag{17.3}$$

where $g_{\mu\nu}$ is $\eta_{\mu\nu}$ or $\delta_{\mu\nu}$ for a Minkowskian signature or Euclidean signature, respectively, and the factor of $\frac{2}{d}$ is fixed by tracing both sides of the equation with $g^{\mu\nu}$. To check what are all the possible solutions for ϵ_μ we differentiate (2.4) twice to find that,

$$[(d-2)\partial_\mu \partial_\nu + g_{\mu\nu}\partial_\alpha \partial^\alpha]\partial_\beta \epsilon^\beta = 0. \tag{17.4}$$

This equation together with (2.4) implies that the third derivatives of ϵ^μ vanish, which means that ϵ^μ can be of order $0, 1, 2$ in x^ν. Obviously the parameters associated with the Poincare group, since they are an isometry and hence do not change the metric, obey (17.1). These transformation parameters together with additional infinitesimal parameters which are linear and quadratic in x^μ are summarized as follows:

$$
\begin{aligned}
\epsilon^\mu &= \epsilon_0^\mu & &\textit{space-time translations} \\
\epsilon^\mu &= \epsilon_0^{\mu\nu} x_\nu & &\textit{Lorentz transformations} \\
\epsilon^\mu &= \epsilon_0 x^\mu & &\textit{scale transformations} \\
\epsilon^\mu &= \tilde{\epsilon}_0^\mu x^2 - 2x^\mu \tilde{\epsilon}_0^\nu x_\nu & &\textit{special conformal transformations} \quad (17.5)
\end{aligned}
$$

where ϵ_0^μ, $\epsilon_0^{\mu\nu}$, ϵ_0, $\tilde{\epsilon}_0^\mu$ are vector, antisymmetric tensor, scalar and vector infinitesimal constants, respectively. The corresponding finite transformations are,

$$x^\mu \rightarrow x^{\mu\prime} = x^\mu + a^\mu$$
$$x^\mu \rightarrow x^{\mu\prime} = a_\nu^\mu x^\nu$$
$$x^\mu \rightarrow x^{\mu\prime} = ax^\mu$$
$$x^\mu \rightarrow x^{\mu\prime} = \frac{x^\mu + \tilde{a}^\mu x^2}{1 + 2\tilde{a}^\mu x_\mu + \tilde{a}^2 x^2}, \tag{17.6}$$

where the various forms of a are the finite parameters of transformation that correspond to the infinitesimal ones above. The last transformation is referred to as the special conformal transformation. It can in fact be decomposed into an inversion transformation $x^\mu \rightarrow -\frac{x^\mu}{x^2}$, a space-time shift transformation and another inversion. The sum of all these transformations is $d + \frac{d(d-1)}{2} + 1 + d = \frac{(d+1)(d+2)}{2}$, which is the dimensions of $SO(2,d)$, the algebra of the conformal group in Minkowski space-time.

Let us analyze now the generators of the conformal transformations and show that indeed they obey the $SO(2,d)$ algebra. The generators are,

$$
\begin{aligned}
P_\mu &= -i\partial_\mu & &\text{space-time translations} \\
L_{\mu\nu} &= i(x_\mu \partial_\nu - x_\nu \partial_\mu) & &\text{Lorentz transformations} \\
D &= -ix^\mu \partial_\mu & &\text{scale transformation} \\
K_\mu &= -i\left[x^2 \partial_\mu - 2x_\mu x^\nu \partial_\nu\right] & &\text{special conformal transformations} \qquad (17.7)
\end{aligned}
$$

Using these expressions for the generators it is a straightforward exercise to realize that they obey the following algebra,

$$
\begin{aligned}
[P_\mu, P_\nu] &= 0 \\
[P_\mu, L_{\nu\rho}] &= i(\eta_{\mu\nu} P_\rho - \eta_{\mu\rho} P_\nu) \\
[L_{\mu\nu}, L_{\rho\sigma}] &= -i(\eta_{\mu\rho} L_{\nu\sigma} - \eta_{\mu\sigma} L_{\nu\rho} + \eta_{\nu\sigma} L_{\mu\rho} - \eta_{\nu\rho} L_{\mu\sigma}) \\
[D, P_\mu] &= -iP_\mu \qquad [D, K_\mu] = iK_\mu \\
[P_\mu, K_\nu] &= 2i(L_{\mu\nu} - \eta_{\mu\nu} D) \\
[D, L_{\mu\nu}] &= 0 \\
[K_\mu, K_\nu] &= 0, \tag{17.8}
\end{aligned}
$$

which is indeed the $SO(2,4)$ algebra. The first three lines constitute the Poincare algebra in four dimensions. It is well known that (17.8) is not the most general form of the $SO(2,4)$ algebra. One can further generalize the construction of the generators by modifying $L_{\mu\nu}$ in the following way,

$$M_{\mu\nu} = L_{\mu\nu} + \Sigma_{\mu\nu}, \tag{17.9}$$

where $\Sigma_{\mu\nu}$ does not act on the space-time points and obeys,

$$[\Sigma_{\mu\nu}, \Sigma_{\rho\sigma}] = -i(\eta_{\mu\rho} \Sigma_{\nu\sigma} - \eta_{\mu\sigma} \Sigma_{\nu\rho} + \eta_{\nu\sigma} \Sigma_{\mu\rho} - \eta_{\nu\rho} \Sigma_{\mu\sigma}). \tag{17.10}$$

Shortly the role of these generators in the conformal transformation of fields will be discussed.

17.2   Conformal invariance of fields, Noether currents and conservation laws

So far we have discussed the conformal transformations as they act on the points in space-time. Now we would like to consider field theories in four dimensions that are classically invariant under the group of conformal transformations. The fields associated with conformal invariant theories may be scalar fields, spinors, vectors or tensors. We denote such a field by $\Phi(x)$.

The transformation of the field under the conformal transformations,

$$\delta\Phi(x) = \delta x^\mu \partial_\mu \Phi(x) + \delta_I \Phi(x), \qquad (17.11)$$

is composed of two parts, the one due to that of the space-time point with δx^μ given in (17.5) and an "internal transformation" $\delta_I \Phi(x)$, which vanishes for space-time translations, while for Lorentz transformations, dilatations and special conformal transformations, takes the form,

$$
\begin{array}{ll}
\delta_I \Phi(x) = -e_{\mu\nu} \Sigma^{\mu\nu} \Phi & \textit{Lorentz transformations} \\
\delta_I \Phi(x) = e\, l\Phi & \textit{scale translations} \\
\delta_I \Phi(x) = 2\tilde{e}_\mu (lx^\mu - x_\nu \Sigma_{\mu\nu})\Phi & \textit{special conformal transformations}
\end{array}
$$
$$(17.12)$$

where l is the conformal dimension of the field and the internal Lorentz generators $\Sigma\mu\nu$ are given by,

$$\Sigma^{\mu\nu} = \frac{i}{4}[\gamma^\mu, \gamma^\nu] \; - \textit{Dirac spinors} \qquad [\Sigma^{\mu\nu}]^\beta_\alpha = \eta^{\mu\beta}\delta^\nu_\alpha - \eta^{\nu\beta}\delta^\mu_\alpha \; - \textit{gauge fields}$$
$$(17.13)$$

Recall that all the parameters of transformations are global, namely space-time independent. To determine the Noether currents associated with the various symmetry transformations, one elevates the transformations into local ones and reads the currents from the variation of the action,

$$\delta S = \int d^4 x J^a_\mu \partial^\mu e_a \qquad (17.14)$$

where e^a is any of the parameters of transformations given in (17.5). The outcome of the Noether procedure are the following conserved currents,

$$
\begin{aligned}
J^{(P)\mu}_\nu &\equiv T^\mu_\nu = \Pi^\mu \partial_\nu \Phi - \delta^\mu_\nu \mathcal{L} \\
J^{(M)\mu}_{\nu\rho} &= x_\nu T^\mu_\rho - x_\rho T^\mu_\nu - \Pi_\mu \Sigma_{\nu\rho}\Phi \\
J^{(D)\mu} &\equiv D^\mu = x_\nu T^{\mu\nu} + l\Pi^\mu \Phi \\
J^{(K)\mu}_\nu &= (2x_\rho x_\nu - \eta_{\nu\rho}x^2)T^{\mu\rho} + 2x_\rho \Pi^\mu (l\delta^\rho_\nu - \Sigma^\rho_\nu)\Phi, \qquad (17.15)
\end{aligned}
$$

where $\Pi^\mu = \frac{\partial \mathcal{L}}{\partial(\partial_\mu \Phi)}$. In fact the variation of the action with respect to dilatations may lead in addition to the divergence of $J^{(D)\mu}$, to another total derivative term Δ_D. However for Lagrangians that are polynomials in the fields and their derivatives, this term vanishes. For the special conformal transformations the additional term is defined by

$$\delta_K S = \int d^4 x e_\mu(x)\left[-\partial_\nu K^{\mu\nu} + 2x^\mu \Delta_D + \Delta_k^\mu\right], \qquad (17.16)$$

with $\Delta_k^\mu = 2\Pi_\nu \Phi(lg^{\mu\nu} + \Sigma^{\mu\nu})\Phi$. For invariance we need $\Delta_D = 0$ and $\Delta_k^\mu = 2\partial_\nu \sigma^{\mu\nu}$. For $l = 1$ and $l = 3/2$, $\sigma_{\mu\nu}$ vanishes, so in these cases $J^{(K)\mu}_\nu$ are really the generators of conformal transformations.

An interesting observation is that all the Noether currents associated with the full conformal group can be expressed in terms of a modified energy-momentum tensor. First note that the energy-momentum tensor defined above in not necessarily symmetric. In fact from the conservation of $J^{(M)\mu}_{\nu\rho}$ the antisymmetric part of $T_{\mu\nu}$ can be determined since,

$$\partial_\mu J^{(M)\mu}_{\nu\rho} = T_{\rho\nu} - T_{\nu\rho} - \partial_\mu(\Pi_\mu \Sigma_{\mu\nu}\Phi) = 0. \qquad (17.17)$$

Using this result it is now easy to define a modified conserved symmetric energy-momentum tensor,

$$T^{(S)}_{\mu\nu} = T_{\mu\nu} + \frac{1}{2}\partial^\rho(\Pi_\rho \Sigma_{\mu\nu}\Phi - \Pi_\mu \Sigma_{\rho\nu}\Phi - \Pi_\nu \Sigma_{\rho\mu}\Phi). \qquad (17.18)$$

The current associated with the Lorentz transformations can be expressed in terms of $T^{(S)}_{\mu\nu}$ as,

$$J^{(M)\mu}_{\nu\rho} = x_\nu T^{(S)\mu}_{\rho} - x_\rho T^{(S)\mu}_{\nu}. \qquad (17.19)$$

One can further modify the symmetric energy-momentum tensor to render it also traceless,

$$T^{(TL)}_{\mu\nu} = T^{(S)}_{\mu\nu} + \frac{1}{2}\partial^\rho \partial^\sigma X_{\rho\sigma\mu\nu}, \qquad (17.20)$$

where $X_{\rho\sigma\mu\nu}$ is defined such that the energy-momentum tensor is traceless and conserved and $\eta^{\mu\nu}\partial^\rho \partial^\sigma X_{\rho\sigma\mu\nu} = 2\partial^\rho \partial^\nu \sigma_{\rho\nu} = -2\eta^{\mu\nu}T^{(S)}_{\mu\nu}$. In terms of this traceless energy-momentum tensor the dilatation current and the current associated with the special conformal transformation are given by,

$$J^{(D)\mu} = x_\nu T^{(TL)\mu\nu} \qquad J^{(K)\mu}_\nu = (2x_\nu x_\rho - x^2 \eta_{\rho\nu})T^{(TL)\mu\rho}. \qquad (17.21)$$

It is thus clear that $J^{(D)\mu}$ and $J^{(K)\mu}_\nu$ are conserved only provided that $T^{(TL)\mu\nu}$ is conserved and traceless.

Note that in the latter form, scale invariance, namely a traceless energy-momentum tensor, implies also conformal invariance.

17.3   Collinear and transverse conformal transformations of fields

Recall that in 2d conformal field theories it is very useful to employ light-cone coordinates (in Minkowski space-time) or holomorphic and anti-holomorphic coordinates (in complex Euclidean space-time). We want to argue now that it is also quite useful to use light-cone coordinates, when analyzing four-dimensional conformal field theories.

Consider the two light-like vectors n_+^μ and n_-^μ,

$$n_+^\mu n_{+\,\mu} = n_-^\mu n_{-\,\mu} = 0 \qquad n_+^\mu n_{-\,\mu} = 1. \tag{17.22}$$

We can then decompose any Lorentz four-vector A_μ as follows,

$$A^\mu = A_- n_+^\mu + A_+ n_-^\mu + A_T^\mu, \tag{17.23}$$

where,

$$A_+ \equiv A_\mu n_+^\mu \qquad A_- \equiv A_\mu n_-^\mu, \tag{17.24}$$

and the transverse part of the four-vector A_T^μ is defined using the transverse part of the metric defined as,

$$\eta_{\mu\nu}^T = \eta_{\mu\nu} - n_-^\mu n_+^\nu - n_+^\mu n_-^\nu \qquad A_T^\mu \equiv \eta_{\mu\nu}^T A_\nu. \tag{17.25}$$

Using this decomposition we find that $A_\mu A^\mu = 2A_+ A_- - A_T^2$.

We can now also consider transformations associated with a subgroup of the full conformal group. In particular consider the special transformations associated with a light-like parameter $\tilde{a}^\mu = \tilde{a} n_-^\mu$. The transformation of x_- takes the form

$$x_- \rightarrow x_-' = \frac{x_-}{1 + 2\tilde{a}x_-}. \tag{17.26}$$

Combining this transformation with the translation along the x_- direction $x_- \rightarrow x_- + a_-$ and scaling $x_- \rightarrow ax_-$ these transformations constitute a subgroup of the full conformal group, the *collinear subgroup* which is an $SL(2, R)$.[2] To verify this group structure we define the following generators,

$$L_+ = -iP_+ \qquad\qquad L_- = \frac{i}{2}K_-$$

$$L_0 = \frac{i}{2}(D + M_{-+}) \qquad E = \frac{i}{2}(D - M_{-+}). \tag{17.27}$$

The generators $L_\pm$ and L_0 obey the algebra,

$$[L_0, L_\pm] = \pm L_\pm \qquad [L_-, L_+] = -2L_0, \tag{17.28}$$

which is indeed the $SL(2, R) \sim SO(2, 1)$ algebra; E commutes with them. It is, actually, the L_0 of the other $SL(2, R)$, the one in the x_+ direction.

---

[2] The use of the $SL(2, R)$ group in applications of conformal symmetry to QCD was introduced in [150] and [83]

The $SL(2, R)$ collinear subgroup is particularly useful for collinear fields, where for instance $\Phi(x)$ takes the form $\Phi(\alpha) \equiv \Phi(\alpha n_+^\mu)$, with α a real number and n_+^μ the light-cone direction defined above. In particular, as will be shown below, this will apply to parton description of quarks. The field $\Phi(\alpha)$ is taken to be an eigenstate of the spin operator Σ_{+-},

$$\Sigma_{+-}\Phi(\alpha) = s\Phi(\alpha), \tag{17.29}$$

so that s is the spin projection to the n_+ direction. The collinear subgroup of the conformal group now acts on the coordinate α as an $SL(2, R)$ transformation,

$$\alpha \rightarrow \alpha' = \frac{a\alpha + b}{c\alpha + d}, \tag{17.30}$$

where a, b, c, d are real numbers with $ad - bc = 1$, and correspondingly the field $\Phi(\alpha)$ transforms as,

$$\Phi(\alpha) \rightarrow (c\alpha + d)^{-2j}\, \Phi\left(\frac{a\alpha + b}{c\alpha + d}\right), \tag{17.31}$$

with,

$$j = \frac{1}{2}(l + s). \tag{17.32}$$

Thus $\Phi(\alpha)$ is a representation of $SL(2, R)$ or an $SL(2, R)$ form of degree j.

The generators of the $SL(2, R)$ group and E act on the collinear field as,

$$
\begin{aligned}
[L_+, \Phi(\alpha)] &= -\partial_\alpha \Phi(\alpha) \\
[L_0, \Phi(\alpha)] &= (\alpha\partial_\alpha + j)\Phi(\alpha) \\
[L_-, \Phi(\alpha)] &= (\alpha^2\partial_\alpha + 2j\alpha)\Phi(\alpha) \\
[E, \Phi(\alpha)] &= \frac{1}{2}(l - s)\Phi(\alpha)
\end{aligned} \tag{17.33}
$$

where $t = (l - s)$ is referred to as the *collinear twist*. In addition $\Phi(\alpha)$ is an eigenstate of the Casimir operator with,

$$\sum_{i=0,1,2} [L_i, [L_i, \Phi(\alpha)]] = j(j - 1)\Phi(\alpha). \tag{17.34}$$

Another subgroup of the conformal group is the transverse subgroup $SL(2, C)$ acting on the transverse coordinates $x_T^\mu = (0, x_1, x_2, 0)$ or in complex coordinates $z = x_1 + ix_2$ and $\bar{z} = x_1 - ix_2$, with fields $\Phi(z, \bar{z})$. This is in fact identical to the $SL(2, C)$ discussed in Chapter 3 where the conformal symmetry of two-dimensional field theories is discussed. In terms of the conformal generators (17.7) the generators of the $SL(2, C)$ are the $P_T^\mu, M_T^{\mu\nu}, D, K_T^\mu$. The coordinate z transforms under the $SL(2, C)$ transformation,

$$z \rightarrow z' = \frac{az + b}{cz + d}, \tag{17.35}$$

which implies the following transformation of $\Phi(z, \bar{z})$,

$$\Phi(z, \bar{z}) \rightarrow (cz + d)^{-2h} \Phi\left(\frac{az + b}{cz + d}, \bar{z}\right), \tag{17.36}$$

where $h = \frac{1}{2}(l + \lambda)$ with λ is the helicity defined by $\Sigma_{z\bar{z}}\Phi = \lambda\Phi$. Similarly for the transformation of $\bar{z}$ and the corresponding transformation of $\Phi(z, \bar{z})$, with $\bar{h} = \frac{1}{2}(l - \lambda)$.

17.4 Collinear primary fields and descendants

In two-dimensional conformal field theories fields were classified into primary fields and descendant ones. The classification was based on their conformal transformations. Correspondingly the states were put into Verma modules each containing a highest weight state and its descendants. Recall the definition of the former,

$$L_0 [\phi(0)|0>] = h [\phi(0)|0>] \quad L_n [\phi(0)|0>] = 0, \quad n > 0. \tag{17.37}$$

In a similar manner the primary operator and the highest weight state of the four-dimensional collinear group are defined [171], [42] as,

$$[L_0, \Phi(0)] = j\Phi(0) \qquad [L_-\phi(0)] = 0 \qquad \Rightarrow$$
$$L_0 [\Phi(0)|0>] = j[\Phi(0)|0>] \qquad L_- [\Phi(0)|0>] = 0, \tag{17.38}$$

$\Phi(0)$ is by definition collinear since it is defined at the origin of the light-cone direction. The fact that in the 2d case the conformal algebra is infinite while in 4d it is finite is manifested by the fact that in the former case there is an infinite set of annihilation operators L_n, $n > 0$ that annihilate the highest weight state, whereas in the latter case it is the single operator L_-.

The descendant fields and correspondingly the descendant states are obtained by repeatedly applying the creation operators, which are L_{-n} in 2d while in 4d it is the operator L_+. So in 4d,

$$\mathcal{O}_n = [L_+, \ldots [L_+, [L_+, \Phi(0)]]] = (-\partial_+)^n \Phi(\alpha)|_{\alpha=0}. \tag{17.39}$$

Note the difference in notation, as it is L_- in 2d while it is L_+ in 4d, both raising.

The descendant operators obey the following commutation relations,

$$[L_0, \mathcal{O}_n] = (j + n)\mathcal{O}_n \qquad [L_+, \mathcal{O}_n] = \mathcal{O}_{n+1} \qquad [L_-, \mathcal{O}_n] = -n(n+2j-1)\mathcal{O}_{n-1}. \tag{17.40}$$

In two-dimensions we discussed the Verma module that includes a highest weight state and all its descendants, and similarly, in four dimensions we consider the so-called conformal tower which also includes the highest weight state and all its descendants. Recall however that there is an essential difference between the two cases due to the fact that in the 2d the algebra is infinite dimensional whereas in 4d it is finite dimensional. In particular the notion of null vectors that played an important role in the 2d case does not exist in four-dimensional CFTs.

We can associate the complete sets of $\Phi(\alpha)$ and $\mathcal{O}_n$ by the following Taylor expansion,

$$\Phi(\alpha) = \sum_{n=0}^{\infty} \frac{(-\alpha)^n}{n!} \mathcal{O}_n. \tag{17.41}$$

An interesting and useful map relates the descendant operators and polynomials. Consider for instance the descendent operator defined in (17.39), which can be re-expressed as,

$$\mathcal{O}_n = \mathcal{P}_n(\partial_\alpha)\Phi(\alpha)|_{\alpha=0} \qquad \mathcal{P}_n(u) = (-u)^n. \tag{17.42}$$

It is straightforward to realize that in terms of these polynomials the operation of $L_0, L_\pm$ takes the form,

$$L_+ \to \tilde{L}_- = -u$$
$$L_- \to \tilde{L}_+ = (u\partial_u^2 + 2j\partial_u)$$
$$L_0 \to \tilde{L}_0 = (u\partial_u + j). \tag{17.43}$$

This correspondence can be viewed as first mapping,

$$\partial_\alpha \to u \qquad \alpha \to \partial_u, \tag{17.44}$$

then interchanging the $+$ and $-$ components, and finally some "normal ordering" of taking the derivatives with respect to u to the right of the factors of u.

The representation in terms of polynomials is referred to as the 'adjoint representation'. Note that since in the original algebra L_- includes a term proportional to α^2, the new algebra includes a second-derivative term ∂_u^2 in $\tilde{L}_+$. This can be avoided by introducing a different argument of the polynomials defined as,

$$\frac{u^n}{\Gamma(n + 2j)} \to \kappa^n, \tag{17.45}$$

so that the action $L_0, L_\pm$ on $\tilde{\mathcal{P}}(\kappa)$ is the same as (17.33) with $\alpha \to \kappa$ and the interchange of L_- and L_+.

We now discuss composite operators built from two "elementary" operators of the form,

$$O(\alpha_1, \alpha_2) = \Phi_{j_1}(\alpha_1)\Phi_{j_2}(\alpha_2), \tag{17.46}$$

with $\alpha_1 \neq \alpha_2$. The operator product expansion with $|\alpha_1 - \alpha_2| \to 0$ is expressed in terms of the composite operators,

$$\mathcal{O}_n(0) = \mathcal{P}_n(\partial_1, \partial_2)\Phi_{j_1}(\alpha_1)\Phi_{j_2}(\alpha_2)|_{\alpha_1=\alpha_2=0}, \tag{17.47}$$

where $\mathcal{P}_n(\partial_1, \partial_2)$ is a homogeneous polynomial of degree n. It can be shown that the complete set of local operators with which one can perform the conformal operator expansion (COPE) takes the form,

$$\mathcal{O}_n^{j_1,j_2}(x) = \partial_+^n \left[\Phi_{j_1}(x) P_n^{(2j_1-1,2j_2-1)} \left(\frac{\overrightarrow{\partial}_+ - \overleftarrow{\partial}_+}{\overrightarrow{\partial}_+ + \overleftarrow{\partial}_+} \right) \Phi_{j_2}(x) \right], \qquad (17.48)$$

$P_n^{(a,b)}(x)$ are the Jacobi polynomials, and we are back to x space here.[3] One can further generalize this construction to a product of three or more operators.

17.5 Conformal operator product expansion

The conformal operator expansion in two dimensions, discussed in Section 3.7.2, was shown to be a very powerful tool in determining correlation functions. Obviously, we anticipate that in four dimensions the COPE will be less powerful. We discuss now the general structure of the COPE.[4] Consider the OPE of two local conformal operators $A(x)B(0)$ with twists and spin projection along the $+$ direction $(t_A, s_A), (t_B, s_B)$, respectively. We perform an expansion for fixed x_- and $x_+, x_T \to 0$, namely $x^2 \to 0$. We want to expand the product in terms of a complete set $\mathcal{O}_{n,n+k}^{j_1,j_2}$ and to leading order in the twist. Such an expansion takes the form,

$$A(x)B(0) = \sum_{n=0}^{\infty} \sum_{k=0}^{\infty} C_{n,k} \left(\frac{1}{x^2} \right)^{1/2(t_A+t_B-t_n)} x_-^{n+k+\Delta} \mathcal{O}_{n,n+k}^{j_1,j_2}(0) + \dots, \quad (17.49)$$

where $\Delta = s_1 + s_2 - s_A - s_B$, $\mathcal{O}_{n,n+k} = (-\partial_+)^k \mathcal{O}_n$, s_1 and s_2 are the spin projections of the constituent fields in the local operators $\mathcal{O}_{n,n+k}^{j_1,j_2}$, $t_n = l_n - n - s_1 - s_2 = l_1 + l_2 - s_1 - s_2$ the twist of the operator $\mathcal{O}_n$, actually independent of n, and the dots refer to higher twist contributions. We want to check to what extent conformal invariance enables us to determine the coefficients $C_{n,k}$. For this purpose we act on the OPE with L_- as,

$$[L_-, A(x)B(0)] = \left(x_-(2j_A + x \cdot \partial_x)A(x) - \frac{1}{2}x^2 \bar{n} \cdot \partial_x A(x) \right) B(0) + \dots \quad (17.50)$$

Inserting (17.49) and taking into account that,

$$[L_-, \mathcal{O}_{n,n+k}^{j_1,j_2}(0)] = -k(k+2j_n-1)\mathcal{O}_{n,n+k-1}^{j_1,j_2}, \qquad (17.51)$$

---

[3] The Jacobi polynomial is given by,

$$P_n^{(a,b)}(z) = \frac{\Gamma(a+n+1)}{n!\Gamma(a+b+n+1)} \sum_{m=0}^{n} \binom{n}{m} \frac{\Gamma(a+b+m+n+1)}{\Gamma(a+m+1)} \left(\frac{z-1}{2} \right)^m.$$

[4] COPE in four dimensions was introduced in [90] and used in QCD in [49], [50], [51].

with $j_n = j_1 + j_2 + n$, we find the following recursion relation for the coefficients $C_{n,k}$,

$$C_{n,k+1} = -\frac{j_A - j_B + j_n + k}{(k+1)(k+2j_n)} C_{n,k}, \tag{17.52}$$

which is solved by,

$$C_{n,k} = (-1)^k \frac{1}{k!} \frac{\Gamma(j_A - j_B + j_n + k)}{\Gamma(j_A - j_B + j_n)} \frac{\Gamma(2j_n)}{\Gamma(2j_n + k)} C_n, \tag{17.53}$$

where $C_n = C_{n,0}$. Plugging this into (17.49) we get the following form for the OPE,

$$A(x)B(0) = \sum_{n=0}^{\infty} C_n \left(\frac{1}{x^2}\right)^{1/2(t_A + t_B - t_n)} \frac{x_-^{n+s_1+s_2-s_A-s_B}}{B(j_A - j_B + j_n, j_B - j_A + j_n)}$$

$$\times \int_0^1 du\, u^{(j_A - j_B + j_n - 1)} (1-u)^{(j_B - j_A + j_n - 1)} O_n^{j_1, j_2}(ux_-), \tag{17.54}$$

where $B(a,b) = \frac{\Gamma(a)\Gamma(b)}{\Gamma(a+b)}$ is the beta function.

17.6   Conformal Ward identities

The application of conformal invariance in determining correlation functions was discussed in Section 2.8 for the case of two-dimensional conformal field theories. It includes both the use of the Ward identities associated with the global $SL(2,C)$ transformation[5] and the full holomorphic conformal transformation. Here in discussing four-dimensional field theories we will encounter two major differences:

(i) Due to the fact that the conformal symmetry group is finite dimensional, there are Ward identities only associated with global transformations.
(ii) When discussing theories with conformal anomalies, there will be modifications of the Ward identities.

Let us start by reminding ourselves of the concept of Ward identities and in particular the conformal ones. Associated with any infinitesimal transformation of a given field $\phi(x) \to \phi(x) + \delta\phi(x)$, the action that describes the system is transformed into,

$$S \to S + \delta S = S + \int d^4x \Delta(\phi, \partial_\mu \phi). \tag{17.55}$$

Associated with this transformation there is a current J_μ such that $\partial^\mu J_\mu = \Delta$. Obviously where $\Delta = 0$ (or a total derivative) the transformation is a symmetry and the corresponding Noether current is conserved. Associated with such a

---

[5] Conformal Ward identities which were studied in [168] are identical to the Callan–Symanzik equation [55] and [204].

transformation of the field there is a constraint on correlation functions of this field. This constraint which is referred to as a *Ward identity* takes the form,

$$\partial_{y^\mu} <TJ_\mu(y)\phi(x_1)...\phi(x_N)> = <T\Delta(y)\phi(x_1)...\phi(x_N)>$$
$$-i\delta^4(x_1 - y) <T\delta\phi(x_1)...\phi(x_N)> ...$$
$$-i\delta^4(x_i - y) <T\phi(x_1)...\delta\phi(x_i)...\phi(x_N)> ...$$
$$-i\delta^4(x_N - y) <T\phi(x_1)...\delta\phi(x_N)> . \qquad (17.56)$$

This relation can be derived straightforwardly using the path integral formulation of correlation functions. The Ward identity takes a simpler form when integrating over y^μ,

$$<T\delta\phi(x_1)...\phi(x_N)> +...+ <T\phi(x_1)...\delta\phi(x_i)...\phi(x_N)> +...$$
$$<T\phi(x_1)...\delta\phi(x_N)> + <Ti\delta S\phi(x_1)...\phi(x_N)> = 0. \qquad (17.57)$$

In particular in analogy with (2.56) the Ward identities associated with dilation and special conformal transformation take the form,

$$\sum_i^N (l_\phi + x_i\partial_i) <T\phi(x_1)...\phi(x_N)> = -i \int d^4x <T\Delta_D(x)\phi(x_1)...\phi(x_N)>$$

$$\sum_i^N (2x_i^\mu(l_\phi + x_i\partial_i) - 2\Sigma_\nu^\mu x_i^\nu - x_i^2\partial_i^\mu) <T\phi(x_1)...\phi(x_N)>$$
$$= -i \int d^4x 2x^\mu <T\Delta_D(x)\phi(x_1)...\phi(x_N)>, \qquad (17.58)$$

where l_ϕ is the canonical dimension, namely that of the free field. Similarly to the way we extracted information about the structure of correlators in 2d CFT in Section 2.9, we can now constrain the form of correlators in 4d. The Ward identities associated with the Poincare transformations imply that any correlation function is in fact not a general function of the N coordinates x_i^μ, but only of the invariants $x_{ij}^2 \equiv (x_i - x_j)^2$.

To understand the implication of the dilatation transformation on the correlation function let us first study the theory at its fixed point, namely at a coupling $g^* = g(\mu^*)$ such that $\beta(g^*) = 0$. Recall that the β function is defined as,

$$\beta(g(\mu)) = \mu\frac{\partial}{\partial_\mu}g(\mu), \qquad (17.59)$$

and hence the vanishing β function implies a fixed point of the coupling constant g. This will be further discussed below for the case of 4d QCD. In this case the dilatation Ward identity takes the form of that of a free theory, like the one given in (17.58), apart from the change of scaling dimension,

$$\sum_i^N (l_\Phi + \gamma(g^*) + x_i\partial_i) <T\phi(x_1)...\phi(x_N)> = 0, \qquad (17.60)$$

where $\gamma(g^*)$ is the anomalous dimension of the filed Φ. As a consequence of this form of the Ward identity, the two-point function of two scalar fields at the fixed point has the form,

$$<\phi(x_1)\phi(x_2)> = N_2(g^*)(\mu^*)^{-2\gamma(g^*)} \left[\frac{1}{(x_1-x_2)^2} \right]^{l_\phi+\gamma(g^*)} . \tag{17.61}$$

For particles with spin s and the same projection on the light-cone,

$$<\phi(x_1)\phi(x_2)> = N_2(g^*)(\mu^*)^{-2\gamma(g^*)} \left[\frac{1}{(x_1-x_2)^2} \right]^{l_\phi+\gamma(g^*)} \left(\frac{(x_1-x_2)_+}{(x_1-x_2)_-} \right)^s , \tag{17.62}$$

where it is assumed that $(x_1-x_2)_T = 0$. At the fixed point, namely $\beta(g^*) = 0$, the Ward identity associated with the special conformal transformation takes the form of that of a free theory with l_ϕ again shifted by the anomalous dimension $l_\phi \to l_\phi + \gamma(g^*)$.

In two dimensions (see Section 2.8) it was found that the three-point function of primary fields is fully determined by the $SL(2,C)$ symmetry and any four-point function of primary fields is determined up to a function of the cross ratio (or anharmonic ratio) $\frac{z_{12} z_{34}}{z_{13} z_{24}}$ and its complex conjugate coordinate. Based on the Poincare, dilation and special conformal transformation in four dimensions, the three-point function is determined here too. For instance, the three-point function of a scalar field is,

$$<\phi(x_1)\phi(x_2)\phi(x_3)> = N_3(g^*)(\mu^*)^{-3\gamma(g^*)} \left[\frac{1}{(x_1-x_2)^2(x_1-x_3)^2(x_2-x_3)^2} \right]^{[l_\phi+\gamma(g^*)]/2} , \tag{17.63}$$

and any correlator of $n > 3$ operators depends only on the ratios $\frac{x_{ij} x_{kl}}{x_{il} x_{jl}}$.

It is worth noting that the Ward identity associated with the dilation (the first equation of (17.58)) is in fact the same as the Callan–Symanzik renormalization group equation. First note that based on dimensional counting and Lorentz invariance the dependence of the N point function on the scale μ takes the form,

$$<T\Phi(x_1)...\Phi(x_N)> = \mu^{Nl_\Phi} G(x_{ik}^2 \mu^2; g(\mu)), \tag{17.64}$$

which means that the following relation holds,

$$\sum_{i=1}^{N}(l_\Phi + x_i \partial_i) <T\Phi(x_1)...\Phi(x_N)> = \mu\frac{\partial}{\partial\mu} <T\Phi(x_1)...\Phi(x_N)> . \tag{17.65}$$

It is easy to realize that the right-hand side of the conformal Ward identity can be rewritten in the form,

$$i\int d^4x <T\Delta_D(x)\Phi(x_1)...\Phi(x_N)> = -M\frac{\partial}{\partial M} <T\Phi(x_1)...\Phi(x_N)>, \tag{17.66}$$

as follows from,

$$\Delta_D(x) = -M\frac{\partial}{\partial M}\mathcal{L}_{\text{eff}}, \tag{17.67}$$

for the cases with no explicit dimension-full parameters. We will show this explicitly for the effective theory of 4d QCD below. On the other hand the dependence of the correlator on M follows from the dependence of the field renormalization factor and the dependence of the coupling constant so that,

$$-M\frac{\partial}{\partial M}<T\Phi(x_1)...\Phi(x_N)> = \left[\beta(g)\frac{\partial}{\partial g} + \sum_{i=1}^{N}\gamma_{\Phi_i}\right]<T\Phi(x_1)...\Phi(x_N)> . \tag{17.68}$$

Combining this together with (17.58) and (17.60) we get the Callan–Symanzik equation,

$$\left[\mu\frac{\partial}{\partial\mu} + \beta(g)\frac{\partial}{\partial g} + \sum_{i=1}^{N}\gamma_{\Phi_i}\right]<T\Phi(x_1)...\Phi(x_N)> = 0. \tag{17.69}$$

17.7 Conformal invariance and QCD_4

So far we have discussed the implications of conformal invariance in general, and in particular the invariance properties under the $SL(2,R)$ collinear group and conformal Ward identities. We are now in the position to examine the application of conformal symmetry to four-dimensional QCD. Recall that the action of four-dimensional $SU(N)$ gauge theory with massless quarks takes the form,

$$\mathcal{L}_{QCD4} = -\frac{1}{4}F_{\mu\nu}^a F^{\mu\nu a} + i\bar{\psi}\slashed{D}\,\psi, \tag{17.70}$$

where $D_\mu = \partial_\mu - igt^a A_\mu^a$ is the covariant derivative, and t^a as usual are the $N \times N$ matrices in the fundamental representation of the $SU(N)$ algebra. It is straightforward to check that the corresponding classical action is invariant under the full set of fifteen transformations associated with the $SO(2,4)$ symmetry group. In particular it is invariant under the scale transformation given by,

$$x_\mu \to \lambda x_\mu \qquad A_\mu(x) \to \lambda A_\mu(\lambda x) \qquad \psi(x) \to \lambda^{3/2}\psi(\lambda x). \tag{17.71}$$

The invariance under these transformations manifests itself in the form of conservation of the corresponding Noether current,

$$D_\mu = x_\nu T^{(TL)\mu\nu} = x_\nu\left[F^{\mu\rho a}F_\rho^{\nu a} + \frac{i}{2}\bar{\psi}(\overleftrightarrow{D})^{(\mu}\gamma^{\nu)}\psi\right], \qquad \partial^\mu D_\mu = 0, \tag{17.72}$$

where $(\overleftrightarrow{D}) \equiv \overrightarrow{D} - \overleftarrow{D}$. The classical invariance is not maintained quantum mechanically. This situation of having classical conformal symmetry but not a corresponding quantum mechanical one is referred to as the *conformal anomaly.*[6] In string theory the two-dimensional conformal symmetry is local. Having an anomaly in a local symmetry renders the theory into an inconsistent one. This implies that (at least in flat space-time) the theory will be defined in a critical dimension where the conformal anomaly vanishes. In the four-dimensional field theories discussed here, like *QCD*$_4$, conformal invariance is a global symmetry and the theory is consistent even when having an anomaly. There are several ways to show that the quantum theory is not scale and hence also not conformal invariant. One may say that the anomaly follows from the fact that the theory has infinities that are cured just by the introduction of a renormalization procedure. The latter involves the introduction of a cutoff scale. Once a scale is introduced the theory is not any more scale invariant.

To see it more explicitly let us consider the low energy effective action of massless *QCD*$_4$. We expand the gluons and quarks in terms of modes and distinguish the low energy (momentum) modes and the high energy modes. Next we integrate the high energy modes to derive the one loop low energy effective action. It takes the form,[7]

$$S_{\mathrm{LE}} = -\frac{1}{4} \int \mathrm{d}^4 x \left[\left(\frac{1}{g_0^2} - \frac{\beta_0}{16\pi^2} \ln\left(\frac{M^2}{\mu^2}\right) \right) F_{\mu\nu}^a F^{\mu\nu a} + ... \right]_{\mathrm{low}}, \qquad (17.73)$$

where M is the UV cutoff, and $\beta_0 = \frac{11}{3} N_c - \frac{2}{3} N_f$ is the coefficient of the one loop beta function. It is easy to check that this one loop renormalized action *is not invariant* under the scale transformations of (17.71). The variation of the action under those transformation reads,

$$\delta S = -\frac{1}{32\pi^2} \beta_0 \ln\lambda \int \mathrm{d}^4 x \left[\frac{1}{g_0^2} F_{\mu\nu}^a F^{\mu\nu a} + ... \right]_{\mathrm{low}}. \qquad (17.74)$$

Thus the quantum mechanically (unlike the classical case) dilatation Noether current is not conserved,

$$\partial_\mu D^\mu \equiv \Delta_D = -\frac{1}{32\pi^2} \left[\beta(g) F_{\mu\nu}^a F^{\mu\nu a} \right]_{\mathrm{low}}, \qquad (17.75)$$

and in deriving the right-hand side of the equation we have used the equations of motion.

The effective action admits also an anomaly with respect to the special conformal transformations.

In (17.46) we discussed the general structure of non-local operators of four-dimensional conformal field theory. In QCD in many cases we encounter a non-local operator built from a quark and an anti-quark at light-like separation, with

[6] The conformal anomaly was introduced in [169] and [6].
[7] The explicit calculation is a one loop perturbative calculation. Since we do not deal with perturbative methods in this book, we do not present here the derivation and refer the reader to references that deal with perturbation theory in *QCD*$_4$.

a line integral connecting them,

$$Q_\mu(\alpha_1, \alpha_2) = \bar{\psi}(\alpha_1)\gamma_\mu P e^{ig \int_{\alpha_2}^{\alpha_1} dt A_+(t)} \psi(\alpha_2) \tag{17.76}$$

where P stands for path ordering. The path integral factor will be denoted $[\alpha_1, \alpha_2]$. In performing the short distance expansion we need now to identify the corresponding conformal operators. To relate the operator ψ to a primary operator we first have to make a spin projection in the following way,

$$\psi_+ = \Gamma_+ \psi \qquad \psi_- = \Gamma_- \psi \qquad \psi = \psi_+ + \psi_-, \tag{17.77}$$

where,

$$\Gamma_+ = \frac{1}{2}\gamma_-\gamma_+, \qquad \Gamma_- = \frac{1}{2}\gamma_+\gamma_-, \qquad \Gamma_- + \Gamma_+ = 1. \tag{17.78}$$

The spin projected parts are,

$$\psi_+(s = +1/2, \ j = 1, \ t = 1) \qquad \psi_-(s = -1/2, \ j = 1/2, \ t = 2). \tag{17.79}$$

With this identification we define the quark anti-quark operators:

$$\begin{aligned} twist - 2 : Q_+ &= \bar{\psi}_+\gamma_+\psi_+ \equiv Q^{(1,1)} \\ twist - 3 : Q_T &= \bar{\psi}_+\gamma_T\psi_- + \bar{\psi}_-\gamma_T\psi_+ \equiv Q^{(1,1/2)} + Q^{(1/2,1)} \\ twist - 4 : Q_- &= \bar{\psi}_-\gamma_-\psi_- \equiv Q^{(1/2,1/2)}. \end{aligned} \tag{17.80}$$

The corresponding local conformal operators are,

$$\begin{aligned} Q_n^{1,1}(\alpha) &= (i\partial_+)^n \left[\bar{\psi}(\alpha)\gamma_+ C_n^{3/2}\left(\overleftrightarrow{D}_+/d_+\right)\psi(\alpha)\right], \\ Q_n^{1,1/2}(\alpha) &= (i\partial_+)^n \left[\bar{\psi}(\alpha)\gamma_+\gamma_T\gamma_- P_n^{1,0}\left(\overleftrightarrow{D}_+/d_+\right)\psi(\alpha)\right], \\ Q_n^{1/2,1/2}(\alpha) &= (i\partial_+)^n \left[\bar{\psi}(\alpha)\gamma_- C_n^{1/2}\left(\overleftrightarrow{D}_+/d_+\right)\psi(\alpha)\right], \end{aligned} \tag{17.81}$$

where

$$\overleftrightarrow{D}_+ = \overrightarrow{D}_+ - \overleftarrow{D}_+ \qquad d_+ = \overrightarrow{D}_+ + \overleftarrow{D}_+, \tag{17.82}$$

and where the Jacobi polynomials with two identical indices were replaced by the Gegenbauer polynomials $P^{(1,1)} \sim C_n^{3/2}$ and $P^{(0,0)} \sim C_n^{1/2}$.

A similar analysis can be carried out for the gluons. The various components of the gluon field have the following properties,

$$\begin{aligned} F_{+T}(s = +1, \ j = 3/2, \ t = 1) &\qquad F_{TT}, F_{+-}(s = 0, \ j = 1, \ t = 2) \\ F_{-T}(s = -1, \ j = 1/2, \ t = 3). & \end{aligned} \tag{17.83}$$

Local operators built from two-gluon fields with leading twist are,

$$G_n^{3/2,3/2}(\alpha) = (i\partial_+)^n \left[F_{+T}(\alpha)C_n^{5/2}\left(\overleftrightarrow{D}_+/d_+\right)F_{+T}(\alpha)\right]. \tag{17.84}$$

Another application of conformal invariance to QCD is the determination of the OPE of two electromagnetic currents $j_\mu^{EM} = \sum_i e_i \bar{\psi}_i\gamma_\mu\psi_i$ where the e_i are the charges of the u, d and s quarks. At the tree level only the transverse components are of interest. The latter have spin $s_j = 0$ and twist $t_j = 3$. The quark operators $\mathcal{Q}_n^{1,1}$ are the relevant basis for the expansion, with conformal spin $j_n = (l_n + 1 + n)/2 = n + 2$ and $t_n = (l_n - 1 - n)/2 = 2$. As $\Delta = 1$ we find

$$J^T(x)J^T(0) \sim$$

$$\sum_{n=0}^{\infty} C_n \left(\frac{1}{x^2}\right)^{(6-t_n)/2} (-ix_-)^{n+1} \frac{\Gamma(2j_n)}{\Gamma(j_n)\Gamma(j_n)} \int_0^1 du[u(1-u)]^{j_n-1} \mathcal{Q}_n^{1,1}(ux_-).$$

$$(17.85)$$

The coefficients C_n can be extracted from deep inelastic scattering via the following matrix element of forward scattering,

$$<P|J^T(x)J^T(0)|P> \sim \sum_{n=0}^{\infty} C_n \left(\frac{1}{x^2}\right)^{(6-t_n)/2} (-ix_-)^{n+1} <P|\mathcal{Q}_n^{1,1}(0)|P> .$$

$$(17.86)$$

Another application of the COPE is the determination of the short-distance expansion of the operator $\mathcal{Q}_+$ (17.76). This case is characterized by $s_A = s_B = s_1 = s_2 = \frac{1}{2}$ so that $\Delta = 0$ and $l_A = l_B = l_1 = l_2 = \frac{3}{2}$ and we find,

$$\mathcal{Q}_+(\alpha_1, \alpha_2) \sim \sum_{n=0}^{\infty} \tilde{C}_n (-i)^n (\alpha_1 - \alpha_2)^n \int_0^1 du\, u^{n+1}(1-u)^{n+1} \mathcal{Q}_n^{1,1}(u\alpha_1 + (1-u)\alpha_2),$$

$$(17.87)$$

where $\tilde{C}_n = \frac{C_n \Gamma(n+2)^2}{\Gamma(2n+4)}$ which can be determined again from forward matrix elements and are found to be $\tilde{C}_n = \frac{2(2n+3)}{(n+1)!}$.

Conformal invariance can be used at short distances to give predictions for the quark distribution amplitudes for flavor non-singlet mesons, namely the wave functions which control the behavior of the exclusive mesons processes at large momentum transfer. Here we discuss as an example the pion distribution amplitude in the leading twist order.

The basic ingredient in computing exclusive reactions including a large momentum transfer to a pion is the matrix element of a quark anti-quark between the vacuum and a one pion state. By using the light-cone gauge $A_+ = 0$ the Wilson line (17.76) is set to unity. We choose a frame where $p_\mu = p_+ n_{-\mu}$ and $x^\mu = x_- n_+^\mu + x_T^\mu$, $x_+ = 0$ so that $x^2 = x_T^2$. The matrix element can then be written as,

$$<0|\bar{d}(0)[0, \infty n]\gamma_+\gamma_5 [\infty n + x, x]u(x)|\pi^+(p)> =$$
$$if_\pi p_+ \int_0^1 dy e^{-iy(p\cdot x)} f(y, \ln x^2) + O(x^2).$$

$$(17.88)$$

This matrix element is the probability amplitude to find the pion in the valence state consisting of a u-quark carrying a momentum y and an anti-d quark of momentum $\bar{y} = 1 - y$ and have a transverse separation x_T. This amplitude is intimately related to the pion electromagnetic form factor for large momentum transfer Q^2 and small separation distance of the order $x_T \sim 1/Q^2$. To approach this limit, one defines the pion distribution amplitude taken at exactly light-like separation where $x_T = 0$. This amplitude reads,

$$<0|\bar{d}(0)[0,\alpha]\gamma_+\gamma_5 u(\alpha)|\pi^+(p)> = if_\pi p_+ \int_0^1 dy e^{-iy(\alpha p_+)} \phi_\pi(y,\mu). \qquad (17.89)$$

The distribution amplitude $\phi_\pi(y,\mu)$ is scale and scheme dependent. In fact the small transverse distance behavior of the valence component of the pion wave function is traded for the scale dependence of the distribution amplitude.

It can be shown that the evolution equation of $\phi_\pi(y,\mu)$ is given by,

$$\mu^2 \frac{d}{d\mu^2}\phi_\pi(y,\mu) = \int_0^1 d\tilde{y} V(y,\tilde{y},\alpha_s(\mu))\phi_\pi(y,\mu), \qquad (17.90)$$

where to leading order in α_s the integral kernel is given by,

$$V_0(y,\tilde{y}) = C_F \left[\frac{1-y}{1-\tilde{y}} \left(1 + \frac{1}{y-\tilde{y}}\right)\theta(y-\tilde{y}) + \frac{\tilde{y}}{y}\left(1 + \frac{1}{y-\tilde{y}}\right)\theta(y-\tilde{y}) \right]_+ , \qquad (17.91)$$

where $]_+$ stands for,

$$[V(\tilde{y},y)]_+ = V(\tilde{y},y) - \delta(y-\tilde{y})\int_0^1 dt V(t,\tilde{y}). \qquad (17.92)$$

Instead of solving this evolution equation one can alternatively proceed by expanding both sides of (17.90) in powers of α. In this way moments of the distribution amplitude are related to matrix elements of renormalized local operators in the following form,

$$<0|\bar{d}(0)\gamma_+\gamma_5(i\overleftrightarrow{D}_+)^n u(0)|\pi^+(p)> = if_\pi(p_+)^{n+1}\int_0^1 dy(2y-1)^n \phi_\pi(y,\mu). \qquad (17.93)$$

This is similar to the leading twist operators that enter the OPE for the unpolarized deep inelastic scattering apart from the flavor, the additional γ_5 factor and the fact that now one has to take into account mixing with operators that contain total derivatives of the form,

$$\mathcal{O}_{n-k,k} = (i\partial_+)^k \bar{d}(0)\gamma_+\gamma_5(i\overleftrightarrow{D}_+)^{n-k}u(0). \qquad (17.94)$$

The mixing matrix is in fact triangular since operators with fewer total derivatives can only mix with operators with more total derivatives but not the other way around. The components of the matrix on the diagonal are true anomalous dimensions, which are identical to those of inelastic scattering,

$$\gamma_n^{(0)} = C_F \left(1 - \frac{2}{(n+1)(n+2)} + 4\sum_{m=2}^{n+1}\frac{1}{m}\right), \qquad (17.95)$$

where,

$$<P|\mathcal{O}_{n,0}(\mu)|P> = <P|\mathcal{O}_{n,0}(\mu_0)|P> \left(\frac{\alpha_s(\mu)}{\alpha_s(\mu_0)}\right)^{\frac{\gamma_n^{(0)}}{\beta_0}} \qquad \beta_0 = \frac{11}{3}N_c - \frac{2}{3}N_f.$$
$$(17.96)$$

Conformal invariance is useful in finding the eigenvectors of the mixing matrix since conformal operators with different conformal spins cannot mix under renormalization to leading order. This happens since to leading order the renormalization is determined by counter terms of the tree level which is conformal invariant. Thus the mixing eigenvector operators are $Q^{1,1}(x)$ defined in (17.80) with the right flavor and γ matrices structure,

$$Q_n^{1,1}(x) = (i\partial_+)^n \left[\bar{d}(x)\gamma_+\gamma_5 C_n^{3/2}\left(\overleftrightarrow{D}_+/d_+\right)u(x)\right].$$
$$(17.97)$$

Note that because of their flavor content these operators cannot mix with operators made out of gluons and they also cannot mix with operators with more fields since they have higher twist. Thus the operators (17.97) are the only relevant ones and they must be multiplicatively renormalized. Comparing (17.93) with (17.97) one concludes that the Gegenbauer moments of the pion distribution amplitudes are given in terms of reduced matrix elements of conformal operators,

$$if_\pi p_+^{n+1} \int_0^1 dy C_n^{3/2}(2y-1)\phi_\pi(y,\mu) = <0|Q_n^{1,1}(0)|\pi^+(p)> .$$
$$(17.98)$$

As was mentioned above these operators are renormalized by a multiplication and the corresponding anomalous dimension is given by (17.95). Thus the final picture is that the distribution amplitude $\phi_\pi(u,\mu)$ can be expanded in a series of Gegenbauer polynomials,

$$\phi_\pi(u,\mu) = 6u(1-u)\sum_{n=0}^{\infty} \phi_n(\mu)C_n^{3/2}(2u-1)$$

$$\phi_n(\mu) = (if_\pi p_+^{n+1})^{-1}\frac{2(2n+3)}{3(n+1)(n+2)}<0|Q_n^{1,1}(0)|\pi^+(p)>$$

$$\phi_n(\mu) = \phi_n(\mu_0)\left(\frac{\alpha_s(\mu)}{\alpha_s(\mu_0)}\right)^{\frac{\gamma_n^{(0)}}{\beta_0}}.$$
$$(17.99)$$

This example demonstrates the application of conformal invariance to solve the problem of operator mixing. There are other applications of conformal symmetry to four-dimensional QCD. We refer the interested reader to [43]. The predictions based on conformal symmetry beyond one loop, for the sector with $\beta = 0$ [51], turned out to be in contradiction with explicit calculations [79]. This paradox was resolved in [166].

18
Integrability in four-dimensional gauge dynamics

Integrability was discussed in Chapter 5 in the context of two-dimensional models. In particular solutions of spin chain models based on the Bethe ansatz approach were described in some detail. Integrable models are characterized by having the same number of conserved charges as the number of physical degrees of freedom. Furthermore, the scattering processes of those models always involve conservation of the number of particles.

A natural question at this point is whether integrability is a property of only two-dimensional models or whether one can also identify systems in four dimensions that admit integrability. Four-dimensional gauge theories have generically infinite numbers of degrees of freedom and their interactions do not conserve the number of particles. Thus four-dimensional gauge theories like the YM theory are not integrable theories. However, it turns out, as will be shown in this chapter, that various sectors of certain four-dimensional gauge theories, which are derived upon imposing certain limits, do admit integrability.

The two-dimensional integrable models discussed in Chapter 5 were non-conformal ones and were characterized by a scale and hence also with particles and an S-matrix. On the other hand the integrable sectors of four-dimensional gauge theories that we are about to describe are conformal invariant. The main idea is that these special conformal invariant sectors can be mapped into two-dimensional spin chains that were described in Section 5.14.

The investigation of this issue is far from complete. Nevertheless, a large body of knowledge has already been accumulated. In recent years this has followed the lines of the AdS/CFT duality [158] which is not covered in this book.[1] The purpose of this section is just to demonstrate the idea of the map between gauge theories and in particular QCD and integrable spin chain models. This will be done by describing the following cases:

(i) $\mathcal{N} = 4$ super YM theory in four dimensions.
(ii) Scale dependence of composite operators in QCD.

$\mathcal{N} = 4$ super YM theory is known to be the maximal global supersymmetric theory in four dimensions. Since supersymmetry is beyond the scope of this book we will not discuss it in the context of the $\mathcal{N} = 4$ SYM. Thus the description

---

[1] For a review of the AdS/CFT the reader can refer to [10].

of the integrable aspects of the theory will be incomplete and will be missing certain essential parts. However, since $\mathcal{N} = 4$ SYM is the simplest interacting four-dimensional non-abelian gauge theory we start with this and then proceed to a certain limit in non-supersymmetric QCD.

There are several review papers on integrability in four-dimensional gauge dynamics. In this chapter we follow [31] about the integrability of $\mathcal{N} = 4$ SYM theory and [34] for the scale dependence of composite operators of QCD.

18.1 Integrability of large N four-dimensional $\mathcal{N} = 4$ SYM

The Lagrangian of $\mathcal{N} = 4$ SYM is given by,

$$\mathcal{L}_{\mathcal{N}=4} = -\frac{1}{4}\mathrm{Tr}\left[F_{\mu\nu}F^{\mu\nu}\right] + \frac{1}{2}\mathrm{Tr}\left[D_\mu\Phi_n D^\mu\Phi^n\right] - \frac{1}{4}g^2\mathrm{Tr}\left[[\Phi^m,\Phi^n]^2\right]$$

$$+\mathrm{Tr}\left[\dot{\psi}_{\dot{\alpha}}^a\sigma_\mu^{\dot{\alpha}\beta}D^\mu\psi_{\beta a}\right] - \frac{i}{2}g\mathrm{Tr}\left[\psi_{\alpha a}\sigma_m^{ab}\epsilon^{\alpha\beta}[\Phi^m,\psi_{\beta b}]\right] - \frac{i}{2}g\mathrm{Tr}\left[\dot{\psi}_{\dot{\alpha}}^a\sigma_{ab}^m\epsilon^{\dot{\alpha}\dot{\beta}}[\Phi_m,\dot{\psi}_{\dot{\beta}}^b]\right],$$

$$(18.1)$$

where $F_{\mu\nu}$ is the field strength associated with an $SU(N)$ gauge group, Φ^m is a set of six $m = 1,\dots,6$ scalar fields, and ψ and $\dot{\psi}$ are doublets of $SU(2) \times SU(2)$. Both the scalars and the spinors are in the adjoint representation of $SU(N)$. The matrices σ^μ and σ^m are the chiral projections of the gamma matrices in four and six dimensions, respectively and ϵ is the totally antisymmetric tensor of $SU(2)$. It is convenient to write the corresponding action as,

$$S = N \int \frac{d^4x}{4\pi^2}\mathcal{L}_{\mathcal{N}=4},\qquad(18.2)$$

where the coupling constant is taken to be $g^2 \equiv \frac{g_{YM}^2 N}{8\pi^2}$.

It is well known that the theory, on top of being invariant under $SU(N)$ gauge symmetry and $SO(6)$ global symmetry, is also conformal invariant and in fact superconformal invariant. The β function of the theory which vanishes to all orders in perturbation theory is believed to vanish also non-perturbatively and hence the theory is assumed to be conformal also in the quantum level. In Section 17.1 we have described the conformal symmetry algebra in four dimensions. Recall the $SO(2,4)^2$ conformal transformations (see 17.7) which are being generated by P^μ, $S^{\mu\nu}$, $\mathcal{D}$, K^μ, the generators of space-time translations, Lorentz transformations, dilation and special conformal transformation, respectively.

A major player in the structure of the $\mathcal{N} = 4$ is the *dilatation operator* $\mathcal{D}$. Whereas the generators of the Poincare group do not get quantum corrections, the dilatation operator does so that in fact,

$$\mathcal{D} = \mathcal{D}_0 + \delta\mathcal{D}(g),\qquad(18.3)$$

---

[2] In fact the $\mathcal{N} = 4$ SYM admits a superconformal algebra of $psu(2,2|4)$ which we do not discuss here.

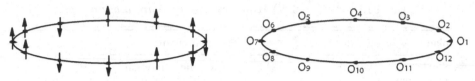

Fig. 18.1. A single trace operator as a spin chain.

where D_0 is the classical operator and δD is the anomalous dilation operator which obviously depends on the gauge coupling g.

The "*states*" of the theory take the form of multi-trace gauge invariant operators,

$$\mathrm{Tr}\,[\mathcal{W}\ldots.\mathcal{W}]\ldots\mathrm{Tr}\,[\mathcal{W}\ldots.\mathcal{W}]\quad \mathcal{W} \in \{D^k\Phi, D^k\Psi, D^k\dot{\Psi}, D^k F\}, \qquad (18.4)$$

where D stands for the covariant derivative and $F \equiv F_{\mu\nu}$ is the field strength. The Hilbert space of states is built, as for any conformal field theory, from Verma modules each characterized by a *highest weight state* or a *primary state*, which were defined in (2.8). An example of a highest weight state is $|\mathcal{K}> = \eta^{mn}\mathrm{Tr}[\Phi_m\,\Phi_n]$. The rest of the ·Verma module includes the descendant states which are derived by acting with lowering operators on primary states. Needless to say the general structure of correlation functions of four-dimensional conformal field theories discussed in Section 17.6 applies also for the case of the $\mathcal{N} = 4$ SYM. In particular recall (2.8) that the two-point function of two operators is given by,

$$<\mathcal{O}(x_1)\mathcal{O}(x_2)> = \frac{M(g)}{|x_1 - x_2|^{2\mathcal{D}(g)}}. \qquad (18.5)$$

The anomalous dimension can be computed perturbatively as a power series in g. As was discussed in Chapter 7 the perturbation expansion becomes much more tractable in the large N limit, namely, in the planar limit. In this limit the dominant diagram has a vanishing Euler number $\chi = 2C - 2G - T = 0$ where C, G, T stand for the number of components, genus, namely the number of handles, and the number of traces, respectively. Since each component requires two traces, one incoming and one outgoing, it implies that the planar limit projects onto diagrams with $G = 0$ and $T = 2C$. This means that only single trace operators are relevant.

We have seen above that in the planar limit we deal with single trace operators. Pictorially, (see Fig. 18.1) a single trace operator looks like a cyclic spin chain. This map can be made precise. Spin chain as integrable models were discussed in Section 5.14. Recall that a spin chain includes a set of L spins with cyclic adjacency property.[3] The spin at each site is a module of the symmetry algebra of the system. The Hilbert space of the whole system is the tensor product of L modules. In Section 5.14 we discussed only chains with a fixed number

---

[3] In Section 5.14 we denoted the number of spins by N. Here to avoid confusion with the rank of the gauge group we will refer to the number of spins as L.

Table 18.1. $\mathcal{N} = 4$ *SYM theory to spin chain dictionary*

planar $\mathcal{N} = 4$ SYM	spin chain
Single trace operator	Cyclic spin chain
Field operator	Spin at a site
Anomalous dilatation operator $g^{-2}\delta\mathcal{D}$	Hamiltonian
Anomalous dimension	Energy eigenvalue
Cyclicity constraint	Zero momentum condition $U = 1$

of spins. One can generalize this situation to incorporate also a *dynamic* spin chain with an unfixed number of spins. In this case the Hilbert space is a tensor product of all Hilbert spaces of a fixed length. In the Heisenberg model each spin has two possible states and the Hilbert space is therefore $\mathcal{C}^{(2^{(L)})}$. In general the spin in the chain can point in more than two directions and in particular also in infinitely many directions, as is the case for the spin chain of the $\mathcal{N} = 4$ SYM theory. In the latter case the spin is mapped into a field operator and the possible spin states to the components of the gauge symmetry multiplet. The cyclicity of the single trace operators maps into a constraint on the spin chain so that states that differ by a trivial shift are identified and hence states with non-trivial momentum are unphysical. In the language of Section 5.14 we have to impose $U = 1$ as a constraint. In the Heisenberg model this renders the Hilbert space into $\frac{\mathcal{C}^{(2^{(L)})}}{Z_L}$. The Hamiltonian of the spin chain model translates into the dilatation operator and the energy eigenvalues to the anomalous dimensions. The full correspondence between the spin chain and the planar limit of the $\mathcal{N} = 4$ SYM theory is summarized in Table 18.1.

Once the correspondence with a spin chain model has been established, one can proceed in a similar way as for the Heisenberg spin chain model. The next step is to write down the algebraic Bethe ansatz which now corresponds to an $SO(6)$ symmetry if one considers operators constructed only from the fields Φ_m or in general the $psu(2,2|4)$ for the full $\mathcal{N} = 4$ SYM theory. The algebraic Bethe ansatz, the analog of (18.6) now reads as follows,

$$\left(\frac{\lambda_k + i/2V_{jk}}{\lambda_k - i/2V_{jk}}\right)^L = \prod_{l \neq k}^{K} \frac{\lambda_k - \lambda_l + iM_{j_k,j_l}}{\lambda_k - \lambda_l - iM_{j_k,j_l}}, \tag{18.6}$$

where L is the size of the chain (N in (5.224)), the total number of excitation is K (l in (5.224)) and where for each of the corresponding Bethe roots λ_k one specifies which of the simple roots is excited by j_k which takes the values of $1, \ldots, \#_{\text{sr}}$ with $\#_{\text{sr}}$ being the number of simple roots which for the $SO(6)$ case is three and for the $psu(2,2|4)$ is seven. M is the Cartan matrix of the algebra (1 in (5.224)) and V are the Dynkin labels of the representation (s in (5.224)).

The condition of zero momentum now reads,

$$1 = U = \prod_{k=1}^{K} \frac{\lambda_k + \frac{i}{2} V_{jk}}{\lambda_k - \frac{i}{2} V_{jk}}. \tag{18.7}$$

The energy of a configuration of roots that satisfies the Bethe equation is,

$$E = \sum_{k=1}^{K} \frac{V_{jk}}{\lambda_k^2 + \frac{1}{4} V_{jk}^2}. \tag{18.8}$$

This is of course the analog of (5.225) and the higher conserved charges are,

$$Q_r = \frac{i}{r-1} \sum_{k=1}^{K} \left(\frac{1}{\left(\lambda_k^2 + \frac{i}{2} V_{jk}\right)^{r-1}} - \frac{1}{\left(\lambda_k^2 - \frac{i}{2} V_{jk}\right)^{r-1}} \right). \tag{18.9}$$

The leading order part of the transfer matrix reads,

$$T(\lambda) = \prod_{k=1}^{K} \frac{\lambda - \lambda_k + \frac{i}{2} V_{jk}}{\lambda - \lambda_k - \frac{i}{2} V_{jk}} + \ldots, \tag{18.10}$$

It was shown that these generalized Bethe equations provide a solution to the planar anomalous dimensions of the $\mathcal{N} = 4$ SYM theory [162]. This is just the tip of the iceberg. The integrable structure of the planar $\mathcal{N} = 4$ has been investigated very thoroughly in recent years. For an early review on the topic the reader can consult [31]. As an epilog let us mention that, as was shown in [106], one can identify in a similar manner with what was done in $\mathcal{N} = 4$ SYM [35], a spin chain structure in gauge theories which are confining and with less or even no supersymmetries. In that case the spin chain Hamiltonian would not correspond to the dilatation operator but was rather associated with the excitation energies of hadrons.

18.2 High energy scattering and integrability

High energy scattering is characterized by the fact that the Mandelstam parameter $s = (p_A + p_B)^2$ is the largest scale of the system, and in the limit of $s \to \infty$ the energy dependence corresponds to a renormalization group flow of the dynamical system that "resides" on the two dimensions transverse to the scattering plane.

It is convenient to study the properties of the high energy scattering amplitude $A(s, t)$ using the Mellin transform,

$$A(s, t) = is \int_{\delta - i\infty}^{\delta + i\infty} \frac{dw}{2\pi i} s^w \tilde{A}(w, t), \tag{18.11}$$

where the integration contour goes to the right of the poles of $A(w, t)$ in the w complex plane. The high energy asymptotic of $A(s, t)$ is determined by the poles of the partial wave amplitudes, namely if $\tilde{A}(w, t) \sim \frac{1}{(w - w_0(t))}$, then $A(s, t) \sim is^{1 + w_0(t)}$. Poles in the w plane are referred to as *reggeons* and the position of the pole is called the *reggeon trajectory*.

The partial wave amplitude $\tilde{A}(w,t)$ can be written using the impact parameter representation as follows,

$$\tilde{A}(w,t) = \int d^2 b_0 e^{i(qb_0)} \int d^2 b_A d^2 b_B \, \Phi_A(\vec{b}_A - \vec{b}_0) T_w(\vec{b}_A, \vec{b}_B) \Phi_B(\vec{b}_B)$$

$$\equiv \int d^2 b_0 e^{i(qb_0)} <\Phi(b_0)|T_w|\Phi(0)>, \qquad (18.12)$$

where the impact factors $\Phi_A(\vec{b}_A)$ and $\Phi_B(\vec{b}_B)$ are the parton distributions which are functions of the transverse coordinates $\vec{b}_A = \vec{b}_A^1, \vec{b}_A^2, \ldots, \vec{b}_A^n$ for the A colliding hadron and $\vec{b}_B = \vec{b}_B^1, \vec{b}_B^2, \ldots, \vec{b}_B^n$ for the B hadron, and $T_w(\vec{b}_A, \vec{b}_B)$ is the scattering (partial wave) amplitude for a given parton configuration. The idea behind this representation of the amplitude is that the transverse coordinates of the partons can be considered as "frozen" during the interaction. It implies that the structure of the poles in the w-plane does not depend on the parton distribution in the colliding hadrons but rather on the general properties of the gluon interaction of the t-channel. It was shown [145], that the propagators of the t-channel gluons develop their own Regge trajectory due to interactions. A t-channel gluon "dressed" by the virtual corrections is referred to as *reggeized gluon*. The reggeized gluons are the relevant degrees of freedom of the high energy scattering. The partial waves $T_w(\vec{b}_A, \vec{b}_B)$ can be classified according to the number of the reggeized gluons propagating in the t-channel. The minimal number required to get a colorless exchange is two gluons. We will discuss here only this case. It can be shown that the amplitude $T_w(\vec{b}_A^1, \vec{b}_A^2 \vec{b}_B^1 \vec{b}_B^2)$ satisfies the so-called BFKL equation that reads [23], [145],

$$w T_w = T_w^{(0)} + \frac{\alpha_s N_c}{\pi} H_{BFKL} T_w, \qquad (18.13)$$

where $T_w^{(0)}$ corresponds to the free exchange of two gluons. Formally one can write the solution as,

$$T_w = \left[w - \frac{\alpha_s N_c}{\pi} H_{BFKL} \right]^{-1} T_w^{(0)}, \qquad (18.14)$$

so that the singularities of T_w are determined by the eigenvalues of the operator,

$$H_{BFKL} \Psi_\alpha(\vec{b}^1, \vec{b}^2) = E_\alpha \Psi_\alpha(\vec{b}^1, \vec{b}^2), \qquad (18.15)$$

where Ψ_α is the eigenstate. The high energy behavior of the scattering amplitude is dominated by the right-most singularity of T_w, namely on the maximal eigenvalue $(E_\alpha)_{\max}$. The equation (18.15) has the interpretation of the two-dimensional Schrödinger equation of two interacting particles. The interacting particles can be identified with reggeized gluons and $\Psi_\alpha(\vec{b}^1, \vec{b}^2)$ is the wavefunction of a colorless bound state of them. Defining the holomorphic and anti-holomorphic coordinates of the reggeized gluons as follows,

$$z_j = x_j + iy_j \quad \bar{z}_j = x_i - iy_i, \tag{18.16}$$

where $\vec{b}_j = (x_j, y_j)$, we can split the BFKL Hamiltonian into a sum of two terms, one that acts only on the holomorphic coordinates and another that acts only on the anti-holomorphic ones, as follow $H_{BFKL} = H + \bar{H}$ where,

$$H = \partial_{z_1}^{-1} \ln(z_{12}) \partial_{z_1} + \partial_{z_2}^{-1} \ln(z_{12}) \partial_{z_2} + \ln(\partial_{z_1} \partial_{z_2}) - 2\psi(1), \tag{18.17}$$

where $z_{12} = z_1 - z_2$, $\psi(x)$ is the Euler digamma function defined by $\psi(x) = \frac{d \ln \Gamma(x)}{dx}$ and in $\bar{H}$ we replace all the z_is by $\bar{z}_i$s.

The BFKL Hamiltonian is further invariant under $SL(2, C)$ transformations. Denoting the $SL(2, C)$ generators (see Section 2.9) by,

$$L_{j-} = -\partial_{z_j} \quad L_{j0} = z_j \partial_{z_j} \quad L_{j+} = z_j^2 \partial_{z_j} \quad L_a = L_{1a} + L_{2a}, \tag{18.18}$$

and similarly for the anti-holomorphic generators, the invariance takes the form,

$$[H_{BFKL}, L_a] = [H_{BFKL}, \bar{L}_a] = 0. \tag{18.19}$$

This implies that H_{BFKL} depends only on the Casimir operators of $SL(2, C)$ algebra of the two particles, namely, with,

$$L_{12}^2 = -(z_1 - z_2)^2 \partial_{z_1} \partial_{z_2} \quad \bar{L}_{12}^2 = -(\bar{z}_1 - \bar{z}_2)^2 \partial_{\bar{z}_1} \partial_{\bar{z}_2}, \tag{18.20}$$

the Hamiltonian must take the form,

$$H = H(L_{12}^2) \quad \bar{H} = \bar{H}(\bar{L}_{12}^2). \tag{18.21}$$

It thus follows that the eigenstates of the Hamiltonian must also be eigenstates of L_{12}^2 and of $\bar{L}_{12}^2$,

$$L_{12}^2 \Psi_{n,\nu} = h(h-1) \Psi_{n,\nu} \quad \bar{L}_{12}^2 \Psi_{n,\nu} = \bar{h}(\bar{h}-1) \Psi_{n,\nu}, \tag{18.22}$$

where the complex dimensions h and $\bar{h}$ are given by,

$$h = \frac{1+n}{2} + i\nu \quad \bar{h} = \frac{1-n}{2} + i\nu. \tag{18.23}$$

The non-negative integer n and the real parameter ν specify the irreducible representation of the $SL(2, C)$ group to which $\Psi_{n,\nu}$ belongs. The wave functions which are eigenstates of the Casimir operators take the form,

$$\Psi_{n,\nu}(\vec{b}) = \left(\frac{z_{12}}{z_{10} z_{20}} \right)^{\frac{1+n}{2} + i\nu} \left(\frac{\bar{z}_{12}}{\bar{z}_{10} \bar{z}_{20}} \right)^{\frac{1-n}{2} + i\nu}, \tag{18.24}$$

where $z_{ij} = z_i - z_j$ and $\vec{b}_0 = (z_0, \bar{z}_0)$ is the center of mass of the state. The conformal dimension of the state is $h + \bar{h} = 1 + 2i\nu$ and the spin $h - \bar{h} = n$. Upon substituting these eigenstates into (18.15) and using the explicit form of the BFKL kernel we find the following eigenvalues,

$$E_{n,\nu} = 2\psi(1) - \psi\left(\frac{1+n}{2} + i\nu \right) - \psi\left(\frac{1+n}{2} - i\nu \right). \tag{18.25}$$

The maximal eigenvalue corresponds to $n = \nu = 0$ or $h = \bar{h} = \frac{1}{2}$ $E_{0,0} = 4\ln 2$. This maximal eigenstate defines the right-most singularity of the partial wave amplitude. This determines the asymptotic behavior of the scattering amplitude in the leading logarithmic approximation,

$$A(s,t) = is^{1+\frac{\alpha_s N}{\pi}4\ln 2} \tag{18.26}$$

which is referred to as the BFKL *Pomeron*. Using the explicit form of the eigenvalue one can reconstruct the operator form of H_{BFKL} acting on the representations of the $SL(2,C)$ group,

$$H_{BFKL} = \frac{1}{2}[H(J_{12}) + H(\bar{J}_{12}) \quad H(j) = 2\psi(1) - \psi(j) - \psi(1-j), \tag{18.27}$$

where $L_{12}^2 = J_{12}(J_{12} - 1)$, and similarly for $\bar{L}_{12}^2$. This is a special case of the Heisenberg spin chain of spin s operators whose Hamiltonian takes the form,

$$H_s = \sum_i^L H(J_{i,i+1}) \quad J_{i,i+1}(J_{i,i+1} + 1) = (\vec{S}_i + \vec{S}_{i+1})^2, \tag{18.28}$$

where $J_{i,i+1}$ is related to the sum of two spins of the neighboring sites, $\vec{S}_i^2 = s(s+1)$ and $H(x)$ is the following harmonic function,

$$H(x) = \sum_{l=x}^{2s-1} \frac{1}{1+l} = \psi(2s+1) - \psi(x+1). \tag{18.29}$$

To connect it to the analysis of Section 5.14 we check this for $s = 1/2$. For this case $J_{i,i+1}$ can take one of the two values $0,1$ for which we have $H(0) = 1$ and $H(1) = 0$, so that the Hamiltonian is a projection into $J_{i,i+1} = 0$ subspace with $H(J_{i,i+1}) = \frac{1}{4} - \vec{S}_i \cdot \vec{S}_{i+1}$ which is identical to (5.158).

One can generalize the exchange of colorless boundstates of two reggeized gluons to exchange of multireggeon boundstates built from N_r reggeized gluons. This is beyond the scope of this book and can be found for instance in [34].

19

Large N methods in QCD_4

We have encountered the large N expansion, or the $\frac{1}{N}$ expansion, in Chapter 7 in the context of two-dimensional field theory, and in particular its application by 't Hooft to solve the mesonic spectrum of QCD in two dimensions [10]. A natural question is thus what does this limit tell us about four dimensional QCD, and in particular can one also solve the mesonic spectrum of QCD in four dimensions in the large N approximation. These questions will be the topics of this chapter. We start with the rules of counting powers of N in four-dimensional QCD, and the relations between Feynman diagrams in the double line notation and Riemann surfaces. We then briefly discuss certain applications of the expansion to the mesonic physics and then follow Witten's seminal analysis of baryons in the planar approximation [222].

The large N technique was introduced by 't Hooft in [122]. Since then there have been many follow-up papers and there is a very rich literature on large N approximation including review papers and books like [223], [66], [165], [46] and [160]. In this chapter we use mainly the latter.

19.1 Large N QCD in four dimensions

Let us remind the reader the basic notations and the classical action of QCD in four dimensions. The two-dimensional ones were presented in (8),

$$ S_{QCD} = \int d^4x \left[-\frac{1}{2} \text{Tr}[F_{\mu\nu} F^{\mu\nu}] + \bar{\Psi}_i (i \not{D} - m_i) \Psi_i \right], \qquad (19.1) $$

where the gauge fields are spanned by $N \times N$ Hermitian matrices T^A such that $A_\mu = A_\mu^A T^A$, $F_{\mu\nu} = \partial_\mu A_\nu - \partial_\nu A_\mu + i\frac{g}{\sqrt{N}}[A_\mu, A_\nu]$, the covariant derivative $D_\mu = \partial_\mu + i\frac{g}{\sqrt{N}} A_\mu$, the fermions Ψ are in the fundamental representation of the color group and $i = 1, \ldots, N_f$ indicates the flavor degrees of freedom. The gauge coupling was chosen to be $\frac{g}{\sqrt{N}}$, to accommodate a large N approximation with g fixed. This can be shown for instance in applying the large N expansion to the β function. The latter, when g is used in the covariant derivatives, is given by (17.73),

$$ \mu \frac{dg}{d\mu} = - \left[\frac{11}{3} N - \frac{2}{3} N_f \right] \frac{g^3}{16\pi^2} + \mathcal{O}(g^5), \qquad (19.2) $$

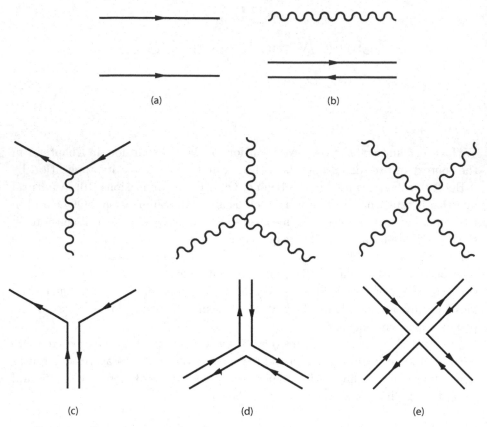

(a)   (b)

(c)   (d)   (e)

Fig. 19.1. Four-dimensional Feynman rules in the usual form and in the double line notation.

which obviously is not suitable for a large N expansion whereas if one replaces $g \to \frac{g}{\sqrt{N}}$ the β function takes the form,

$$\mu \frac{dg}{d\mu} = - \left[\frac{11}{3} - \frac{2N_f}{3N} \right] \frac{g^3}{16\pi^2} + \mathcal{O}(g^5). \qquad (19.3)$$

The rules of the Feynman diagrams in two-dimensional QCD (see Fig. 10.1) include the fermion propagator, the gluon propagator and the quark gluon vertex, all expressed in the double line notation. Using the light-cone in two dimensions one eliminates the three- and four-gluon vertices. In four dimensions due to the transverse directions the gluon vertices cannot be eliminated by choosing a gauge. Thus all together the four-dimensional Feynman rules are expressed in Fig. 19.1.

The figures a,b,c are identical to the two-dimensional ones (Fig. 10.1) whereas figures d and e are the three- and four-gluon vertices. The quark propagator (19.1a) is given by,

$$<\psi^a (x)\bar{\psi}^b (y)> = S(x - y)\delta^{ab}. \qquad (19.4)$$

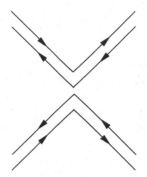

Fig. 19.2. Color flow in the double line notation associated with two traces.

The $SU(N)$ gluon propagator in the double line notation reads,

$$<(A_\mu)_b^a(x)(A_\nu)_d^c(y)> = D_{\mu\nu}(x-y)\frac{1}{2}\left(\delta_d^a\delta_b^c - \frac{1}{N}\delta_b^a\delta_d^c\right). \qquad (19.5)$$

For a $U(N)$ gauge group the propagator does not include the term which is proportional to $\frac{1}{N}$. One can view the $SU(N)$ color indices as those of a $U(N)$ theory plus an additional "ghost" $U(1)$ gauge field that cancels the contribution of the $U(1)$ gauge field in the $U(N)$ gauge group. For many applications the distinction between the $U(N)$ and $SU(N)$ in the $1/N$ expansion is sub-leading. The three and four gluon vertices (see Fig. 19.1 d,e) emerge obviously from the $\text{Tr}[F_{\mu\nu}F^{\mu\nu}]$ term in the action. Note that it is a single trace operator and hence the four-gluon vertex is the one depicted in (19.1 e) and not in Fig. 19.2 which corresponds to the color flow of a two-trace operator.

To compute the N dependence of Feynman diagrams it is convenient to re-scale the gluon field and the quark field as follows,

$$\frac{gA_\mu}{\sqrt{N}} \to \hat{A}_\mu \quad \psi \to \sqrt{N}\hat{\psi} \qquad (19.6)$$

In terms of the re-scaled field the QCD action reads,

$$\mathcal{L} = N\left[-\frac{1}{2g^2}\text{Tr}[\hat{F}_{\mu\nu}\hat{F}^{\mu\nu}] + \sum_{i=1}^{N_f}\bar{\hat{\psi}}_i(i\slashed{D} - m_i)\hat{\psi}_i\right]. \qquad (19.7)$$

From this Lagrangian we can read off the powers of N and $\lambda \equiv g^2 = N$ of each part of a Feynman diagram. The vertex operator scales like N, the propagator as $\frac{1}{N}$ and every color index loop gives a factor of N. If we combine the dependence on λ we find for the gluon, that the vertex behaves as $\frac{N}{\lambda}$, the propagator as $\frac{\lambda}{N}$ and the color loop as N, while for the fermions neither the propagator nor the quark-gluon vertex depend on the coupling.

Thus a connected vacuum diagram with V vertices, E propagators, namely edges and F loops, namely faces, is of order (see (7.7)),

$$N^{V-E+F}\lambda^{(E_{(G)}-V_{(G)})} = N^{\chi}\lambda^{(E_{(G)}-V_{(G)})}, \qquad (19.8)$$

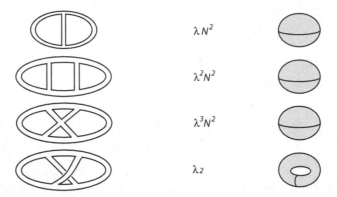

λN^2

$\lambda^2 N^2$

$\lambda^3 N^2$

$\lambda 2$

Fig. 19.3. Examples of Feynman diagrams with only gluons.

where,

$$\chi \equiv V - E + F = 2 - 2h - b, \tag{19.9}$$

is the Euler character of the surface, h is the genus, namely, the number of handles and b is the number of boundaries. $E_{(G)}$ and $V_{(G)}$ are the appropriate qualities for gluons. For instance the sphere has $\chi = 2$ since it has no handles and no boundaries, the disk has $\chi = 1$ since it has no handles and one boundary and the torus has one handle and no boundaries and hence it has $\chi = 0$. Thus the Feynman diagrams look like triangulated two-dimensional surfaces. In fact all possible gluon exchange may fill the holes of the triangulated structure forming a smooth surface with no boundaries for gluon only diagrams, and with boundaries for diagrams that include quark loops. It was conjectured that the two-dimensional surface is the world sheet of a string theory which is dual to QCD. There has been tremendous progress in this string/gauge duality in recent years following the seminal AdS/CFT duality of Maldacena [158]. This is beyond the scope of this book and we refer the reader to the relevant literature, for instance the review [10].

To further demonstrate the determination of the order of a diagram consider first the diagrams that involve only gluons which appear in Fig. 19.3. The diagram in (a) has $V = 2, E = 3, F = 3$ and thus it behaves as $N^2\lambda$. Similarly in (b) and (c) $V = 4, E = 6, F = 4$ and $V = 5, E = 8, F = 5$ so that they behave as $N^2\lambda^2$ and $N^2\lambda^3$, respectively. The three diagrams (a), (b) and (c) are all planar diagrams and have a topology of a sphere. Note however that diagram (d) which is non-planar behaves as $N^0\lambda^2$, namely of genus one. In the large N limit this last diagram is obviously suppressed.

So far we have only discussed diagrams with gluons. Quarks propagators are represented (see Fig. 19.1a) by a single line. A closed quark loop is a boundary and hence using (19.9) it contributes to the diagram a factor of $\frac{1}{N}$. Consider for example the diagram drawn in Fig. 19.4.

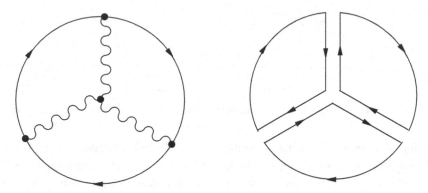

Fig. 19.4. Four-dimensional Feynman rules in the usual form and in the double line notation.

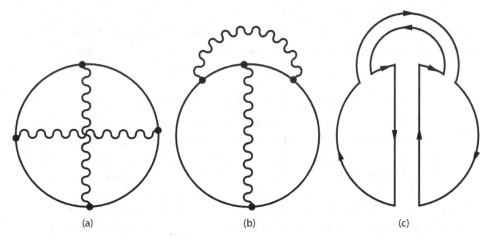

Fig. 19.5. Non-planar diagram with gluons and quarks.

It is a diagram of order N. This follows trivially from the fact that it has zero genus, $h = 0$ and one boundary $b = 1$. Alternatively we have one gluon vertex, three gluon propagators and three loops and hence $N^{1-3+3} = N$. Obviously this is also the result when one uses the unrescaled operators where each vertex contributes $\frac{1}{\sqrt{N}}$ and each index loop N, so that we get $(\frac{1}{\sqrt{N}})^4 \times N^3 = N$. Similar to the non-planar gluon diagram (19.3d), Fig. 19.5 describes a non-planar diagram that includes both gluons and quarks. This diagram scales like $\frac{1}{N}$ since there is no gluon vertex, two gluon propagators and one index loop $N^{0-2+1} = \frac{1}{N}$.

As was mentioned above the difference between the $SU(N)$ case versus the $U(N)$ can be accounted by adding a ghost $U(1)$ gauge field whose role is to cancel the extra $U(1)$ part of the $U(N)$. The $U(1)$ commutes with the $U(N)$ gauge fields and therefore does not interact with them and hence one has to incorporate only the coupling of the quark fields to the $U(1)$ gauge field. When we consider a connected diagram with gluons and $U(1)$ ghost gauge fields the contribution to the

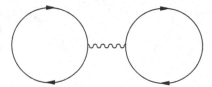

Fig. 19.6. $U(1)$ ghost propagator connecting two otherwise disconnected diagrams.

counting of orders of N due to the gluons is not affected, whereas each $U(1)$ ghost field contributes a factor of $\frac{1}{N^2}$. The latter follows from the fact that the ghost $U(1)$ propagator contributes a factor of $1/N$ and another $1/N$ factor due to the two coupling constants at the end of the propagator. This can easily be seen from the unrescaled action (19.1). For diagrams that are connected with the ghost field and otherwise disconnected as is depicted in Fig. 19.6 the situation is different. For instance that diagram is of order N^0 since it has $N \times N \times \frac{1}{N^2}$. Note however that even in this case there is a difference of order $\frac{1}{N^2}$ between the $SU(N)$ and $U(N)$ cases, (actually, in Fig. 19.6 the contribution of $SU(N)$ vanishes).

19.1.1 Counting rules for correlation functions

So far we have described the counting rules for vacuum diagrams. We now proceed to the counting rules of correlation functions of gluons and quarks which are vacuum expectation values of gauge invariant operators made out of gluon and quark fields. The latter should be color singlets, not necessarily local, that cannot be split into color singlet pieces. Thus operators like $\bar\psi\psi, Tr[F_{\mu\nu}F^{\mu\nu}], \bar\psi(y)e^{-i\int_x^y A_\mu(z)dz^\mu}\psi(x)$ are allowed whereas $(\bar\psi\psi)^2$ is not. As usual the procedure to compute correlation functions of such operators is to add appropriate source terms to the action and differentiate the generating function with respect to sources that correspond to the operators. If we denote by $\hat{\mathcal{O}}_i$ a gauge invariant operator made out of the rescaled fields, we shift the Lagrangian density as follows, $\mathcal{L}_0 \to \mathcal{L}_0 + \sum_i NJ_i\mathcal{O}_i$, where $\mathcal{L}_0$ is the Lagrangian density without the sources and J_i is the source that corresponds to $\mathcal{O}_i$. Thus any correlator can be determined as follows,

$$<\hat{\mathcal{O}}_1....\hat{\mathcal{O}}_n> = \frac{1}{iN}\frac{\partial}{\partial J_1}\cdots\frac{1}{iN}\frac{\partial}{\partial J_n}W(J). \tag{19.10}$$

In terms of N counting, for correlation functions of only gluon fields $W(J)$ is of order N^2 and hence the correlator is of order N^{2-n}. The leading order of $W(J)$ in the case where quarks fields are also involved is of order N which means that the correlator is of order N^{1-n}. Let us denote by $\hat{G}_i$ and $\hat{M}_i$ glueball and meson gauge invariant operators, respectively built from the rescaled fields $\hat{A}_\mu$ and $\hat\psi, \bar{\hat\psi}$. The leading order in N of the various correlators are summarized in the following table. The operator $\sqrt{N}\hat{M}$ is the operator that creates a meson

Table 19.1. *The N counting for glueball and*
meson correlators

Correlator	N counting
$<\hat{G}_1 \hat{G}_2>_c$	N^0
$<\hat{G}_1 \dots \hat{G}_{n_g}>_c$	N^{2-n_g}
$<\hat{M}_1 \hat{M}_2>_c$	N^{-1}
$<\sqrt{N}\hat{M}_1 \sqrt{N}\hat{M}_2>_c$	N^0
$<\sqrt{N}\hat{M}_1 \sqrt{N}\hat{M}_{n_h}>_c$	$N^{1-\frac{n_h}{2}}$
$<\hat{G}_1 \dots \hat{G}_{n_g} \sqrt{N}\hat{M}_1 \sqrt{N}\hat{M}_{n_h}>_c$	$N^{1-n_g-\frac{n_h}{2}}$

with a unit amplitude. In particular we read from the table that the glueball meson interaction is of order $\frac{1}{\sqrt{N}}$.

19.2 Meson phenomenology

The picture that emerges from N counting is that of mesons and glueballs interacting weakly with a coupling of $\frac{1}{\sqrt{N}}$. At the tree level the singularities are poles. At one loop, namely at order $\frac{1}{N}$ the singularities are two particle cuts, at two loops three-particle cuts and so on. We now describe certain phenomena of meson physics that are accounted for by $\frac{1}{N}$ arguments and quite often cannot be explained in any other way.

- The spectrum at low energies of QCD in the large N limit include infinitely many narrow glueball and meson resonances. The fact that the number of resonances is infinite follows from the need to reproduce the logarithmic running of QCD correlation functions. A meson two-point function can be written as a sum of resonances,

$$\int d^4 x e^{iqx} <M(x)M(0)>_c = \sum \frac{Z_i}{q^2 - m_i^2}, \tag{19.11}$$

since single meson exchange dominates in the large N limit. The logarithmic dependence on q^2 of the left-hand side can be recast only provided that the sum on the right-hand side includes infinitely many terms. The resonances are narrow since their decay width goes to zero in the large N limit. This follows from the fact that the phase space factor is N independent and the coupling constant behaves like $\frac{1}{\sqrt{N}}$.

- In Chapter 17 we encountered the pion decay constant f_π. Let us check how it scales with N. Recall its definition $<0|\bar{\psi}\gamma_5 T^A \psi|\pi^b(p)> = i f_\pi p^\mu \delta^{ab}$. The corresponding gauge invariant correlator is $<N\hat{M}_1 \sqrt{N}\hat{M}_2>$, where the first operator $N\hat{M}_1$ corresponds to the axial current and the second $\sqrt{N}\hat{M}_2$ to the

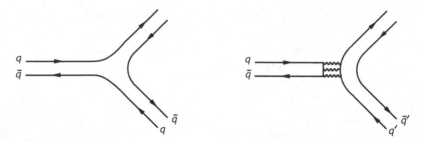

Fig. 19.7. Zweig's rule for the decay of a meson into two mesons.

pion produced from the vacuum with a unit amplitude. This correlator scales like $\sqrt{N}$ and hence,

$$f_\pi \sim \sqrt{N}. \tag{19.12}$$

- The suppression of exotic states of the form $q\bar{q}q\bar{q}$ and the fact that the meson is almost a pure $q\bar{q}$ state with little impact of the $q\bar{q}$ sea are straightforward consequences of large N. Since a quark loop, as we have seen above, is suppressed by a factor of $\frac{1}{N}$ the $q\bar{q}$ sea is irrelevant. Since in the leading order the mesons are non interacting in large N, there is no interaction that will bind two mesons into a $q\bar{q}q\bar{q}$ exotic state.
- Consider the two diagrams of Fig. 19.7.
  Using the counting rules it is obvious that the right-hand diagram is $\frac{1}{N}$ suppressed in comparison to the one on the left. Correspondingly the meson will preferably decay into two mesons of the left-hand side of the figure, what is referred to as Zweig rule conserving decay, and not to the two mesons on the right which is a Zweig rule suppressed decay. In this sense large N predicts the Zweig rule. The same mechanism is in charge of the fact that there is almost flavor singlet and octet degeneracy. In the leading order in large N the whole nonet is degenerate since the diagrams that split singlets from octets involve a $q\bar{q}$ annihilation which is order of $\frac{1}{N}$. In the large N for instance the vector mesons (ρ, w, ϕ, K^*) are degenerate.
- It is known that meson decay proceeds mainly via decay into two body states and not into states of more mesons. Large N tells us that the decay into two mesons behaves as $\frac{1}{\sqrt{N}}$, whereas a decay into four mesons is of order $\frac{1}{N^{3/2}}$.
  This can also be compared to the decay of a meson via creation of a quark anti-quark pair in the mesonic flux tube [60].
- The N counting rules tell us also that meson scattering amplitudes are given by an infinite sum of tree diagram of exchange of physical mesons. This fits nicely the so-called Regge phenomenology, where strong interactions are interpreted as an infinite sum of tree diagrams with hadron exchange.
- Another very important phenomenon is related to the axial $U(1)$, the theta term and the mass of the η'. This will be described in detail in Section 22.5, but here we present the picture in the large N.

19.2.1 Axial U(1) and the mass of the η′

Consider the full action of four-dimensional YM theory, which includes also the θ term,

$$S_{YM} = \int \mathrm{d}^4 x \left[-\frac{1}{4}\mathrm{Tr}[F_{\mu\nu}F^{\mu\nu}] + \frac{\theta g^2}{16\pi^2 N}\mathrm{Tr}[F_{\mu\nu}\tilde{F}^{\mu\nu}] \right]. \tag{19.13}$$

The normalization of the θ term is such that θ is an angular variable, namely, under the shift of $\theta \to \theta + 2\pi$ the action is shifted by $2\pi n$ where n is some integer, so that e^{iS} is unchanged. This result follows from the quantization of the θ term,

$$\int \mathrm{d}^4 x \frac{g^2}{16\pi^2 N}\mathrm{Tr}[F_{\mu\nu}\tilde{F}^{\mu\nu}] = \text{integer}. \tag{19.14}$$

We would like to explore the dependence on θ of the vacuum energy in the pure YM theory and in the theory with massless quarks. In particular we would like to determine $\frac{\mathrm{d}^2 E}{\mathrm{d}\theta^2}|_{\theta=0}$.

Using the path integral formulation we find that,

$$\frac{\mathrm{d}^2 E}{\mathrm{d}\theta^2}|_{\theta=0} = \frac{1}{N^2}\left(\frac{g^2}{16\pi^2}\right)^2 \int \mathrm{d}^4 x <T(\mathrm{Tr}[F_{\mu\nu}\tilde{F}^{\mu\nu}(x)]\mathrm{Tr}[[F_{\mu\nu}\tilde{F}^{\mu\nu}])(0)> . \tag{19.15}$$

Let us introduce an IR cutoff and take,

$$\frac{\mathrm{d}^2 E}{\mathrm{d}\theta^2}|_{\theta=0} = \left(\frac{g^2}{16\pi^2 N}\right)^2 \lim_{k\to 0} U(k)$$

$$U(k) = \int \mathrm{d}^4 x e^{(\nu k x)} <T(\mathrm{Tr}[F_{\mu\nu}\tilde{F}^{\mu\nu}](x)\mathrm{Tr}[[F_{\mu\nu}\tilde{F}^{\mu\nu}(0)])> . \tag{19.16}$$

It is easy to check that in perturbation theory $U(k)$ is of order N^2 due to the contribution of the N^2 degrees of freedom of the gluons. However, perturbatively, $\lim_{k\to 0} U(k) = 0$ since $F\tilde{F}$ is a total derivative. One concludes that perturbatively the vacuum energy is θ independent. To better understand (19.16) we rewrite $U(k)$ in terms of a sum over intermediate single particle states,

$$U(k) = \sum_{gb} \frac{N^2 (a_{gb})_n^2}{k^2 - (m_{gb})_n^2} + \sum_{mes} \frac{N(a_{mes})_n^2}{k^2 - (m_{mes})_n^2}, \tag{19.17}$$

where gb stands for glueball and mes for meson. $N a_{gb}$ and $\sqrt{N}a_{mes}$ are the amplitudes for $\mathrm{Tr}[F\tilde{F}]$ to create a glueball and meson state, respectively. This result again follows from the N counting rules,

$$<0|\mathrm{Tr}[F\tilde{F}]|mes>\sim \sqrt{N} \qquad <0|\mathrm{Tr}[F\tilde{F}]|gb>\sim N. \tag{19.18}$$

The fact that only single states and not multi-states are taken in the intermediate states is since the latter are suppressed in the large N. In the pure YM without quarks the first term vanishes and hence $U(0) \sim N^2$ and $\frac{\mathrm{d}^2 E}{\mathrm{d}\theta^2}|_{\theta=0} \sim 1$. In the presence of massless quarks we know that there could not be any θ dependence and thus we should be able to show that the first term is cancelled out. However, it seems that there is no way that the second term can cancel the first.

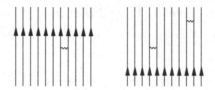

Fig. 19.8. Perturbative correction to the free propagator due to an exchange of one gluon (left) and two gluons (right).

In fact it is possible if there is one meson state with mass $m_{\text{mes}} \sim \frac{1}{\sqrt{N}}$ and if the two terms have opposite sign. This obviously can cancel the $k = 0$ term in $U(k)$ and does not cancel for non-trivial k, but this is exactly what enters (19.16). The opposite sign follows from the fact that an additional equal time commutator term has to be added to (19.15) (see the appendix of [221]). Assuming such a state with mass $m_{\text{mes}} \sim \frac{1}{\sqrt{N}}$ the form of $U(0)$ is,

$$U(0) = N \frac{a_{\eta'}^2}{M_{\eta'}^2}. \tag{19.19}$$

Using the axial anomaly equation which will be further discussed in (22.5),

$$\partial_\mu J_5^\mu = N_f \frac{g^2}{4\pi^2 N} \text{Tr}[F\tilde{F}], \tag{19.20}$$

we get,

$$\frac{g^2}{8\pi^2} <0|\text{Tr}[F\tilde{F}]|\eta'> = \frac{N}{2N_f} <0|\partial_\mu J_5^\mu|\eta'> = \frac{N}{2N_f} f_{\eta'} M_{\eta'}^2, \tag{19.21}$$

From this relation we get the Veneziano–Witten formula or the mass of the η',

$$M_{\eta'}^2 = \left(\frac{2N_{(f)}}{f_{(\pi)}}\right)^2 \frac{d^2 E}{d\theta^2}\Big|_{\theta=0}. \tag{19.22}$$

The picture that emerges from this discussion is that the η' is a Goldstone boson in the large N limit. It has a mass of the order of $M_{\eta'} \sim \frac{1}{\sqrt{N}}$. The dependence on η' of nonzero amplitudes can be obtained from the dependence on θ in the theory without quarks by the following replacement,

$$\theta \rightarrow \theta + \left(\frac{2N_{(f)}}{f_{(\eta')}}\right)\eta'. \tag{19.23}$$

Note that $f_{(\eta')} = f_{(\pi)}$ to leading order in $\frac{1}{N}$.

19.3 Baryons in the large N expansion

Whereas we have seen that the large N expansion is very useful in discussing mesons, it may seem that it is not the case for baryons. Baryonic diagrams depend on N both via the combinatorial factors associated with the diagrams, as well as the fact that the diagrams themselves include N quarks.

The problem is clearly demonstrated when computing the perturbative correction to the free propagator of an N quarks state. The correction occurs due to an exchange of a gluon between two quarks (see Fig. 19.8). The gluon exchange diagram scales as $\frac{1}{N}$. However since there are $\frac{1}{2}N(N-1)$ possible pairs the net effect is of order N. In a similar manner the exchange of two gluons is of order $(\frac{1}{\sqrt{N}})^4 N^4 \sim N^2$, where the first factor comes from the four vertices and the second from the number of ways to choose the four quarks. Higher-order exchange diagrams will have higher order divergence in large N. We will now show, that in spite of this fatal obstacle, there is a large N approximation to the problem of the baryons. The idea is to divide the problem into two parts, in the first one uses diagrammatic methods to study the problem of n quark interaction, and then the effect of these forces on an N quark state. Let us first apply this approach for determining the dependence of the mass of the baryon on N in the quark model. Assuming that the mass gets contributions from the quark masses, quark kinetic energy and quark–quark potential energy, the mass of the baryon reads,

$$M_{\rm B} = N \left[m_{\rm q} + T_{\rm q} + \frac{1}{2}V_{\rm q} \right],$$
(19.24)

where $m_{\rm q}$ is the quark mass, $T_{\rm q}$ is the kinetic energy of the quark and $V_{\rm q}$ is the quark–quark potential energy. Thus we observe that the mass of the baryon scales as N. This result will be shown to hold even beyond the quark model. Again we have made use of the fact that the potential energy is combined from the N^2 combinatorial factor and the $\frac{1}{N}$ factor that comes from the vertices, or gluon propagator.

Leaving aside the quark model, we want to address first the baryonic system made out of very heavy quarks.

19.3.1 The Hartree approximation

In the baryons, the quarks are anti-symmetric in color. Thus they are symmetric in flavor, space and spin combined. Hence they act like bosons. A natural framework to analyze such a system of bosonic charged particles that are subjected to a central potential is the Hartree approximation, in which each particle moves independently of the others in a potential which is determined self consistently by the motion of all the other particles. The justification of the use of this approximation is the large N limit which renders the interactions to be weak. Therefore neglecting the fact that the particle trajectory affects the state of all the other particles and hence the potential that it feels, is justified. Also it is obvious that taking the potential created by all particles and not the one created by all the particles apart from the one we consider, is a $\frac{1}{N}$ effect. Since, as mentioned above, the particles in the non-color degrees of freedom are bosons it implies that in the ground state of the baryon all the particles will sit in the ground state of the Hartree potential.

Let us take the Hamiltonian of the system to be,

$$H = \frac{1}{2m} \sum_a |\vec{p}_a|^2 + \frac{1}{2N} \sum_{a \neq b} V^2(\vec{r}_a, \vec{r}_b) + \frac{1}{6N^2} \sum_{a \neq b \neq c} V^3(\vec{r}_a, \vec{r}_b, \vec{r}_c) + \cdots,$$
(19.25)

where we have suppressed the flavor and spin degrees of freedom, V^n stands for the n body interaction and is independent of N and its strength is of order N^{1-n}, as explained above. In fact since the number of clusters of n quarks is of order N^n, each term in the Hamiltonian is proportional to N.

Next one takes for the ground state wave function a product of the wave functions of each of the particles, namely,

$$\psi(\vec{r}_a, \dots \vec{r}_N) = \prod_a \phi(\vec{r}_a) \tag{19.26}$$

where the particle wave functions are determined by a variational method. The expectation value of the Hamiltonian,

$$<\psi|H|\psi> = N \left[\frac{1}{2m} \int d^3\vec{r}|\nabla\phi|^2 + \frac{1}{2} \int d^3\vec{r}_1 d^3\vec{r}_2 V^2(\vec{r}_1, \vec{r}_1)|\phi(r_1)\phi(r_2)|^2 + \right.$$
$$\left. + \frac{1}{6} \int d^3\vec{r}_1 d^3\vec{r}_2 d^3\vec{r}_3 V^3(\vec{r}_1, \vec{r}_2, \vec{r}_3)|\phi|(r_1)\phi(r_2)\phi(r_3)|^2 + \dots \right], \tag{19.27}$$

has to be minimized with respect to $\phi(r)$ subjected to the constraint that,

$$\int d^3\vec{r}|\phi|^2 = 1. \tag{19.28}$$

The minimization translates to,

$$\left[-\frac{\nabla^2}{2m} + V(\vec{r}) \right]\phi = \epsilon\phi, \tag{19.29}$$

where ϵ is the Lagrange multiplier associated with the constraint, and the Hartree potential is,

$$V = \left[\int d^3\vec{r}_1 V^2(\vec{r}, \vec{r}_1)|\phi(r_1)|^2 + \frac{1}{2} \int d^3\vec{r}_1 d^3\vec{r}_2 V^3(\vec{r}, \vec{r}_1, \vec{r}_2)|\phi(r_1)\phi(r_2)|^2 + \dots \right]. \tag{19.30}$$

We will now treat first the case of heavy quarks and subsequently will address, in a less rigorous manner, the case of light quarks.

19.3.2 Baryons made out of heavy quarks

A non relativistic Schrodinger equation is an adequate framework to deal with very heavy baryons. In this setup for short distances the quark–quark potential is an attractive Coulomb potential so that the Hamiltonian takes the form,

$$H = Nm_{\mathrm{q}} - \sum_{i=1}^{N} \frac{\partial_i^2}{2m_{\mathrm{q}}} - \frac{g^2}{N} \sum_{i<j} \frac{1}{|x_i - x_j|}. \tag{19.31}$$

The system is effectively that of N bosons with a Coulomb interaction with a strength of $\frac{1}{N}$.

In spite of the fact that the interaction potential behaves like $\frac{1}{N}$, we cannot treat this term as a perturbation since each quark interacts with N other quarks and hence the total quark–quark interaction of each quark is of order one. This situation calls for a Hartree approximation where, as explained above, the quark is exposed to an average effective potential. The fluctuations of the effective potential are negligible and hence we can consider a background c-number potential.

For heavy quarks where the potential is taken to be a Coulomb potential we find,

$$<\psi|H - N\epsilon|\psi> = N[M + \tfrac{1}{2m}\int d^3\vec{r}|\nabla\phi|^2 - \tfrac{g^2}{2}\int d^3\vec{r}\int d^3\vec{r}\,'\frac{|\phi(r)|^2|\phi(r')|^2}{|r-r'|}$$
$$-\epsilon\int d^3\vec{r}|\phi(r)|^2]. \tag{19.32}$$

The main point here is that each of the terms is proportional to N and hence the result of the minimization is N independent. The variation with respect to ϕ^* results in the following Schrodinger equation,

$$-\frac{\nabla^2\phi}{2m} - g^2\phi\int d^3\vec{r}\,'\frac{|\phi(r')|^2}{|r-r'|} = \epsilon\phi. \tag{19.33}$$

One can convert this integro-differential equation into a fourth-order differential equation

$$-\frac{1}{2m}\nabla^2\left(\frac{\nabla^2\phi}{\phi}\right) + 4\pi g^2|\phi|^2 = 0. \tag{19.34}$$

For radial solutions, for instance, the ground state of this equation takes the form,

$$-\frac{1}{2m}\left[\frac{d^2}{dr^2} + \frac{2}{r}ddr\right]\left(\frac{1}{\phi}\left[\frac{d^2}{dr^2} + \frac{2}{r}ddr\right]\phi\right) + 4\pi g^2|\phi|^2 = 0, \tag{19.35}$$

which is derived by dividing (19.33) by ϕ and acting with ∇^2.

Even without solving this equation, it is clear that the mass of the baryon is linear with N and that the charge distribution of the baryon which implies in particular its size is N independent.

19.3.3 Baryons made out of light quarks

Up to this point we have used a non relativistic Hartree approximation which is valid only for heavy quarks. However, phenomenologically, one is more interested in baryons made out of light quarks and in scattering processes that involve relativistic quarks. We will show now that even for the light quark baryons, a Hartree-like approximation, namely that each quark moves independently of the others in a potential which is determined self consistently by the motion of all the other particles, is still justified. Moreover, it will be argued that just as for the

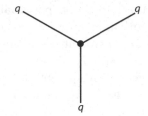

Fig. 19.9. Baryon stringy configuration for $N = 3$.

heavy quark baryons, also for the baryons made out of light quarks, the mass is linear with N, whereas the size and the shape of the baryon are N independent.

A major difference between the case of heavy quarks versus that of light ones is that in the latter case one has to introduce on top of the two-body interaction also a three-body interaction and, in general, n-body interactions. In addition one has to use a relativistic analog of the Hartree approximation. In two dimensions one can solve the relativistic Hartree approximation. Unfortunately, the four-dimensional analog is not known. Let us first discuss a non relativistic Hartree approximation with n-body interactions and then argue about the relativistic analog. The Hamiltonian for the case with any n-body interaction takes the form,

$$H = \frac{1}{2m} \sum_a |\vec{p}_a|^2 + \frac{1}{2N} \sum_{a \neq b} V^2(\vec{r}_a, \vec{r}_b) + \frac{1}{6N^2} \sum_{a \neq b \neq c} V^3(\vec{r}_a, \vec{r}_b, \vec{r}_c) + \cdots,$$

(19.36)

where we have suppressed the flavor and spin degrees of freedom, V^n stands for the n-body interaction and is independent of N. The strength of V^n is of order N^{1-n} since breaking the n quark line costs a factor of N^{-n} and since the baryon is in a totally antisymmetric representation, each quark line carries a different color index.

We now substitute this Hamiltonian into $<\psi|H|\psi>$ and use a variational method as above. Since for each V^n term there are N^n ways to choose a set of n quarks, here again the expectation value of the Hamiltonian is linear in N.

Next we have to introduce a four-dimensional relativistic Hartree approximation. In two dimensions in the large N limit the Hartree approximation is exact. The generalization to four dimensions, however, is not known and hence one can make only the qualitative statement that even in this case the mass is linear in N and the size and shape are independent of N.

The Hartree approximation of light quarks moving in an effective potential can be also related to a string model of the baryon. In this model, the N quarks are attached to a common junction as can be seen in Fig. 19.9 for the case of $N = 3$.[1] In the large N approximation the junction can be regarded as a heavy object and its motion can be ignored. The interaction of the quarks with the fixed junction can be thought of as an interaction with an effective Hartree potential.

---

[1] The modern picture of the latter is that of a wrapped D brane.

19.3.4 Baryonic excited states

Low-lying excitations of the baryons are described by wave functions where out of N single particle states, a number $n_k \ll N$ of states are placed in the kth excited state of the Hamiltonian. The corresponding mass of the excited baryon is $M = M_0 + \sum_{n_k} n_k \epsilon_k$ where ϵ_k is the energy of the kth excited state.

Highly excited states have a finite fraction of single particle excited states. Denote by p this fraction, namely there are $(1-p)N$ particles in the ground state and pN ones in excited states which we take to be ϕ_1 so that the baryon wave function is,

$$\psi(\vec{r}_a, \dots \vec{r}_N) = \sum (-1)^P \prod_a^{pN} \phi_1(\vec{r}_a) \prod_a^{(1-p)N} \phi_0(\vec{r}_a). \tag{19.37}$$

Inserting this ansatz into the expectation value of the Hamiltonian one gets a set of two coupled nonlinear equations for ϕ_0 and ϕ_1. This structure can obviously be generalized to states with higher single particle excited states.

Another approach to studying excited states is to apply a time-dependent Hartree approximation. It is easy to check that starting with the Hartree ansatz for the wave function but now with single particle wave functions that are also time dependent, one finds instead of (19.33) the following time-dependent Schrodinger equation,

$$-\frac{\nabla^2 \phi}{2m} - g^2 \phi \int d^3 \vec{r}' \frac{|\phi(r')|^2}{|r - r'|} = i\partial_t \phi(\vec{r}, t). \tag{19.38}$$

This equation is solved by $\phi(t, \vec{r}) = e^{-i\epsilon t} \phi(\vec{r})$ where $\phi(\vec{r})$ is a solution of the time-independent equation. By Galilean boosting along, for instance the x direction, a static baryon solution, we find the solution,

$$\phi(\vec{r}, t) = \phi(x - vt) e^{i(Mvx - \epsilon t - \frac{1}{2}Mv^2 t)}, \tag{19.39}$$

which is a baryon travelling with a constant velocity. This is an additional solution to the time-dependent equation. In fact starting with any function $\phi(\vec{r}, 0)$ and substituting it into (19.38) a new solution will be generated. These solutions will generically be excited states, but not in energy eigenstates, since they will not have a harmonic form.

To generate excited baryon solutions which are in eigenstates of the energy, we make use of the DHN procedure discussed in Section 5.5.1 in the context of two-dimensional field theories. The idea is to look for solutions which are periodic in time and to quantize them by requiring that the action during a period will obey,

$$\int_0^T dt <\psi|H - i\partial_t|\psi> = 2\pi n, \tag{19.40}$$

where n is some integer number. Recall that this condition follows from the fact that the solutions are invariant under time translations, so from $\psi(t)$ we can also generate a solution $\psi(t - t_0)$ for any t_0 and also any linear combinations of them, and in particular a harmonic varying solution $\int_0^T dt_0 e^{-it_0 E} \psi(t - t_0)$.

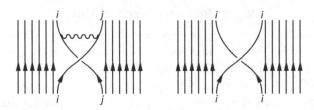

Fig. 19.10. Baryon–baryon scattering. Exchange of a quark on the right, while on the left, such an exchange plus an exchange of a gluon.

In analogy to the discussion in Section 5.5.1 here as well one can introduce a non-abelian flavor group, namely construct baryons made out of several flavors. In this case one introduces into the Hartree wave function a separate single particle wave function for each flavor.

For very heavy quarks one can neglect the spin-dependent forces, and hence anticipate that the baryons are spherically symmetric. However, for less heavy quarks this is no longer the case. For a baryon made out of a single flavor, in the ground state all the spins are aligned and hence the total spin is $\frac{1}{2}N$. Due to the fact that the total spin is very large, the effect of the coupling of this large spin to the orbit is significant and hence the ground state will no longer be spherically symmetric. If one takes the large N analog of the baryon to be composed of $\frac{N+1}{2}$ quarks of one flavor and $\frac{N+1}{2} - 1$ of the other flavor, then the net spin will be $\frac{1}{2}$ since the spin–spin interaction will align the spin of the different flavors in an antiparallel way. Unlike the one flavor case where the spin is $\frac{N}{2}$, the spin $\frac{1}{2}$ will be too small to affect the spherically symmetric ground state via spin orbit interaction, and hence for that case it will remain symmetric.

19.4 Scattering processes

In Section 19.2 it was shown that in the leading order of the large N there is no meson–meson scattering. The same applies also for meson–glueball and glueball–glueball scattering. Let us now address the question of baryon–baryon and baryon–meson scattering.

Baryon–baryon scattering is dominated by an interchange of one quark between two baryons. Whether the process involves only an interchange or also in addition an exchange of a gluon, as is shown in Fig. 19.10, the amplitude is of order N. In the case of no gluon exchange, there is a choice of the interchanging quark in one of the baryon, which goes like N. Once a quark in one baryon is chosen it can be interchanged only with a quark in the second baryon that carries exactly the same color index, hence there is no additional N dependence. Thus altogether the amplitude is order N. Note also that the diagram (19.10) comes with a factor of (-1). The amplitude for an interchange that is accompanied with an exchange of a gluon is also of order N, which follows from the fact that there is a factor of N from choosing the quark in the first baryon,

Fig. 19.11. Annihilation of a quark coming from the baryon and anti-quark coming from the anti-baryon.

another factor of N from the other baryon and a factor of $\frac{1}{N}$ from the quark gluon vertices.

This fact that the amplitude is order N is behind why there is a smooth large N limit to the baryon–baryon scattering. Recall that the mass of the baryon and hence also the non-relativistic kinetic energy of the baryon are linear in N. Thus the total Hamiltonian can be written as $H = N\hat{H}$ where $\hat{H}$ is N independent. The eigenvectors of $\hat{H}$ and hence of the scattering process are N independent.

Quantitatively one addresses the question of baryon–baryon scattering using a non relativistic time-dependent Schrodinger equation for a system of $2N$ quarks. Due to the exclusion principle the total wave function should be a product of two orthonormal space and spin wave functions $\phi_i(x,t)$ where $i = 1,2$ in the following way,

$$\psi(x_1,\ldots,x_{2N},t) = \sum_P (-1)^P \prod_{i=1}^N \phi_1(x_i,t) \prod_{j=1}^N \phi_2(x_j,t). \qquad (19.41)$$

Using again the time-dependent variational principle, we find that in the case where all the spins of the quarks are parallel so we can ignore them,

$$i\partial_t \phi_1(x,t) = \frac{\nabla^2}{2M} \phi_1(x,t) - g^2 \phi_1(x,t) \int \frac{dy \phi_1^*\phi_1(y,t)}{|x-y|}$$
$$-g^2 \phi_1(x,t) \int \frac{dy \phi_2^*\phi_1(y,t)}{|x-y|}, \qquad (19.42)$$

and another equation where $\phi_1 \leftrightarrow \phi_2$. Apart from the last term this equation is identical to the one describing a single baryon (19.38), hence the last term obviously describes the interaction between the two baryons. To describe baryon–baryon scattering we start with inital conditions where the wave functions ϕ are localized at two far away regions of space, but heading for a collision. When the two wave functions overlap the interaction term is important and determines the scattering via (19.42).

The baryon anti-baryon scattering is dominated by an annihilation of a quark coming from the baryon and an anti-quark from the anti-baryon. The amplitude of this process is of order N since choosing one quark is order N, choosing an anti-quark is order N and the coupling is order $\frac{1}{N}$ (see Fig. 19.11). Again this is like the scaling of the kinetic term and hence there is a smooth limit. The

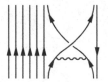

Fig. 19.12. Exchange of a quark and a gluon in meson–baryon scattering.

variational procedure now involves a wave function composed of N quark and N anti-quark wave functions, namely,

$$\psi(\vec{r}_a, \ldots, \vec{r}_N) = \prod_a \phi(\vec{r}_a) \prod_b \bar{\phi}(\vec{r}_b), \tag{19.43}$$

where $\bar{\phi}(\vec{r}_b)$ is the wave function of a single anti-quark. The minimization now yields a pair of coupled equations for ϕ and $\bar{\phi}$.

The meson–baryon scattering is described in a diagram like Fig. 19.12. The diagram is of order N^0 since there is a factor of $\frac{1}{N}$ from the coupling and N from the number of ways to choose the quark from the baryon. Recall that the baryon kinetic energy is order N and that of the meson is order one. Hence the interaction term is negligible from the point of view of the baryon and it does not feel the meson but the meson motion is affected by the interaction and thus there is a meson baryon non-trivial scattering. Denoting again the wave function of a quark of the baryon as $\phi(x, t)$ and that of the meson as $\phi_{\mathrm{M}}(x_{\mathrm{M}}, y_{\mathrm{M}}, t)$, the trial many body wave function reads,

$$\psi(\vec{r}_a, \ldots, \vec{r}_N, x_{\mathrm{M}}, y_{\mathrm{M}}, t) = \prod_a \phi(\vec{r}_a, t) \phi_{\mathrm{M}}(x_{\mathrm{M}}, y_{\mathrm{M}}, t). \tag{19.44}$$

Again we substitute this wave function into the variational principle $\int dt <\psi|H - i\partial_t|\psi>$. The solution for ϕ of the corresponding equations is identical to the solution of the baryon and hence indeed the baryon is not affected by the presence of the meson. On the other hand the equation for ϕ_M is affected by the presence of ϕ. The equation will be that of a free meson plus two additional terms describing the interaction that take the form,

$$H_{\mathrm{int}} = -g^2 \phi(x) \left[\int \frac{dz \phi^*(z, t) \phi_{\mathrm{M}}(z, y_{\mathrm{M}}, t)}{|x_{\mathrm{M}} - z|} + \int \frac{dz \phi^*(z, t) \phi_{\mathrm{M}}(z, y_{\mathrm{M}}, t)}{|z - y_{\mathrm{M}}|} \right]. \tag{19.45}$$

Thus the interaction term and hence the whole equation is linear in ϕ_{M}.

20

From 2d bosonized baryons to 4d Skyrmions

20.1 Introduction

Low energy effective actions associated with four-dimensional QCD, and in particular the Skyrme model have been very thoroughly studied with an emphasis on both formal aspects such as anomalies as well as phenomenological ones like the spectrum of baryons. This chapter is devoted to the Skyrme model. We first derive the various terms of the Skyrme action. These include the sigma term, the WZ term, the mass term and the Skyrme term. The first three terms we have encountered already in the two-dimensional analog, the bosonized QCD (Chapter 13) whereas the fourth one, the Skyrme term, shows up as a stabilization term only in the four-dimensional case. We then construct the classical soliton solution, the Skyrmion, of the corresponding equations of motion. Next we determine the classical mass and radius of the baryon. In a similar manner to the procedure taken in the two-dimensional model, we quantize the system semiclassically. This yields mass splitting between the nucleons and the delta particles and the axial coupling of the nucleons. Most of the discussion will be done for $SU(N_f = 2)$ but we will also discuss certain properties of the three-flavor case. The Skyrme model was introduced in [195], [196], [197] and [91].

The topic of baryons as Skyrmions was discussed and reviewed in several books [53], [157] and reviews [22], [186], [231]. In several sections of this chapter we follow the latter review.

20.2 The Skyrme action

In two dimensions using the bosonized version of QCD, we were able to integrate in the strong coupling limit the color degrees of freedom and derive the low energy effective action of the flavor degrees of freedom. The latter took the form of a WZW action for the group $U(N_f)$ of level N_c for massless QCD and modified flavored WZW action with a mass term for massive QCD. The main point there was that the action derived was exact. In four dimensions the situation is quite different. For once we do not have a bosonized version of QCD which enables us (at least in the massless case) to decouple the flavor and color degrees of freedom and then integrate over the latter. However, due to confinement, the low energy degrees of freedom of four-dimensional QCD are also only flavor degrees of freedom and hence it is natural to use an $N_f \times N_f$ group element $g(x^\mu)$ of the flavor group $U(N_f)$. Since we do not have a systematic way to derive the corresponding action, we will now consider various terms that eventually construct the full Skyrme model.

20.2.1 The Sigma term

Recall that the group element $g(x)$ transforms under left and right transformations as follows,

$$g(x) \rightarrow Ug(x) \quad g(x) \rightarrow g(x)U. \tag{20.1}$$

As we have shown in two dimensions the sigma term is the term with the lowest number of derivatives. In two dimensions it takes the form,

$$S_{2d} = \frac{1}{12\pi} \int d^2x \, \text{Tr} \left[\partial_\mu g \partial^\mu g^{-1} \right]. \tag{20.2}$$

It is easy to see that the analog in four dimensions has the form,

$$S = \frac{1}{16} f_\pi^2 \int d^4x \, \text{Tr} \left[\partial_\mu g \partial^\mu g^{-1} \right], \tag{20.3}$$

where f_π has dimensions of mass (it will be shown below that by comparison to experimental data, for $N_f = 3$, it has to be taken to be $\sim 93MeV$). This coefficient is needed since our group element still does not carry classically any conformal dimension. The sigma term which is clearly the one with the lowest number of derivatives can also be expressed as,

$$S = \frac{1}{16} f_\pi^2 \int d^4x \, \text{Tr} \left[L_\mu L^\mu \right] = \frac{1}{16} f_\pi^2 \int d^4x \, \text{Tr} \left[R_\mu R^\mu \right], \tag{20.4}$$

where,

$$L_\mu = g^{-1} \partial_\mu g \quad R_\mu = g \partial_\mu g^{-1}. \tag{20.5}$$

It is important to note that by construction the L_μ obey the so-called Maurer–Cartan equation,

$$\partial_\mu L_\nu - \partial_\nu L_\mu + [L_\mu, L_\nu] = 0. \tag{20.6}$$

and so does R_μ. Note that they are both like pure gauges in a non-abelian gauge theory, and thus they have $F_{\mu\nu} = 0$.

Before proceeding to the WZ term let us check the symmetries of this action in comparison with the known symmetries of QCD. It is easy to check that it is invariant under global $SU(N_f)_L \times SU(N_f)_R \times U(1)_B$ transformations. It is further invariant under the following three discrete transformations,

$$\begin{aligned} Transpose: \quad & g \leftrightarrow g^T & \vec{x} \leftrightarrow \vec{x}, \, t \leftrightarrow t \\ P_0: \quad & g \leftrightarrow g & \vec{x} \leftrightarrow -\vec{x}, \, t \leftrightarrow t \\ (-1)^{N_B}: \quad & g \leftrightarrow g^{-1} & \vec{x} \leftrightarrow \vec{x}, \, t \leftrightarrow t. \end{aligned} \tag{20.7}$$

The second transformation P_0 is a parity transformation and the third is the number of bosons modulo two. To check whether these discrete symmetries are also shared by QCD we first expand $g(x)$ around unity,

$$g(x) = 1 + \frac{2i}{f_\pi} \sum_1^{N_f^2-1} T^a \pi^a(x), \tag{20.8}$$

in terms of the Goldstone bosons $\pi(x)$ and then consider $N_f = 3$. It turns out that P_0 and $(-1)^{N_B}$ are not conserved separately but only the combination $P = P_0(-1)^{N_B}$. This is demonstrated by the process $K^+K^- \to \pi^+\pi^0\pi^-$. Obviously the number of bosons modulo two is not conserved as well as the parity P_0 since the π^i are pseudo scalars. It is thus clear that the action (20.3) cannot describe the effective action of QCD and another term that does conserve P and not P_0 and $(-1)^{N_B}$ separately. It is well known that the parity transformation P_0 is violated by a term which is proportional to the Levi Civita tensor which in four dimensions reads $\epsilon_{\mu\nu\rho\sigma}$. However, it is very easy to verify that the only term proportional to $\epsilon_{\mu\nu\rho\sigma}$, namely $\epsilon^{\mu\nu\rho\sigma} \operatorname{Tr}[g^{-1}\partial_\mu g g g^{-1}\partial_\nu g g^{-1}\partial_\rho g g^{-1}\partial_\sigma g]$ vanishes due to the antisymmetric nature of $\epsilon_{\mu\nu\rho\sigma}$ and the cyclicity of the trace.

20.2.2   The WZ term

Experienced with the two-dimensional WZ term it is clear that this situation naturally calls for a four-dimensional WZ term: [225], [226]. Recall that the two-dimensional WZ term was written as a three-dimensional integral over a three-dimensional ball or a three disk whose boundary is the two-dimensional space-time. In a similar manner we can construct a term defined on a five-dimensional disk $D5$ whose boundary is the four-dimensional space-time and has the form,

$$S_{WZ} = -i\frac{\lambda_S}{240\pi^2} \int_{D^5} d^5x \, \epsilon^{ijklm} \operatorname{Tr}[g^{-1}\partial_i g g^{-1}\partial_j g g^{-1}\partial_k g g^{-1}\partial_l g g^{-1}\partial_m g], \tag{20.9}$$

where now i, j, k, l, m denote coordinates on $D5$ and λ_S is a coefficient that has to be determined. Extending the map $g(x^\mu)$ from the four-dimensional space-time to the $SU(N)$ group manifold into a map from $D5$ to the group manifold is based on the fact that $\pi_4(SU(N)) = 0$ and $\pi_1(SU(N)) = 0$. Now let us check whether there are any constraints on λ_S. The analogous two-dimensional case tells us that $\lambda_S = N_c$. We will now verify this result in two steps. First we show that on general topological grounds it has to be an integer and then by relating the action to QCD we show that this integer has to be equal to the number of colors. To understand the topological nature of the WZ term it is convenient to use a compactified Euclidean four space of a topology $S^1 \times S^3$ where the S^1 factor corresponds to a compactified time direction. Now the five-dimensional disk D^5 can be taken now to be $S^3 \times D^2$. However as is clear from Fig. 20.1 there are in fact two options of choosing the disk D^2 namely D_n^2 and D_s^2. Requiring that the independence of the physics on choice translates into,

$$e^{iS_W^n z} = e^{iS_W^s z} \to \int_{(D_n^2+D_s^2)\times S^3} w_5^0 = \int_{S^2 \times S^3} w_5^0 = 2\pi \text{ integer}, \tag{20.10}$$

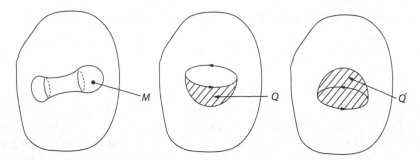

Fig. 20.1. Two options of choosing the disk D^2.

where we have used the fact that the sum of the two disks $(D_n^2 + D_s^2)$ is topologically equivalent to S^2, that the five cycles in the group manifold $SU(N)$ of the topology $S^2 \times S^3$ can be represented by the cycles of topology S^5 and that $\pi_5(SU(N)) = Z$ and hence any S^5 in $SU(N)$ is topologically a multiple of the basic S^5 on which w can be normalized such that $\int_{S_0^5} w_5^0 = 2\pi$.

We thus conclude that the coefficient λ_S has to be an integer. As we mentioned above in two dimensions we have shown that this integer has to be N_c. We will show below when discussing the phenomenology of the Skyrme model that this is the case also in four dimensions. Thus we will take from here on that $\lambda_S = N_c$.

20.2.3 The Skyrme term

Baryons were described in the context of the bosonized theory of two-dimensional QCD in terms of soliton solutions of the WZW theory in flavor space (see Chapter 13). In a similar manner we anticipate that also in four dimensions solitons are intimately related to baryons. However, in Section 5.3 it was shown that according to Derrick's theorem, there are no stable solitons in the space dimension larger that one. To recapitulate this theorem let us analyze the scaling behavior of the energy of a soliton solution in four dimensions. It is easy to realize that in D space dimensions the energy that corresponds to the action (20.3) reads,

$$E = \int d^D x \frac{f_\pi^2}{4} \operatorname{Tr}[L_\mu L^\mu]. \tag{20.11}$$

where a change in normalization was made.

Under a scaling of $x \to \lambda x$, $g(x) \to g(\lambda x)$ since as mentioned above $g(x)$ has a zero scaling dimension, and the energy scales as,

$$E_\lambda = \lambda^{2-D} E. \tag{20.12}$$

Thus for $D = 3$ the energy of the system can shrink to zero by large scaling and hence the solutions are not stable against scale transformations. To avoid this problem we add a term which is quartic in L_μ so that the total Lagrangian is,

$$\mathcal{L} = \mathcal{L}_2 + \mathcal{L}_4 = \frac{f_\pi^2}{4} \operatorname{Tr}[L_\mu L^\mu] + \frac{1}{4} e^2 Tr([L_\mu, L_\nu]^2). \tag{20.13}$$

Under scale transformation the energy scales as,

$$E_\lambda = \lambda^{2-D} E_{(2)} + \lambda^{4-D} E_{(4)}. \tag{20.14}$$

It is easy to see that for $D \geq 3$ there is a minimum at $\lambda = 1$ and with

$$\frac{dE_\lambda}{d\lambda} = 0 \rightarrow \quad \frac{E_{(2)}}{E_{(4)}} = -\frac{4-D}{2-D}$$

$$\frac{d^2 E_\lambda}{d\lambda^2} > 0 \rightarrow \quad 2(D-2)E_{(2)} > 0. \tag{20.15}$$

In fact the energy is bounded from below by the topological charge. The energy associated with the action (20.13) takes the following form for a static configuration,

$$E = \int d^3 x \left\{ -\frac{f_\pi^2}{4} \operatorname{Tr}[L_i L_i] - \frac{1}{4} e^2 \operatorname{Tr}([L_i, L_j]^2) \right\}, \tag{20.16}$$

which can be rewritten as,

$$E = -\frac{f_\pi^2}{4} \int d^3 x \operatorname{Tr}\left[L_i L^i + \frac{e^2}{f_\pi^2} (\sqrt{2} \epsilon_{ijk} L^j L^k)^2 \right] \geq$$

$$-\frac{f_\pi^2}{4} \int d^3 x \left| \operatorname{Tr}\left[\left(\frac{2\sqrt{2}e}{f_\pi} \epsilon_{ijk} L^i L^j L^k \right) \right] \right| = 12\sqrt{2}\pi^2 e f_\pi |B|. \tag{20.17}$$

This is referred to as the Bogomol'ny bound. It is interesting to note that unlike other cases like instantons (see Section 22.1) there is no configuration that saturates the bound. The configuration that does saturates the bound has the form $L_i = \frac{\sqrt{2}e}{f_\pi} \epsilon_{ijk} L_j L_k$, however it is easy to see that it does not obey the Maurer–Cartan equation (20.6).

20.2.4 A mass term

In two dimensions the mass term was shown to be a key ingredient to having soliton solutions of strong coupling of QCD_2. In four dimensions this is not the case. Stable soliton solutions do not require a mass term. However, to incorporate the fact that the pions are not massless, one adds a mass term to the full Lagrangian. The mass term has the same form as that of the two-dimensional theory, namely,

$$S_m = \frac{m_\pi^2 f_\pi^2}{16} \int d^4 x \operatorname{Tr}[g + g^{-1} - 2]. \tag{20.18}$$

Upon substitution for $g(x)$ in the ansatz (20.8) this term takes the form of a mass term for the π fields,

$$S_m = \int d^4 x \frac{1}{2} m_\pi^2 (\pi^a)^2. \tag{20.19}$$

For $N_f > 2$ one can use a mass term that breaks flavor symmetry by assigning different masses to different flavors. One can also generalize the mass term by using general functions of g which in the limit of $g \to 1$ approach g.

20.2.5 Gauging the Skyrme action

In Section 9.3.1 we discussed the gauging of the WZW action. In that case we were interested in gauging the diagonal $SU_D(N_c) \in SU_L(N_c) \times SU_R(N_c)$. We presented there two methods for the gauging procedure: (i) Noether trial and error method, (ii) covariantization of the associated currents. Here in the case of the Skyrme action we would like to gauge the $U(1)$ diagonal global abelian symmetry that corresponds to electromagnetism, as well as the full $SU(N_f) \times SU(N_f)$ global symmetry of the Skyrme action. Let us first identify the diagonal abelian symmetry that we want to gauge to incorporate EM gauge fields. In the particular case of $N_f = 3$ the EM charge matrix of the u, d, s quarks is given by,

$$Q = \begin{pmatrix} 2/3 & & \\ & -1/3 & \\ & & -1/3 \end{pmatrix}. \tag{20.20}$$

Thus a local transformation that corresponds to the EM gauge transformation is,

$$g(x) \to U g(x) U^{-1} \quad U \sim 1 + i\epsilon(x)[Q, g]. \tag{20.21}$$

As usual the local transformation can be a symmetry of the action only provided we add to the action gauge fields that transform under the EM gauge transformation as $A_\mu \to A_\mu - \frac{1}{e}\partial_\mu \epsilon(x)$ where e is a unit EM charge. For the sigma term and the Skyrme term it is obvious that gauge invariance is achieved by replacing the ordinary derivative with covariant ones, namely $\partial_\mu \to D_\mu = \partial_\mu + ie\partial_\mu$. The gauging of the WZ term $\Gamma \equiv S_{WZ}$ is more subtle and as was done for the two-dimensional case; we use a trial and error method. First we compute the variation of the term under the global $U(1)$ symmetry. We find that,

$$\Gamma \to \Gamma - \int d^4x \, \partial_\mu \epsilon(x) J^\mu$$

$$J^\mu = \frac{1}{48\pi^2} \epsilon^{\mu\nu\rho\sigma} \{ \mathrm{Tr}\left[-Q(R_\nu R_\rho R_\sigma) \right]$$
$$+ \mathrm{Tr}\left[Q(L_\nu L_\rho L_\sigma) \right] \} \tag{20.22}$$

where Q is defined in (20.20). The next step in gauging the WZ term is to replace the original term with,

$$\Gamma \to \Gamma - e \int d^4x \, A_\mu J^\mu. \tag{20.23}$$

It turns out that the action after this replacement is still not gauge invariant but it is invariant with the following addition,

$$S = \frac{f_\pi^2}{16} \int d^4x \, \text{Tr} \left[D_\mu g D^\mu g^{-1} \right] + N_c \tilde{\Gamma},$$

$$\tilde{\Gamma}(g, A_\mu) = \Gamma(g) - e \int d^4x A_\mu J^\mu + \frac{ie}{24\pi^2} \int d^4x \epsilon^{\mu\nu\rho\sigma} \partial_\mu A_\nu A_\rho$$

$$\times \text{Tr} \left[Q^2 (L_\sigma - R_\sigma) - QgQg^{-1} R_\sigma \right]. \tag{20.24}$$

In a similar manner one can gauge the full global symmetry or its subgroups. Since we will not need it in this chapter we refer the reader to references, for instance [231].

20.3 The baryon as a Skyrmion

The two-dimensional solitonic baryon was analyzed at two levels, firstly the classical configuration and then at the semi-classical level. At both levels properties of the baryon such as its mass and conserved charges were computed. In this section we present the analogous calculations for the 4d Skyrmion and then we compare the four- and two-dimensional results.[1]

20.3.1 The classical Skyrmion

The Skyrmion is by definition a solitonic solution of the Skyrme action. Soliton solutions in two dimensions were discussed in general in Section 5.3 and in particular the solitons of the low energy effective action of QCD_2 in the strong coupling limit in Chapter 13. In fact the classical solitonic baryons were solutions of a sine-Gordon equation that was derived from an action that included a sigma term and a mass term since for static configuration the WZ vanishes. The latter property holds also in four dimensions so the relevant action now includes the sigma term and the Skyrme term. In fact we will describe here the case of two flavors and as mentioned above the WZ term vanishes for the $SU(2)$ group manifold anyhow. The equation of motion derived by computing the variation of the action with respect to $g^{-1}\delta g$ is,

$$\partial^\mu L_\mu - 2\frac{e^2}{f_\pi^2} \partial^\mu \left[L_\nu, [L_\mu, L_\nu] \right] = 0. \tag{20.25}$$

Obviously, an equivalent equation can be written by replacing $L_\mu \to R_\mu$. Parameterizing the general static configuration as,

$$g(x) = e^{i\vec{\tau}\cdot\hat{r}F(r)} = \cos(F(r)) + i\vec{\tau}\cdot\hat{r}\sin(F(r)). \tag{20.26}$$

---

[1] The classical properties of the $SU(2)$ baryonic Skyrmion were analyzed in [133], [5] and afterwards in many other papers.

For this ansatz the equation of motion reads,

$$F'' + \frac{2}{r}F' - \frac{\sin(2F)}{r^2} + 8\frac{e^2}{f_\pi^2}\left[\frac{\sin(2F)\sin^2 F}{r^4} - \frac{F'^2\sin(2F)}{r^2}\frac{2F''\sin^2 F}{r^2}\right] = 0.$$
(20.27)

The boundary conditions are taken to be,

$$F(r=0) = \pi \quad F(r \to \infty) = 0.$$
(20.28)

The mass of the classical Skyrmion is derived by substituting a solution of the equation of motion into (20.16) getting,

$$M_s = 4\pi \int_0^\infty r^2\,\mathrm{d}r\,\frac{f_\pi^2}{2}\left[F'^2 + \frac{2\sin^2 F}{r^2}\right] + 4e^2\frac{\sin^2 F}{r^2}\left[2F'^2 + \frac{\sin^2 F}{r^2}\right]. \quad (20.29)$$

Using the virial property this reduces to,

$$M_s = 4\pi\sqrt{2}e\int_0^\infty x^2\,\mathrm{d}x\left[\left(\frac{\mathrm{d}F}{\mathrm{d}x}\right)^2 + \frac{2\sin^2 F}{r^2}\right],$$
(20.30)

where $x = \frac{f_\pi}{e\sqrt{2}}r$ is dimensionless. The value of the integral is ~ 11.7. One can either use the mass of the proton combined with the mass of the delta, to determine f_π and the coefficient of the Skyrme term, or use the experimental values of f_π and the axial coupling to be discussed shortly.

Let us now analyze the radial profile of the soliton $F(r)$. Asymptotically for $r \to \infty$ only the terms inside the square brackets can be neglected leading to a solution of the form $F(r) \to \frac{16e^2}{f_\pi^2}\frac{A^2}{r^2}$, where again A can be determined by comparing to experimental data and is found to be $A \sim 1.08$. On the other limit around the origin it is easy to see that the equation is solved by $F(r) \sim n\pi - ar$. The numerical solution of $F(r)$ that interpolates between these two boundary conditions is drawn in Fig. 20.2.

In addition to the mass, we have also extracted in two dimensions from the classical soliton the flavor properties and baryon number. For the Skyrmion we should also be able to determine these properties as well as its spin. When we insert the classical soliton solution in the baryon density we get,

$$B^0 = \frac{i}{24\pi^2}\epsilon^{ijk}L_iL_jL_k = \frac{1}{2\pi^2}\sin^2 F\frac{F'}{r^2},$$
(20.31)

so that the baryonic charge is,

$$B = 4\pi\int_0^\infty \mathrm{d}r\,r^2\,B^0(r) = \frac{1}{\pi}(F(0) - F(\infty)) + 12\pi[\sin(2F(\infty)) - \sin(2F(0))] = 1.$$
(20.32)

where we have used the boundary conditions of $F(r)$ specified above. Using the distribution of the baryonic charge, one can define the rms radius of the baryon as follows,

$$r_{\mathrm{rms}} = \frac{e}{\pi f_\pi}\left(-\int_0^\infty \mathrm{d}x x^2\sin^2 FF'\right)^{1/2},$$
(20.33)

which is of the order of 0.48 for the $B = 1$.

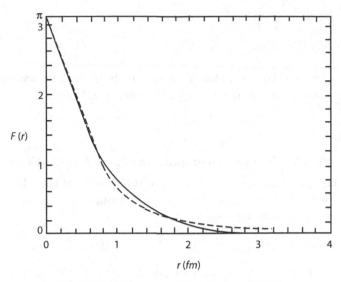

F (r)

r (fm)

Fig. 20.2. The numerical solution of $F(r)$.

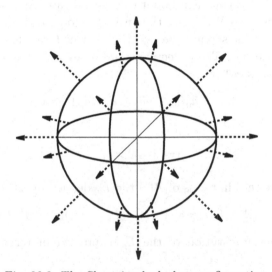

Fig. 20.3. The Skyrmion hedgehog configuration.

The hedgehog configuration, see Fig. 20.3, used as an ansatz for the Skyrmion, is by construction invariant under the operation of,

$$K \equiv J + I = (L + S) + I, \qquad (20.34)$$

where L, S and J are the orbital angular momentum, the spin and the total angular momentum, and I is the isospin.

It follows from,

$$[K, g(x)] = i \sin F \left[[(r \times -\nabla), \vec{\tau} \cdot \hat{r}] + \left[\frac{\vec{\tau}}{2}, \vec{\tau} \cdot \hat{r} \right] \right] = 0. \tag{20.35}$$

Hence the Skyrmion carries a charge of $K = 0$. It is straightforward to notice that it is also an invariant under the parity operator defined in (20.7), so that altogether it is $K = 0^+$ state.

20.3.2 Semiclassical quantization of the soliton

Recall that the quantization of the collective coordinates of the two-dimensional soliton was performed by elevating the static group element to a space-time dependent one in the following way,

$$g(x) \to g(x, t) = A(t)g(x)A^{-1}(t) \quad A(t) \in U(N_f). \tag{20.36}$$

Nothing in this prescription is two dimensional and hence we now use the same ansatz also for the four-dimensional soliton. In two dimensions we discussed the general N_f case, here we start with the simplest case of $N_f = 2$ and then we discuss the $N_f = 3$ and comment about the general case. As discussed above for $SU(N_f = 2)$ there is no WZ term, thus we have to substitute (20.36) into the action that includes the sigma term and the Skyrme term (for simplicity we do not add the mass term). The collective coordinates $A(t)$ can be parameterized in the following ways either,

$$A(t) = a_0(t) + i\vec{a}(t) \cdot \vec{\sigma} \quad (a_\mu)^2 = 1, \tag{20.37}$$

or,

$$A^{-1}\dot{A} \equiv \frac{i}{2}\vec{\sigma} \cdot \vec{w}. \tag{20.38}$$

It is easy to verify that in terms of $A(t)$ the L_μ defined in (20.5),

$$L_0 = A(t)g^{-1}(x)(A^{-1}\dot{A})(t)g(x)A^{-1}(t) \quad L_i = A(t)g^{-1}\partial_i g(x)A^{-1}(t). \tag{20.39}$$

The result of the substitution of the L_μ expressed in terms of w into the Lagrangian is,

$$\mathcal{L} = \mathcal{L}_{cl} + \frac{1}{2}\alpha^2 w^2, \tag{20.40}$$

where the constant of proportionality α^2 is computed as a spatial integral over the chiral angle to be $\frac{53.3}{e^3 f_\pi}$. In terms of spin and isospin operators the Hamiltonian can be rewritten in terms of a second Casimir operator,

$$\mathcal{H} = E_{cl} + \frac{1}{2\alpha^2}J^2 = E_{cl} + \frac{1}{2\alpha^2}I^2. \tag{20.41}$$

In terms of the a_μ variables the Lagrangian density takes the form,

$$\mathcal{L} = \mathcal{L}_{cl} + \lambda(\dot{a}_\mu)^2, \tag{20.42}$$

where λ is the moment of inertia given by,

$$\lambda = \frac{8\pi}{3} f_\pi^2 \int_0^\infty dr r^2 \sin^2 F \left[1 + \frac{8e^2}{f_\pi^2} \left(F^2 + \frac{\sin^2 F}{r^2} \right) \right]. \tag{20.43}$$

The corresponding Hamiltonian is therefore,

$$\mathcal{H} = \pi_\mu \dot{a}_\mu - \mathcal{L} = E_{cl} + \frac{1}{8\lambda} \pi_\mu^2 = E_{cl} + \frac{1}{8\lambda} \left(-\frac{\partial^2}{\partial a_\mu^2} \right), \tag{20.44}$$

where in the last step we introduced the canonical quantization namely,

$$[a_\mu, \pi_\nu] = i\delta_{\mu\nu} \quad \pi_\mu = -\frac{i\partial}{\partial a_\mu}, \tag{20.45}$$

subjected to the constraint $(a_\mu)^2 = 1$.

The Noether charges associated with the angular momentum and isospin expressed in terms of a_μ take the form,

$$I_k = \frac{i}{2} \left(a_0 \frac{\partial}{\partial a_k} - a_k \frac{\partial}{\partial a_0} - \epsilon_{klm} a_l \frac{\partial}{\partial a_m} \right)$$

$$J_k = \frac{i}{2} \left(a_k \frac{\partial}{\partial a_0} - a_0 \frac{\partial}{\partial a_k} - \epsilon_{klm} a_l \frac{\partial}{\partial a_m} \right). \tag{20.46}$$

Choosing a fermionic rather than bosonic wave function, namely, odd under changing $A \rightarrow -A$, implies that the wave function of baryons of $I = J = \frac{1}{2}$ is linear in a_μ, for instance the proton wave function is $|p> = \frac{1}{\pi}(a_1 + ia_2)$ and those of $I = J = \frac{3}{2}$ cubic in a_μ like $|\Delta^{++}> = \frac{\sqrt{2}}{\pi}(a_1 + ia_2)^2$. It is thus obvious that the mass difference between the Δ and the nucleon is,

$$M_\Delta - M_N = \frac{3}{2\alpha^2}. \tag{20.47}$$

Expectation values of flavor charges can be computed in the semi-classical approximation by expressing the Noether currents and charges in terms of the $a_\mu(t)$ and their corresponding momenta $\pi_\mu = -\frac{i\partial}{\partial a_\mu}$ (20.46) and sandwiching these operators in between quantum states, like N and Δ. For instance the space components of the Baryonic current is given by,

$$B^i = i \frac{e^{ijk}}{2\pi^2} \frac{\sin^2 F}{r} \hat{r}_k \text{Tr} [\dot{A}^{-1} A\tau_j]. \tag{20.48}$$

We use this expression to compute the isoscalar magnetic moment, defined by,

$$\mu_{I=0} = \frac{1}{2} \int d^3 r \vec{r} \times \vec{B}, \tag{20.49}$$

of the proton as follows,

$$(\mu_{I=0})_3 = -\frac{2i}{3\pi} \int dr r^2 F <p, 1/2| \text{Tr} (\tau_3 \dot{A}_{-1} A)|p, 1/2>$$

$$\frac{i}{3} <r^2>_{I=0} <p, 1/2| \text{Tr} (\tau_3 \dot{A}_{-1} A)|p, 1/2> = \frac{1}{6I} <r^2>_{I=0}. \tag{20.50}$$

In a similar manner one can compute the isovector magnetic moment.

Another important property of nucleons that can be extracted from the description of baryons as semiclassical solitons is the axial coupling g_A defined via,

$$<N'(p_2)|J_\mu^{A^a}|N(p_1)> = <\bar{u}(p_2)\frac{\tau^a}{2}[g_A(q^2)\gamma_\mu\gamma_5 + h_A(q^2)q_\mu\gamma_5]u(p_1)>, \quad (20.51)$$

where $J_\mu^{A^a}$ is the axial current and $q = p_2 - p_1$. In the chiral limit $h_A(q^2)$ has a pion pole whose residue is $-2f_\pi g_{\pi NN}$ namely,

$$h_A(q^2) = \frac{d_A(q^2)}{q^2} \quad d_A(0) = -2f_\pi g_{\pi NN}, \quad (20.52)$$

where $g_{\pi NN}$ is the pion nucleon coupling. Current conservation implies that,

$$2M_N g_A(q^2) + q^2 h_A(q^2) = 0. \quad (20.53)$$

In the nucleon rest frame the nonrelativistic limit $q \to 0$, taken in a symmetric form $q_i q_j \to \frac{1}{3}\delta_{ij}q^2$ yields,

$$\lim_{q \to 0} <N'(p_2)|J_i^A a|N(p_1)> = \lim_{q \to 0} < u(p_2)\frac{\tau^a}{2}[g_A(0)\sigma_i$$

$$+h_A(q^2)\vec{\sigma}\cdot\hat{\vec{q}}\hat{q}_i]u(p_1)> =$$

$$= \lim_{q \to 0} g_A(0)(\delta_{ij} - \hat{q}_i\hat{q}_j) <N'(p_2)|\sigma_i\frac{\tau^a}{2}|N(p_1)>$$

$$= \frac{2}{3}g_A(0) <N'(p_2)|\sigma_i\frac{\tau^a}{2}|N(p_1)>, \quad (20.54)$$

where we have made use of the Goldberger–Triman relation,

$$g_A(0) = \frac{g_{\pi NN}f_\pi}{M_N}. \quad (20.55)$$

In the Skyrme model we can extract the axial coupling $g_A(0)$ in the following way. First we compute the space integral over the axial current,

$$\int d^3x J^{A^a}_i(x) = -\frac{1}{2}d\,\text{Tr}\,[\tau_i A^{-1}\tau^a A], \quad (20.56)$$

where d is the space integral over a function that depends only on the classical soliton configuration. We then sandwich this operator in between nucleon states to find,

$$\lim_{q \to 0} \int d^3x e^{i\vec{q}\cdot\vec{x}} <N'|J^{A^a}_i(x)|N> = \frac{2}{3}d <N'|\sigma_i\frac{\tau^a}{2}|N> . \quad (20.57)$$

Equating the last expression with (20.54) it was found that the Skyrme model value of $g_A(0) = 0.61$ whereas experientially it is equal to 1.33.

20.3.3 The Skyrme model and large N_c QCD

In Section 19.2 it was shown that the scattering amplitude or quadrilinear coupling of mesons in the large N_c limit behaves like $\frac{1}{N_c}$. Recall that in this limit we take $N_c \to \infty$ while keeping $\lambda_{t'\text{Hooft}} = g^2 N_c$ finite. The amplitude that is proportional to g^2 behaves like $\frac{1}{N_c}$. By comparing this result to the quadrilinear coupling of the Skyrme model we are able to determine the dependence of the Skyrme coefficients on N_c. Let us start by expanding the sigma term in terms of the pion fields,

$$\mathcal{L}_\sigma = \frac{f_\pi^2}{4} \text{Tr}\left[\partial_\mu g \partial^\mu g^{-1}\right] \sim \frac{1}{2}\partial_\mu \vec{\pi} \cdot \partial^\mu \vec{\pi} + \frac{1}{6f_\pi^2}\left[(\vec{\pi} \cdot \partial^\mu \vec{\pi})^2 - \vec{\pi}^2 \partial_\mu \vec{\pi} \cdot \partial^\mu \vec{\pi}\right] + \mathcal{O}(\pi^6).$$
$$(20.58)$$

The quadrilinear coupling behaves like $\frac{1}{f_\pi^2}$. If we expand the Skyrme term in a similar manner we find that in that case the coupling behaves like $\frac{1}{e^2 f_\pi^4}$, and hence we conclude in agreement with (19.12) that,

$$f_\pi \sim \sqrt{N_c} \quad e \sim \frac{1}{\sqrt{N_c}}. \qquad (20.59)$$

This enables us to check the N_c dependence of the classical Skyrmion mass and its semi-classical extension,

$$M_{cl} \sim \frac{f_\pi}{e} \sim N_c \quad M_{sc} \sim \frac{1}{\alpha^2} \sim e^3 f_\pi \sim \frac{1}{N_c}. \qquad (20.60)$$

Recall for comparison that the two-dimensional solitonic baryons were shown to have classical mass which is also order N_c, but the semi-classical correction term behaves like N_c^0 and not $\frac{1}{N_c}$.

20.4 The Skyrme model for $N_f = 3$

Phenomenologically we should obviously be interested in the case of $N_f = 3$ rather than only two flavors. Moreover, to let the WZ term play a role we also have to go beyond $N_f = 2$. So we have two reasons to discuss now the $U(N_f = 3)$ classical solitons and their semi-classical quantization. The action is now the sum of a sigma term (20.3), the Skyrme term (20.13) and the WZ term (20.9). We can further add a mass term (20.18). The latter can be used to introduce an explicit breaking of the flavor symmetry by assigning different masses to the different flavor degrees of freedom. In fact one can add additional flavor symmetry breaking terms which, to simplify the treatment, we would not do.

We first need to choose a parametrization for the $U(N_f = 3)$ group element. The analog of (20.8) including the $U(1)$ factor now reads,

$$g(x) = e^{i\frac{\sqrt{2}}{\sqrt{3}f_\pi}\eta^0(x)} e^{i\Phi(x)} = e^{i\frac{\sqrt{2}}{\sqrt{3}f_\pi}\eta^0} e^{i\sum_{a=1}^{a=8}\frac{\sqrt{2}}{f_\pi}\lambda^a \phi^a}, \qquad (20.61)$$

where λ^a are the $SU(3)$ Gell–Mann matrices. In terms of the pions, kaons and η we have,

$$\Phi \equiv \begin{pmatrix} \frac{1}{\sqrt{2}}\pi^0 + \frac{1}{\sqrt{6}}\eta^8 & \pi^+ & K^+ \\ \pi^- & -\frac{1}{\sqrt{2}}\pi^0 + \frac{1}{\sqrt{6}}\eta^8 & K^0 \\ K^- & \bar{K}^0 & \frac{2}{\sqrt{6}}\eta^8 \end{pmatrix}. \tag{20.62}$$

Next we have to choose an ansatz for the static classical configuration $g_0(x)$. Recall that in the two-dimensional case we took an embedding of the $U(1)$ in $U(N_f)$ of the form $g_0(x) = \text{Diag}(1, 1, \ldots, e^{-i\sqrt{\frac{4\pi}{N_c}}\phi(x)})$. In analogy in the four-dimensional case we embed the $SU(2)$ hedgehog configuration in the $SU(3)$ group element as follows,

$$g_0(x) = \begin{pmatrix} & & 0 \\ e^{iF(r)\vec{\tau}\cdot\hat{r}} & & 0 \\ 0 & 0 & 1 \end{pmatrix}. \tag{20.63}$$

Since the WZ term vanishes for an $SU(2)$ group the solution for $F(r)$ is identical to that discussed in Section 20.3 and hence the elevation to $N_f = 3$ shows up basically only in the semi-classical quantization of the collective coordinates. Recall that the latter are introduced via $g_0(x) \to A(t)g_0(x)A^{-1}(t)$. In two dimensions we parameterized the quantum fields $A(t)$ in terms of the $Z_i, i = 1, \ldots, N_f$ variables which was adequate for the CP^{N_f-1} that the collective coordinates span in that case. Clearly in the present case since the $g_0(x) \in SU(2)$ a different ansatz is required. The most straightforward one is in terms of the angular velocities w_a that generalize those of (20.38) as follows,

$$A^{-1}(t)\dot{A}(t) = \frac{i}{2}\sum_{a=1}^{8}\lambda^a w_a. \tag{20.64}$$

When substituting this into the Lagrangian one finds,

$$L_{SU(3)} = L_{cl} + \frac{1}{2}\alpha^2 \sum_{a=1}^{3} w_a^2 + \frac{1}{2}\beta^2 \sum_{a=4}^{7} w_a^2 - N_c\frac{B}{2\sqrt{3}}w_8. \tag{20.65}$$

Note that the WZ term which is proportional to N_c and linear in the angular velocity w_8 associated with λ_8 that commutes with the classical ansatz $[\lambda_8, g_0(x)]$. This is very reminiscent of the structure of the WZ term in two dimensions that is also linear in the angular velocity associated with the hypercharge with N_c as a coefficient.

The quantization of the system is performed as if it is an "$SU(3)$ rigid top" namely one defines the right generators,

$$\mathcal{R}_a = -\frac{\partial L_{SU(3)}}{\partial w_a} = -\alpha^2 w_a = -J_a, \quad a = 1, 2, 3$$
$$-\beta^2 w_a, \qquad a = 4, .., 7 \tag{20.66}$$
$$\frac{N_c B}{2\sqrt{3}}, \qquad a = 8,$$

and imposes the quantization condition,

$$[\mathcal{R}_a, \mathcal{R}_b] = -i f_{abc} \mathcal{R}_c, \tag{20.67}$$

where f_{abc} are the structure constants of $SU(3)$.

The Hamiltonian of the system takes the form,

$$H = E_{cl} + \frac{1}{2}\left[\frac{1}{\alpha^2} - \frac{1}{\beta^2}\right] J^2 + \frac{1}{2\beta^2} C_2 - \frac{3}{8\beta^2}. \tag{20.68}$$

Again this form of the Hamiltonian is similar to the one we found in two dimensions which is also proportional to the second Casimir operator. One can now apply the Hamiltonian on states associated with representations of the $SU(3)$ and compute the corresponding masses. Using the eigenvalues of the second Casimir operator of the representations 8, 10, $\bar{1}0$, 27 which are given by 3, 6, 6, 8, respectively we can determine the masses of the various Skyrme hadrons. The full analysis of the $N_f = 3$ baryons is beyond the scope of this book. We refer the interested reader to literature. For a review of the topic see for instance [231].

21

From two-dimensional solitons to four-dimensional magnetic monopoles

21.1 Introduction

Solitons play an important role in non-perturbative two-dimensional fields as we have seen in the first part of this book. They are intimately related to non-trivial topology, they are an essential ingredient in integrable models, and they enable the phenomenon of fermion boson duality-bosonization. When passing to four-dimensional field theories the topology may be even richer and thus we would anticipate having topological solitons as static solutions also in four-dimensional space-time. As we have seen in Section 5.3, Derrik's theorem does not permit the existence of solitons of scalar field theory in space dimensions higher than one, however, they are not prohibited in theories that include higher spin fields, in particular in theories of scalar fields coupled to non-abelian gauge fields. Indeed as we will see in this section certain theories of this type that admit spontaneous symmetry breaking, admit soliton solutions. These configurations carry a conserved topological charge which guarantees their stability against decay to the vacuum. As it will turn out this charge is in fact a magnetic charge and hence these solitons are magnetic monopoles, or in the more general case dyons with both magnetic and electric charge. The construction of dyons from static solutions will be the analog process of building up breathers from two-dimensional solitons.

In the next section we present the basics of the Yang–Mills Higgs theory. We then show the relation between magnetic monopoles and topological solitons both for the simplest case of $SU(2)$ (and $SO(3)$) as well as for a general non-abelian gauge group. The next topic is the seminal solution of 't Hooft and Polyakov. Then we discuss zero modes, time-dependent solutions and dyons. In the following section we discuss the very important limit of BPS. We then describe the construction of multi-monopole solutions that was proposed by Nahm. We show its application to the construction of BPS monopoles of charge one and two. The next topic is the moduli space of monopoles. We determine the metric on this space for the case of widely separated monopoles.

The topic of magnetic monopoles and dyons has been covered by several review papers, proceedings of meetings and books, for instance [21], [214], [67], [7], [182] and [193], respectively. Here in this chapter we made use of mainly the former two references.

Monopoles and dyons play an important role in $\mathcal{N} = 2$ SYM. They admit a mathematical non-perturbative structure that can be determined exactly. Since we do not discuss supersymmetry in this book the monopoles and dyons of that theory will not be addressed in this chapter. We refer the reader to [191] and [192].

21.2 The Yang–Mills Higgs theory – basics

Consider the Yang–Mills Higgs system described by the following Lagrangian density,

$$\mathcal{L} = -\frac{1}{2}\text{Tr}[F_{\mu\nu}F^{\mu\nu}] + \text{Tr}[D_\mu\Phi D^\mu\Phi] - V(\Phi), \tag{21.1}$$

where the (non-abelian) gauge group is G, Φ is in the adjoint representation of the group, namely,

$$A_\mu = T^a A_\mu^a \quad \Phi = T^a \Phi^a, \tag{21.2}$$

where T^a are the generators of G (see Section 3.1), the covariant derivative reads,

$$D_\mu = \partial_\mu\Phi + ie[A_\mu, \Phi], \tag{21.3}$$

and $V(\Phi)$ is given by,

$$V(\Phi) = -\mu^2\text{Tr}[\Phi^2] + \lambda(\text{Tr}[\Phi^2])^2, \tag{21.4}$$

where λ is taken to be positive so that the energy is bounded from below, and we also take $\mu^2 > 0$. In general one can discuss a similar system where Φ is in any representation of G but here we consider only the case of the adjoint representation.

Let us start with the simplest case where $G = SU(2)$ and Φ is in the triplet (adjoint) representation. For such a case the vacuum solution can be put in the form,

$$\Phi(x) = v\frac{\sigma_3}{2} \equiv \Phi_0 \quad v \equiv \sqrt{\frac{\mu^2}{\lambda}}$$
$$A_\mu(x) = 0. \tag{21.5}$$

In this case the vacuum expectation value of the Higgs field breaks the $SU(2)$ symmetry down spontaneously to a $U(1)$ symmetry along the $a = 3$ direction. The physical fields will be denoted as follows,

$$\mathcal{A}_\mu = A_\mu^3 \quad W_\mu = \frac{A_\mu^1 + iA_\mu^2}{\sqrt{2}} \quad \varphi = \Phi^3, \tag{21.6}$$

which associate with the massless "photon", pair of mesons W, W^* with a mass of eV and charges $\pm e$ and an electrically neutral scalar boson with mass $m_H = \sqrt{2}\mu$, respectively.

For a general group G which we take to be a simple Lie group of rank r (for the basic definitions see Section 3.1). In this case the expectation value of $\phi = \Phi_0$ can be taken to lie in the Cartan subalgebra of the group G. Using the notation $\vec{H}$ for the r-dimensional vector of the elements of the Cartan subalgebra, Φ_0 is characterized by a vector $\vec{h}$ such that,

$$\Phi_0 = \vec{h} \cdot \vec{H}. \tag{21.7}$$

The generators of the unbroken subgroup are those generators of G that commute with Φ_0. These are all the generators of the Cartan subalgebra together with ladder operators associated with roots orthogonal to $\vec{h}$. If none of the $\vec{\gamma}$ are orthogonal to h the unbroken symmetry is $U(1)^r$, whereas if there are some roots $\vec{\gamma}$ such that $\vec{\gamma} \cdot \vec{h} = 0$ then the unbroken symmetry is $U(1)^{r-r'} \times K$ where K is of rank r' and it has $\vec{\gamma}$ as its root diagram.

21.3 Topological solitons and magnetic monopoles

In Section 5.2 when discussing two dimensional solitons, we identified a topological conserved current and an associated topological charge. Configurations that carry a non-trivial value with respect to this charge cannot, due to charge conservation, decay to vacuum. These configurations were shown to be stable solutions of the equations of motion and to have finite energy. Thus we anticipate that also in four-dimensional field theories, and in particular in the Yang–Mills Higgs theory we discuss here, there should be solutions of the equations of motion that associate with non-trivial topological charges and one can determine their existence even without solving the equations of motion. It is easy to verify that for $G = SU(2)$ the following current,

$$k_\mu = \frac{1}{8\pi} \epsilon_{\mu\nu\rho\sigma} \epsilon^{abc} \partial^\nu \hat{\Phi}^a \partial^\rho \hat{\Phi}^b \partial^\sigma \hat{\Phi}^c, \tag{21.8}$$

where $\hat{\Phi}^a = \frac{\Phi^a}{|\Phi|}$, is conserved for any configuration whether it solves the equations of motion or not. It follows trivially from the total anti-symmetry of $\epsilon_{\mu\nu\rho\sigma}$ that $\partial^\mu k_\mu = 0$. The corresponding charge is,

$$Q = \int d^3x k_0 = \frac{1}{8\pi} \int d^3x \epsilon_{ijk} \epsilon^{abc} \partial^i \hat{\Phi}^a \partial^j \hat{\Phi}^b \partial^k \hat{\Phi}^c$$

$$= \frac{1}{8\pi} \int d^3x \epsilon_{ijk} \epsilon^{abc} \partial^i (\hat{\Phi}^a \partial^j \hat{\Phi}^b \partial^k \hat{\Phi}^c) = \frac{1}{8\pi} \int d^2 S_i x \epsilon_{ijk} \epsilon^{abc} \hat{\Phi}^a \partial^j \hat{\Phi}^b \partial^k \hat{\Phi}^c. \tag{21.9}$$

This topological charge is the winding number associated with the map,

$$\Phi_0(\infty): \quad S_2^{\text{space}} \to S_2^{G/H}, \tag{21.10}$$

where S_2^{space} is the boundary of the space at $r = \infty$ and where the coset space G/H is in our example $G/H = SU(2)/U(1) = S_2$. It is thus an integer charge $Q = n$ and as such it must be invariant under smooth deformations of the surface of integration that do not cross any of the zeros of Φ. In fact a configuration with $Q = n$ must have at least $|n|$ zeros of the Higgs field. If we distinguish between a $+$ zero and a $-$ anti-zero then the net number is precisely n. Obviously if we consider Higgs configurations of $Q = n = 1$ there must be one with minimum energy. This cannot be smoothly deformed to a vacuum since the winding number is quantized. Hence such a configuration must be a local minimum of the energy and therefore a static classical solution.

The next question we want to address is what is the connection between these non trivial soliton solutions and magnetic monopoles? The field strength of the abelian gauge field A_μ defined in (21.6), $\mathcal{F}_{\mu\nu} = \partial_\mu A_\nu - \partial_\nu A_\mu$ is the outcome of,

$$\tilde{\mathcal{F}}_{\mu\nu} = \hat{\Phi}^a F^a_{\mu\nu} - \frac{1}{g}\epsilon^{abc}\hat{\Phi}^a D_\mu \hat{\Phi}^b D_\nu \hat{\Phi}^c, \tag{21.11}$$

when we take $\hat{\Phi} = (0, 0, 1)$, namely $\Phi = \varphi$. What distinguishes $\tilde{\mathcal{F}}_{\mu\nu}$ from an ordinary abelian field strength is that it does not obey the Biachi identity,

$$*d\tilde{\mathcal{F}} = \frac{1}{2}\epsilon_{\mu\nu\rho\sigma}\partial^\nu \tilde{\mathcal{F}}^{\rho\sigma} = ek_\mu = \frac{4\pi}{g}k_\mu, \tag{21.12}$$

where g is the magnetic charge which will be shown to be equal to $4\pi/e$. Defining now the magnetic field B_i associated with $\tilde{\mathcal{F}}_{\mu\nu}$ as usual as,

$$B_i \equiv \frac{1}{2}\epsilon_{ijk}\tilde{\mathcal{F}}^{jk}, \tag{21.13}$$

we find that,

$$\nabla \cdot \vec{B} = \frac{4\pi}{g}k_0 \quad Q_M = \frac{1}{g}\int d^3x\, k_0 = \frac{4\pi}{e}Q = \frac{4\pi}{e}n. \tag{21.14}$$

We have thus realized that the non-trivial soliton configurations carry a magnetic charge and hence are magnetic monopoles. We can further determine the classical mass of the monopole since the total energy of such a solution is,

$$E = \int d^3x \left[\text{Tr}[E_i^2] + \text{Tr}[(D_0\Phi)^2] + \text{Tr}[B_i^2] + \text{Tr}[(D_i\Phi)^2] + V(\Phi) \right]. \tag{21.15}$$

For a static configuration that does not carry electric charge, the first two terms are expected to vanish. Then one can show that the form of the mass has to be,

$$M = \frac{4\pi v}{e}f\left(\frac{\lambda}{e^2}\right), \tag{21.16}$$

where $f(\frac{\lambda}{e^2})$ should be of order one.

So far we have discussed the topological charges of the group $G = SU(2)$ case. Let us now address the general case. Instead of the map (21.10), the asymptotic Higgs field constitutes in the general case a map,

$$\Phi(\infty): \quad \partial\mathcal{M} \to \frac{G}{H}, \tag{21.17}$$

where $\partial\mathcal{M}$ is the boundary of the space which for ordinary flat Minkowski space-time is S_2 and G/H is the coset of the unbroken symmetry group H and the original group G, which in the discussion above was $SU(2)/U(1) = S_2$. These maps from the boundary of space to the coset, fall into equivalent classes which form the homotopy group $\pi_2(G/H)$. For simply connected group G, namely with $\pi_1(G) = 0$ the classification of the maps is in fact done by $\pi_1(H)$ since for this type of G,

$$\pi_2(G/H) = \pi_1(H). \tag{21.18}$$

This follows from the following exact sequence[1]

$$\ldots \to \pi_2(G) \quad \to \quad \pi_2(G/H) \quad \to \quad \pi_1(H) \quad \to \quad \pi_1(G) \quad \to \ldots. \tag{21.19}$$

The image of a given homomorphism equals the kernel of the next one in the sequence. It is well known that for any semi-simple group G, $\pi_2(G) = 0$ and hence,

$$\pi_2(G/H) \cong \text{Ker}[\pi_1(H) \quad \to \pi_1(G)]. \tag{21.20}$$

Now since for a simply connected group $\pi_1(G) = 0$ we find (21.18). Let us describe now several cases of physical interest:

- The 't Hooft–Polyakov solution that will be discussed in the next section, is slightly different since in that case $G = SO(3)$ which is not simply connected $\pi_1(SO(3)) = \mathcal{Z}_2$ and hence only the even elements of $\pi_1(H = U(1))$ are in the kernel of the homomorphism of (21.20). This is the source of the fact that the quantization condition is twice the one given by Dirac (see (21.27)) even though in both cases $H = U(1)$.
- For a simply connected G of rank r and with H which is the full Cartan subalgebra, namely $H = U(1)^r$, the homotopy group that classifies the magnetic monopoles is $\pi_1(H = U(1)^r) = \mathcal{Z}^r$.
- In the spontaneous breaking of the electro-weak theory we have $G = SU(2) \times U(1)$ and $H = U(1)$ such that $G/H \cong S^3$. Since $\pi_2(S^3) = 0$ magnetic monopoles are excluded in this theory.
- On the other hand a wide class of grand unified theories do admit magnetic monopoles. The most prominent example is the $G = SU(5)$ grand unified model with $H = SU(3) \times SU(2) \times U(1)$. This is an example of the case that $H = U(1) \times K$ where K is a semi-simple and simply connected group. In this case there is only a single component of the magnetic charge that is topologically conserved.
- Another interesting scenario is the case where the group G twice undergoes a spontaneous symmetry breaking namely, $G \to H_1 \subset G \to H_2 \subset H_1$. This is relevant to an evolution of the universe where at as early stage magnetic

---

[1] The reader who is not familiar with the notion of an exact sequence can refer to any text book on topology or alternatively to the book of Coleman [66] where an elegant proof of this theorem is presented.

monopoles associated with $\pi_2(G/H_1)$ are being created and then the question is what is their fate when the universe undergoes a second phase transition to H_2? This can be determined by an exact sequence similar to (21.19).

21.4 The 't Hooft–Polyakov magnetic monopole solution

Equipped with knowledge based on the topological arguments of above, that magnetic monopoles do exist for theories associated with non-trivial $\pi_2(G/H)$ we want to proceed to the determination of explicit configurations of the non-trivial topological solutions. It turns out that this is a non-trivial task and only for a limited set of cases can it be accomplished analytically. We start with the simplest case where $G = SU(2)$, or $G = SO(3)$ as was done in the original solution of 't Hooft and Polyakov [123], [175]. We will see in the next section an explicit solution using a special limit of the theory. Without this limit one can simplify the procedure by searching for spherically symmetric solutions. This implies that the fields must be invariant under a combination of rotations and a compensating gauge transformation. The latter can be space independent by using the so called "hedgehog" gauge where rotational invariance requires that the fields be invariant under a combined rotation and global internal $SU(2)$ transformation. Stating it differently, one looks for a configuration which is symmetric under a mixed angular momentum,

$$\vec{J} = \vec{L} + \vec{I}, \tag{21.21}$$

where $\vec{L} = -i\vec{r} \times \vec{\nabla}$ is the ordinary spatial part of the angular momentum and $\vec{I}$ are the generators of the $SU(2)$ gauge group. With this definition of $\vec{J}$ the ansatz should obey,

$$[J_i, \Phi] = 0 \quad [J_i, A_j] = i\epsilon_{ijk} A_k. \tag{21.22}$$

Using this gauge 't Hooft and Polyakov suggested the following ansatz for the fields,

$$A_i^a = \epsilon_{iam} \hat{r}^m \left[\frac{1 - u(r)}{er} \right] \quad \Phi^a = \hat{r}^a h(r). \tag{21.23}$$

To write down the equations that determine $u(r)$ and $h(r)$ one can substitute these expressions into the Lagrangian density (21.1) and vary with respect to $u(r)$ and $h(r)$. This is easier than the usual procedure of substituting (21.23) into the equations of motion derived from (21.1). The resulting equations are,

$$u'' - \frac{(u^2 - 1)u}{r^2} + e^2 u h^2 = 0$$

$$h'' + \frac{2}{r} h' - \frac{2u^2 h}{r^2} + \lambda(v^2 - h^2)h = 0. \tag{21.24}$$

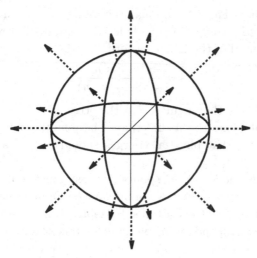

Fig. 21.1. The hedgehog configuration of the Higgs field. The orientation in isospace is aligned with the position vector.

The primes denote derivatives with respect to r. Finiteness of the energy associated with the solution requires that,

$$\Phi_0(\infty) = 0 \quad \rightarrow \quad u(\infty) = 0$$
$$D_i\Phi_0(\infty) = 0 \quad \rightarrow \quad h(\infty) = v. \tag{21.25}$$

Similarly requiring that the solutions are non-singular at the origin implies that,

$$u(0) = 1 \quad h(0) = 0. \tag{21.26}$$

Qualitatively, the profile of the Higgs field is that of a hedgehog, as can be seen in Fig. 21.1. The orientation in isospace is aligned with the position vector.

Analytic solutions of these equations will be derived in the next section using a special (BPS) limit. In general one has to solve these equations numerically. The physical picture that comes out from these calculations is that there is a central core of radius $R_{\text{core}} \sim \frac{1}{ev}$, outside of which $u(r)$ and $|h - u|(r)$ decrease exponentially. The mass of the monopole takes the form $M = \frac{4\pi v}{e} f(\frac{\lambda}{e^2})$ where $f(0) = 1$ and $f(\infty) = 1.787$.

21.5 Charge quantization

In his seminal paper on magnetic monopoles in quantum mechanics [78] Dirac found out that the magnetic charge g and the electric charge e must be related via the famous charge quantization condition,

$$eg = 2\pi n, \tag{21.27}$$

where n is an integer. This implies that the existence of a magnetic monopole explains the observed quantization of all electric charges. He proposed a solution of a magnetic potential of the following form,

$$e\vec{A} = \frac{eg}{4\pi} \frac{\hat{r} \times \hat{n}}{r(1 - \hat{r} \cdot \hat{n})}, \tag{21.28}$$

which has in addition to the singularity at the origin also a singularity extending from the origin out along the $\hat{n}$ direction. The quantization condition follows from the requirement that physical charges should not be able to detect the string.

Yang–Mills Higgs models with spontaneous symmetry breaking such that $\pi_2(G/H) \neq 0$, as we have seen above, admit magnetic monopole solutions associated with each element of the Cartan subalgebra of the unbroken group H. For large distance the corresponding magnetic fields take the form,

$$eA_i = \frac{eg}{4\pi}(1 + \cos\theta)\partial_i\phi + \mathcal{O}\left(\frac{1}{r^2}\right). \tag{21.29}$$

The generalization of the quantization argument of Dirac to the non-abelian case is straightforward. This follows from the demand that the electrically charged fields of the theory are single valued if acted upon by a group element e^{eg}. This quantization condition can be solved in terms of the simple roots $\vec{\gamma}_i$ where $i = 1, \ldots, r$ where r is the rank of H, of the root system of H. From the simple roots one constructs a convenient basis (H_i) for the elements of the Cartan subalgebra with the property that each element has half-integer eigenvalues when acting on the basis vector of any representation. This is achieved by taking,

$$H_i \equiv \frac{\vec{\gamma}_i}{|\vec{\gamma}_i|^2}\vec{H}. \tag{21.30}$$

In this basis the solution of the quantization condition is,

$$\frac{eg}{4\pi} = \sum_{i=1}^{r} n_i H_i. \tag{21.31}$$

This solution can be represented as an r-dimensional lattice dual to the weight lattice of the group (see Section 3.1). For the simple example of $SO(3)$ the rank is one and thus one gets $eg = 4\pi n$ twice as the condition of Dirac due, as was explained above, to the fact that $SO(3)$ is not simply connected. For the group $SU(3)$ which is of rank $r = 2$ the charge lattice is drawn in Fig. 21.2. In general it was shown that the charge lattice is the weight lattice of the dual gauge group.

21.6 Zero modes, time-dependent solutions and dyons

From the static $SU(2)$ monopole solution discussed above we can generate obviously (infinitely) more solutions by applying gauge transformations. This of course will be avoided by fixing a gauge. However even in that case there is

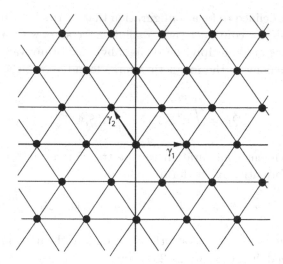

Fig. 21.2. The charge lattice of the $SU(3)$ monopole.

one unfixed global gauge $U(1)$ phase. This adds up to the parameters associated with the translation of the monopole. Infinitesimal transformations of this set of four parameters generate field variations δA_i and $\delta\Phi$ that preserve the equations of motion and leave the energy unchanged. These variations in general will be referred to as *zero modes*.

Time-dependent excitations of the translational zero modes can be derived by substituting $\vec{r} \to \vec{r} - \vec{v}t$ into the static solution. This has to be done together with ensuring the Gauss law constraint,

$$D_i E^i = ie[\Phi, D^0\Phi], \qquad (21.32)$$

which also implies that for most choices of gauge $A^0 \neq 0$. Substituting the solutions of the Gauss law into the expression of the energy (21.15) one can show that the change of energy for non-relativistic velocities is as one expects, given by,

$$\Delta E = \frac{1}{2}M|\vec{v}|^2. \qquad (21.33)$$

Next we want to describe the excitation of the zero mode associated with the fourth parameter, that of the global gauge phase. The Noether charge associated with this transformation is the electric charge that corresponds to the unbroken $U(1)$ gauge symmetry. In terms of the physical variables defined in (21.6) this takes the form,

$$Q_E = -ie\int \mathrm{d}^3x[W^{j*}\mathcal{D}_0(W_j - \mathcal{D}_j W_0) - W^j\mathcal{D}_0(W_j^* - \mathcal{D}_j W_0^*)], \qquad (21.34)$$

where the $U(1)$ covariant derivative is defined by $\mathcal{D}_\mu W_\nu = (\partial_\mu - ie\mathcal{A}_\mu)W_\nu$ and where a string gauge is used where the Higgs field direction is uniform.

A dyon [134] is defined to be a configuration that carries both magnetic as well as electric charges. To construct such a solution we start with a static solution in the string gauge and multiply the W field by a uniformly varying $U(1)$ phase factor e^{iwt}. In analogy to the magnetic charge found above, the electric charge has the form,

$$Q_E = \int d^2 S_i \, \tilde{\mathcal{F}}_{0i} = \int d^2 S_i \, \hat{\Phi}^i \, E_i^a . \tag{21.35}$$

The time-dependent configuration can be transformed into a static solution by a $U(1)$ transformation of the form,

$$W_i \rightarrow e^{-iwt} W_i \quad \mathcal{A}_i \rightarrow \mathcal{A}_i \quad \mathcal{A}_0 \rightarrow \bar{\mathcal{A}}_0 = \mathcal{A}_0 - \frac{w}{e} . \tag{21.36}$$

In this static form we can transform the solution into the non-singular hedgehog gauge with A_i^a and Φ^a given by (21.23) and,

$$A_0^a = \hat{r}^a \, j(r) = \hat{r}^a \, \bar{\mathcal{A}}_0 . \tag{21.37}$$

For this case the static field equations, which are the analog of (21.24) become,

$$h'' + \frac{2}{r} h' - \frac{2u^2 h}{r^2} + \lambda(v^2 - h^2) h = 0,$$

$$u'' - \frac{(u^2 - 1)u}{r^2} - e^2 u(h^2 - j^2) = 0,$$

$$j'' + \frac{2}{r} j' - \frac{2u^2 j}{r^2} = 0, \tag{21.38}$$

where the first equation is identical to the one in (21.24), in the second there is a $2e^2 u j^2$ addition and the third equation is the Gauss law.

The dependence of Q_E on w can be determined by substituting the ansatz into (21.34) and recalling that $W_0 = 0$ we find,

$$Q_E = \frac{8\pi w}{e} \int dr u(r)^2 \frac{j(r)}{j(\infty)} \equiv Iw. \tag{21.39}$$

The integral can be estimated since $u(r)$ falls off exponentially outside a region of radius $\sim 1/v$ so that,

$$I = \frac{4\pi}{e^2 v} \bar{I}, \tag{21.40}$$

where $\bar{I}$ is of order unity. In analogy to (21.33) one can show that the correction to the energy is,

$$\Delta E = \frac{Q_E^2}{2eI} = \left(\frac{e}{4\pi}\right)^2 \frac{Q_E^2 M}{2\bar{I} f} \sim \frac{Q_E^2}{2Q_M^2} M. \tag{21.41}$$

21.7 BPS monopoles and dyons

A special limit of the monopole configuration occurs when one takes the limit of,

$$\lambda \to 0 \quad \mu^2 \to 0 \quad v^2 = \frac{\mu^2}{\lambda} \text{ fixed.} \tag{21.42}$$

In this limit the energy of the system,

$$\begin{aligned} E &= \int d^3x \text{Tr}[\vec{E} \cdot \vec{E} + \vec{B} \cdot \vec{B} + (D_0\Phi)^2 + (\vec{D}\Phi)^2] \\ &= \int d^3x \text{Tr}[(\vec{E} \pm \sin\alpha \vec{D}\Phi)^2 + (\vec{B} \pm \cos\alpha \vec{D}\Phi)^2 + (D_0\Phi)^2] \\ &\pm 2 \int d^3x [\cos\alpha \text{Tr}[\vec{B} \cdot \vec{D}\Phi] + \sin\alpha \text{Tr}[\vec{E} \cdot \vec{D}]]. \end{aligned} \tag{21.43}$$

The passage from the first line to the rest is of course an identity for arbitrary α. Next we perform an integration by parts in the last line and use the Gauss law $\vec{D} \cdot E - ie[\Phi, D^0\Phi] = 0$ and $\vec{D} \cdot \vec{B} = 0$ we find,

$$\begin{aligned} E &= \int d^3x \text{Tr}[(\vec{E} \pm \sin\alpha \vec{D}\Phi)^2 + (\vec{B} \pm \cos\alpha \vec{D}\Phi)^2 + (D_0\Phi)^2 \\ &\pm \cos\alpha \mathcal{Q}_M \pm \sin\alpha \mathcal{Q}_E \\ E &\geq \pm \cos\alpha \mathcal{Q}_M \pm \sin\alpha \mathcal{Q}_E, \end{aligned} \tag{21.44}$$

where the magnetic and electric charges $\mathcal{Q}_M = vQ_M$ and $\mathcal{Q}_E = vQ_E$ are given by (21.14) and (21.35)

$$\mathcal{Q}_M = 2 \int d^2\vec{S} \cdot \text{Tr}[\Phi\vec{B}] \quad \mathcal{Q}_E = 2 \int d^2\vec{S} \cdot \text{Tr}[\Phi\vec{E}]. \tag{21.45}$$

Recall that so far α is arbitrary. The most stringent bound is achieved when one takes $\tan\alpha = \frac{\mathcal{Q}_E}{\mathcal{Q}_M}$, for which the bound reads,

$$E \geq \sqrt{\mathcal{Q}_M^2 + \mathcal{Q}_E^2}. \tag{21.46}$$

It is easy to realize that the bound is saturated if,

$$\vec{E} = \cos\alpha \vec{D}\Phi \quad \vec{B} = \sin\alpha \vec{D}\Phi \quad D_0\Phi = 0. \tag{21.47}$$

These first-order equations are referred to as the Bogomolny Prasad Sommerfeld equations or BPS equations ([41] and [180]). The configurations that obey these equations have the minimal value of energy,

$$E = \sqrt{\mathcal{Q}_M^2 + \mathcal{Q}_E^2}, \tag{21.48}$$

for given magnetic and electric charges $\mathcal{Q}_M, \mathcal{Q}_E$ respectively and hence are also solutions of the (second-order) equations of motion of the system. In particular the magnetic monopole which carries no electric charge and is static and hence $Q_E = 0$ and $D_0\Phi = 0$ (in the $A_0 = 0$ gauge), obeys the Bogomolny equation,

$$\vec{B} = \vec{D}\Phi. \tag{21.49}$$

The BPS limit seems to be unnatural and artificial since we have introduced a potential to give a non-trivial expectation value to the field Φ and then we tuned the potential to zero keeping the expectation value. It turns out that in certain suspersymmetric models the BPS equations follow from the requirement of the invariance under supersymmetry. Since supersymmetry is not discussed in this book we refer the reader to the literature, for instance [214].

If we go back to the equations of motion and substitute $\lambda = 0$ the equations take the form,

$$h'' + \frac{2}{r}h' - 2\frac{u^2 h}{r^2} = 0,$$

$$u'' - \frac{u(u^2 - 1)}{r^2} - e^2 u h^2 = 0. \tag{21.50}$$

The solution of this set of equations is given by,

$$u(r) = \frac{evr}{\sinh(evr)} \qquad h(r) = v\coth(evr) - \frac{1}{er} \tag{21.51}$$

The fact that Φ falls off as $1/r$ and not exponentially is due to the fact that in the BPS limit it has a vanishing mass and hence associates with a long range force.

The BPS equations can also be solved for the case of a dyon, namely a configuration that carries both a magnetic as well as an electric charge. In that case the solution reads,

$$u(r) = \frac{e\tilde{v}r}{\sinh(e\tilde{v}r)}$$

$$h(r) = \frac{\sqrt{Q_M^2 + Q_E^2}}{Q_M}\left[\tilde{v}\coth(e\tilde{v}r) - \frac{1}{er}\right]$$

$$j(r) = -\frac{Q_M}{Q_E}\left[\tilde{v}\coth(e\tilde{v}r) - \frac{1}{er}\right]. \tag{21.52}$$

21.8 Montonen Olive duality

In two dimensions we have seen an equivalence between a soliton configuration and an elementary field. This was the essence of bosonization manifested for instance in the equivalence between the sine-Gordon theory and the Thirring model (see Section 6.2). Montonen and Olive [163] conjectured an analogous duality between the spectrum of states created from the elementary field and from those created by the solitons. Since the former are electrically charged and the latter are magnetically charged, the duality is in a sense a non-abelian

Table 21.1. *The spectrum of the* $SU(2)$
YM Higgs theory in the BPS limit

	Mass	Q_E	Q_M
photon	0	0	0
ϕ	0	0	0
$W^\pm$	ev	$\pm e$	0
Monopole	$\frac{4\pi v}{e}$	0	$\pm \frac{4\pi}{e}$

generalization of electric–magnetic duality of the Maxwell equations. To understand this duality notice that under the following operation,

$$Q_{\rm E} \leftrightarrow Q_{\rm M} \qquad e \leftrightarrow \frac{4\pi}{e}, \tag{21.53}$$

the entries associated with the W mesons and those of the magnetic monopoles in Table 21.1 are interchanged.

It turns out that on top of the self-duality of the spectrum, there is a similar duality also in the low energy scattering. It is a well-known property of BPS states in general and in particular the magnetic monopoles that there is no net force between them. This follows up from an exact cancellation between the magnetic repulsion and the attraction due to an exchange of a Higgs scalar. The zero velocity limit of the scattering amplitude of two W bosons also admits a no force behavior. The exchange of a single photon is cancelled out by the exchange of a Higgs boson.

In the non-supersymmetric YM Higgs theory the duality of the spectrum and the scattering amplitudes cannot be lifted to a duality of the full theory. A simple indication of this is the fact that the W boson carries a spin one, whereas the quantum state built from a spherical monopole carries spin 0. In supersymmetric analogs of the YM Higgs theory this difficulty may be overcome since both the W bosons as well as the magnetic monopoles are members of supersymmetric multiplets that contain states with several different spins. It turns out that the YM theory with 16 supercharges, the so-called $\mathcal{N} = 4$ SYM admits a complete invariance under the Olive Montonen duality. Since we do not discuss supersymmetry in the book we refer the interested reader to the review papers mentioned above, for instance [214].

21.9 Nahm construction of multimonopole solutions

This construction [167] maps the Bogomolny equation in three variables into a nonlinear equation in one variable. We present the construction for the $SU(2)$ case. To simplify the notations we set the coupling constant $e = 1$. It can be restored when needed.

The construction of an $SU(2)$ k monopole is built from three steps:

1. We look for a quartet of Hermitian $k \times k$ matrices $T_\mu(s)$, where $\mu = 0, i = 1, 2, 3$ which satisfy the Nahm equation,

$$\frac{dT_i}{ds} + i[T_0, T_i] + \frac{i}{2}\epsilon_{ijk}[T_j, T_k] = 0, \tag{21.54}$$

where s is an auxiliary variable that takes its value in the interval $-v/2 \leq s \leq v/2$ and where v is the vacuum expectation value of the Higgs field. For $k = 1$ since the commutators vanish we get that T_i are constants. In fact due to the ordinary gauge invariance one can choose a gauge where $T_0 = 0$ and hence the equation reads,

$$\frac{dT_i}{ds} + \frac{i}{2}\epsilon_{ijk}[T_j, T_k] = 0. \tag{21.55}$$

The boundary condition that one should impose for the multimonople case is that the $T_i(s)$ have poles at the boundaries of the form,

$$T_i(s) = -\frac{L_i^\pm}{s \mp \frac{v}{2}} + O(1). \tag{21.56}$$

The Nahm equations implies that the $L_i^\pm$ form a k-dimension representation of the $SU(2)$ algebra,

$$[L_i^\pm, L_j^\pm] = i\epsilon_{ijk}L_k^\pm. \tag{21.57}$$

These representations should be irreducible, namely, must be equivalent to the $(k-1)/2$ representation.

2. The next step is to solve the construction equation for the $2k$ component vector $w(s, \vec{r})$,

$$\Delta^\dagger w(s, \vec{r}) \equiv \left[-\frac{d}{ds} - T_i \otimes \sigma_i + r_i I_k \otimes \sigma_i \right] w(s, \vec{r}) = 0. \tag{21.58}$$

We denote by $w_a(s, \vec{r})$ a completely linearly independent set of normalizable solutions that obey the orthonormality condition,

$$\int_{-v/2}^{v/2} dsw_a^\dagger(s, \vec{r})w_b(s, \vec{r}) = \delta_{ab}. \tag{21.59}$$

3. It can be proved that for the $SU(2)$ case there are only two normalizable solutions $w_a(s, \vec{r})$. The space-time fields are given in terms of these as follows,

$$\Phi^{ab}(\vec{r}) = \int_{-v/2}^{v/2} dsw_a^\dagger(s, \vec{r})w_b(s, \vec{r}),$$

$$A_j^{ab}(\vec{r}) = \int_{-v/2}^{v/2} dsw_a^\dagger(s, \vec{r})\partial_j w_b(s, \vec{r}). \tag{21.60}$$

We do not bring here the the details of the proof of this construction (see for instance [71]) we just mention that it includes the following elements: (i) Using the expressions for Φ^{ab} and A_i^{ab} given above in (21.60) one can show that B_i^{ab} and $D_i \Phi^{ab}$ are identical and hence the Bogomolny equation is obeyed. (ii) Showing that the solutions lie in $SU(2)$ and (iii) computing the long-range behavior and showing that the solutions indeed have a magnetic charge equal to k. We refer the interested reader to, for instance, [214] for the detailed proof. Instead we proceed now to a demonstration of the application of the method both for $k = 1$ and $k = 2$. As was mentioned above for the former case the T_i are constants independent of s. In fact these constant values of $\vec{T}$ enter the construction equation as $(\vec{r} - \vec{T}) \cdot \vec{\sigma}$ namely, they are the coordinates of the center of mass of the monopole and thus by shifting to a frame of coordinates which is centered at the monopole center $\vec{T} = 0$. The construction equation (21.58) takes the form,

$$\frac{dw}{ds} = \vec{r} \cdot \vec{\sigma} w. \tag{21.61}$$

The two normalizable solutions of this equation are,

$$w_a(s, \vec{r}) = \mathcal{N}(r) e^{s\vec{r} \cdot \vec{\sigma}} \eta_a, \tag{21.62}$$

where $\mathcal{N}$ is a normalization factor and η_a are orthonormal constant vectors. From the orthonormality condition we find that,

$$\mathcal{N}(r) = \sqrt{\frac{r}{\sinh(vr)}}, \tag{21.63}$$

where we have made use of the first of the following integrals,

$$\int_{-v/2}^{v/2} e^{2s\vec{r} \cdot \vec{\sigma}} ds = \frac{\sinh(vr)}{r} I_2$$

$$\int_{-v/2}^{v/2} s e^{2s\vec{r} \cdot \vec{\sigma}} ds = \frac{\vec{r} \cdot \vec{\sigma}}{r^3} [vr \cosh(vr) - \sinh(vr)]. \tag{21.64}$$

Using the integral we find that the Higgs field and the corresponding gauge field are given by,

$$\Phi_{ab} = \frac{1}{2}\left(v \coth(vr) - \frac{1}{r}\right) \eta_a^\dagger \frac{\vec{r} \cdot \sigma}{r} \eta_b,$$

$$\vec{A}_{ab} = -i\eta_a^\dagger \partial_i \eta_b - i\epsilon_{ijk} \hat{r}_j \eta_a \sigma_k \eta_b \left(\frac{1}{2r} - \frac{v}{2\sinh(vr)}\right). \tag{21.65}$$

Upon setting $\eta_1^t = (1, 0)$ and $\eta_2^t = (0, 1)$ we retrieve the hedgehog solution of 21.51).

21.9.1 *SU(2) two-monopole solutions*

The $k = 2$ solutions are characterized by a priori eight parameters out of which one corresponds to the $U(1)$ phase and as for the case of $k = 1$ does not enter the Nahm data. Three parameters relate to the translation and three to the rotations of the solution. Thus we end up with one non-trivial parameter. Intuitively this parameter should relate to the separation distance between the two centers of the solution. Let us see if the Nahm construction verifies this intuition and to what extent one can write an explicit solution for this case. The $T_i(s)$ which now are not constants can be decomposed into the following form,

$$T_i(s) = \frac{1}{2}\vec{T}_i^v(s) \cdot \vec{\sigma} + T_i^s(s)I_2. \tag{21.66}$$

Substituting this into the Nahm equation implies that the $\vec{T}_i^v(s)$ have to obey,

$$\frac{d\vec{T}_i^v(s)}{ds} = \frac{1}{2}\epsilon_{ijk}\vec{T}_j^v(s) \times \vec{T}_k^v(s). \tag{21.67}$$

It is easy to realize that there is a relation between the $\vec{T}_k^v(s)$, namely the following matrix,

$$T_{ij} = \vec{T}_i^v(s) \cdot \vec{T}_j^v(s) - \frac{1}{3}\delta_{ij}\vec{T}_k^v(s) \cdot \vec{T}_k^v(s), \tag{21.68}$$

is s independent. After some tedious algebra one can show that the most general form of $\vec{T}_k^v(s)$ takes the form,

$$\vec{T}_i^v(s) = \frac{1}{2}\sum_i A_{ij}f_j(s + v/2, \kappa, d)\tau_j + T_i^s I_2, \tag{21.69}$$

where A_{ij} is an orthogonal matrix of constants that diagonalize T_{ij}, $f(s + v/2, \kappa, d)$ are the Euler top functions[2] and the parameter d can be shown for widely separated monopoles to be the inter-monopole distance as the original intuition taught us.

21.10 Moduli space of monopoles

As we have seen above the monopole configurations are parameterized by certain moduli. One defines a space of these moduli, the moduli space. The properties of this space are intimately related to the low energy behavior of monopoles and dyons. In Section 22.3 we will describe the moduli space of YM instantons. Denoting the collective coordinates that parameterize a monopole configuration

---

[2] The Euler top function can be expressed in terms of the elliptic functions $SN_\kappa(x), CN_\kappa$ and $DN_\kappa(x)$ as follows:

$$f_1(x, \kappa, D) = -D\frac{CN_\kappa(Dx)}{SN_\kappa(Dx)}$$
$$f_2(x, \kappa, D) = -D\frac{DN_\kappa(Dx)}{SN_\kappa(Dx)}$$
$$f_3(x, \kappa, D) = -D\frac{1}{SN_\kappa(Dx)} \tag{21.70}$$

by z_r, the Lagrangian of the system is approximated by,

$$\mathcal{L} = -(\text{total rest mass of monopoles}) + \frac{1}{2} g_{rs}(z) \dot{z}_r \dot{z}_s. \qquad (21.71)$$

The metric on the moduli space $g_{rs}(z)$ can be determined from the background zero models of the gauge fields as follows,

$$g_{rs}(z) = 2 \int d^3 x \text{Tr}[\delta_r A_i \delta_s A_i + \delta_r \Phi \delta_s \Phi] = 2 \int d^3 x \text{Tr}[\delta_r A_a \delta_s A_a], \qquad (21.72)$$

where a takes the values $a = 1, \ldots, 4$ with $A_4 = \Phi$, and where,

$$\delta_r A^a = \frac{\partial (A^{cl})^a}{\partial z_r} - D^a \epsilon_r, \qquad (21.73)$$

and where ϵ_r is defined via $A_0 = \dot{z}_r \epsilon_r$ which follows from the Gauss law, $D_a F^{a0} = 0$.

In the case of a single monopole, as we have seen above, there are four zero modes associated with the location of the center of the monopole $\vec{r}_{cm}$ and the global $U(1)$ phase so that,

$$g_{rs}(z) \dot{z}_r \dot{z}_s = M(\dot{\vec{r}}_{cm})^2 + \frac{I}{e} \dot{\alpha}^2, \qquad (21.74)$$

where M is the mass of the monopole and I is defined via (21.39) $Q_E = Iw$.

The moduli space of BPS monopoles is hyper-Kähler. This property that implies that there are three almost complex structures with correspondingly three closed Kähler forms will be discussed in detail in Chapter 22, so we will not discuss it here for the BPS monopoles.

An important part of the structure of the moduli space is encoded in its isometries, namely the symmetries that preserve the form of the metric. Naturally since the underlying space where the monopoles reside is an R^3 the isometries include three translations of the center of mass of the collective coordinates. The same applies to the spatial rotation of the monopole. It takes a monopole solution to another monopole solution and hence it maps one point in the moduli space into another one. Another type of isometries is those associated with the unbroken $U(1)$ gauge groups. These isometries, unlike the rotational isometry preserve the complex structure. One can choose a coordinate basis where the gauge transformations act as translations of the angular variables ξ^A, with the corresponding Killing vectors being $K_A = \frac{\partial}{\partial \xi^A}$. Denoting by y^p the rest of the coordinates, the Lagrangian associated with the moduli space approximation can be written as,

$$L = \frac{1}{2} g_{pq}(y) \dot{y}^p \dot{y}^q + \frac{1}{2} \tilde{g}_{AB}(y)[\dot{\xi}^A + \dot{y}^p w_p^A(y)][\dot{\xi}^B + \dot{y}^q w_q^B(y)]. \qquad (21.75)$$

Thus the coordinates ξ^A are cyclic coordinates and their conjugate momenta are conserved. In fact the latter are the electric charges of the dyonic cores.

For monopoles that are separated by large distances the task of determining the metric of the moduli is much easier. Consider a system of N fundamental monopoles all well separated from each other. In such a layout only abelian interactions are relevant and there is an enhanced gauge symmetry. Instead of having a conserved electric charge for each unbroken $U(1)$, there is an effective conserved charge for each monopole core. The moduli space is spanned in this case by $3N$ coordinates of the positions of the cores and N angles ξ_j, $j = 1, \ldots, N$. The enhance symmetry is the translation along each of the ξ_j. The approximated Lagrangian then takes the form,

$$L = \frac{1}{2} M_{ij}(x)\vec{\dot{x}}^i \cdot \vec{\dot{x}}^j + \frac{1}{2}\tilde{g}_{ij}(x)[\dot{\xi}^i + w_k^i(x)\vec{\dot{x}}^k][\dot{\xi}^j + w_l^i(x)\vec{\dot{x}}^l]. \tag{21.76}$$

By computing the pairwise interactions between the separated dyons, one can determine the functions $M_{ij}(x), \tilde{g}_{ij}(x)$ and $w_j^i(x)$. We refer the reader to [214] for the derivation and we cite here the results,

$$M_{ij} = m_i - \sum_{k \neq 1} \frac{4\pi \vec{\alpha}_i^* \cdot \vec{\alpha}_k^*}{e^2 r_{ik}}, \quad i = j \tag{21.77}$$

$$M_{ij} = \sum_{k \neq 1} \frac{4\pi \vec{\alpha}_i^* \cdot \vec{\alpha}_j^*}{e^2 r_{ij}}, \quad i \neq j \tag{21.78}$$

$$\vec{W}_{ij} = -\sum_{k \neq 1} \vec{\alpha}_i^* \cdot \vec{\alpha}_k^* \vec{w}_{ik}, \quad i = j \tag{21.79}$$

$$\vec{W}_{ij} = \vec{\alpha}_i^* \cdot \vec{\alpha}_i^* w_{ij}, \quad i \neq j \tag{21.80}$$

and $K = \frac{(4\pi)^2}{e^4} M^{-1}$. It can be further shown that the metric of the moduli space of a two monopole BPS solution can be determined exactly and it takes the form of a Taub–Nut metric or Atiya–Hitchin metric [19] depending whether the monopoles are distinct or the same. This is beyond the scope of this book and we refer the reader to for instance the review [214].

22

Instantons of QCD

In Chapter 5 we saw that solutions of the classical equations of motion, which are characterized by a topological number, play an important role in two-dimensional QFT. Derick's theorem (5.36) forbids scalar field soliton solutions in higher than two-dimensional space-time. However, for gauge fields one can bypass the theorem, and indeed, as we have seen in Chapter 21, there are solitons in the form of magnetic monopoles in four-dimensional gauge theories. The topic of this chapter will be solutions of the Yang–Mills theory defined on a Euclidean space-time which have finite action and are topological in their nature, the *instantons*. We will start with a description of the basic properties of one instanton solution including the topological charge that characterizes it. We then describe the construction of multi-instanton solutions and the moduli space of instantons including its dimension, complex nature, singularities and symmetries. When Wick rotated to Minkowski space-time the instanton describes a tunneling process between different vacua. We will elaborate on this phenomenon in the context of the four-dimensional YM theory. Various properties of QCD and hadron physics were thought to be related to instantons. In certain cases like confinement, the relation to instantons is still a mystery. One case where the role of instantons is clear is the $U(1_A)$ problem. This will be described in the last section of this chapter.

The one instanton solution was derived in [32]. The basic properties of instantons were worked out by many authors including [125], [57]. The instantons of $SU(N)$ gauge symmetry were derived in [218]. There are several review papers such as [65], [208] and [189] and books [182] and [188] that describe the basic instanton solutions.

22.1 The basic properties of the instanton

The action of the four-dimensional YM theory in Euclidean space-time can be rewritten in the following form,

$$S = \frac{1}{2g^2} \int \mathrm{d}^4 x \, \mathrm{Tr} \, [F^{\mu\nu} F_{\mu\nu}] = \frac{1}{4g^2} \int \mathrm{d}^4 x F^{a\,\mu\nu} F^a_{\mu\nu}$$

$$S = \frac{1}{2g^2} \int \mathrm{d}^4 x \left[\pm \mathrm{Tr} \, [F^{\mu\nu \, *} F_{\mu\nu}] + \frac{1}{2} \mathrm{Tr} \, [(F_{\mu\nu} \mp \,^* F_{\mu\nu})(F^{\mu\nu} \mp^* F^{\mu\nu})] \right], \quad (22.1)$$

where as usual $F_{\mu\nu} = F_{\mu\nu}^a T^{a\,1}$ and $^*F_{\mu\nu} = \frac{1}{2}\epsilon_{\mu\nu\rho\sigma} F_{\rho\sigma}$ is the dual field strength, and g is the YM coupling. The first term in the second line is a topological invariant, or a topological charge[2] since,

$$Q = \frac{1}{16\pi^2} \int \mathrm{d}^4 x \, \mathrm{Tr}\,[F^{\mu\nu\,*}F_{\mu\nu}] = \int \mathrm{d}^4 x \, \partial_\mu K_\mu = \oint \mathrm{d}\sigma_\mu K_\mu, \tag{22.2}$$

where,

$$K_\mu = \frac{1}{8\pi^2}\epsilon_{\mu\nu\rho\sigma}\,\mathrm{Tr}\,\left[A_\nu\left(\partial_\rho A_\sigma + \frac{3}{2}A_\rho A_\sigma\right)\right]. \tag{22.3}$$

In fact Q is the Pontryagin index or the winding number of maps from the sphere at space-time infinity to the $SU(2)$ group manifold which is also the three sphere, namely $S_s^3 \to S_g^3$. This topological invariant is the homotopy $\pi_3(S^3)$ which is an integer $\pi_3(S^3) \in \mathcal{Z}$. This can be shown as follows. Since the self-dual field is asymptotically a pure gauge, namely on σ_μ $A_\mu = U\partial_\mu U^{-1}$ and $F_{\mu\nu} = 0$ hence,

$$Q = \frac{1}{24\pi^2}\oint \mathrm{d}\sigma_\mu\epsilon_{\mu\nu\rho\sigma}\,\mathrm{Tr}\,[A_\nu A_\rho A_\sigma]$$

$$Q = \frac{1}{24\pi^2}\oint \mathrm{d}\sigma_\mu\epsilon_{\mu\nu\rho\sigma}\,\mathrm{Tr}\,[(\partial_\nu U)U^{-1}(\partial_\rho U)U^{-1}(\partial_\sigma U)U^{-1}]. \tag{22.4}$$

If we take for U the following ansatz that will be shown below to correspond to the one instanton solution,

$$U(x) = \hat{x}_\mu\sigma_\mu = x_0 + ix_i\sigma_i, \tag{22.5}$$

we found,

$$Q = \oint \mathrm{d}\sigma_\mu K_\mu = -\frac{1}{24\pi^2}\int \mathrm{d}\sigma_\mu\left(\frac{-12x_\mu}{|x^4|}\right)$$

$$= \frac{1}{2\pi^2}\int \mathrm{d}\Omega x_\mu |x|^2\frac{x_\mu}{|x|^4} = \frac{1}{2\pi^2}\int \mathrm{d}\Omega = 1. \tag{22.6}$$

Thus we have shown that Q is indeed the winding number that measures how many times we wind S_g^3 when we integrate over S_s^3.

Let us now return to (22.1). Since it is a sum of a topological charge and a positive semi-definite quantity, it is clear that it is minimized when the latter vanishes namely,

$$F_{\mu\nu} = \pm^*F_{\mu\nu}. \tag{22.7}$$

The corresponding gauge fields A_μ (with a $+$ sign) will be referred to as *instanton* or *self-dual* gauge field and those with a $-$ sign as *anti instanton* or *anti self-dual*

---

[1] In this chapter we denote the $SU(N)$ adjoint indices with $a = 1, \ldots, N^2 - 1$ whereas in Chapter 19 we used A and not a.

[2] Recall in analogy the topological charge defined in two-dimensional scalar field theories (5.3).

gauge field. It is straightforward to show that the self-duality condition implies the equation of motion,

$$D_\mu F_{\mu\nu} = 0. \tag{22.8}$$

A solution of the equation of motion is not necessary self-dual but it can be shown that the non-self-dual configurations are saddle points and not local minima of the action.

Comparing the expression of the action (22.1) and the topological charge (22.2) it is clear that a (anti) self-dual configuration that carries an instanton number $(Q = -k), Q = k$ has an action of,

$$S = \frac{8\pi^2}{g^2}|k|. \tag{22.9}$$

One can add the topological charge as an additional term to the action. To be more precise one adds a θ term,

$$S_\theta = i\frac{\theta}{16\pi^2}\int d^4x \, \text{Tr}\left[F^{\mu\nu}{}^*F_{\mu\nu}\right] = i\theta k. \tag{22.10}$$

Whereas the ordinary YM action is the same for the instanton and anti-instanton, the θ term obviously distinguishes between them by assigning opposite charges to them. We will further discuss the *theta* term in Section 22.5. For the self-dual solution up to a constant the action is equal to the topological charge which by definition does not depend on the metric. This exhibits the topological nature of the instanton. Another indication of this nature is the fact that it has vanishing energy-momentum tensor as follows from,

$$T_{\mu\nu} = -\frac{2}{g^2}\,\text{Tr}\left[F_{\mu\rho}F_{\nu\rho} - \frac{1}{4}\delta_{\mu\nu}F_{\alpha\beta}F^{\alpha\beta}\right] = 0. \tag{22.11}$$

This clearly implies that instantons do not curve the space-time they reside in.

The one instanton solution for the $SU(2)$ can be constructed from $U(x)$ given in (22.5) via,

$$A_\mu^a(x) = U^{-1}\partial_\mu U\frac{\rho^2}{(x-X)^2 + \rho^2}, \tag{22.12}$$

which yields the explicit form,

$$A_\mu^a(x) = 2\frac{\eta_{\mu\nu}^a(x-X)^\nu}{(x-X)^2 + \rho^2}, \tag{22.13}$$

where $\eta_{\mu\nu}^a$ is the 't Hooft antisymmetric symbol defined by,

$$\begin{aligned} \eta_{\mu\nu}^a &= \epsilon_{\mu\nu}^a \quad \mu,\nu = 1,2,3 \quad \eta_{\mu 4}^a = -\eta_{4\mu}^a = \delta_\mu^a \\ \bar{\eta}_{\mu\nu}^a &= \epsilon_{\mu\nu}^a \quad \mu,\nu = 1,2,3 \quad \bar{\eta}_{\mu 4}^a = -\bar{\eta}_{4\mu}^a = -\delta_\mu^a. \end{aligned} \tag{22.14}$$

It is easy to check that $\eta^a_{\mu\nu}$ and $\bar{\eta}^a_{\mu\nu}$ are self-dual and anti self-dual respectively. The corresponding field strength takes the form,

$$F^a_{\mu\nu} = -4\eta^a_{\mu\nu} \frac{\rho^2}{[(x-X)^2+\rho^2]^2}. \tag{22.15}$$

Obviously since $\eta^a_{\mu\nu}$ is self-dual so is $F^a_{\mu\nu}$.

The one instanton solution is characterized by eight parameters, four correspond to the center of the instanton X_μ, one to the size of the instanton ρ and three to three global $SU(2)$ gauge transformations. Recall that fixing a guage we fix only the local gauge transformations. The space of parameters of the k instanton solutions will be further addressed in Section 22.3 where it will be shown that in general for $SU(N)$ the dimension of the moduli space is $4kN$.

The instanton solution (22.13) falls off asymptotically as $\frac{1}{x}$ and hence it contributes a finite amount to the integral of the topological charge (winding number) which is of the form $\int A^3 x^3 d\Omega$. The field strength falls off as $1/x^4$ such that indeed the corresponding action is finite. However due to the $\frac{1}{x}$ asymptotic behavior it is difficult to form square integrable expressions that contain it. For that purpose one can use the following singular gauge transformation,

$$U = \frac{\sigma^\dagger_\mu (x-X)^\mu}{|x-X|}, \tag{22.16}$$

which renders the instanton to have a $\frac{1}{x^3}$ fall off as can be seen from,

$$A^a_\mu = \frac{1}{g} \frac{2\rho^2 (x-X)^\nu \bar{\eta}^a_{\nu\mu}}{(x-X)^2 [(x-X)^2+\rho^2]}. \tag{22.17}$$

The singular instanton is obviously singular at the location of the instanton $x_\mu = X_\mu$. This singularity is not physical and can be removed by a gauge transformation or by puncturing the Euclidean space with the singular point being removed.

Instantons of $SU(N)$ gauge theory can be constructed by embedding $SU(2)$ instantons in $SU(N)$ for instance,

$$A^{SU(N)}_\mu = \begin{pmatrix} 0 & 0 \\ 0 & A^{SU(2)}_\mu \end{pmatrix}, \tag{22.18}$$

where the instanton is the 2×2 matrix on the lower right. The most general $SU(N)$ one instanton configuration can be derived from (22.18) by the following transformation,

$$A^{SU(N)}_\mu = U^\dagger \begin{pmatrix} 0 & 0 \\ 0 & A^{SU(2)}_\mu \end{pmatrix} U \quad U \in \frac{SU(N)}{SU(N-2) \times U(1)}. \tag{22.19}$$

Operating with $U \in SU(N-2) \times U(1)$ obviously leaves the basic configuration invariant and thus only transformations with elements of the coset are relevant. This is in accordnace with the fact that for a k instanton solution the stability group is $S(U(N-2k) \times U1))$ as will be shown in Section 22.3. The dimension of the coset is $N^2 - 1 - ((N-2)^2 = 4N - 5$. Together with the 5 parameters of the location and the size we have $4N$ collective coordinates. Indeed in Section 22.3 we will see that in general the dimension of the moduli space is $4Nk$ for instanton number equals k solution. To demonstrate this counting consider the case of $SU(3)$ for which the generators are the Gell–Mann matrices $\{\lambda^a\}$, $a = 1, \ldots, 8$. The first three generators λ_a, $a = 1, 2, 3$ form the $SU(2)$ $k = 1$ instanton. $\lambda_4, \ldots, \lambda_7$ form two doublets under this $SU(2)$ so they can generate new solutions while λ_8 commutes with the $SU(2)$ and hence leaves the basic $SU(2)$ solution invariant.

One can express the instanton solution in terms of quaternionic notation. This will turn out to be convenient for the ADHM construction of multi-instanton solutions (see Section 22.2). The idea is to make use of the representations of the covering group $SU(2)_L \times SU(2)_R$ of the Lorentz group of four-dimensional Euclidean space-time rather than the $SO(4)$ Lorentz group itself. In particular we represent any four vector of $SO(4)$ as a $(2, 2)$ representation of $SU(2)_L \times SU(2)_R$, for instance the four vector x_μ is denoted by $x_{\alpha\dot\alpha}$ or $x^{\dot\alpha\alpha}$ defined as follows,

$$x_{\alpha\dot\alpha} = x_\mu \sigma^\mu_{\alpha\dot\alpha} \quad x^{\dot\alpha\alpha} = x_\mu \bar\sigma^{\dot\alpha\alpha}_\mu, \tag{22.20}$$

where $\sigma^\mu_{\alpha\dot\alpha}$ is a 2×2 matrix of $(i\vec\sigma, 1)$ and $\bar\sigma^\mu_{\dot\alpha\alpha} = (\sigma^\mu_{\alpha\dot\alpha})^\dagger$. In terms of the quaternionic notation the one instanton solution (22.13) for $SU(2)$ gauge theory is given by,

$$A_\mu = \frac{1}{g} \frac{2(x-X)^\nu \Sigma_{\nu\mu}}{(x-X)^2 + \rho^2}, \tag{22.21}$$

where $\Sigma_{\mu\nu}$, which were introduced in Section 17.1, are the part of the Lorentz generators that do not act on the space-time coordinates but only on the internal degrees of freedom. Here using the $SU(2)_L \times SU(2)_R$ notation we define them as follows,

$$\Sigma_{\mu\nu} = \frac{1}{4}(\sigma_\mu \bar\sigma_\nu - \bar\sigma_\nu \sigma_\mu) \quad \bar\Sigma_{\mu\nu} = \frac{1}{4}(\bar\sigma_\mu \sigma_\nu - \sigma_\nu \bar\sigma_\mu). \tag{22.22}$$

The self-duality property of the instanton configuration follows trivially from the fact that $\Sigma_{\mu\nu}$ is self-dual, namely,

$$\Sigma_{\mu\nu} = \frac{1}{2}\epsilon_{\mu\nu\lambda\rho}\Sigma^{\lambda\rho} \quad \bar\Sigma_{\mu\nu} = -\frac{1}{2}\epsilon_{\mu\nu\lambda\rho}\bar\Sigma^{\lambda\rho}. \tag{22.23}$$

The corresponding field strength reads,

$$F_{\mu\nu} = \frac{1}{g} \frac{4\rho\Sigma_{\mu\nu}}{((x-X)^2 + \rho^2)^2}. \tag{22.24}$$

As was discussed above, the instanton solution (22.21) falls off asymptotically as $\frac{1}{x}$ and hence it is difficult to form square integrable expressions that contain it. The solution in the singular gauge now reads,

$$A_\mu = \frac{1}{g} \frac{2\rho^2 (x-X)^\mu \bar\Sigma_{\nu\mu}}{((x-X)^2(x-X)^2 + \rho^2)}. \tag{22.25}$$

22.2 The ADHM construction of instantons

The vacua of the YM theory is given by pure gauge configurations (which can be written in terms of a double index notation,

$$A_{ij}^{\mu} = \frac{1}{g} U^{\dagger l}_{i} \partial^{\mu} U_{lj}, \tag{22.26}$$

where $i, j, l = 1, \ldots, N$. It is straightforward to check that this gauge field obeys the self-duality condition. The idea of the ADHM construction[3] is to generalize this configuration also to the k instanton case by taking now the matrices U to be of the form $U_{\mathcal{I}i}$ where $\mathcal{I} = 1, \ldots, N + 2k$ with the orthonormality condition,

$$U^{\dagger \mathcal{I}}_{i} U_{\mathcal{I}j} = \delta_{ij}. \tag{22.27}$$

The U matrices are the basis vectors of a null space,

$$\Delta^{\dagger \dot{\alpha} \mathcal{I}}_{I} U_{\mathcal{I}i} = 0 = U^{\dagger \mathcal{I}}_{i} \Delta_{\mathcal{I}I\dot{\alpha}}, \tag{22.28}$$

where $I = 1, \ldots, k$ and $\Delta_{\mathcal{I}I\dot{\alpha}}$ is a $(N + 2k) \times 2k$ complex valued matrix which is taken to be linear in the space-time coordinate x_{μ}, namely takes the form,

$$\Delta_{\mathcal{I}I\dot{\alpha}} = a_{\mathcal{I}I\dot{\alpha}} + b^{\alpha}_{\mathcal{I}I} x_{\alpha\dot{\alpha}} \qquad \Delta^{\dagger \dot{\alpha} \mathcal{I}}_{I} = a^{\dagger \dot{\alpha} \mathcal{I}}_{I} + x^{\dot{\alpha}\alpha} b^{\mathcal{I}}_{I\alpha}, \tag{22.29}$$

and with $\Delta^{\dagger \dot{\alpha} \mathcal{I}}_{I} \equiv (\Delta_{\mathcal{I}I\dot{\alpha}})^{*}$.

The ADHM k instanton solution of the form (22.26) is self-dual if one further requires that $\Delta_{\mathcal{I}I\dot{\alpha}}$ obeys the following condition,

$$\Delta^{\dagger \dot{\alpha} \mathcal{I}}_{I} \Delta_{\mathcal{I}J\dot{\beta}} = \delta^{\dot{\alpha}}_{\dot{\beta}} (f^{-1})_{IJ}, \tag{22.30}$$

where f is an arbitrary x-dependent $k \times k$ dimensional Hermitian matrix. Since $f_{IJ}(x)$ is arbitrary there are three "ADHM constraints" on a, $a^{\dagger}$, b and $b^{\dagger}$,

$$a^{\dagger \dot{\alpha} \mathcal{I}}_{I} a_{\mathcal{I}J\dot{\beta}} = \left(\frac{1}{2} a^{\dagger} a\right)_{IJ} \delta^{\dot{\alpha}}_{\dot{\beta}}$$

$$a^{\dagger \dot{\alpha} \mathcal{I}}_{I} b_{\mathcal{I}I\dot{\beta}} = b^{\dagger \beta \mathcal{I}}_{I} a^{\dot{\alpha}}_{\mathcal{I}J}$$

$$b^{\dagger \mathcal{I}}_{\alpha I} b^{\beta}_{\mathcal{I}I} = \left(\frac{1}{2} a^{\dagger} a\right)_{IJ} \delta^{\beta}_{\alpha}. \tag{22.31}$$

It is straightforward to realize that the ADHM construction is invariant under,

$$\Delta \to \mathcal{A}\Delta\mathcal{B}^{-1} \qquad U \to \mathcal{A}U \qquad f \to \mathcal{B}f\mathcal{B}^{\dagger}, \tag{22.32}$$

where $\mathcal{A} \in U(N + 2k)$ and $\mathcal{B} \in GL(k, \mathcal{C})$. Thus by construction there is a redundancy in a and b. One can choose a simple canonical form for b and a as follows,

$$b^{\beta}_{\mathcal{I}J} = \begin{pmatrix} 0 \\ \delta^{\beta}_{\alpha} \delta_{IJ} \end{pmatrix} \qquad b^{\dagger \mathcal{I}}_{\beta J} = (0, \delta^{\alpha}_{\beta} \delta_{ji})$$

$$a_{\mathcal{I}J\dot{\alpha}} = \begin{pmatrix} \hat{a}_{iJ\dot{\alpha}} \\ (a'_{\alpha\dot{\alpha}})_{IJ} \end{pmatrix} \qquad a^{\dagger \dot{\alpha} \mathcal{I}}_{J} = (\hat{a}^{\dot{\alpha}}_{Ji}, (a'^{\dagger \dot{\alpha}\alpha})_{IJ}). \tag{22.33}$$

where $\mathcal{I} = i + I\alpha$.

---

[3] Multi-instanton solutions were presented in [220], [132] and other papers. Our discussion of the construction of multi-instanton is based on the paper of ADHM [20]. This approach was further discussed in [70]. We follow the description of the construction given in [81].

In this parametrization the third ADHM constraint is automatically obeyed
while the other two take the form,

$$\bar{\sigma}^{\dot{\alpha}}_{\dot{\beta}}(a^{\dagger\dot{\beta}}a_{\dot{\alpha}}) = \bar{\sigma}^{\dot{\alpha}}_{\dot{\beta}}((a^{\dagger})^{\dot{\beta}}\hat{a}_{\dot{\alpha}} + a'^{\dagger\dot{\beta}}a'_{\dot{\alpha}}) = 0, \tag{22.34}$$

where we made use of the fact that a' must be Hermitian. In this canonical
parametrization the matrix f reads,

$$f = 2((\hat{a}^{\dagger})^{\dot{\alpha}}\hat{a}_{\dot{\alpha}} + (a'_{\mu} + x_{\mu}1_{[k]\times[k]})^2)^{-1}. \tag{22.35}$$

The field strength $F_{\mu\nu}$ that corresponds to the ADHM k instanton configuration (22.26) can be written in the following form,

$$F_{\mu\nu} = \partial_{\mu}A_{\nu} - \partial_{\nu}A_{\mu} + g[A_{\mu}, A_{\nu}] = \frac{1}{g}\partial_{[\mu}(U^{\dagger}\partial_{\nu]}U) + \frac{1}{g}(U^{\dagger}\partial_{[\mu}U)(U^{\dagger}\partial_{\nu]}U)$$

$$= \frac{1}{g}\partial_{[\mu}U^{\dagger}(1 - U^{\dagger}U)\partial_{\nu]}U) = \frac{1}{g}\partial_{[\mu}U^{\dagger}\Delta f\Delta^{\dagger}\partial_{\nu]}U$$

$$= \frac{1}{g}U^{\dagger}\partial_{[\mu}\Delta f\partial_{\nu]}\Delta^{\dagger}U = \frac{1}{g}U^{\dagger}b\sigma_{[\mu}\bar{\sigma}_{\nu]}fb^{+}U = 4\frac{1}{g}U^{\dagger}b\sigma_{\mu\nu}fb^{+}U, \tag{22.36}$$

where we have made use of,

$$\mathcal{P}^{\mathcal{J}}_{\mathcal{I}} \equiv U_{\mathcal{I}i}U^{\dagger\mathcal{J}} = \delta^{\mathcal{J}}_{\mathcal{I}} - \Delta_{\mathcal{I}I\dot{\alpha}}f_{IJ}\Delta^{\dagger\dot{\alpha}\mathcal{J}}_{J}. \tag{22.37}$$

To get an explicit expression for A_{μ} we make use of the decomposition

$$U_{\mathcal{I}i} = \begin{pmatrix} \hat{U}_{ij} \\ (U'_{\alpha})_{Ii} \end{pmatrix} \qquad \Delta_{IJ\dot{\alpha}} = \begin{pmatrix} \hat{a}_{iJ\dot{\alpha}} \\ (\Delta'_{\alpha\dot{\alpha}})_{IJ} \end{pmatrix}. \tag{22.38}$$

From the completeness condition (22.37) $\hat{U}$ can take the form,

$$U = \sqrt{(i_{[N]\times[N]} - \hat{a}_{\dot{\alpha}}f(\hat{a})^{\dagger\dot{\alpha}})} \quad U' = \Delta'_{\dot{\alpha}}f(\hat{a})^{\dagger\dot{\alpha}}(\hat{u})^{\dagger-1}. \tag{22.39}$$

We next show that for the particular case of $k = 1$ the ADHM solution (22.26)
is identical to (22.25). For $k = 1$ we have to drop the indices I, J. One can verify
that in that case the parameters a'_{μ} can be identified with the center of the
instanton X_{μ}. From the ADHM constraint (22.34) we get that,

$$(\hat{a}^{\dagger})^{\dot{\beta}}\hat{a}_{\dot{\alpha}} = \rho^2\delta^{\dot{\beta}}_{\dot{\alpha}}, \tag{22.40}$$

and,

$$f = \frac{1}{(x_{\mu} - X_{\mu})^2 + \rho^2}, \tag{22.41}$$

where ρ will naturally be the size of the instanton.
From the relation (22.39) we deduce that,

$$\hat{U} = 1_{[N]\times[N]} + \frac{1}{\rho^2}\left(\sqrt{\frac{(x - X)^2}{(x - X)^2 + \rho^2}} - 1\right)(\hat{a}^{\dagger})^{\dot{\beta}}\hat{a}_{\dot{\alpha}},$$

$$U' = -\frac{(x - X)_{\alpha\dot{\alpha}}(\hat{a}^{\dagger})^{\dot{\alpha}}}{|x - X|\sqrt{(x - X)^2 + \rho^2}}. \tag{22.42}$$

Plugging these expressions into U we finally get the singular form of the one instanton solution (22.25).

In general for $k \neq 1$ one can show that the gauge configuration given in (22.36) indeed carries an instanton of charge k. To derive this result one makes use of the following relation,

$$g^2 \, \text{Tr} \, [F_{\mu\nu}^2] = \partial_\mu \partial^\mu tr_k [\log f]. \tag{22.43}$$

This relation can be proven by expanding the two sides of the equations using the explicit expression for $F_{\mu\nu}$ (22.36). Upon integrating (22.43) over the whole Euclidean space-time divided by $\frac{1}{16\pi^2}$ and making use of the fact that asymptotically $f(x) \to \frac{1}{x^2}$, we find that indeed it is equal to the instanton charge k.

22.3 On the moduli space of instantons

The moduli space of instantons[4] $\mathcal{M}$ is the space of inequivalent self-dual Yang–Mills configurations. The notion of moduli space of solutions is an important tool in general and in particular for instantons. We have encountered it in the context of magnetic monopoles in (21.10). We will elaborate in this section about the basic properties of the moduli space of instantons such as its dimension, complex structure, metric, symmetries, and singularities.

Consider a small fluctuation $\delta A_\mu(x)$ around an instanton solution $A_\mu(x)$ which is also a self-dual solution of the YM equation, namely, it obeys to linear in $\delta A_\mu(x)$,

$$D_\mu \delta A_\nu - D_\nu \delta A_\mu = \epsilon_{\mu\nu\rho\sigma} D^\rho \delta A^\sigma, \tag{22.44}$$

where D_μ is the covariant derivative in the instanton background. In terms of the quateronic notation this equation reads,

$$\bar{\sigma}_\beta^{\dot\alpha} \slashed{D}^{\dagger\,\beta\alpha} \delta A_{\alpha\dot\alpha} = 0, \tag{22.45}$$

where $\slashed{D} = \sigma^\mu D_\mu$. Next we would like to guarantee that the fluctuation is not a local gauge transformation. This can be done by requiring that the fluctuation be orthogonal to any gauge transformation, namely,

$$\int d^3x \, \text{Tr}[\delta A_\mu D^\mu \Lambda] = 0 \;\; \to \;\; D^\mu \delta A_\mu = 0, \tag{22.46}$$

where we have made use of an integration by parts to derive the last expression. In the quaternionic notation this condition takes the form of $\slashed{D}^{\dagger\,\dot\alpha\alpha} \delta A_{\alpha\dot\alpha} = 0$ which combined with (22.45) is given by,

$$\slashed{D}^{\dagger\,\dot\alpha\alpha} \delta A_{\alpha\dot\beta} = 0. \tag{22.47}$$

[4] The properties of the moduli space of instantons were discussed by many authors. In particular [155], [141] and [80]. The review about the moduli space that we are using is [81].

The fluctuation that obeys this equation is referred to as a *zero mode* since it is a physical fluctuation that does not change the action. Note that this is exactly the equation of motion of a Weyl spinor in the background of the instanton $A_\mu(x)$. The zero modes defined by (22.47) are the collective coordinates of the instantons.

Since the YM instantons are characterized by the topological charge defined in (22.2) so is the corresponding moduli space. We thus discuss the moduli space of instantons of charge k which we denote by $\mathcal{M}_k$.

It can be proven that the moduli space of instantons is a manifold. In fact we will see below that $\mathcal{M}_k$ has some conical singular points associated with zero size instantons. The coordinates on the moduli space are the collective coordinates that were just shown to be equivalent to the zero model (22.47). We denote by X_n the collective coordinates where $n = 1, \ldots, \dim \mathcal{M}_k$. A trivial set of coordinates are the space-time coordinates of the center of the instanton X_μ accordingly the moduli space is a product of the form,

$$\mathcal{M}_k = \mathcal{R}^4 \times \hat{\mathcal{M}}_k. \tag{22.48}$$

The collective coordinates X_μ follow from the fact that the instanton solution breaks the symmetry of the action under space-time translations. There are other collective coordinates that associate with symmetries of the theory that the instanton configuration breaks. However, not all symmetries yield non equivalent collective coordinates and not all the coordinates associate with broken symmetries.

From the ADHM construction it follows that the moduli space is identified with the variable $a_{\dot\alpha}$ subject to the ADHM constraints (22.31) quotiented by the residual $U(K)$ symmetry transformation (22.32) with,

$$\mathcal{A} = \begin{pmatrix} 1_{[N]\times[N]} & 0 \\ 0 & \mathcal{C}1_{[2]\times[2]} \end{pmatrix}, \quad \mathcal{B} = \mathcal{C} \quad \mathcal{C} \in U(k), \tag{22.49}$$

which preserve the canonical form of b (22.33) and transform a as follows,

$$\hat{a}_{i I \dot\alpha} \rightarrow \hat{a}_{\dot\alpha} \mathcal{C} \quad a'_\mu \rightarrow \mathcal{C}^\dagger a'_\mu \mathcal{C}. \tag{22.50}$$

Thus the dimension of the moduli space is,

$$\dim \mathcal{M}_k = 4k(N + K) - 4 \dim \ U(k) = 4k(N + k) - 4k^2 = 4NK. \tag{22.51}$$

This result can be derived also by using an index theorem that counts the zero modes at a point in the moduli space. In fact as we will see shortly the space $\mathcal{M}_k$ is a hyper-Kahler quotient of the flat space $\mathcal{R}^{4k(k+N)}$ by the $U(k)$ group. The one instanton solution of $SU(2)$ is indeed characterized by the four coordinates of its center, its size and three global $SU(2)$ gauge transformations.

The moduli space is a complex manifold. A complex manifold is an even-dimensional manifold that admits a complex structure **I** a linear map of the tangent space to itself such that $\mathbf{I}^2 = 0$. There are always local holomorphic coordinates $(Z^i, \bar{Z}^i), i = 1, \ldots, n$ (see Section 1.1) for which,

$$\mathbf{I} = \begin{pmatrix} i\delta^{\bar{i}}_{\bar{j}} & 0 \\ 0 & -i\delta^i_j \end{pmatrix} \quad g = g_{i\bar{j}} dZ^i d\bar{Z}^{\bar{j}} \quad w = i g_{i\bar{j}} dZ^i \wedge d\bar{Z}^{\bar{j}}. \tag{22.52}$$

where g is an Hermitian metric and w is referred to as the fundamental 2 form. In the case that the fundamental 2 form is closed namely $dw = 0$ it is called the *Kähler form* and the associated manifold is a Kähler manifold. The latter is also characterized by the fact that the complex structure is covariantly constant and the Kähler metric can be derived from a *Kähler potential*,

$$\nabla_\mu \mathbf{I} = 0 \quad g_{i\bar{j}} = \partial_i \partial_{\bar{j}} K. \tag{22.53}$$

The moduli space of instanton is not only a Kähler manifold but in fact a *hyper Kähler manifold* which means that it admits three linearly independent complex structures, $\mathbf{I}^{(c)}, c = 1, 2, 3$ that satisfies the algebra

$$\mathbf{I}^{(c)} \mathbf{I}^{(d)} = -\delta^{cd} + \epsilon^{cde} \mathbf{I}^{(e)}. \tag{22.54}$$

The four-dimensional Euclidean space $\mathcal{R}^4$ is hyper Kähler and the three complex structures are,

$$\mathbf{I}^{(c)}_{\mu\nu} = -\eta^c_{\mu\nu} \quad (\mathbf{I} \cdot x)_{\alpha\dot\alpha} = i x_{\alpha\dot\beta} \vec{\sigma}^{\dot\beta}_{\dot\alpha}, \tag{22.55}$$

where $\eta^c_{\mu\nu}$ is the 't Hooft η symbol defined in (22.14) and the expression in the left-hand side is the quaterionic formulation. Now recall that by the definition of the zero modes (22.47), if $\delta_n A_{\alpha\dot\alpha}$ is a zero mode so is also $\delta_n A_{\alpha\dot\alpha} C^{\dot\beta}_{\dot\alpha}$ for any constant matrix $\mathbf{C}$ and in particular also to $\vec{\sigma}$ and hence if $\delta_n A_{\alpha\dot\alpha}$ so is also $(\mathbf{I} \cdot \delta_n A)_{\alpha\dot\alpha} = i\delta_n A_{\alpha\dot\alpha} \vec{\sigma}^{\dot\beta}_{\dot\alpha}$. Since the zero modes form a complete set there must exist $\vec{\mathbf{I}}^n_m$ such that,

$$(\vec{\mathbf{I}} \cdot \delta_m A)_{\alpha\dot\alpha} = \delta_n A_{\alpha\dot\alpha} \vec{\mathbf{I}}^n_m, \tag{22.56}$$

from which it implies that $\vec{\mathbf{I}}^n_m$ satisfies the algebra (22.54).

The Kähler potential which is common to the three complex structures of the moduli space of instantons takes the form,

$$K = -\frac{g^2}{4} \int d^4 x \, x^2 \, \mathrm{Tr} F_{\mu\nu} F^{\mu\nu}. \tag{22.57}$$

Using the form of $\mathbf{I}^{(c)}$ on $\mathcal{R}^4$ given in (22.52) for instance $I^{(3)}$ associated with the complex coordinates on $\mathcal{R}^4$ $ix^3 + x^4$ and $ix^1 - x^2$, we find that,

$$(\mathbf{I}^{(c)} \cdot \partial_{Z^i} A)_{\alpha\dot\alpha} = i\partial_{Z^i} A_{\alpha\dot\alpha}, \tag{22.58}$$

for instance for $\mathbf{I}^{(3)}$ we get $\partial_{Z^i} A_{\alpha 2} = \partial_{\bar{Z}^i} A_{\alpha 1} = 0$. Furthermore the derivative with the respect to the holomorphic and anti-holomorphic coordinates of the gauge fields obey the equations of zero modes namely,

$$\delta_i A_\mu \equiv \partial_{Z^i} A_\mu \quad \bar{\delta}_{\bar{i}} A_\mu \equiv \partial_{\bar{Z}^{\bar{i}}} A_\mu. \tag{22.59}$$

It further follows that $\partial_{\bar{z}^j}\partial_{z^i}A_\mu = 0$ and hence,

$$\partial_{\bar{z}^j}\partial_{z^i}\operatorname{Tr}[F_{\mu\nu}^2] = \partial_\mu\partial^\mu\operatorname{Tr}[\delta_i A_\mu\bar{\delta}_j A_\mu - 2\partial_\mu\partial_\nu]\operatorname{Tr}\delta_i A_\mu\bar{\delta}_j A_\nu. \qquad (22.60)$$

Upon integrating by parts twice we find the metric on the moduli space,

$$\partial_{\bar{z}^j}\partial_{z^i}K = -2g^2\int d^4x\operatorname{Tr}\delta_i A_\mu\bar{\delta}_j A_\mu = g_{i\bar{j}}. \qquad (22.61)$$

Next let us now discuss the symmetries of the moduli space, in particular the realization of symmetries of the gauge theory which are broken by the instanton configuration. We start with the four-dimensional conformal group (see Section 17). In the quaternionic formulation the basic variable of the ADHM construction Δ is transformed as follows,

$$x \to x' = (Ax + B)(cX + D)^{-1} \quad \det\begin{pmatrix} AB \\ CD \end{pmatrix} = 1$$

$$\Delta(x;a,b) \to \Delta(x';a,b) = \Delta(x;aD + bB, aC + bA)(Cx + D)^{-1} \quad (22.62)$$

In fact the term $(Cx + D)^{-1}$ in the right-hand side of the last equation is irrelevant since the gauge field depends on U and $U^\dagger$ defined in (22.26) is redundant.

We can now use transformations of (22.32) that keeps the canonical structure of b (22.33). Upon applying this transformation a goes into,

$$a \to A(aD + bB)B^{-1}. \qquad (22.63)$$

A particular example of transformations which belong to the conformal group are the translations. For this case,

$$\Delta(x;a,b) \to \Delta(x;a + b\epsilon, b), \qquad (22.64)$$

from which it follows that,

$$a'_\mu \to a'_\mu + \epsilon_\mu \mathbb{1}_{[k]\times[k]} \quad \hat{a}_{\dot{\alpha}} \to \hat{a}_{\dot{\alpha}}. \qquad (22.65)$$

It is thus clear that indeed the components a'_μ are proportional to the coordinates of the center of the instanton,

$$\operatorname{tr}_k a'_\mu = kX_\mu. \qquad (22.66)$$

Global gauge transformations act non trivially on the ADHM variables if $N \le 2k$, while if $N \ge 2k$ there are transformations that leave the instantons fixed. This is the stability group of the instanton. One can embed the k instanton solution in an $SU(2k)$ subgroup of $SU(N)$ and show that the stability group is $S(U(N - 2k) \times U(1))$.

The moduli space $\hat{\mathcal{M}}_k$ is in fact not a smooth manifold due to certain singularities. However, these singularities do not signal any pathology of the moduli space and integrals over the moduli space are well defined. It can be shown that $\hat{\mathcal{M}}_k$ is a cone. For the moduli space of single instanton $k = 1$ the apex of the cone is the point $\rho = 0$ where the instanton has a zero size. This structure can be generalized also to the $k \ne 1$ instantons.

Topological characteristics of the moduli space can be described by a topological field theory where the observables of the theory are the topological invariants. This is beyond the scope of this book and we refer the interested reader to the list of references for this chapter.

22.4 Instantons and tunneling between the vacua of the YM theory

The vacua of the YM theory in Minkowsi space-time are defined to be the gauge configurations for which the energy vanishes. Using the temporal gauge $A_0 = 0$, the Hamiltonian of the theory is

$$H = \frac{1}{2g^2} \int d^3x \, \text{Tr} \, [E^2 + B^2]. \tag{22.67}$$

Thus a classical vacuum has a vanishing field strength,

$$F_{\mu\nu} = 0 \quad \rightarrow A_i(x,t) = iU(x,t)\partial_i U(x,t)^\dagger. \tag{22.68}$$

Thus the vacuum gauge configuration is that of a pure gauge. Prior to a discussion of how to tunnel between two vacuum states, we have to classify and enumerate the vacua namely following (22.68) the group elements $U(x)$. This translates to the equivalence classes of maps from S^3 to the $SU(N)$ group manifold. This is done by the topological charge or winding number or Pontryagin number defined in (22.2). Since this step is very essential in the discussion of the tunneling let us clarify this point. Let us analyze the tunneling between a vacuum state $A_i(x,t_1)$ at $t = t_1$ into another vacuum state $A_i(x,t_2)$ at $t = t_2$. On top of fixing $A_0(x,t) = 0$ we can use the residual gauge symmetry to set $A_i(x,t_1) = 0$. Next we consider a path in the space of gauge configurations that connects the two vacua points and has a finite energy H (in Minkowski space-time). Finite energy implies that for large $|\vec{x}| \rightarrow \infty$ it has to be a pure gauge,

$$A_\mu \rightarrow_{|x|\rightarrow\infty} U\partial_\mu U^\dagger. \tag{22.69}$$

Since $A_0 = 0$, $U(x,t) = U(x)$ is time independent. Because $U(x,t_1) = 1$ we obtain that for all t and $|x| \rightarrow \infty$, $U = 1$ and hence also $A_i(x,t_2) \longrightarrow_{|x|\rightarrow\infty} 0$. The fact that asymptotically in $|\vec{x}|$ all $A_i = 0$ allows us to compactify the spacelike hypersurface at fixed t into S^3. The following Fig. 22.1 describes the situation.

On the boundary of hyper-cylinder the gauge fields A_i vanish apart from on the hyper-disk at $t = t_2$ where A_i is a pure gauge. Consider now the topological charge which we have seen (22.2) is in fact a surface term. Since on the boundary $F_{\mu\nu} =$ the contribution to the surface integral takes the form,

$$Q = \frac{1}{24\pi^2} \epsilon^{\mu\nu\rho\sigma} \oint d\sigma_\mu \text{Tr} \, [A_\nu A_\rho A_\sigma]$$

$$= \frac{1}{24\pi^2} \epsilon^{0ijk} \int d^3x r[U\partial_i U^\dagger U\partial_j U^\dagger U\partial_k U^\dagger], \tag{22.70}$$

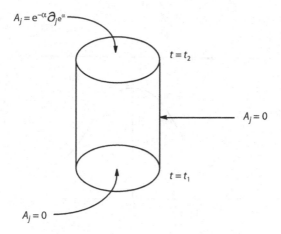

Fig. 22.1. Compactification of the space coordinates on S^3 at fixed t.

where in the last expression the integration is over the three sphere at $t = t_2$ since at all other parts of the surface of the hyper-cylinder $A_i = 0$. Thus the configurations in Minkowski space-time that connect a vacuum state at $t = t_1$ to one at $t = t_2$ are classified by the winding number of the maps of S^3(space) $\rightarrow S^3$(group) just as the maps of instanton in Euclidean space-time.[5] In the latter case S^3(space) is the boundary of R^4 whereas in the former it is a compactification of R^3 at $t = t_2$.

It is easy to realize that there is no way to interpolate a vacuum at $t = t_1$ of zero winding number with a one at $t = t_2$ with non vanishing winding number with a configuration of zero energy. The latter corresponds to a pure gauge configuration which has $F_{\mu\nu} = 0$ everywhere and hence also vanishing topological number. Thus the energy of the tunneling configuration as a function of time should look as in Fig. 22.2.

To identify the configuration that has the largest tunneling rate we consider a family of gauge configurations characterized by the collective coordinates associated with a coordinate transformation from t to $\lambda(t)$ such that,

$$A_i^{(\lambda)}(x, t) = A_i(x, \lambda(t)), \tag{22.71}$$

with the requirement that $\lambda(t_1) = t_1$ and $\lambda(t_2) = t_2$. Next we compute the electric and magnetic fields,

$$E_i = F_{i0} = -\partial_0 A^{(\lambda)}(x, t) = \frac{\partial A_i}{\partial \lambda}(x, \lambda(t))\dot{\lambda}$$

$$B_i = \frac{1}{2}\epsilon_{ijk}F_{jk} = \frac{1}{2}\epsilon_{ijk}(\partial_j A_k(x, \lambda(t)) + A_j(x, \lambda(t)(A_k(x, \lambda(t)) - (j \leftrightarrow k) \tag{22.72}$$

---

[5] The role of instantons in tunneling between different vacua was proposed in [131]. It was also discussed in [26], [40] and [44]. This topic is reviewed in [185] and in [209]. We follow the latter.

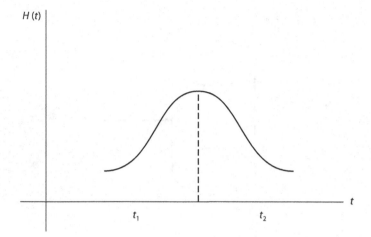

Fig. 22.2. The energy of the tunneling configuration as a function of time.

and substitute them into the Lagrangian $L = \int d^3x \mathcal{L} = \int d^3x[-\frac{1}{g^2}\,\mathrm{Tr}\,[E^2 - B^2]]$ which can be written in the following form,

$$L = -\frac{1}{2}m(\lambda)(\dot\lambda)^2 - V(\lambda),$$

$$m(\lambda) = \frac{2}{g^2}\int d^3x\,\mathrm{Tr}\left(\frac{\partial A_i}{\partial\lambda}\right)^2 \geq 0,$$

$$V(\lambda) = -\frac{1}{g^2}\int d^3x\,\mathrm{Tr}\,(B^i)^2 \geq 0. \tag{22.73}$$

The Lagrangian (22.73) is the Lagrangian of a particle that moves from one vacuum at $t = t_1$ where $V(\lambda) = m(\lambda) = 0$ to a vacuum at $t = t_2$, where again $V(\lambda) = m(\lambda) = 0$. The quantum mechanical tunneling rate is proportional to e^{-2R} where R is given by,

$$R = \int_{\lambda_1}^{\lambda_2} d\lambda\sqrt{2m(\lambda)(V(\lambda) - E)}$$

$$= \int_{\lambda_1}^{\lambda_2} d\lambda\sqrt{\left[\left(\frac{1}{g^2}\int d^3x\,\mathrm{Tr}\left(\frac{\partial A_i}{\partial\lambda}\right)^2\right)\left(\frac{1}{g^2}\int d^3x\,\mathrm{Tr}\,(B_i)^2\right)\right]}$$

$$= \frac{2}{g^2}\int_{t_1}^{t_2} dt\sqrt{\left(\int d^3x\,\mathrm{Tr}\,(E_i)^2\right)\left(\frac{1}{g^2}\int d^3x\,\mathrm{Tr}\,(B_i)^2\right)}. \tag{22.74}$$

Using the triangle inequality we can relate the tunneling rate to the winding or instanton number as follows,

$$R \geq \frac{2}{g^2}|\int_{t_1}^{t_2} d^4x\,\mathrm{Tr}\,[E^i B^i]| = \frac{8\pi^2}{g^2}|k|, \tag{22.75}$$

where the instanton number is $Q = k$. Thus we see that the tunneling rate is bounded by,

$$e^{-R} \le e^{-\frac{8\pi^2}{g^2}|k|}, \tag{22.76}$$

and the bound is saturated for instanton configurations. To be more precise the most probable tunneling paths are given by a Minkowski gauge configuration with $\vec{E} = \pm\vec{B}$ which when viewed as a configuration in Euclidean space are instantons. Conversely given an instanton $A^E_\mu(x,t)$ in Euclidean space one can construct a set of paths in Minkowski space-time $A^M_\mu(x,\lambda(t))$ such that $A^{M,(\lambda)}_i(x,t) = A^E_\mu(x,\lambda(t))$ and $A^{M,(\lambda)}_0(x,t) = A^E_4(x,\lambda(t))$.

22.5 Instantons, theta vacua and the $U_A(1)$ anomaly

It was shown in the previous section that the instantons connect different vacua. This means that the vacuum of the YM theory cannot be described by any of the states of zero energy and a specific topological charge, but instead has to be a superposition of all these states, namely,

$$|\theta> = \sum_k e^{ik\theta}|k> . \tag{22.77}$$

The generator of large gauge transformation that changes the winding number by one unit, namely, $T|k> = |k+1>$ has to be a symmetry generator that commutes with the Hamiltonian so that $T|vac> = e^{i\varphi}|vac>$ for some phase φ. Indeed for the θ vacuum we get $T|\theta> = \sum_k e^{ik\theta}|k+1> = e^{-i\theta}|\theta>$.

The energy associated with the θ vacua given by,

$$E(\theta) = -2K\cos(\theta)e^{-S}. \tag{22.78}$$

This follows from the following steps. Consider the amplitude to tunnel from a vacuum $|i>$ to a vacuum $|j>$ is given by,

$$<j|e^{-Ht}|i> = \sum N_\pm \frac{\delta_{N_+ - N_- - (j-i)}}{N_+!N_-!}(Kte^{-S})^{N_+ + N_-}, \tag{22.79}$$

when the instantons are sufficiently dilute and where K is the pre-exponential factor in the tunneling amplitude, and $N_\pm$ are the number of instantons and anti-instantons. We introduce the parameter θ via a representation of the Kroneker delta function,

$$\delta_{N_+ - N_+ + (i-j)} = \frac{1}{2\pi}\int_0^{2\pi} e^{i\theta(N_+ - N_+ + (i-j))}. \tag{22.80}$$

Upon performing the summations over N_+ and N_- we get,

$$<j|e^{-Ht}|i> = \int_0^{2\pi} e^{i\theta(i-j)} e^{2Kt\cos(\theta)e^{-S}}, \tag{22.81}$$

which implies that the energy of the θ vacuum is as given in (22.78). Note that this does not imply that the YM theory has a continuous spectrum without mass gap since the θ parameter is fixed for a given theory and it cannot be changed. Fixing the value of θ can be achieved by adding a θ term to the action (22.10),

$$S_{\text{YM}} = \frac{1}{2g^2} \int d^4x \, \text{Tr} \, F_{\mu\nu} F^{\mu\nu} \rightarrow \frac{1}{2g^2} \int d^4x \, \text{Tr} \, F_{\mu\nu} F^{\mu\nu} + \frac{\theta}{16\pi^2} \int d^4x \, \text{Tr} \, [F_{\mu\nu}{}^* F^{\mu\nu}].$$

$$(22.82)$$

The additional θ term is a surface term and hence does not affect the equations of motion, however it is not invariant under CP or T transformations.[6] As will be discussed below, with no massless quarks indeed the θ term implies a strong CP violation. The most severe restriction on CP violation comes from the electric dipole moment of the neutron. This sets the upper bound to theta to be,

$$\theta < 10^{-9}. \qquad (22.83)$$

The puzzle of why θ is so tiny is referred to as the strong CP problem. One proposal to handle this problem is the introduction of the axion, χ, a pseudo scalar field with a coupling of the form $\chi \, \text{Tr} \, [F_{\mu\nu}{}^* F^{\mu\nu}]$ so that the effective θ is the sum of $\sqrt{<\chi>}$ and the θ term. As will be discussed below there is in fact an even simpler mechanism to resolve the strong CP problem and that is having a massless u quark. This brings us to the next topic which is the incorporation of light quarks to the game.

   In the presence of light quarks there is a simple physical observable that distinguishes between the different topological vacua, the axial current. Recall that for N_f massless quarks the theory is classically invariant under global $U_L(N_f) \times U_R(N_f) \equiv SU_L(N_f) \times SU_R(N_f) \times U_B(1) \times U_A(1)$ symmetry. The $SU(N_f) \times U_V(1)$ symmetry group factors are realized in nature also quantum mechanically. The invariance under the axial $SU(N_f)$ transformations is broken spontaneously and there are $N_f^2 - 1$ Goldstone bosons. For $N_f = 2$ these are the pions. The $U_A(1)$ axial symmetry is not conserved quantum mechanically. In analogy to the anomaly of the axial symmetry in two dimensions discussed in Section 9.1, in four dimensions as well one can show using various different methods that,

$$\partial^\mu J_\mu^5 = \frac{N_f}{8\pi^2} \, \text{Tr} \, [F_{\mu\nu}{}^* F^{\mu\nu}], \qquad (22.84)$$

where $J_\mu^5 = \sum_i \bar{\psi}_i \gamma_\mu \gamma_5 \psi_i$ is the axial current and $\psi_i \; i = 1, \ldots, N_f$ are the fields of the various flavored quarks. This resolves the so-called $U_A(1)$ puzzle, namely the absence of the fourth Goldstone boson for $N_f = 2$ or the ninth one for $N_f = 3$. Indeed for the former case one could associate the η pseudo-scalar meson with the fourth Goldstone boson, however it has a mass of $478 MeV$ whereas a current

---

[6] A proposal for resolving the strong CP problem was proposed in [172]. The $U_A(1)$ problem has been resolved by 't Hooft. The mass of the η' was proposed by Witten [221] and by Veneziano [216]. Our discussion of this topic follows the review [185].

algebra theorem states that it has to be lighter than $\sqrt{3}m_\pi$. The right-hand side of (22.84) is proportional to the divergence of the topological current (22.2) $\partial_\mu K_\mu$ so one can define a modified conserved axial current $\tilde{j}^5_\mu = j^5_\mu - N_f K_\mu$. However, unlike the topological charge, the topological current is not gauge invariant. A massless pole in the correlator of K_μ does not necessarily correspond to a massless particle. One may wonder also about the fact that the right-hand side of (22.84) is a surface term and hence cannot have a physical significance. However, as was emphasized above due to instantons the surface term is relevant. Let us see this explicitly. We start by computing the change in the axial charge,

$$\Delta Q_5 = Q_5(t = +\infty) - Q_5(t = -\infty) = \int d^4x \partial^\mu J^5_\mu$$

$$= N_f \int d^4x \partial^\mu \operatorname{Tr}\left[S(x,x)\gamma_\mu\gamma_5\right], \qquad (22.85)$$

where $S(x,y)$ is the fermion propagator $S(x,y) = <x|(i\not{D})^{-1}|y>$ that can be determined from the eigenfunction equation $i\not{D}\,\psi_\lambda = \lambda\psi_\lambda$ in the form $S(x,y) = \sum_\lambda \frac{\psi_\lambda(x)\psi^\dagger_\lambda(y)}{\lambda}$. Substituting this expression we get,

$$\Delta Q_5 = N_f \int d^4x \partial^\mu \operatorname{Tr}\left(\sum_\lambda \frac{\psi_\lambda(x)\psi^\dagger_\lambda(y)}{\lambda}2\lambda\gamma_5\right) = 2N_f(n_L - n_R), \qquad (22.86)$$

where we have used the fact that ψ_λ and $\gamma_5\psi_\lambda$ are orthogonal so only the $n_L(n_R)$ left (right) zero modes contribute.

Integrating the left-hand side of (22.84) we get the topological charge Q which is thus related to ΔQ_5. The latter counts the number of left-handed zero modes minus the number of right-handed zero modes. This is obviously associated with instantons. Each instanton contributes one unit to the topological charge and has a left-handed zero mode, whereas an anti-instanton has a right-handed zero mode and $Q = -1$. This is the way the instantons contribute to the axial anomaly and hence to the resolution of the $U_A(1)$ problem. For the case of $N_f = 3$ this implies that this would be the ninth Goldstone boson, the η' is massive even if the quark masses vanish. It was shown that the mass of the η' is related to the topological susceptibility in the following form,

$$\frac{2N_f}{f^2_\pi}\chi_{\text{top}} = \frac{2N_f}{f^2_\pi}\int d^4x <Q(x)Q(0)> = m^2_\eta + m^2_{\eta'} - 2m^2_K. \qquad (22.87)$$

The combination of masses on the right-hand side corresponds to the part of the η' mass which is not due to the strange quark mass.

23

Summary, conclusions and outlook

23.1 General

Relativistic quantum field theory has been very successful in describing strong, electromagnetic and weak interactions, in the region of small couplings by perturbation theory, within the framework of the standard model.

However, the region of strong coupling, like the hadronic spectrum and various scattering phenomena of hadrons within QCD, is still largely unsolved.

A large variety of methods have been used to address this question, including lattice gauge simulations, light-cone quantization, low energy effective Lagrangians like the Skyrme model and chiral Lagrangians, large N approximation, techniques of conformal invariance, the integrable model approach, supersymmetric models, string theory approach, QCD sum rules, etc. In spite of this major effort the gap between the phenomenology and the basic theory has only been partially bridged, and the problem is still open.

The goals of this book are to provide a detailed description of the tool box of non-perturbative techniques, to apply them on simplified systems, mainly of gauge dynamics in two dimensions, and to examine the lessons one can learn from those systems about four-dimensional QCD and hadron physics.

The study of two-dimensional problems to improve the understanding of four-dimensional physical systems was found to be fruitful. This follows two directions, one is the utilization of non-perturbative methods on simpler setups and the second is extracting the physical behavior of hadrons in one space dimension.

Obviously, physics in two dimensions is simpler than that of the real world since the underlying manifold is simpler and since the number of degrees of freedom of each field is smaller. There are some additional simplifying features in two-dimensional physics. In one space dimension there is no rotation symmetry and no angular momentum. The light-cone is disconnected and is composed of left moving and right moving branches. Therefore, massless particles are either on one branch or the other. These two properties are the basic building blocks of the idea of transmutation between systems of different statistics. Also, the ultra-violet behavior is more convergent in two dimensions, making for instance QCD_2 a superconvergent theory.

In this summary chapter we go over several notions, concepts and methods with emphasis on the comparison between the two- and four-dimensional worlds and what one can deduce about the latter from the former. In particular we deal with conformal invariance, integrability, bosonization, solitons and topological charges, confinement versus screening and finally the hadronic spectrum and scattering.

23.2 Conformal invariance

From the outset there is a very dramatic difference between conformal invariance in two and four dimensions. The former is characterized by an infinite-dimensional algebra, the Virasoro algebra, whereas the latter is associated with the finite-dimensional algebra of $SO(4,2)$. This basic difference stems from the fact that whereas the conformal transformations in four dimensions are global, in two dimensions the parameters of conformal transformations are holomorphic functions (or anti-holomorphic), see Section 17.5 versus 2.1. Nevertheless there are several features of conformal invariance which are common to the two cases. We will now compare various aspects of conformal invariance in two and four dimensions:

- The notion of a primary field and correspondingly a highest weight state is used both in two-dimensional conformal field theories as well as for the four-dimensional collinear algebra. It is expressed in the former as (17.38),

$$L_0[\phi(0)|0>] = h[\phi(0)|0>] \qquad L_n[\phi(0)|0>] = 0, \qquad n > 0 \qquad (23.1)$$

and for the latter,

$$L_0[\Phi(0)|0>] = j[\Phi(0)|0>] \qquad L_-[\Phi(0)|0>] = 0. \qquad (23.2)$$

The difference is of course the infinite set of annihilation operators L_n versus the single annihilation operator L_- in four dimensions.
- The COPE, the conformal operator product expansion has a compact form in two dimensional CFT (Section 2.12)

$$\mathcal{O}_i(z, \bar{z})\mathcal{O}_j(w, \bar{w}) \sim \sum_k C_{ijk}(z - w)^{h_k - h_i - h_j}(\bar{z} - \bar{w})^{\bar{h}_k - \bar{h}_i - \bar{h}_j}\mathcal{O}_k(w, \bar{w}),$$

$$(23.3)$$

where C_{ijk} are the *product coefficients*, while in four dimensions it reads,

$$A(x)B(0) = \sum_{n=0}^{\infty} C_n \left(\frac{1}{x^2}\right)^{1/2(t_A + t_B - t_n)} \frac{x_-^{n + s_1 + s_2 - s_A - s_B}}{B(j_A - j_B + j_n, j_B - j_A + j_n)}$$

$$\times \int_0^1 du u^{(j_A - j_B + j_n - 1)}(1 - u)^{(j_B - j_A + j_n - 1)}\mathcal{O}_n^{j_1, j_2}(ux_-), \qquad (23.4)$$

where the definitions of the various quantities are in Chapter 17. Again there is a striking difference between the simple formula in two dimensions and the complicated one in four dimensions.

- As an example let us compare the OPE of two currents. Recall from the discussion of Chapter 3 the expression in two dimensions reads,

$$J^a(z)J^b(w) = \frac{k\delta^{ab}}{(z-w)^2} + i\frac{f_c^{ab}J^c(w)}{(z-w)} + \text{finite terms}, \qquad (23.5)$$

for any non-abelian group, and in particular for the abelian case the second term on the right-hand side is missing. For comparison the OPE of the transverse components of the electromagnetic currents given in Chapter 17 takes the form,

$$J^T(x)J^T(0) \sim$$

$$\sum_{n=0}^{\infty} C_n \left(\frac{1}{x^2}\right)^{(6-t_n)/2} (-ix_-)^{n+1} \frac{\Gamma(2j_n)}{\Gamma(j_n)\Gamma(j_n)} \int_0^1 du[u(1-u)]^{j_n-1} Q_n^{1,1}(ux_-). \qquad (23.6)$$

- The conformal Ward identity associated with the dilatation operator in four dimensions (17.60),

$$\sum_i^N (l_\phi + \gamma(g^*) + x_i\partial_i) <T\phi(x_1)...\phi(x_N)>= 0, \qquad (23.7)$$

where l_ϕ is the canonical dimension and $\gamma(g^*)$ is the anomalous dimension, seems very similar to the one in two dimensions,

$$\sum_i (z_i\partial_i + h_i) <0|\phi_1(z_1,\bar{z}_1)...\phi_n(z_n,\bar{z}_n)|0>= 0. \qquad (23.8)$$

In both cases one has to determine the full quantum conformal dimensions of the various operators. However, as was shown in Section 2.7, in certain CFT models, like the unitary minimal models, there are powerful tools based on unitarity which enable us to determine exactly the dimensions h_i of all the primary operators and hence all the operators of the model. On the other hand, it is a non-trivial task to determine the anomalous dimensions in other models in two dimensions, and of course four-dimensional operators. In certain supersymmetric theories there are operators whose dimension is protected, but generically one has to use perturbative calculations to determine the anomalous dimensions of gauge theories to a given order in the coupling constant.

Using the Ward identity one can extract the form of the two-point function of operators of spin s in four dimensions. It is given by,

$$<\phi(x_1)\phi(x_2)>= N_2(g^*)(\mu^*)^{-2\gamma(g^*)} \left[\frac{1}{(x_1-x_2)^2}\right]^{l_\phi+\gamma(g^*)} \left(\frac{(x_1-x_2)_+}{(x_1-x_2)_-}\right)^s. \qquad (23.9)$$

The corresponding two-point function in two dimensions, which depends only on the conformal dimension of the operator h, reads,

$$G_2(z_1,\bar{z}_1,z_2,\bar{z}_2) \equiv <0|\phi_1(z_1,\bar{z}_1)\phi_1(z_2,\bar{z}_2)|0> = \frac{c_2}{(z_1-z_2)^{2h_1}(\bar{z}_1-\bar{z}_2)^{2\bar{h}_1}}. \qquad (23.10)$$

- As for higher point functions, we have seen in Section 2.9 that one can use the local Ward identities together with Virasoro null vectors to write down partial differential equations. The result for a four-point function (2.63) was later used to determine the four-point function of the Ising model (2.94). Two dimensional conformal field theories are further invariant under affine Lie algebra transformations, and as we have shown in Section 3.6 those can be combined with null vectors to derive the so-called Knizhnik–Zamolodchikov equations (3.69), which were later used to solve for the four-point function of the $SU(N)$ WZW model in Section 4.4. These types of differential equations that fully determine correlation functions are obviously absent in four-dimensional interacting conformal field theories.

23.3 Integrability

Integrability was discussed in Chapter 5 in the context of two-dimensional models and in Chapter 18 in four-dimensional gauge theories. For systems with a finite number of degrees of freedom, like spin chain models, there is a finite number of conserved charges, equal of course to the number of degrees of freedom. For integrable field theories there is an infinite countable number of conserved charges. Furthermore, the scattering processes of those models always involve a conservation of the number of particles.

In two dimensions we have encountered continuous integrable models like the sine-Gordon model as well as discretized ones like the XXX spin chain model. The integrable sectors of gauge dynamical systems discussed in Chapter 18 are based on identifying an exact map between certain properties of the systems and a spin chain structure. In two dimensions the spin chain models follow from a discretization of the space coordinate, by placing a spin variable on each site that can take several values and imposing periodicity. In the four-dimensional $\mathcal{N} = 4$ super YM theory discussed in Section 18.1 the spin chain corresponds to a trace of field operators and in the process of high-energy scattering of Section 18.2 it is a "chain" of reggeized gluons exchanged in the t-channel of a scattering process. A summary of the comparison among the basic two-dimensional spin chain, the "spin chains" associated with the planar $\mathcal{N} = 4$ SYM, and the high-energy scattering in QCD, is given in Table 23.1. A powerful method to solve all these spin chain models is the use of the algebraic Bethe ansatz. This was discussed in detail for the the $XXX_{1/2}$ model in Section 5.14. The solutions of the energy eigenvalues needed for the high-energy scattering process was based on generalizing this method to the case of spin s Heisenberg model (see Section 18.2) and for the $\mathcal{N} = 4$ to the case of an $SO(6)$ invariance.

There is one conceptual difference between the spin chains of the two-dimensional models and those associated with the $\mathcal{N} = 4$ SYM in four dimensions. In the former the models are non conformal, involving a scale, and hence

Table 23.1. *Spin chain structure of the two-dimensional model and the four-dimensional gauge systems of $\mathcal{N} = 4$ SYM and of high-energy behavior of scattering amplitudes in QCD.*

Spin chain	Planar $\mathcal{N} = 4$ SYM	High energy scattering in QCD
Cyclic spin chain	Single trace operator	Reggeized guons in t-channel
Spin at a site	Field operator	$SL(2)$ spin
Number of sites	Number of operators	Number of gluons
Hamiltonian	Anomalous dilatation operator	H_{BFKL}
Energy eigenvalue	Anomalous dimension $g^{-2}\delta\mathcal{D}$	$\sim \frac{1}{\lambda}\frac{\log A}{\log s}$
Evolution time	Global time	The total rapidity $\log s$
Zero momentum $U = 1$	Cyclicity constraint	

also with particles and an S-matrix. The integrable sectors of four-dimensional gauge theories, however, are conformal invariant.

The study of integrable models in two dimensions is quite mature, whereas the application of integrability to four-dimensional systems is at an infant stage. The concepts of multi-local charges described in Section 5.11 and of quantum groups discussed in Section 5.13 have been applied only slightly to gauge dynamical systems in four dimensions.

23.4 Bosonization

Bosonization is the formulation of fermionic systems in terms of bosonic variables and fermionization is just the opposite process. The study of bosonized physical systems offers several advantages:

(1) It is usually easier to deal with commuting fields rather than anti-commuting ones.
(2) In certain examples, like the Thirring model, the fermionic strong coupling regime turns into the weak coupling one in its bosonic version, the sine-Gordon model (see Section 6.2).
(3) The non-abelian bosonization, especially in the product scheme (see Section 6.3.4), offers a separation between colored and flavored degrees of freedom, which is very convenient for analyzing low lying spectrum.
(4) Baryons composed of N_C quarks are a many-body problem in the fermion language, while simple solitons are in the boson language.
(5) One loop fermionic computations involving the currents turn into tree level consideration in the bosonized version. The best-known example of the latter are the chiral (or axial) anomalies (see Section 9.1).

In four dimensions, spin is obviously non-trivial and one cannot constitute generically a bosonization equivalence. However, in certain circumstances a four-dimensional system can be described approximately by fields that depend only on the time and on one space direction. In those cases one can apply the bosonization technique. Examples of such scenarios are monopole induced proton decay, and fractional charges induced on monopoles by light fermions. In these cases the relevant degrees of freedom are in an s-wave and hence taken to depend only on the time and the radial direction. This enables one to use the corresponding bosonized field. There is a slight difference with two dimensions, as the radial coordinate goes from zero to infinity, so "half" a line. Appropriate boundary conditions enable us to use a reflection, so as to extend to a full line.

23.5 Topological field configurations

- The topological charges in any dimensions are conserved regardless of the equations of motion of the corresponding systems. In two dimensions it is very easy to write down a current which is conserved without the use of the equations of motion. This is referred to as a topological conservation. Consider a scalar field ϕ or its non-abelian analog ϕ^a that transforms in the adjoint representation of a group, then the following currents are abelian and non-abelian conserved currents,

$$J_\mu = \epsilon_{\mu\nu}\partial^\nu\phi \qquad J_\mu^a = \epsilon_{\mu\nu}\partial^\nu\phi^a. \qquad (23.11)$$

Recall that for a system that admits, for instance an abelian case, also a current $J_\mu = \partial_\mu\phi$ that is conserved upon the use of the equations of motion, one can then replace the two currents with left and right conserved currents $J_\pm = \partial_\pm\phi$ or $J = \partial\phi$ and $\bar{J} = \bar{\partial}\phi$, as was discussed in Chapter 1. The charge associated with the topological conserved current is given by,

$$Q_{\text{top}} = \int dx\phi' = [\phi(t,+\infty) - \phi(t,-\infty)] \equiv \phi_+ - \phi_-, \qquad (23.12)$$

where the space dimension is taken to be $\mathcal{R}$. For a compactified space dimension, namely an S^1 this charge vanishes, except for cases where the field is actually an angle variable, in which case the charge is 2π. The latter appears in the case of $U(1)$ gauge theory in two dimensions, where there is a winding number.

- Obviously one cannot have such topologically conserved currents and charges in four dimensions. However, for theories that are invariant under a non-abelian group, one can construct also in four dimensions a topological current and charge, as for the cases of Skyrmions, magnetic monopoles and instantons. For the Skyrmions the topological current is given by,

$$J_{\text{skyre}}^\mu = \frac{i\epsilon^{\mu\nu\rho\sigma}}{24\pi^2} \text{Tr}\,[L_\nu L_\rho L_\sigma]. \qquad (23.13)$$

Table 23.2. *Topological classical field configurations in two and four dimensions*

classical field	dim.	map	topological current
soliton	two	[1]	$\epsilon^{\mu\nu}\partial_\nu\phi$
baryon	two	$S^1 \to S^1$	$\epsilon^{\mu\nu}Tr[g^{-1}\partial_\nu g]$
Skyrmion	four	$S^3 \to S^3$	$\frac{i\epsilon^{\mu\nu\rho\sigma}}{24\pi^2}Tr[L_\nu L_\rho L_\sigma]$
monopole	four	$S^2_{space} \to S^2_{G/H}$	$\frac{1}{8\pi}\epsilon_{\mu\nu\rho\sigma}\epsilon^{abc}\partial^\nu\hat\Phi^a\partial^\rho\hat\Phi^b\partial^\sigma\hat\Phi^c$
instanton	four	$S^3_s \to S^3_g$	$\frac{i\epsilon^{\mu\nu\rho\sigma}}{16\pi^2}Tr[A_\nu\partial_\rho A_\sigma + \frac{2}{3}A_\nu A_\rho A_\sigma]$

- The topological charges, for compact spaces, are the winding numbers of the corresponding topological configurations. For a compact one space dimension, we have the map of $S^1 \to S^1$ related to the homotopy group $\pi_1(S^1)$. In two space dimensions, the windings are associated with the map $S_2 \to S_2^{G/H}$, as for the magnetic monopoles. For three space dimensions, it is $S^3 \to S^3$ for the Skyrmions at $N_f = 2$, and the non-abelian instantons for the gauge group $SU(2)$. The topological data of the various models is summarized in Table 23.2.
- According to Derrick's theorem (see Section 5.3), for a theory of a scalar field with an ordinary kinetic term with two derivatives, and any local potential at $D \geq 2$, the only non singular time-independent solutions of finite energy are the vacua. However, as we have seen in Chapters 20, 21 and 22, there are solitons in the form of Skyrmions and monopoles and instantons. Those configurations bypass Derrick's theorem by introducing higher derivative terms or including non-abelian gauge fields.
- As was emphasised in Chapter 20, the extraction of the baryonic properties in the Skyrme model is very similar to the one for the baryons in the bosonized theory in two dimensions. Unlike the latter which is exact in the strong coupling limit, one cannot derive the former starting from the underlying theory. Another major difference between the two models is of course the existence of angular momentum only in the four-dimensional case.
- A non-trivial task associated with topological configurations is the construction of configurations that carry multipole topological charge, for instance a multi-baryon state both of the bosonized QCD_2 as well as of the Skyrme model, a multi-monopole solution and a multi-instanton solution. For the two-dimensional baryons (as discussed in Section 13.6) the construction is a straightforward generalization of the configuration of baryon number one. For the multi-monopole solutions we presented Nahm's construction, and for the multi-instantons the ADHM construction. These constructions, which are in fact related, are much more complicated than that for the two-dimensional muti-baryons.

---

[1] Depends on the type of the soliton. See Section 5.3.

- A very important phenomenon that occurs in both two and four dimensions
  is the strong-weak duality, and the duality between a soliton and an elementary field. In two dimensions we have encountered this duality in the relation
  between the Thirring model and the sine-Gordon model, where the coupling
  of the latter β is related to that of the former g as (6.27),

$$\frac{\beta^2}{4\pi} = \frac{1}{1 + \frac{g}{\pi}}. \tag{23.14}$$

This also relates the elementary fermion field of the Thirring model with the
soliton of the sine-Gordon model. In particular for $g = 0$ corresponding to
$\beta^2 = 4\pi$, the Thirring model describes a free Dirac fermion, while the soliton
of the corresponding sine-Gordon theory is the same fermion in its bosonization
disguise. An analog in four dimensions is the Olive–Montonen duality discussed
in Section 21.8, which relates the electric charge e with the magnetic one
$e_M = \frac{4\pi}{e}$, where the former is carried by the elementary states $W^\pm$ and the
latter by the magnetic monopoles.

23.6 Confinement versus screening

Naive intuition tells us that dynamical quarks in the fundamental representation can screen external sources in the fundamental representation, dynamical
adjoint quarks can screen adjoint sources, but that dynamical adjoint cannot
screen fundamentals. The picture that emerged from our two dimensional calculations (Chapter 14) showed that this was not the case. We found that massless
adjoint quarks could screen an external source in the fundamental representation. Moreover we have seen that any massless dynamical field will necessarily
be in the screening phase. The argument for that was that in all cases we have
considered we have found that the string tension is proportional to the mass of
the dynamical quarks,

$$\sigma \sim mg, \tag{23.15}$$

where m is the mass of the quark and g is the gauge coupling, and hence for the
massless case it vanishes. This was shown in Chapter 14 based on performing a
chiral rotation that enabled us to eliminate the external sources and computing
the string tension as the difference between the Hamiltonian of the system with
the external sources and the one without them namely (14.12),

$$\sigma = <H> - <H_0>. \tag{23.16}$$

It seems as though the situation in two dimensions is very different from that in
four dimensions. From the onset there is a dramatic difference between two and
four dimensions relating to the concept of confining theory. In two dimensions
both the Coulomb abelian potential and the non-abelian one are linear with
the separation distance L, whereas obviously in four dimensions the Coulomb
potential between two particles behaves as $1/L$. The confining potential is linear

with L in both two and four dimensions. However, that does not explain the difference between two and four dimensions, it merely means that in two dimensions the coulomb and confining potentials behave in the same manner.[2] The determination of the string tension in two dimensions cannot be repeated in four dimensions. The reason is that in the latter case the anomaly is not linear in the gauge field and thus one cannot use the chiral rotation to eliminate the external quark anti-quark pair. That does not imply that the situation in four dimensions differs from the two-dimensional one, it just means that one has to use different methods to compute the string tension in four dimensions.

What are the four-dimensional systems that might resemble the two-dimensional case of dynamical adjoint matter and external fundamental quarks? A system with external quarks in the fundamental representation in the context of pure YM theory seems a possible analog since the dynamical fields, the gluons, are in the adjoint representation, though they are vector fields and not fermions. An alternative is the $\mathcal{N} = 1$ SYM where in addition to the gluons there are also gluinos which are Majorana fermions in the adjoint representation. Both these cases should correspond to the massless adjoint case in two dimensions. The latter admits a screening behavior where as the four-dimensional models seem to be in the confining phase. This statement is supported by several different types of calculations in particular for the non supersymmetric case this behavior is found in lattice simulations.

At this point we cannot provide a satisfactory intuitive explanation why the behavior in two and four dimensions is so different. There is also no simple picture of how the massless adjoint dynamical quarks in two dimensions are able to screen external charges in the fundamental representation.

It is worth mentioning that there is ample evidence that four-dimensional hadronic physics is well described by a string theory. This is based for instance on realizing that mesons and baryons in nature admit Regge trajectory behavior which is an indication of a stringy nature. Any string theory is by definition a two-dimensional theory and hence a very basic relation between four-dimensional hadron physics and two-dimensional physics.

In addition to the ordinary string tension which relates to the potential between a quark and anti-quark in the fundamental representation, one defines the k string that connects a set of k quarks with a set of k anti-quarks. This object has been examined in four-dimensional YM as well as four-dimensional $\mathcal{N} = 1$ SYM. These two cases seem to be the analog of the two-dimensional QCD theory with adjoint quarks and with external quarks in a representation that is characterized by k boxes in the Young tableau description. In Chapter 14 we have derived an expression for the string tension as a function of the representation of the external and dynamical quarks and in particular for dynamical adjoint

---

[2] Note that the linear potential in two dimensions is already there at lowest order, while obviously in four dimensions it is a highly non-perturbative effect.

fermions and external quarks in the k representation. If there is any correspondence between the four-dimensional adjoint matter field and the two-dimensional
adjoint quarks it must be with massive adjoint quarks since for the massless case,
as was mentioned above, the two-dimensional string tension vanishes whereas the
four-dimensional one does not. Thus one may consider a correspondence for a
softly broken $\mathcal{N} = 1$ case where the gluinos are massive.

In two dimensions for the pure YM case we found that the string tension
behaves like $\sigma \sim g^2 k_{\text{ext}}^2$ whereas a Wilson line calculation yields $\sigma \sim g^2 C_2(R)$
where $C_2(R)$ is the second Casimir operator in the R representation of the
external quarks. For the QCD$_2$ case of general k external charges and adjoint
dynamical quarks, one can derive from (14.49) that,

$$\sigma_k^{2d} \sim \sin^2\left(\frac{\pi k}{N_c}\right),\tag{23.17}$$

whereas in four dimensions it is believed that for general k, the string tension
either follows a Casimir law or a sinusoidal rule as follows,

$$\sigma_k^{\text{cas}} \sim \frac{k(N_c - k)}{N_c} \qquad \sigma_k^{\text{sin}} \sim \sin\left(\frac{\pi k}{N_c}\right).\tag{23.18}$$

It is an open problem which of these holds.

As expected all these expressions are invariant under $k \to N - k$ which corresponds for antisymmetric representations to replacing a quark with an anti-
quark.

23.7 Hadronic phenomenology of two dimensions versus four dimensions

QCD_2 was addressed first in the fermionic formulation. In his seminal work 't
Hooft deduced the mesonic spectrum in the large N_C limit as is described in
Chapter 10. We further presented three additional approaches to the hadronic
spectra in two dimensions, the currentization method for massless quarks for the
entire plane of N_C and N_f, the DLCQ approach to extract the mesonic spectrum
for the case of fundamental as well as adjoint quarks and finally the bosonized
formulation in the strong coupling limit to determine the baryonic spectrum. As
for the four-dimensional hadronic spectrum we described the use of the large N_C
planar limit and the analysis of the baryonic world using the Skyrme model. It
is worth mentioning again that whereas in the four-dimensional case the Skyrme
approach is only an approximated model derived by an "educational guess", in
two dimensions the action in the strong coupling regime is exact.

23.7.1 Mesons

As was just mentioned the two-dimensional mesonic spectrum was extracted
using the large N_C approximation in the fermionic formulation for $N_f = 1$

('t Hooft model), the currentization for massless quarks and the DLCQ approach for both cases of quarks in the fundamental and the adjoint representation. For the particular region of $N_C \gg N_f$ and $m = 0$ the fermionic large N_c and the currentization treatments yielded identical results. In fact this result is achieved also using the DLCQ method for adjoint fermions upon a truncation to a single parton and replacing g^2 with $2g^2$ (see (12.42)). For massive fundamental quarks the DLCQ results match very nicely those of lattice simulations and the large N_c calculations as can be seen from Figs (12.1) and (12.2).

In all these methods the corresponding equations do not admit exact analytic solutions for the whole range of parameters and thus one has to resort to numerical solutions. However, in certain domains one can determine the analytic behavior of the wavefunctions and masses.

The spectrum of mesons in two dimensions is characterized by the dependence of the meson masses M_{mes} on the gauge coupling g, the number of colors N_c, the number of flavors N_f, the quark mass m_q and the excitation number n. In four-dimensional QCD the meson spectra depend on the same parameters apart from the fact that Λ_{QCD}, the QCD scale, is replacing the two-dimensional gauge coupling and of course some additional quantum numbers. The following lines summarize the properties of the spectrum

- The highly excited states $n \gg 1$, are characterized by,

$$M_{mes}^2 \sim \pi g^2 N_c n. \tag{23.19}$$

This seems to fit the behaviors of mesons in nature. This behavior is referred to as a Regge trajectory and it follows easily from a bosonic string model of the mesons. Following this analogy, the role of the string tension in a two-dimensional model is played by $g^2 N_c$. This seems to be in contradiction with the statement that the string tension is proportional to $m_q g$, as seen in the discussion of screening versus confinement.

It is very difficult to derive the Regge trajectory behavior from direct calculations in four-dimensional QCD.

- The opposite limit of low-lying states and in particular the ground state can be deduced in the limit of large quark masses $m_q \gg g$ and small quark masses $g \gg m_q$. For the ground state in the former limit we find,

$$M_{mes}^0 \cong m_{q_1} + m_{q_2}, \tag{23.20}$$

where m_{q_i} are the masses of the quark and anti-quark. In the opposite limit of $m_q \ll g$

$$(M_{mes}^0)^2 \cong \frac{\pi}{3} \sqrt{\frac{g^2 N_c}{\pi}} (m_1 + m_2). \tag{23.21}$$

For the special case of massless quarks we find a massless meson. This is very reminiscent of the four-dimensional picture for the massless pions. For small masses this is similar to the pseudo-Goldstone boson relation where,

$$m_\pi^2 \sim \frac{<\bar\psi\psi>}{f_\pi^2}(m_1 + m_2). \qquad (23.22)$$

Note that in two dimensions the massless mesons decouple.

- The 't Hooft model cannot be used to explore the dependence on N_{f} the number of flavors. This can be done from the 't Hooft-like equations derived in Chapter 11. It was found that for the first massive state there is a linear dependence of the meson mass squared on N_{f}

$$M_{\mathrm{mes}}^2 \sim N_{\mathrm{f}}. \qquad (23.23)$$

We are not aware of a similar behavior of the mesons in four dimensions.

- The 't Hooft model (Chapter 10) provides the solution of the meson spectrum in the planar limit in two dimensions. The planar, namely large N_{c} limit, in four dimensions is too complicated to be similarly solved. As we have seen in Chapter 19 one can extract the scaling in N_{c} dependence of certain hadronic properties like the mass the size and scattering amplitude but the full determination of the hadronic spectrum and scattering is still an unresolved mystery.

- Tremendous progress has been made in the understanding of the supersymmetric theory of $\mathcal{N} = 4$ partly by demonstrating that certain sectors of it can be described by integrable spin chain models (Section 18.1).

- As was demonstrated in Chapter 12 the DLCQ method has been found very effective to address the spectrum of mesons of two-dimensional QCD. This raises the question of whether one can use the DLCQ method to handle the spectrum of four-dimensional QCD. This task is clearly much more difficult. On route to the extraction of the hadronic spectrum of QCD_4 an easier system has been analyzed. It is that of the collinear QCD (see Chapter 17) where in the Hamiltonian of the system one drops off all interaction terms that depend on the transverse momenta. In this effective two-dimensional setup the transverse degrees of freedom of the gluon are retained in the form of two scalar fields. This system which was not described in the book has been solved in [14] where a complete bound and continuum spectrum was extracted as well as the Fock space wavefunctions.

23.7.2 Baryons

In Chapter 13 we have described the spectrum of baryons in multiflavor two-dimensional QCD in the strong coupling limit $\frac{m_q}{e_c} \to 0$. The four-dimensional baryonic spectrum was discussed in the large N_{c} limit in Chapter 19 and using the Skyrme model approach in Chapter 20. We would like now to compare these spectra and to investigate the possibility of predicting four-dimensional baryonic properties from the simpler two-dimensional model. In the former case the mass is a function of the QCD scale Λ_{QCD}, the number of colors N_{c} and the number

Table 23.3. *Scaling of baryon masses with N_c in two and four dimensions*

	two dimensions	four dimensions
Classical baryon mass	N_c	N_c
Quantum correction	N_c^0	N_c^{-1}

of flavors N_f and in the latter it is a function of e_c, N_c and N_f. Thus it seems that the dimensionful gauge coupling in two dimensions is the analog of Λ_{QCD} in four dimensions.

- In two dimensions the mass of the baryon was found to be,

$$E = 4m\sqrt{\frac{2N_c}{\pi}} + m\sqrt{2}\sqrt{\left(\frac{\pi}{N_c}\right)^3\left[C_2 - N_c^2\frac{(N_f-1)}{2N_f}\right]}, \qquad (23.24)$$

where the classical mass m is given by

$$m = \left[N_c c m_q \left(\frac{e_c\sqrt{N_f}}{\sqrt{2\pi}}\right)^{\Delta_c}\right]^{\frac{1}{1+\Delta_C}}, \qquad (23.25)$$

with $\Delta_c = \frac{N_c^2-1}{N_c(N_c+N_f)}$. Due to the fact that in two dimensions there is no spin, the structure of the spectrum with respect to the flavor group is obviously different in two and four dimensions. For instance the lowest allowed state for $N_c = N_f = 3$ is in two dimensions the totally symmetric representation **10**, whereas it is the mixed representation **8** in four dimensions.
- Let us discuss now the scaling with N_c in the large N_c limit. The classical term behaves like N_c, while the quantum correction like 1. This classical result is in accordance with four dimensions, derived when the large N expansion is applied to the baryonic system (see Chapter 19), that the baryon mass is linear in N_c and with the Skyrmion result (see Chapter 20). However, whereas in two dimensions the quantum correction behaves like N_c^0, namely suppressed by a factor of $\frac{1}{N_c}$ compared to the classical term, in four dimensions it behaves like $\frac{1}{N_c}$ namely a suppression of $\frac{1}{N_c^2}$. This is summarized in Table 23.3.
- In terms of the dependence on the number of flavors, it is interesting to note that both in two dimensions and in four dimensions, the contribution to the mass due to the quantum fluctuations has a term proportional to the second Casimir operator associated with the representation of the baryonic state under the $SU(N_f)$ flavor group (compare (23.24) with (20.68)).
- Another property of the baryonic spectrum that can be compared between the two- and four-dimensional cases is the flavor content of the various states. In Chapter 13 we have computed the $\bar{u}u, \bar{d}d$ and $\bar{s}s$ content for the Δ^+ and Δ^{++} states. Recall that in the two-dimensional model for $N_c = N_f = 3$ we do not have a state in the **8** representation but only in the **10** so strictly speaking there

Table 23.4. *Flavor content of two-dimensional and four-dimensional baryons*

	two dimensions state	value	four dimensions state	value
$\langle \bar{u}u \rangle$	Δ^+	$\frac{1}{2}$	p	$\frac{2}{5}$
$\langle \bar{d}d \rangle$	Δ^+	$\frac{1}{3}$	p	$\frac{11}{30}$
$\langle \bar{s}s \rangle$	Δ^+	$\frac{1}{6}$	p	$\frac{7}{30}$
$\langle \bar{s}s \rangle$	Δ^{++}	$\frac{1}{6}$	Δ	$\frac{7}{24}$
$\langle \bar{s}s \rangle$			Ω^-	$\frac{5}{24}$

is no exact analog of the proton. Instead we take the charge $= +1\Delta^+$ as the two-dimensional analog of the proton. In the Skyrme model one can compute in a similar manner the flavor content of the four-dimensional baryons. The two- and four-dimensional states compare as is summarized in Table 23.4.

23.8 Outlook

We can imagine future developments associated with the topics covered in the book in three different directions: Further progress in the application of the methods discussed in the book to unravel the mysteries of gauge dynamics in nature; applications of the methods in other domains of physics not related to four-dimensional gauge theories; and improving our understanding of the strong interaction and hadron physics due to other non-perturbative techniques that are not discussed in the book. Let us now briefly fantasize on hypothetical developments in those three avenues.

23.8.1 Further progress in the application of the methods discussed in the book

- A lesson that follows from the book is that the exploration of physical systems on one space dimension is both simpler to handle and sheds light on the real world so there are plenty of other unresolved questions that could be explored first in two dimensions. This includes exploration of the full standard model and the physics beyond the standard model including supersymmetry and its dynamical breaking, large extra dimensions, compositeness etc.
- There has been tremendous development in recent years in applying methods of integrable models and in particular of spin chains, like the thermal Bethe ansatz, to $\mathcal{N} = 4$ SYM theory, namely, in the context of supersymmetric conformal gauge theory. We have no doubt that there will be further development in computing the anomalous dimensions of gauge invariant operators and correlators.

- Moreover, one can identify in a similar manner to $\mathcal{N} = 4$ SYM theory spin chain structures in gauge theories which are confining and with less or even no supersymmetries. In that case the spin chain Hamiltonian would not correspond to the dilatation operator but rather be associated with the excitation energies of hadrons.
- It is plausible that the full role of magnetic monopoles and of instantons has not yet been revealed. They have already had several reincarnations and there may be more. For instance there was recently a proposal to describe baryons as instantons which are solitons of a five-dimensional flavor gauge theory in curved five dimensions.

23.8.2 Applications to other domains

- A very important application of two-dimensional conformal symmetry has been to superstring theories. A great part of the developments in superstring theories is attributed to the infinite-dimensional conformal symmetry algebra. In fact it went in both directions and certain progress in understanding the structure of conformal invariance has emerged from the research of string theories. A similar symbiotic evolution took place with regard to the affine Lie algebras.
- String theories and in particular the string theory on $AdS_5 \times S^5$ have recently been analyzed using the tools of integrable models like mapping to spin chains, using the Bethe ansatz equations, identifying a set of infinitely many conserved charges and using structure of Yangian symmetry.
- Spin chain models have been suggested to describe systems of "real" spins in condensed matter physics. As was discussed in this book the application of the corresponding tools to field theory systems has been quite fruitful. The opposite direction will presumably also take place and the use of properties of integrability that were understood in field theories will shed new light on certain condensed matter systems.
- The application of conformal invariance to condensed matter systems at criticality has a long history. There has been recently an intensive effort to further develop the understanding of systems like various superconductors, the fractional Hall effect and other systems using modern conformal symmetry techniques.

23.8.3 Developments in gauge dynamics due to other methods

- An extremely important framework for analyzing gauge theories has been supersymmetry. Regardless if it is realized in nature or not, it is evident that there are more tools to handle supersymmetric gauge theories and hence they are much better understood than non supersymmetric ones. One can gain novel insight about non supersymmetric theories by introducing supersymmetry breaking terms to well understood supersymmetric models. For instance one can start with the Seiberg Witten solution of $\mathcal{N} = 2$ [192] where the structure of vacua is known and extract confinement behavior in $\mathcal{N} = 1$ and non supersymmetric theories.

- A breakthrough in the understanding of gauge theories in the strong coupling regime took place with the discovery by Maldacena of the AdS/CFT holographic duality [158]. The strongly coupled $\mathcal{N} = 4$ in the large N and large 't Hooft parameter λ is mapped into a weakly curved supergravity background. Thousands of research papers that followed develop this map in many different directions and in particular also in relation to the pure YM theory and QCD in four dimensions. There is very little doubt that further exploration of the duality will shed new light on QCD and on hadron physics.

- String theory has been born as a possible theory of hadron physics. It then underwent a phase transition into a candidate for the theory of quantum gravity and even a unifying theory for everything. In recent years, mainly due to the AdS/CFT duality there is a renaissance of the idea that hadrons at low energies should be described as strings. This presumably combined with the duality seems to be a useful tool that will improve our understanding of gauge dynamics.

- The computations of scattering amplitudes in gauge theories has been boosted in recent years due to various developments including the use of techniques based on twistors, on a novel T-duality in the context of the Ads/CFT duality and on a conjectured duality between Wilson lines and scattering amplitudes. One does not need a wild imagination to foresee further progress in the industry of computing scattering amplitudes.

To summarize, non-perturbative methods have always been very important tools in exploring the physical world. We have no doubt that they will continue to be a very essential ingredient in future developments of science in general and physics in particular.

References

[1] E. Abdalla and M. C. B. Abdalla, "Updating QCD in two-dimensions," *Phys. Rept.* **265**, 253 (1996) [arXiv:hep-th/9503002].

[2] E. Abdalla, M. C. B. Abdalla and K. D. Rothe, *Nonperturbative Methods in Two-Dimensional Quantum Field Theory,* Singapore: World Scientific (2001).

[3] A. Abrashkin, Y. Frishman and J. Sonnenschein, "The spectrum of states with one current acting on the adjoint vacuum of massless QCD2," *Nucl. Phys.* B **703**, 320 (2004) [arXiv:hep-th/0405165].

[4] C. Adam, "Charge screening and confinement in the massive Schwinger model," *Phys. Lett.* B **394**, 161 (1997) [arXiv:hep-th/9609155].

[5] G. S. Adkins, C. R. Nappi and E. Witten, "Static properties of nucleons in the Skyrme model," *Nucl. Phys.* B **228**, 552 (1983).

[6] S. L. Adler, J. C. Collins and A. Duncan, "Energy momentum tensor trace anomaly in spin 1/2 quantum electrodynamics'" *Phys. Rev.* D **15**, 1712 (1977).

[7] I. Affleck, "On the realization of chiral symmetry in (1+1)-dimensions," *Nucl. Phys.* B **265**, 448 (1986).

[8] O. Aharony, O. Ganor, J. Sonnenschein and S. Yankielowicz, "On the twisted G/H topological models," *Nucl. Phys.* B **399**, 560 (1993) [arXiv:hep-th/9208040].

[9] O. Aharony, O. Ganor, J. Sonnenschein, S. Yankielowicz and N. Sochen, "Physical states in G/G models and 2-d gravity," *Nucl. Phys.* B **399**, 527 (1993) [arXiv:hep-th/9204095].

[10] O. Aharony, S. S. Gubser, J. M. Maldacena, H. Ooguri and Y. Oz, "Large N field theories, string theory and gravity," *Phys. Rept.* **323**, 183 (2000) [arXiv:hep-th/9905111].

[11] A. Y. Alekseev and V. Schomerus, "D-branes in the WZW model," *Phys. Rev.* D **60**, 061901 (1999) [arXiv:hep-th/9812193].

[12] D. Altschuler, K. Bardakci and E. Rabinovici, "A construction of the c < 1 modular invariant partition function," *Commun. Math. Phys.* **118**, 241 (1988).

[13] L. Alvarez-Gaume, G. Sierra and C. Gomez, "Topics in conformal field theory," contribution to the Knizhnik Memorial Volume, L. Brink, *et al.*, World Scientific. In Brink, L. (ed.) *et al.*: *Physics and Mathematics of Strings* 16-184 (1989). Singapore.

[14] F. Antonuccio and S. Dalley, "Glueballs from (1+1)-dimensional gauge theories with transverse degrees of freedom," *Nucl. Phys.* B **461**, 275 (1996) [arXiv:hep-ph/9506456].

[15] A. Armoni, Y. Frishman and J. Sonnenschein, "The string tension in massive QCD(2)," *Phys. Rev. Lett.* **80**, 430 (1998) [arXiv:hep-th/9709097].

[16] A. Armoni, Y. Frishman and J. Sonnenschein, "The string tension in two dimensional gauge theories," *Int. J. Mod. Phys.* A **14**, 2475 (1999) [arXiv:hep-th/9903153].

[17] A. Armoni, Y. Frishman and J. Sonnenschein, "Massless QCD(2) from current constituents," *Nucl. Phys.* B **596**, 459 (2001) [arXiv:hep-th/0011043].

[18] A. Armoni and J. Sonnenschein, "Mesonic spectra of bosonized QCD in two-dimensions models," *Nucl. Phys.* B **457**, 81 (1995) [arXiv:hep-th/9508006].

[19] M. F. Atiyah and N. J. Hitchin, "The geometry and dynamics of magnetic monopole. M.B. Porter lectures" Princeton, USA: Princeton University Press (1988) 133p.

[20] M. F. Atiyah, N. J. Hitchin, V. G. Drinfeld and Yu. I. Manin, "Construction of instantons," *Phys. Lett.* A **65**, 185 (1978).

[21] F. A. Bais, "To be or not to be? Magnetic monopoles in non-abelian gauge theories," in 't Hooft, Hackensack, G. ed. *Fifty years of Yang–Mills Theory*, New Jersey World Scientific, C 2005.

[22] A. P. Balachandran, "Solitons in nuclear and elementary physics. Proceedings of the Lewes workshop," World Scientific, 1984.

[23] I. I. Balitsky and L. N. Lipatov, "The Pomeranchuk singularity in quantum chromodynamics," *Sov. J. Nucl. Phys.* **28**, 822 (1978) [*Yad. Fiz.* **28**, 1597 (1978)].

[24] V. Baluni, "The Bose form of two-dimensional quantum chromodynamics," *Phys. Lett.* B **90**, 407 (1980).

[25] T. Banks, "Lectures on conformal field theory," presented at the Theoretical Advanced Studies Institute, St. John's College, Santa Fe, N. Mex., Jul 5 – Aug 1, 1987. Published in Santa Fe: TASI 87:572.

[26] T. I. Banks and C. M. Bender, "Anharmonic oscillator with polynomial self-interaction," *J. Math. Phys.* **13**, 1320 (1972).

[27] K. Bardakci and M. B. Halpern, "New dual quark models," *Phys. Rev.* D **3**, 2493 (1971).

[28] A. Bassetto, G. Nardelli and R. Soldati, *Yang-Mills Theories in Algebraic Noncovariant Gauges: Canonical Quantization and Renormalization,* Singapore World Scientific (1991).

[29] R. J. Baxter, *Exactly Solved Models in Statistical Mechanics,* London: Academic press (1989).

[30] K. Becker, M. Becker and J. H. Schwarz, *String Theory and M-Theory: A modern introduction,* Cambridge, UK: Cambridge University Press (2007).

[31] N. Beisert, "The dilatation operator of N = 4 super Yang-Mills theory" *Phys. Rept.* **405**, 1 (2005) [arXiv:hep-th/0407277].

[32] A. A. Belavin, A. M. Polyakov, A. S. Shvarts and Yu. S. Tyupkin, "Pseudoparticle solutions of the Yang-Mills equations," *Phys. Lett.* B **59**, 85 (1975).

[33] A. A. Belavin, A. M. Polyakov and A. B. Zamolodchikov, "Infinite conformal symmetry in two-dimensional quantum field theory," *Nucl. Phys.* B **241**, 333 (1984).

[34] A. V. Belitsky, V. M. Braun, A. S. Gorsky and G. P. Korchemsky, "Integrability in QCD and beyond," *Int. J. Mod. Phys.* A **19**, 4715 (2004) [arXiv:hep-th/0407232].

[35] D. E. Berenstein, J. M. Maldacena and H. S. Nastase, "Strings in flat space and pp waves from N = 4 super Yang Mills," *JHEP* **0204**, 013 (2002) [arXiv:hep-th/0202021].

[36] M. Bernstein and J. Sonnenschein, "A comment on the quantization of the chiral bosons" *Phys. Rev. Lett.* **60**, 1772 (1988).

[37] J. D. Bjorken and S. D. Drell, *Relativistic Quantum Field Theory,* New York: McGraw-Hill (1965), ISBN 0-07-005494-0.

[38] G. Bhanot, K. Demeterfi and I. R. Klebanov, "(1+1)-dimensional large N QCD coupled to adjoint fermions," *Phys. Rev.* D **48**, 4980 (1993) [arXiv:hep-th/9307111].

[39] D. Bernard and A. Leclair, "Quantum group symmetries and nonlocal currents in 2-D QFT," *Commun. Math. Phys.* **142**, 99 (1991).

[40] K. M. Bitar and S. J. Chang, "Vacuum tunneling of gauge theory in minkowski space," *Phys. Rev.* D **17**, 486 (1978).

[41] E. B. Bogomolny, "Stability of classical solutions," *Sov. J. Nucl. Phys.* **24**, 449 (1976) [*Yad. Fiz.* **24**, 861 (1976)].

[42] V. M. Braun, S. E. Derkachov, G. P. Korchemsky and A. N. Manashov, "Baryon distribution amplitudes in QCD," *Nucl. Phys.* B **553**, 355 (1999) [arXiv:hep-ph/9902375].

[43] V. M. Braun, G. P. Korchemsky and D. Mueller, "The uses of conformal symmetry in QCD," *Prog. Part. Nucl. Phys.* **51**, 311 (2003) [arXiv:hep-ph/0306057].

[44] E. Brezin and J. L. Gervais, "Nonperturbative aspects in quantum field theory. Proceedings of Les Houches Winter Advanced Study Institute, March 1978," *Phys. Rept.* **49**, 91 (1979).

[45] E. Brezin, C. Itzykson, J. Zinn-Justin and J. B. Zuber, "Remarks about the existence of nonlocal charges in two-dimensional models," *Phys. Lett.* B **82**, 442 (1979).

[46] E. Brezin and S. R. Wadia, "*The large N expansion in quantum field theory and statistical physics: From spin systems to two-dimensional gravity,*" Singapore, Singapore: World Scientific (1993) 1130 p

[47] S. J. Brodsky, "Gauge theories on the light-front," *Braz. J. Phys.* **34**, 157 (2004) [arXiv:hep-th/0302121].

[48] S. J. Brodsky, H. C. Pauli and S. S. Pinsky, "Quantum chromodynamics and other field theories on the light cone," *Phys. Rept.* **301**, 299 (1998) [arXiv:hep-ph/9705477].

[49] S. J. Brodsky, Y. Frishman, G. P. Lepage and C. T. Sachrajda, "Hadronic wave functions at short distances and the operator product expansion," *Phys. Lett.* B **91**, 239 (1980).

[50] S. J. Brodsky, Y. Frishman and G. P. Lepage, "On the application of conformal symmetry to quantum field theory," *Phys. Lett.* B **167**, 347 (1986).

[51] S. J. Brodsky, P. Damgaard, Y. Frishman and G. P. Lepage, "Conformal symmetry: exclusive processes beyond leading order," *Phys. Rev.* D **33**, 1881 (1986).

[52] R. C. Brower, W. L. Spence and J. H. Weis, "Effects of confinement on analyticity in two-dimensional QCD," *Phys. Rev.* D **18**, 499 (1978).

[53] G. E. Brown, 'Selected papers, with commentary, of Tony Hilton Royle Skyrme," Singapore: World Scientific (1994) 438 p. (World Scientific series in 20th century physics, 3).

[54] R. N. Cahn, *Semisimple Lie Algebras and their Representations,* Menlo Park, USA: Benjamin/Cummings (1984) 158 p. (Frontiers In Physics, 59)

[55] C. G. Callan, "Broken scale invariance in scalar field theory," *Phys. Rev.* D **2**, 1541 (1970).

[56] C. G. Callan, N. Coote and D. J. Gross, "Two-dimensional Yang-Mills theory: a model of quark confinement," *Phys. Rev.* D **13**, 1649 (1976).

[57] C. G. Callan, R. F. Dashen and D. J. Gross, "Toward a theory of the strong interactions," *Phys. Rev.* D **17**, 2717 (1978).

[58] J. L. Cardy, "Boundary conditions, fusion rules and the Verlinde formula," *Nucl. Phys.* B **324**, 581 (1989).

[59] J. L. Cardy, "Conformal invariance and statistical mechanics," Les Houches Summer School 1988:0169-246.

[60] A. Casher, H. Neuberger and S. Nussinov, "Chromoelectric flux tube model of particle production," *Phys. Rev.* D **20**, 179 (1979).

[61] A. Chodos, "Simple connection between conservation laws in the Korteweg-De Vries and sine-Gordon systems," *Phys. Rev.* D **21**, 2818 (1980).

[62] E. Cohen, Y. Frishman and D. Gepner, "Bosonization of two-dimensional QCD with flavor," *Phys. Lett.* B **121**, 180 (1983).

[63] S. R. Coleman, "Quantum sine-Gordon equation as the massive Thirring model," *Phys. Rev.* D **11**, 2088 (1975).

[64] S. R. Coleman, "More about the massive Schwinger model," *Annals Phys.* **101**, 239 (1976).

[65] S. R. Coleman, "The uses of instantons," *Subnucl. Ser.* **15**, 805 (1979).

[66] S. Coleman, *Aspects of Symmetry,* selected Erice lectures of Sidney Coleman, Cambridge, UK, Cambridge Univ. Press (1985).

[67] S. R. Coleman, The magnetic monopole fifty years later, In the *Unity of Fundamental Interactions,* ed. A. Zichichi: New York, Plenum (1983).

[68] S. R. Coleman, R. Jackiw and L. Susskind, "Charge shielding and quark confinement in the massive Schwinger model," *Annals Phys.* **93**, 267 (1975).

[69] S. Cordes, G. W. Moore and S. Ramgoolam, "Large N 2-D Yang-Mills theory and topological string theory," *Commun. Math. Phys.* **185**, 543 (1997) [arXiv:hep-th/9402107].

[70] E. Corrigan, D. B. Fairlie, S. Templeton and P. Goddard, "A Green's function for the general selfdual gauge field," *Nucl. Phys.* B **140**, 31 (1978).

[71] E. Corrigan and P. Goddard, "Construction of instanton and monopole solutions and reciprocity," *Annals Phys.* **154**, 253 (1984).

[72] S. Dalley and I. R. Klebanov, "String spectrum of (1+1)-dimensional large N QCD with adjoint matter," *Phys. Rev.* D **47**, 2517 (1993) [arXiv:hep-th/9209049].

[73] R. F. Dashen and Y. Frishman, "Four fermion interactions and scale invariance," *Phys. Rev.* D **11**, 2781 (1975).

[74] R. F. Dashen, B. Hasslacher and A. Neveu, "Nonperturbative methods and extended hadron models in field theory. 2. Two-dimensional models and extended hadrons," *Phys. Rev.* D **10**, 4130 (1974).

[75] G. Date, Y. Frishman and J. Sonnenschein, "The spectrum of multiflavor QCD in two-dimensions'" *Nucl. Phys.* B **283**, 365 (1987).

[76] G. F. Dell-Antonio, Y. Frishman and D. Zwanziger, "Thirring model in terms of currents: solution and light cone expansions," *Phys. Rev.* D **6**, 988 (1972).

[77] P. Di Francesco, P. Mathieu, D. Senechal *Conformal Field Theory*, Series:Graduate Texts in Contemporary Physics. New York: Springer-Verlag (1997).

[78] P. A. M. Dirac, "Theory Of Magnetic Monopoles," *In *Coral Gables 1976, Proceedings, New Pathways In High-energy Physics, Vol. I*, New York: Plenum Press, 1976, 1-14.*

[79] F. M. Dittes and A. V. Radyushkin, "Two loop contribution to the evolution of the pion wave function," *Phys. Lett.* B **134**, 359 (1984); M. H. Sarmadi, *Phys. Lett.* B **143**, 471 (1984); S. V. Mikhailov and A. V. Radyushkin, *Nucl. Phys.* B **254**, 89 (1985); G. R. Katz, *Phys. Rev.* D **31**, 652 (1985).

[80] N. Dorey, T. J. Hollowood, V. V. Khoze, M. P. Mattis and S. Vandoren, "Multi-instanton calculus and the AdS/CFT correspondence in N = 4 *Nucl. Phys.* B **552**, 88 (1999) [arXiv:hep-th/9901128].

[81] N. Dorey, T. J. Hollowood, V. V. Khoze and M. P. Mattis, "The calculus of many instantons," *Phys. Rept.* **371**, 231 (2002) [arXiv:hep-th/0206063].

[82] M. R. Douglas, K. Li and M. Staudacher, "Generalized two-dimensional QCD," *Nucl. Phys.* B **420**, 118 (1994) [arXiv:hep-th/9401062].

[83] A. V. Efremov and A. V. Radyushkin, "Factorization and asymptotical behavior of pion form-factor in QCD," *Phys. Lett.* B **94**, 245 (1980).

[84] T. Eguchi and H. Ooguri, "Chiral bosonization on a Riemann surface," *Phys. Lett.* B **187**, 127 (1987).

[85] S. Elitzur and G. Sarkissian, "D-branes on a gauged WZW model," *Nucl. Phys.* B **625**, 166 (2002) [arXiv:hep-th/0108142].

[86] J. R. Ellis, Y. Frishman, A. Hanany and M. Karliner, "Quark solitons as constituents of hadrons," *Nucl. Phys.* B **382**, 189 (1992) [arXiv:hep-ph/9204212].

[87] J. R. Ellis, Y. Frishman and M. Karliner, "Meson baryon scattering in QCD(2) for any coupling," *Phys. Lett.* B **566,** (2003) 201 [arXiv:hep-ph/0305292].

[88] L. D. Faddeev, "How algebraic Bethe ansatz works for integrable model," arXiv:hep-th/9605187.

[89] B. L. Feigin and D. B. Fuchs "Skew symmetric differential operators on the line and Verma modules over the Virasoro algebra" *Funct. Anal. Prilozhen* **16**, (1982) 47.

[90] S. Ferrara, A. F. Grillo and R. Gatto, "Improved light cone expansion," *Phys. Lett.* B **36**, 124 (1971) [*Phys. Lett.* B **38**, 188 (1972)].

[91] D. Finkelstein and J. Rubinstein, "Connection between spin, statistics, and kinks," *J. Math. Phys.* **9**, 1762 (1968).

[92] R. Floreanini and R. Jackiw, "Selfdual fields as charge density solitons," *Phys. Rev. Lett.* **59**, 1873 (1987).

[93] D. Friedan, "Introduction To Polyakov's string theory," Published in Les Houches Summer School 1982:0839.

[94] D. Friedan, E. J. Martinec and S. H. Shenker, "Conformal invariance, supersymmetry and string theory," *Nucl. Phys.* B **271**, 93 (1986).

[95] D. Friedan, Z. a. Qiu and S. H. Shenker, "Conformal invariance, unitarity and two-dimensional critical exponents," *Phys. Rev. Lett.* **52**, 1575 (1984).

[96] Y. Frishman, A. Hanany and J. Sonnenschein, "Subtleties in QCD theory in two-dimensions," *Nucl. Phys.* B **429**, 75 (1994) [arXiv:hep-th/9401046].

[97] Y. Frishman and M. Karliner, "Baryon wave functions and strangeness content in QCD in two-dimensions," *Nucl. Phys.* B **344**, 393 (1990).

[98] Y. Frishman and M. Karliner, "Scattering and resonances in QCD(2)," *Phys. Lett.* B **541**, 273 (2002). Erratum-ibid. B **562**, 367, (2003). [arXiv:hep-ph/0206001].

[99] Y. Frishman and J. Sonnenschein, "Bosonization of colored-flavored fermions and QCD in two-dimenstions," *Nucl. Phys.* B **294**, 801 (1987).

[100] Y. Frishman and J. Sonnenschein, "Gauging of chiral bosonized actions'" *Nucl. Phys.* B **301**, 346 (1988).

[101] Y. Frishman and J. Sonnenschein, "Bosonization and QCD in two-dimensions," *Phys. Rept.* **223**, 309 (1993) [arXiv:hep-th/9207017].

[102] Y. Frishman and W. J. Zakrzewski, "Multibaryons in QCD in two-dimensions," *Nucl. Phys.* B **328**, 375 (1989).

[103] Y. Frishman and W. J. Zakrzewski, "Explicit expressions for masses and bindings of multibaryons in QCD(2)," *Nucl. Phys.* B **331**, 781 (1990).

[104] S. Fubini, A. J. Hanson and R. Jackiw, "New approach to field theory," *Phys. Rev.* D **7**, 1732 (1973).

[105] O. Ganor, J. Sonnenschein and S. Yankielowicz, "The string theory approach to generalized 2-d Yang-Mills theory," *Nucl. Phys.* B **434**, 139 (1995) [arXiv:hep-th/9407114].

[106] E. G. Gimon, L. A. Pando Zayas, J. Sonnenschein and M. J. Strassler, "A soluble string theory of hadrons," *JHEP* **0305**, 039 (2003) [arXiv:hep-th/0212061].

[107] D. Gepner, "Nonabelian bosonization and multiflavor QED and QCD in two-dimensions," *Nucl. Phys.* B **252**, 481 (1985).

[108] D. Gepner and E. Witten, "String theory on group manifolds," *Nucl. Phys.* B **278**, 493 (1986).

[109] P. H. Ginsparg, "Applied conformal field theory," Published in Les Houches Summer School 1988:1-168, arXiv:hep-th/9108028.

[110] P. Goddard, A. Kent and D. I. Olive, "Virasoro algebras and coset space models," *Phys. Lett.* B **152**, 88 (1985).

[111] P. Goddard and D. I. Olive, "Kac-Moody and Virasoro algebras in relation to quantum physics," *Int. J. Mod. Phys.* A **1**, 303 (1986).

[112] D. Gonzales and A. N. Redlich, "The low-energy effective dynamics of two-dimensional gauge theories," *Nucl. Phys.* B **256**, 621 (1985).

[113] M. B. Green, J. H. Schwarz and E. Witten, "Superstring theory. Vol. 2: Loop amplitudes, anomalies and phenomenology," Cambridge, UK: Univ. Pr. (1987) 596 P. (Cambridge Monographs On Mathematical Physics)

[114] C. Gomez, G. Sierra and M. Ruiz-Altaba, "Quantum groups in two-dimensional physics," *Cambridge, UK: Univ. Pr. (1996) 457 p*

[115] D. J. Gross, "Two-dimensional QCD as a string theory," *Nucl. Phys.* B **400**, 161 (1993) [arXiv:hep-th/9212149].

[116] D. J. Gross, I. R. Klebanov, A. V. Matytsin and A. V. Smilga, "Screening vs. Confinement in 1+1 Dimensions," *Nucl. Phys.* B **461**, 109 (1996) [arXiv:hep-th/9511104].

[117] D. J. Gross and A. Neveu, "Dynamical symmetry breaking in asymptotically free field theories," *Phys. Rev.* D **10**, 3235 (1974).

[118] D. J. Gross and W. Taylor, "Two-dimensional QCD is a string theory," *Nucl. Phys.* B **400**, 181 (1993) [arXiv:hep-th/9301068].

[119] D. J. Gross and W. Taylor, "Twists and Wilson loops in the string theory of two-dimensional QCD," *Nucl. Phys.* B **403**, 395 (1993) [arXiv:hep-th/9303046].

[120] I. G. Halliday, E. Rabinovici, A. Schwimmer and M. S. Chanowitz, "Quantization of anomalous two-dimensional models," *Nucl. Phys.* B **268**, 413 (1986).

[121] M. B. Halpern, "Quantum solitons which are SU(N) fermions," *Phys. Rev.* D **12**, 1684 (1975).

[122] G. 't Hooft, "A planar diagram theory for strong interactions," *Nucl. Phys.* B **72**, 461 (1974).

[123] G. 't Hooft, "Magnetic monopoles in unified gauge theories" *Nucl. Phys.* B **79**, 276 (1974).

[124] G. 't Hooft, "A two-dimensional model for mesons," *Nucl. Phys.* B **75**, 461 (1974).

[125] G. 't Hooft, "Computation of the quantum effects due to a four-dimensional pseudoparticle," *Phys. Rev.* D **14**, 3432 (1976) [Erratum-ibid. D **18**, 2199 (1978)].

[126] P. Horava, "Topological strings and QCD in two-dimensions," hep-th/9311156, talk given at NATO Advanced Research Workshop on New Developments in String Theory, Conformal Models and Topological Field Theory, Cargese, France, 12–21 May 1993.

[127] K. Hornbostel, "The application of light cone quantization to quantum chromodynamics in (1+1)-dimensions," Ph.D. thesis, SLAC-R-333, Dec 1988.

[128] K. Hornbostel, S. J. Brodsky and H. C. Pauli, "Light cone quantized QCD in (1+1)-dimensions," *Phys. Rev.* D **41**, 3814 (1990).

[129] C. Imbimbo and A. Schwimmer, "The Lagrangian formulation of chiral scalars," *Phys. Lett.* B **193**, 455 (1987).

[130] C. Itzykson and J. B. Zuber, *Quantum Field Theory, New York, USA:* McGraw-Hill (1980) 705 P.(International Series In Pure and Applied Physics)

[131] R. Jackiw and C. Rebbi, "Vacuum periodicity in a Yang-Mills quantum theory," *Phys. Rev. Lett.* **37**, 172 (1976).

[132] R. Jackiw, C. Nohl and C. Rebbi, "Conformal properties of pseudoparticle configurations", *Phys. Rev.* D **15**, 1642 (1977).

[133] A. D. Jackson and M. Rho, "Baryons as chiral solitons," *Phys. Rev. Lett.* **51**, 751 (1983).

[134] B. Julia and A. Zee, "Poles with both magnetic and electric charges in nonabelian gauge theory," *Phys. Rev.* D **11**, 2227 (1975).

[135] N. Ishibashi, "The boundary and crosscap states in conformal field theories," *Mod. Phys. Lett.* A **4**, 251 (1989).

[136] V. G. Kac, "Simple graded algebras of finite growth," *Funct. Anal. Appl.* **1**, 328 (1967).

[137] L. P. Kadanoff, "Correlators along the line of two dimensional Ising model," *Phys. Rev.* **188**, 859 (1969).

[138] M. Kaku, *Strings, Conformal Fields, and M-Theory,* New York, USA: Springer (2000) 531 p.

[139] V. A. Kazakov, "Wilson loop average for an arbitrary contour in two-dimensional U(N) gauge theory," *Nucl. Phys.* B **179**, 283 (1981).

[140] S. V. Ketov, *Conformal Field Theory,* Singapore, Singapore: World Scientific (1995) 486 p.

[141] V. V. Khoze, M. P. Mattis and M. J. Slater, "The instanton hunter's guide to supersymmetric SU(N) gauge theory," *Nucl. Phys.* B **536**, 69 (1998) [arXiv:hep-th/9804009].

[142] E. Kiritsis, *String Theory in a Nutshell,* Princeton, USA: Univ. Pr. (2007) 588 p.

[143] V. G. Knizhnik and A. B. Zamolodchikov, "Current algebra and Wess-Zumino model in two dimensions," *Nucl. Phys.* B **247**, 83 (1984).

[144] W. Krauth and M. Staudacher, "Non-integrability of two-dimensional QCD," *Phys. Lett.* B **388**, 808 (1996) [arXiv:hep-th/9608122].

[145] E. A. Kuraev, L. N. Lipatov and V. S. Fadin, "The Pomeranchuk singularity in nonabelian gauge theories," *Sov. Phys. JETP* **45**, 199 (1977) [Zh. Eksp. Teor. Fiz. **72**, 377 (1977)]. "Multi-reggeon processes in the Yang-Mills theory," *Sov. Phys. JETP* **44**, 443 (1976) [*Zh. Eksp. Teor. Fiz.* **71**, 840 (1976)].

[146] D. Kutasov, "Duality off the critical point in two-dimensional systems with nonabelian symmetries," *Phys. Lett.* B **233**, 369 (1989).

[147] D. Kutasov, "Two-dimensional QCD coupled to adjoint matter and string theory," *Nucl. Phys.* B **414**, 33 (1994) [arXiv:hep-th/9306013].

[148] D. Kutasov and A. Schwimmer, "Universality in two-dimensional gauge theory," *Nucl. Phys.* B **442**, 447 (1995) [arXiv:hep-th/9501024].

[149] T. D. Lee and Y. Pang, "Nontopological solitons," *Phys. Rept.* **221**, 251 (1992).

[150] G. P. Lepage and S. J. Brodsky, "Exclusive processes in quantum chromodynamics: evolution equations for hadronic wave functions and the form-factors of mesons," *Phys. Lett.* B **87**, 359 (1979).

[151] L. N. Lipatov, *The Creation of Quantum Chromodynamics and the Effective Energy,* Bologna, Italy: Univ. Bologna (1998) 367 p.

[152] J. H. Lowenstein and J. A. Swieca, "Quantum electrodynamics in two-dimensions," *Annals Phys.* **68**, 172 (1971).

[153] M. Luscher, "Quantum nonlocal charges and absence of particle production in the two-dimensional nonlinear Sigma model," *Nucl. Phys.* B **135**, 1 (1978).

[154] D. Lust and S. Theisen, "Lectures on string theory," *Lect. Notes Phys.* **346**, 1 (1989).

[155] A. Maciocia, "Metrics on the moduli spaces of instantons over Euclidean four space," *Commun. Math. Phys.* **135**, 467 (1991).

[156] G. Mack and A. Salam, "Finite component field representations of the conformal group," *Annals Phys.* **53**, 174 (1969).

[157] V. G. Makhankov, Y. P. Rybakov and V. I. Sanyuk, *The Skyrme model: Fundamentals, methods, applications,* Berlin, Germany: Springer (1993) 265 p. (Springer series in nuclear and particle physics).

[158] J. M. Maldacena, "The large N limit of superconformal field theories and supergravity," *Adv. Theor. Math. Phys.* **2**, 231 (1998) [*Int. J. Theor. Phys.* **38**, 1113 (1999)] [arXiv:hep-th/9711200].

[159] S. Mandelstam, "Soliton operators for the quantized sine-Gordon equation," *Phys. Rev.* D **11**, 3026 (1975).

[160] A. V. Manohar, "Large N QCD," arXiv:hep-ph/9802419, Published in *Les Houches 1997, Probing the standard model of particle interactions, Pt. 2* 1091-1169.

[161] A. A. Migdal, "Recursion equations in gauge field theories," *Sov. Phys. JETP* **42**, 413 (1975) [*Zh. Eksp. Teor. Fiz.* **69**, 810 (1975)].

[162] J. A. Minahan and K. Zarembo, "The Bethe-ansatz for N = 4 super Yang-Mills," *JHEP* **0303**, 013 (2003) [arXiv:hep-th/0212208].

[163] C. Montonen and D. I. Olive, "Magnetic monopoles as gauge particles?," *Phys. Lett.* B **72**, 117 (1977).

[164] R. V. Moody, "Lie algebras associated with generalized Cartan matrices," *Bull. Am. Math. Soc.* **73**, 217 (1967).

[165] M. Moshe and J. Zinn-Justin, "Quantum field theory in the large N limit: A review," *Phys. Rept.* **385**, 69 (2003) [arXiv:hep-th/0306133].

[166] D. Mueller, "Constraints for anomalous dimensions of local light cone operators in phi**3 in six-dimensions theory," *Z. Phys.* C **49**, 293 (1991); *Phys. Rev.* D **49**, 2525 (1994).

[167] W. Nahm, "A simple formalism for the BPS monopole," *Phys. Lett.* B **90**, 413 (1980).

[168] N. K. Nielsen, "Gauge invariance and broken conformal symmetry," *Nucl. Phys.* B **97**, 527 (1975).

[169] N. K. Nielsen, "The energy momentum tensor in a nonabelian quark gluon theory," *Nucl. Phys.* B **120**, 212 (1977).

[170] S. P. Novikov, "Multivalued functions and functionals, An analogue to Morse theory," *Sov. Math. Dock.* **24**, 222 (1981)

[171] T. Ohrndorf, "Constraints from conformal covariance on the mixing of operators of lowest twist," *Nucl. Phys.* B **198**, 26 (1982).

[172] R. D. Peccei and H. R. Quinn, "Constraints imposed by CP conservation in the presence of instantons," *Phys. Rev.* D **16**, 1791 (1977).

[173] M. E. Peskin and D. V. Schroeder, *An Introduction To Quantum Field Theory,* Reading, USA: Addison-Wesley (1995) 842 p.

[174] J. Polchinski, *String Theory.* Vol. 2: *Superstring Theory and Beyond,* Cambridge, UK: Univ. Pr. (1998) 531 p.

[175] A. M. Polyakov, "Particle spectrum in quantum field theory," *JETP Lett.* **20**, 194 (1974) [*Pisma Zh. Eksp. Teor. Fiz.* **20**, 430 (1974)].

[176] A. M. Polyakov, "Hidden symmetry of the two-dimensional chiral fields," *Phys. Lett.* B **72**, 224 (1977).

[177] A. M. Polyakov, "Quantum geometry of bosonic strings," *Phys. Lett.* B **103**, 207 (1981).

[178] A. M. Polyakov, *Gauge Fields and Strings,* Chur, Switzerland: Harwood (1987) 301 p. (Contemporary concepts in Physics, 3).

[179] A. M. Polyakov and P. B. Wiegmann, "Goldstone fields in two-dimensions with multi-valued actions," *Phys. Lett.* B **141**, 223 (1984).

[180] M. K. Prasad and C. M. Sommerfield, "An exact classical solution for the 't Hooft monopole and the Julia-Zee dyon," *Phys. Rev. Lett.* **35**, 760 (1975).

[181] E. Rabinovici, A. Schwimmer and S. Yankielowicz, "Quantization in the presence of Wess-Zumino terms," *Nucl. Phys.* B **248**, 523 (1984).

[182] R. Rajaraman, *An Introduction to Solitons and Instantons in Quantum Field Theory,* Amsterdam, Netherlands: North-holland (1982) 409p.

[183] C. Rebbi and G. Soliani, *Solitons and particles,* World Scientific 1984.

[184] B. E. Rusakov, "Loop averages and partition functions in U(N) gauge theory on two-dimensional manifolds," *Mod. Phys. Lett.* A **5**, 693 (1990).

[185] T. Schafer and E. V. Shuryak, "Instantons in QCD," *Rev. Mod. Phys.* **70**, 323 (1998) [arXiv:hep-ph/9610451].

[186] J. Schechter and H. Weigel, "The Skyrme model for baryons," arXiv:hep-ph/9907554.

[187] T. D. Schultz, D. C. Mattis and E. H. Lieb, "Two-dimensional Ising model as a soluble problem of many fermions," *Rev. Mod. Phys.* **36**, 856 (1964).

[188] M. A. Shifman, *Instantons in Gauge Theories*, Singapore, Singapore: World Scientific (1994) 488 p.

[189] E. V. Shuryak, "The role of instantons in quantum chromodynamics. 1. Physical vacuum," *Nucl. Phys.* B **203**, 93 (1982).

[190] J. S. Schwinger, "Gauge invariance and mass. 2," *Phys. Rev.* **128**, 2425 (1962).

[191] N. Seiberg and E. Witten, "Monopoles, duality and chiral symmetry breaking in N=2 supersymmetric QCD," *Nucl. Phys.* B **431**, 484 (1994) [arXiv:hep-th/9408099].

[192] N. Seiberg and E. Witten, "Monopole condensation, and confinement in N=2 supersymmetric Yang-Mills theory," *Nucl. Phys.* B **426**, 19 (1994) [Erratum-ibid. B **430**, 485 (1994)] [arXiv:hep-th/9407087].

[193] Y. M. Shnir, *Magnetic Monopoles*, Berlin, Germany: Springer (2005) 532 p.

[194] W. Siegel, "Manifest Lorentz invariance sometimes requires nonlinearity," *Nucl. Phys.* B **238**, 307 (1984).

[195] T. H. R. Skyrme, "A Nonlinear theory of strong interactions," *Proc. Roy. Soc. Lond.* A **247**, 260 (1958).

[196] T. H. R. Skyrme, "Particle states of a quantized meson field," *Proc. Roy. Soc. Lond.* A **262**, 237 (1961).

[197] T. H. R. Skyrme, "A unified field theory of mesons and baryons," *Nucl. Phys.* **31**, 556 (1962).

[198] Smirnov, F. A. "Form factors in completely integrable models of quantum field theory," Adv. Ser. Math. Phys. **14**:1–208 (1992).

[199] J. Sonnenschein, "Chiral bosons" *Nucl. Phys.* B **309**, 752 (1988).

[200] M. Spiegelglas and S. Yankielowicz, "G/G topological field theories by cosetting G(K)," *Nucl. Phys.* B **393**, 301 (1993) [arXiv:hep-th/9201036].

[201] P. J. Steinhardt, "Baryons and baryonium in QCD in two-dimensions," *Nucl. Phys.* B **176**, 100 (1980).

[202] M. Stone, *Bosonization*, Singapore: World Scientific (1994) 539 p.

[203] H. Sugawara, "A Field theory of currents," *Phys. Rev.* **170**, 1659 (1968).

[204] K. Symanzik, "Small distance behavior in field theory and power counting," *Commun. Math. Phys.* **18**, 227 (1970).

[205] W. E. Thirring, "A soluble relativistic field theory," *Annals Phys.* **3**, 91 (1958).

[206] C. B. Thorn, "Computing the Kac determinant using dual model techniques and more about the no-ghost theorem," *Nucl. Phys.* B **248**, 551 (1984).

[207] S. B Treiman, R. Jackiw and D. Gross *Current Algebra and its Applications* (Princeton, University Press, New Jersey, 1972).

[208] A. I. Vainshtein, V. I. Zakharov, V. A. Novikov and M. A. Shifman, "ABC of instantons," *Sov. Phys. Usp.* **25**, 195 (1982) [*Usp. Fiz. Nauk* **136**, 553 (1982)].

[209] S. Vandoren and P. van Nieuwenhuizen, "Lectures on instantons," arXiv:0802.1862 [hep-th].

[210] E. P. Verlinde, "Fusion rules and modular transformations in 2d conformal field theory," *Nucl. Phys.* B **300**, 360 (1988).

[211] E. P. Verlinde and H. L. Verlinde, "Chiral bosonization, determinants and the string partition function," *Nucl. Phys.* B **288**, 357 (1987).

[212] M. S. Virasoro, "Subsidiary conditions and ghosts in dual resonance models," *Phys. Rev.* D **1**, 2933 (1970).

[213] M. Wakimoto, "Fock representations of the affine Lie algebra A1(1)," *Commun. Math. Phys.* **104**, 605 (1986).

[214] E. J. Weinberg and P. Yi, "Magnetic monopole dynamics, supersymmetry, and duality," *Phys. Rept.* **438**, 65 (2007) [arXiv:hep-th/0609055].

[215] S. Weinberg, *The Quantum Theory of Fields.* Vol. 1: *Foundations,* Cambridge, UK: Univ. Pr. (1995) 609 p.

[216] G. Veneziano, "U(1) Without instantons," *Nucl. Phys.* B **159**, 213 (1979).

[217] J. Wess and B. Zumino, "Consequences of anomalous Ward identities," *Phys. Lett.* B **37**, 95 (1971).

[218] F. Wilczek, "Inequivalent embeddings of SU(2) and instanton interactions," *Phys. Lett.* B **65**, 160.

[219] K. G. Wilson, "Nonlagrangian models of current algebra," *Phys. Rev.* **179**, 1499 (1969).

[220] E. Witten, "Some exact multipseudoparticle solutions of classical Yang-Mills theory," *Phys. Rev. Lett.* **38**, 121 (1977).

[221] E. Witten, "Instantons, the Quark model, and the 1/N expansion," *Nucl. Phys.* B **149**, 285 (1979).

[222] E. Witten, "Baryons in the 1/N expansion," *Nucl. Phys.* B **160**, 57 (1979).

[223] E. Witten, "Large N chiral dynamics," *Annals Phys.* **128**, 363 (1980).

[224] E. Witten, "Nonabelian bosonization in two dimensions," *Commun. Math. Phys.* **92**, 455 (1984).

[225] E. Witten, "Global aspects of current algebra," *Nucl. Phys.* B **223**, 422 (1983).

[226] E. Witten, "Current algebra, baryons, and quark confinement," *Nucl. Phys.* B **223**, 433 (1983).

[227] E. Witten, "On holomorphic factorization of WZW and coset models," *Commun. Math. Phys.* **144**, 189 (1992).

[228] E. Witten, "Two-dimensional gauge theories revisited," *J. Geom. Phys.* **9**, 303 (1992) [arXiv:hep-th/9204083].

[229] C. N. Yang and R. L. Mills, "Conservation of isotopic spin and isotopic gauge invariance," *Phys. Rev.* **96**, 191 (1954).

[230] C. N. Yang, "Some exact results for the many body problems in one dimension with repulsive delta function interaction," *Phys. Rev. Lett.* **19**, 1312 (1967).

[231] I. Zahed and G. E. Brown, "The Skyrme model," *Phys. Rept.* **142**, 1 (1986).

[232] A. B. Zamolodchikov, "Renormalization group and perturbation theory near fixed points in two-dimensional field theory," *Sov. J. Nucl. Phys.* **46**, 1090 (1987) [*Yad. Fiz.* **46**, 1819 (1987)].

[233] A. B. Zamolodchikov, "Exact solutions of conformal field theory in two-dimensions and critical phenomena," *Rev. Math. Phys.* **1**, 197 (1990).

[234] A. B. Zamolodchikov, "Thermodynamic Bethe anzatz in relativistic models, scaling three state Potts and Lee-Yang models," *Nucl. Phys.* B **342**, 695 (1990).

[235] A. B. Zamolodchikov and A. B. Zamolodchikov, "Factorized S-matrices in two dimensions as the exact solutions of certain relativistic quantum field models," *Annals Phys.* **120**, 253 (1979).

[236] A. B. Zamolodchikov, A. B. Zamolodchikov and I. M. Khalatnikov, "Physics Reviews, Vol. 10, pt. 4: Condormal field theory and critical phenomena in two-dimensional systems," London, UK: Harwood (1989) 269–433. (Soviet scientific reviews, Section A, 10.4)

[237] B. Zwiebach, *A First Course in String Theory,* Cambridge, UK: Univ. Pr. (2004) 558 p

Index

Printed in the United States
by Baker & Taylor Publisher Services